Impact of Societal Norms on Safety, Health, and the Environment

Impact of Societal Norms on Safety, Health, and the Environment

Case Studies in Society and Safety Culture

Lee T. Ostrom
University of Idaho
Idaho Falls, USA

Registered Office
John Wiley & Sons, Inc., 111 River Street, Hoboken, NJ 07030, USA

Editorial Office
111 River Street, Hoboken, NJ 07030, USA

For details of our global editorial offices, customer services, and more information about Wiley products visit us at www.wiley.com.

Wiley also publishes its books in a variety of electronic formats and by print-on-demand. Some content that appears in standard print versions of this book may not be available in other formats.

Library of Congress Cataloging-in-Publication Data applied for

Cover Design: Wiley
Cover Image: © Thatree Thitivongvaroon/Getty Images

Set in 9.5/12.5pt STIXTwoText by Straive, Chennai, India

SKY10035922_090222

Contents

Preface

Accidental deaths are the fourth most common cause of death in the United States, claiming approximately 200,000 people each year. Deaths from falls account for approximately 42,000 deaths and deaths from automobile accidents accounts for approximately 40,000 deaths. 4764 fatal occupational injuries occurred in 2020 and this represents the lowest annual number since 2013. A worker died every 111 minutes from a work-related injury in 2020. Compare this with the approximately 16,000 occupational deaths in 1940. This decrease in the number of deaths is due to the increased focus on safety in workplace and more recently to the concentration on the importance of safety culture in organizations. However, horrific accidents still occur every day. Every organization has a safety culture. There are numerous organizations that have very positive safety cultures and there are many that have very negative safety cultures. This book contains case studies on accidents and the safety culture attributes associated with the events. Please use the descriptions of the safety culture problems in the case studies to help improve your organizations.

Abbreviations

%	percent
°	degrees
°C	degrees celsius
°F	degrees fahrenheit
3D	three-dimensional
65NJ	Helo Kearny Heliport
AA	aluminum association
AAO	acetaldehyde oxime
AAR	Association of American Railroads (United States)
AAR	Association of American Railroads
AB	able bodied seaman
AC	advisory circular
AC	air conditioning
ACC	American Chemistry Council
agl	Above ground level
AHJ	Authority Having Jurisdiction
AIChE	American Institute of Chemical Engineers
AIS	automatic identification system
AIS	abbreviated injury scale
Amtrak	National Railroad Passenger Corporation
AN	ammonium nitrate
ANFO	ammonium nitrate/fuel oil
ANSI	American National Standards Institute
AP	anterior–posterior
API	American Petroleum Institute
ARA	Agricultural Retailers Association
ARPA	automatic radar plotting aid
ARS	air rescue systems
ARSST	Advanced Reactive System Screening Tool

AS	ammonium sulfate
ASLRRA	American Short Line and Regional Railroad Association (United States)
ASSE	American Society of Safety Engineers
ASTM	American Society for Testing Materials
ATC	air traffic control
ATC	automatic train control
ATF	Bureau of Alcohol, Tobacco, Firearms, and Explosives (United States)
ATT	airframe total time
b/d	barrels per day
BC	borden chemical
BEA	Bureau d'Enquêtes et d'Analyses pour la Sécurité de l'Aviation Civile
BFI	Browning-Ferris Industries
BLET	Brotherhood of Locomotive Engineers and Trainmen
BLEVE	boiling liquid expanding vapor explosion
BOV	bottom outlet valve
BRM	bridge resource management
BRM/BTM	bridge resource management/bridge team management hp
CAL	confirmatory action letter
CalOSHA	California Division of Occupational Safety and Health
CAN	certified nursing assistant
CANUTEC	Canadian Transport Emergency Centre
CCPS	Center for Chemical Process Safety
CCTV	closed circuit television
CDC	Centers for Disease Control and Prevention
CDP	Center for Domestic Preparedness
CDRH	Center for Devices and Radiological health, FDA
CDT	Central Daylight Time
CEO	chief executive officer
CEPP	Chemical Emergency Preparedness Program
CERCLA	Comprehensive Environmental Response, Compensation, and Liability Act
CFATS	Chemical Facility Anti-Terrorism Standards
CFM	cubic feet per minute
CFR	Code of Federal Regulations (United States)
CMU	concrete masonry unit
CN	Canadian National
CNCG	Concentrated Non-Condensable Gas
COI	chemical of interest

Conrail	consolidated rail corporation
COO	chief operating officer
CPR	Canadian Pacific Railway
CRM	crew resource management
CROR	Canadian Rail Operating Rules
CSA	Canadian Standards Association
CSAT	Chemical Security Assessment Tool
CSB	Chemical Safety and Hazard Investigation Board (United States)
CSCD	Chemical Security Compliance Division
CT	computed tomography
CTA	Canadian Transportation Agency
CTA	CTA Acoustics, Inc.
CTC	centralized traffic control
CTG	Continuing Training Grant
CTPS	computerized treatment plan
CWR	continuous welded rail
CX	customer experience
DCS	distributed control system
DG	dangerous good
DHS	Department of Homeland Security (United States)
DIERS	Design Institute for Emergency Relief Systems
Diglyme	diethylene glycol (dimethyl) ether
DO	director of operations
DOL	Department of Labor (United States)
DOT	Department of Transportation (United States)
DWC	Division of Workers' Compensation
ECDIS	Electronic Chart Display and Information System
ECP	Electronically Controlled Pneumatic (Braking System)
ECR	engine control room
EDT	eastern daylight time
EHS	extremely hazardous substance
EMPG	Emergency Management Performance Grant
EMS	emergency medical services
EMT	emergency medical technician
EO	executive order
EOC	emergency operations center
EOT	engine order telegraph
EPA	Environmental Protection Agency (United States)
EPCRA	Emergency Planning and Community Right-to-Know Act
ERAP	emergency response assistance plan
ERDC	Engineer Research and Development Center

ERG	Emergency Response Guidebook
ERP	Emergency Response Plan
ERT	Emergency Response Team
ESC	Electrostatic Charging Tendency
eTicketing	electronic ticketing
EU	European Union
FAA	Federal Aviation Administration
FAST	firefighter assist and search team
FAST	Fixing America's Surface Transportation
FBU	Fluoroproducts Business Unit
FDA	Food and Drug Administration
FDEP	Florida Department of Environmental Protection
FDNY	Fire Department of the City of New York
FEMA	Federal Emergency Management Agency (United States)
FFCL	fuel flow control lever
FFFIPP	Fire Fighter Fatality Investigation and Prevention Program
FGAN	fertilizer grade ammonium nitrate
FOM	flight operations manual
FOV	field of view
FP&S	Fire Prevention and Safety
fpm	feet per minute
FR	Federal Register
FRA	Federal Railroad Administration (United States)
FRS	Facility Registry Service
FSDO	flight standards district office
FSOL	fuel shutoff lever
g/cm^3	grams per cubic centimeter
GAO	Government Accounting Office
GAO	Government Accountability Office (United States)
GDC	General Duty Clause
GE	General Electric Company
GHS	Globally Harmonized System of Classification and Labeling of Chemicals
GM	General Motors
GMP	Good Manufacturing Practices
GOI	General Operating Instructions
GPCC	Greater Pittsburg Cancer Center
GPD	Grant Programs Directorate
GPN	Graduate Practical Nurse
GRT	gross register tons
GSI	General Special Instructions

HAZCOM	OSHA's Hazard Communication
HAZMAT	hazardous material
HAZOP	Hazard and operability study
HAZWOPER	Hazardous Waste Operations and Emergency Response
HCS	Hazard Communication Standard
HDR	high dose rate
HHC	highly hazardous chemical
HSE	Health and Safety Executive (United Kingdom)
HSNTP	Homeland Security National Training Program
HUD	Department of Housing and Urban Development (United States)
HVAC	heating, ventilation, and air conditioning
HVLC	high volume low concentration
IAFC	International Association of Fire Chiefs
IAFF	International Association of Fire Fighters
IAP	Incident Action Plan
IBC	International Building Code
IC	Incident Commander
ICA	instructions for continued airworthiness
ICAO	International Civil Aviation Organization
ICBO	International Conference of Building Officials
ICC	International Code Council
ICHEME	Institution of Chemical Engineers
ICS	Incident Command System
ICT	Insurance Council of Texas
ICWUC	International Chemical Workers Union Council
IDLH	Immediately Dangerous to Life or Health
IEC	International Electrotechnical Commission
IFC	International Fire Code
IIS	Inspection Information System (TC)
IIT	Incident Investigation Team
IME	Institute of Makers of Explosives
IMIS	Integrated Management Information System
IMO	International Maritime Organization
IOSHA	Indiana Occupational Safety and Health Administration
IP	Inspection Procedure
IRCC	Indiana Regional Cancer Center
Irving	Irving Oil Ltd.
ISA	International Society of Automation
ISO	Insurance Services Office
ISO	International Organization for Standardization
IST	inherently safer technology

JFRD	Jacksonville Fire and Rescue Department
JSO	Jacksonville Sheriff's Office
kip	kilopound (1 kip = 1000 lb)
KJRB	Downtown Manhattan/Wall Street Heliport
km/h	kilometers per hour
K-Mag	potassium magnesium
KSt	Dust Deflagration Index
kts	knots
kW	kilowatts
KYOSHA	Kentucky Department of Labor, Office of Occupational Safety and health
LE	locomotive engineer
LED	light-emitting diode
LEL	lower explosive limit (also known as lower flammable limit)
LEP	Local Emphasis Programs
LEPC	Local Emergency Planning Committee
LER	locomotive event recorder
LFL	lower flammable limit (also known as lower explosive limit)
LGA	LaGuardia Airport
LLC	Limited Liability Company
LNG	liquefied natural gas
LOA	letter of authorization
LOC	limiting oxidant (oxygen) concentration
LOPA	Layers of Protection Analysis
LPBC	Local Performance Based Compensation Program
LPG	liquid propane gas
LPN	licensed practical nurse
LS	Lumbar-Sacral
LVHC	low volume high concentration
m	meters
MACT	Maximum Achievable Control Technology
MARC	Maryland Area Regional Commuter
MAWP	maximum allowable working pressure
MC	manual chapter
MCI	mass casualty incident
MCMT	methylcyclopentadienyl manganese tricarbonyl
MCPD	methylcyclopentadiene
MEC	Minimum Explosive Concentration
MeSH	methyl mercaptan
MG	miscellaneous guidance
MIC MOC	methyl isocyanate management of change

MIE	minimum ignition energy
MIOSHA	Michigan Occupational Safety and Health Administration
MLLW	mean lower low water
mm	millimeters
MMA	Montreal, Maine and Atlantic Railway
MOC	management of change
MOU	memorandum of understanding
MP	milepost
mph	miles per hour
MSDS	Material Safety Data Sheet
msl	mean sea level
NAICS	North American Industry Classification System
NASFM	National Association of State Fire Marshals
NBSR	Southern New Brunswick Southern Railway
NCG	non-condensable gas
NCOSHA	North Carolina Department of Labor, Occupational Safety and Health Division
NEC	National Electric Code
NEIC	National Earthquake Information Center
NEP	National Emphasis Program
NERRTC	National Emergency Response and Rescue Training Center
NESHAP	National Emissions Standards for Hazardous Air Pollutants
NEW	net explosive weight
NFA	National Fire Academy
NFPA	National Fire Protection Association
NIMS	National Incident Management System
NIOSH	National Institute for Occupational Safety and Health
nm	nautical miles
NMSS	Nuclear Material Safety and Safeguards
NOAA	National Oceanic and Atmospheric Administration OS
NORAC	Northeast Operating Rules Advisory Committee (Operating Rulebook)
NO_x	nitrogen oxide
NPD	National Preparedness Directorate
NPRM	notice of proposed rulemaking
NRC	Nuclear Regulatory Commission
NRC	National Research Council of Canada
NRS	Northwest River Supplies
NTED	National Training and Education Division
NTSB	National Transportation Safety Board (United States)
NVFC	National Volunteer Fire Council

NYC	New York City
NYPD	New York City Police Department
OAG	Office of the Auditor General
OB	Operating Bulletin
OEM	Office of Emergency Management (Philadelphia)
OIG	Office of Inspector General
OpSpec	operations specification
ORIS	Oak Ridge Institute for Sciences and Education
OSC	Oncology Services Corporation
OSH Act	Occupational Safety and Health Act of 1970 (United States)
OSHA	Department of Labor Occupational Safety and Health Administration (United States)
OSHA	Occupational Safety and Health Administration
OSHRC	Occupational Safety and Health Review Commission (United States)
OTI	OSHA Training Institute
OTIS	Operational Tests and Inspections Program
OTSC	Office of the Texas State Chemist
PAI	Permit Authorizing Individual
PAI	principal avionics inspector
PCA	Packaging Corporation of America
PCDS	personnel carrying device system
PDD	proximity detection device
PEL	Permissible exposure limit
PES	programmable Electronic System
PFD	personal flotation device
PFD	Philadelphia Fire Department
PG	packing group
PHA	process hazard analysis
PHMSA	Pipeline and Hazardous Materials Safety Administration (United States)
PHMSA	Pipeline and Hazardous Materials Safety Administration
PIC	pilot-in-command
PMI	principal maintenance inspector
POI	principal operations inspector
ppb	parts per billion
PPC	Public Protection Classification
PPD	Philadelphia Police Department
PPE	Personal Protective Equipment
ppm	parts per million
PPU	Portable pilot unit

PRD	pressure relief device
PRV	pressure relief valve
psi	pounds per square inch
PSI	Process Safety Information
psia	pounds per square inch absolute
psig	pound-force per square inch gauge
psig	pounds per square inch gauge
PSM	OSHA Process Safety Management Standard
PSM	Process Safety Management
PTC	positive train control
PVC	polyvinyl chloride
Q&A	question and answer
QA	quality assurance
QC	quality control
QM	quality management
QNS&L	Quebec North Shore and Labrador Railway
QRB	quick release brake (valve)
QSR	Quebec Southern Railway
RAC	Railway Association of Canada
RAGAGEP	Recognized and Generally Accepted Good Engineering Practice
RBPS	Risk Based Process Safety
RCMS	Responsible Care Management System®
RCRA	Resource Conservation and Recovery Act
RCS	relative culture strength
RDPC	Rural Domestic Preparedness Consortium
REAC/TS	Radiation Emergency Assistance Center/Training Site
REL	recommended exposure limit
RFI	request for information
RFM	rotorcraft flight manual
RFMS	rotorcraft flight manual supplement
RMP	Risk Management Plan Rule
RMR	Reactivity Management Roundtable
RN	registered nurse
RODS	Rail Occurrence Database System (TSB)
rpm	revolutions per minute
RSA	Railway Safety Act
RSC	reset safety control
RSI	railway safety inspector
RSO	Radiation Safety Officer
RTC	rail traffic controller
RTR	Registered Technologist Radiographer

RTT	Registered Therapy Technician
RWI	Rail World, Inc.
S/N	serial number
SAA	State Administrative Agency
SAChE	Safety and Chemical Engineering Education Committee
SAF	Swedish Air Force
SAFER	Staffing for Adequate Fire and Emergency Response
SARA	Superfund Amendments and Reauthorization Act
SB	service bulletin
SBA	Small Business Administration
SBU	sense and braking unit
SCBA	self-contained breathing apparatus
SCBA	self-contained breathing apparatus
SD	secure digital
SDS	safety data sheet
SEP	Special Emphasis Programs
SEPTA	Southeastern Pennsylvania Transportation Authority
SERC	State Emergency Response Commission
SFFMA	State Firefighters' and Fire Marshals' Association
SFMO	State Fire Marshal's Office
SHI	Substance Hazards Index
SHIB	Safety and Health Information Bulletin
SHM	Scenery Hill Manner
SIBU	Standard Insecticide Business Unit
SIC	Standard Industry Code
SIC	Standard Industrial Classification
SIS	Safety Instrumented System
SMS	Safety Management System
SMS	Manual Safety Management System
SOLAS	International Convention for the Safety of Life at Sea
SOP	standard operating procedure
SOR	Southern Ontario Railway
SPCC	spill prevention, control, and countermeasures
SPRS	supplemental passenger restraint system
SPTO	single-person train operations
SQ	Sûreté du Québec
SSO	Safety Systems Overview
SST	Strobel Starostka Transfer, LLC
STC	supplemental type certificate
STD	start-to-discharge (pressure)
TAC	Texas Administrative Code

TAPPI	Technical Association of Pulp and Paper Industry
TC	Transport Canada
TCDS	type certificate data sheet
TCEQ	Texas Commission on Environmental Quality
TCFP	Texas Commission on Fire Protection
TDG	transportation of dangerous goods
TDI	Texas Department of Insurance
TEEX	Texas A&M Engineering Extension Service
TFI	The Fertilizer Institute
TGAN	technical grade ammonium nitrate
TIESB	Texas Industrial Emergency Services Board
TIP	Technical Information Paper
TNT	trinitrotoluene
TOPS	Tour Operators Program of Safety
TR	technical report
TRANSCAER	Transportation Community Awareness and Emergency Response
Tranz Rail	Tranz Rail Holdings Limited (New Zealand)
TRI	Toxics Release Inventory
TRS	Total Reduced Sulfur
TSB	Transportation Safety Board of Canada
TSO	technical standard order
TSR	Track Safety Rules
TWA	time-weighted average
TX	Texas
UK	United Kingdom
UEL	upper explosive limit (also known as upper flammable limit)
UFC	Uniform Fire Code California Division of Occupational Safety and Health
UFCW	United Food and Commercial Workers
UFL	upper flammable limit (also known as upper explosive limit)
UN	United Nations (product code)
UNECE	United Nations Economic
USACE	US Army Corps of Engineers
USC	United States Code
USFA	Fire Administration (United States)
USGS	Geological Survey (United States)
USS	United States Ship
VDR	voyage data recorder
VFD	volunteer fire department
VFR	visual flight rules

VHF	very high frequency
VIA	VIA Rail Canada, Inc.
VSP2	Vent Sizing Package 2
VTS	Vessel traffic service
WC	Wisconsin Central
WFC	West Fertilizer Company
WFD	West Fire Department
WFSI	World Fuel Services, Inc.
WIS	West Intermediate School
WISD	West Independent School District
WMS	West Middle School
WVFD	West Volunteer Fire Department

1

Safety Culture Concepts

1.0   Introduction

Summer blockbuster movies always have some huge disaster as a major part of the plot. These events include events like airplanes crashing, volcanoes erupting, explosions, ships sinking, fires, tornadoes, earthquakes, railroad trains crashing, or alien invasions. In the movies all the bloodshed is fake. In reality, these types of disastrous events cause real people to die, to become severely injured and destroy families' homes and businesses. Events like tornadoes, earthquakes, and volcanoes can't be prevented. However, airplanes crashing, industrial explosions, ships sinking, trains crashing, and fires can be prevented. Alien invasions, well we don't know yet (or maybe we do).

Safety and health professionals dedicate their lives trying to prevent people from being killed or injured and to prevent property damage. Safety and health professionals include:

- Safety engineers
- Industrial hygienists
- Fire inspectors
- Fire fighters
- Health physicist
- Radiation safety officers
- Ergonomists
- Risk analysts
- Human factors practitioners

   In addition to:

- Professional engineers of all types
- Chemists
- Biologists

Impact of Societal Norms on Safety, Health, and the Environment: Case Studies in Society and Safety Culture, First Edition. Lee T. Ostrom.

- Ecologists
- Physicists
- Medical professionals
- Police and Military

It is quite a list of professionals who work to keep us safe. Some of these professionals are dedicated to designing safe products, some to ensuring safe working conditions, some working to ensure we live healthy lives, and some ensuring our physical security.

I wrote this book because of my passion about safety and helping people come home safe every day from work. The case studies in this book represent a broad range of events that have happened and could happen tomorrow, if precautions are not taken. Safety culture is the focus of the book because disastrous events can be prevented if the safety culture of organizations is improved.

The book is organized into 10 chapters. Chapter 1 discusses the concepts of safety culture; Chapter 2 discusses aspects of the evolution of safety and safety culture through the centuries and decades, Chapters 3–9 discuss a wide range of accident case studies and the associated safety culture attributes. Chapter 10 discusses methods of assessing safety culture.

The case studies presented in the book should inspire employers to ensure their facilities and processes are engineered and maintained to be safe, employees are properly trained, and organizations place a high value on the lives of their employees.

I want to thank the investigators form the Chemical Safety Board (CSB), National Transpiration Safety Board (NTBS), Canadian Transportation Safety Board (CTBS), Nuclear Regulatory Commission (NRC), the Department of Energy (DOE), and local fire and police departments for their integrity in performing the detailed accident investigations that I have used in this book.

1.1 Culture

The American Heritage Dictionary defines culture as "The totality of socially transmitted behavior patterns, arts, beliefs, institutions, and all other products of human work and thought characteristic of a community or population." A culture is comprised of behavioral norms, patterns of perceptions, language/speech, and even building design features that make the culture what it is. It is difficult to understand a culture in total, but it is possible to study and understand individual norms. A social norm is defined as an unspoken rule of behavior that, if not followed, will result in sanctions. In an organization, a norm might be that managers must business attire.

In this organization, a manager who arrives at a meeting in casual clothes might be teased or reprimanded. If he or she consistently failed to wear the appropriate clothing, might be considered unprofessional, not reflecting the company image, and face severe sanctions, including loss of his or her position.

Every organization, as does every country, has a culture. Even within a country, cultures vary widely. The cultures within the United States vary greatly. We all know how different the culture in a northeast state varies from the deep South. Consider also that a city culture varies from a culture just outside a city. We recently watched a movie entitled "Into the White." The movie was based on a true story about a German and a British air crew stranded in the high Norwegian plateau during World War II after shooting each other down. The interesting part about this movie to me is when they were out of food and starving a British and a German airman went hunting for food. They shot a rabbit and brought it back to their cabin and all five looked at the dead rabbit and had no clue how to prepare it to eat. I grew up in the mountains of the west and I was hunting at 12 and cleaning what I killed. These airmen had been denied that skill because of the cultures they grew up in.

Cultures vary widely in countries and areas within countries. Consider the languages in the small country of Dagestan. The country is only about $20,000\,mi^2$ (about $50,000\,km^2$) and has only 3.1 million residents. There are more than 30 ethnic groups and 81 nationalities. There are 14 official languages, and 12 ethnic groups constituting more than 1% of its total population. In addition, there are over 40 languages (Charles Rivers 2019). So, how did all these cultures come about? Dagestan borders the Caspian Sea and is in the Caucasus Mountains. It was at a crossroads during ancient times and many villages were high in the mountains. Adjacent villages were separated enough that distinct dialects developed. Also, the invasions by various groups brought languages and customs to Dagestan as well.

So, what is the point of this? It is that within companies and organizations safety cultures vary widely, just as cultures range widely in cities, counties, states, and countries.

A safety culture is composed of safety norms within a company or an organization. A safety norm can be positive or negative (Ostrom, Wilhelmsen, and Kaplan 1993). A positive norm is that a lab worker always dons their safety glasses and a fire-resistant lab coat every time they enter a lab. A negative norm is when an electrician fails to use proper lock and tag procedures when working on electrical circuits. The case studies will present the results of numerous case studies involving negative safety norms.

Pidgeon (1991) writings from 1991 are still true today. He wrote that a "good" safety culture is hard to define. Part of the reason for this is that each organization's culture is somewhat unique. Culture can be influenced by the nation or region, by the technologies and tools it uses, and by the history of success and failure the company/organization has achieved. Safety culture of an organization may be influenced by the marketplace and regulatory setting in which it operates. Safety culture may be influenced by the vision, values, and beliefs of its leaders as well. All these influences make it difficult to say what a "good" safety culture will look like in a particular setting. Despite differences, good safety cultures do have things in common:

- Good safety cultures have employees with particular patterns of attitudes toward safety practice.
- Because it is impractical to establish formal, explicit rules for all foreseeable hazards, norms within the organization are required to provide guidance in particular circumstances.
- In a "good" safety culture employees might be alert for unexpected changes and ask for help when they encounter an unfamiliar hazard.
- They would seek and use available information that would improve safety performance. In a "good" safety culture, the organization rewards individuals who call attention to safety problems and who are innovative in finding ways to locate and assess workplace hazards.
- All groups in the organization participate in defining and addressing safety concerns, and one group does not impose safety on another in a punitive manner.
- Organizations with a "good" safety culture are reflexive on safety practices.
- They have mechanisms in place to gather safety-related information, measure safety performance, and bring people together to learn how to work more safely.
- They use these mechanisms not only to support solving immediate safety problems but also to learn how to better identify and address those problems on a day-to-day basis.
- What is acceptable in a company regarding safety must be defined and practiced if a corporate culture that values safety is to be created. Ideally, employees should know all the risks associated with their jobs, what is required for safety, and take responsibility for themselves. In other words, develop a norm in which employees are aware of all the risks in their workplace or are continually on the lookout for risks.

The result is an overall positive attitude toward safety.

1.2 Safety and Health Pioneers

People have been trying to understand and control the factors that lead to occupational illnesses and accidents for two millennia. Some of the first leaders in occupational safety and health were Hippocrates, Pliny the Elder, Galen, Agricola, and Bernardino Ramazzini. The Greek physician Hippocrates identified lead as a hazardous material in the mining industry in about 400 BCE (BC). He helped develop rules for working in mines. The Roman Pliny the Elder in about 100 CE (AD) identified zinc fumes and sulfur vapors as hazardous. He developed a face mask made from animal bladders to help protect chemical workers. Galen was another Greek physician who characterized the pathology of lead poisoning and the hazards of working copper miners who were exposed to acid mists. Agricola was a German scholar and a very early industrial hygienist and described the diseases of miners. He also developed preventative measures to avoid diseases associated with mining.

Bernardino Ramazzini made a huge impact on safety and health. Most safety and health professionals consider him to be the founder of occupational medicine. Those of us who work also in ergonomics consider him to have been a pioneer in the study of musculoskeletal injuries.

He was born on October 4, 1633, in the small town of Capri. This town is located in the duchy of Modula, Italy. In his lifetime he established the field of occupational medicine. In 1682 Duke Francesco II of Modena assigned him to establish a medical department at the University of Modena. His title was professor "Medicinae Theoricae."

He was appointed chair of practical medicine in Padua, Republic of Venice, in 1700. This was the premier medical faculty in Italy. That same year he wrote the seminal book on occupational diseases and industrial hygiene, De Morbis Artificum Diatriba (Diseases of Workers).

He is best known for his work on exposure to toxic materials. His pioneering effort in musculoskeletal illnesses included linking occupations to specific disorders (Franco and Fusetti 2004). Ramazzini was one of the first to observe that common musculoskeletal illnesses could develop due to common stresses associated with poor ergonomics, for instance, prolonged stationary postures or of unnatural postures. Just as today, people working in awkward postures like bakers, scribes, weavers, or washer women could develop illnesses. People in professions that required prolonged static postures like workers who stand or are required to sit for long periods of time can develop problems as well. In addition, workers who are required to perform tasks that require heavy muscular performance are at risk of injury.

1.3 The Evolution of Accident Causation Models

The next step in the evolution of safety was the development of accident causation models. Accident causation models have been evolving for about 100 years. Heinrich's Domino Theory developed in the 1930s was based on the premise that a social environment conducive to accidents was the first of five dominos to fall

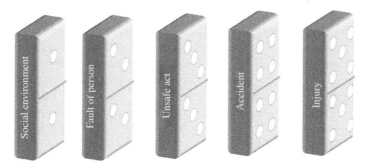

Figure 1.1 Heinrich's theory of accident causation. Source: Redrawn by Ostrom (2022).

Figure 1.2 Heinrich's theory of an accident sequence. Source: Redrawn by Ostrom (2022).

in an accident sequence (Figures 1.1 and 1.2) (Heinrich 1931; Heinrich, Peterson, and Roos 1980). The social environment in this case is associated with the culture the worker grew up in. Included in this were what Heinrich called, ancestry traits, like stubbornness, greed, and recklessness. The other four dominos in sequence were fault of person (personal traits), unsafe acts/mechanical issues/facility, accident, and injury/property damage. If a domino is removed, then the accident sequence is stopped. This theory is now 90-plus years old and focuses on the inherent traits of the person, instead of all the other cultural influences on safety, within an organization. However, his theory does support the concept that accidents occur because of a sequence of events (Figure 1.3).

Heinrich's accident process is:

1. The environment is where and how a person was raised and educated, therefore the culture of his upbringing.
2. Faults of persons are inherited or acquired because of their social environment or acquired by ancestry.
3. Personal and mechanical hazards exist only through the fault of careless persons or poorly designed or improperly maintained equipment.
4. An accident occurs only because of a personal or mechanical hazards.
5. A personal injury (the final domino) occurs only because of an accident.

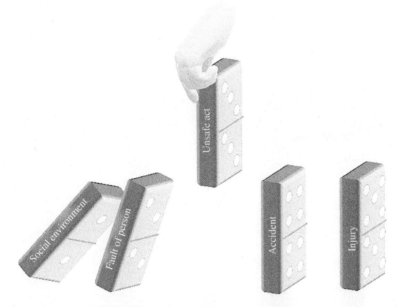

Figure 1.3 Removing a domino in the sequence. Source: Redrawn by Ostrom (2022).

Heinrich believed that the unsafe act or mechanical/physical hazard should be examined and corrected first to be able to prevent accidents.

A major development in the accident causation process by Heinrich is that an accident is any unplanned, uncontrolled event that could result in personal injury or property damage. For example, if a person gets their hand caught in a piece of equipment an accident has occurred even if no injury resulted. Heinrich developed the initial accident hierarchy triangle (Figure 1.4). This triangle shows his idea of how many near miss accidents there were (300) to minor injury accidents (29) to major injury accidents/deaths (1).

Frank Bird (Bird and Germain 1996) modified the dominos to reflect his theory of accident causation in the late 1960s and early 1970s (Figure 1.5). His ratio of near miss accidents to incidents to serious incidents to accidents is shown in Figure 1.6.

Bird's theory was that accidents occur because of:

1. Lack of management control or oversight
   The items he felt management should control are planning, organizing, directing, controlling, and coordinating, job analysis, personal communication, selection and training, "standards" in each work activity identified measuring performance by standards and correcting performance by improving the existing program.

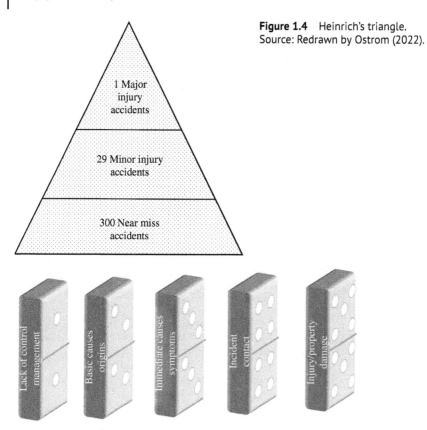

Figure 1.4 Heinrich's triangle. Source: Redrawn by Ostrom (2022).

Figure 1.5 Frank Bird's theory of accident causation. Source: Ostrom (2022).

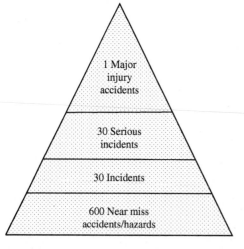

Figure 1.6 Frank Bird's accident triangle. Source: Redrawn by Ostrom (2022).

2. Origins or basic causes

   The origins fall into two categories:

   Personal factors include lack of knowledge or skill, improper motivation and physical or mental problems.

   Job factors include inadequate work standards, design, maintenance, purchasing standards, abnormal usage, and others.

   These basic causes and conditions and failure to identify them permit the second domino to fall. This initiates the possibility of further chain reaction.

3. Immediate causes

   Immediate causes are only symptoms of the underlying problems in an organization. These underlying conditions manifest in unsafe acts and unsafe conditions.

4. Accident

   The accident results because of the unsafe acts or conditions. There are ways to mitigate the results of unsafe acts and conditions using personal protective equipment, for example.

5. Injury/damage

   Injury is the most important item of loss and second, comes property damage.

The Swiss cheese model of accident causation was developed by James Reason (1990, 1997). Figure 1.7 shows a depiction of the model.

In the Swiss cheese model defenses, barriers, and safeguards are integral to ensuring the safety of a system. Modern complex systems have many defensive layers. These include engineered components like alarms, physical barriers, automatic shutdowns, and interlocks. Other controls rely on the human. These are the key operating personnel in the system and include operators, pilots, medical personnel, and maintenance personnel. Procedures and other administrative control are also an important component of safety.

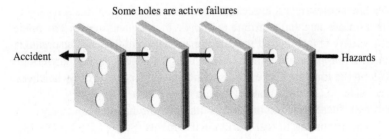

Figure 1.7 James Reason's Swiss cheese model. Source: Redrawn by Ostrom (2022).

As Reason explains; "In an ideal world each defensive layer would be intact. However, they are more like slices of Swiss cheese, having many holes – though unlike in the cheese, these holes are continually opening, shutting, and shifting their location. The presence of holes in anyone "slice" does not normally cause a bad outcome. Usually, this can happen only when the holes in many layers momentarily line up to permit a trajectory of accident opportunity, bringing hazards into damaging contact with victims."

The holes in the defenses, according to Reason, come about for two reasons: active failures and latent conditions. Adverse events involve a combination of these two sets of factors.

Active failures are the unsafe acts committed by people. Unsafe acts include slips, lapses, fumbles, mistakes, and procedural violations. These types of failures have their influence on defenses for a short period of time.

Latent conditions are those that, as the name implies, are hidden in a system until a set of conditions occur that triggers this type of failure to manifest itself. These types of conditions are, for example, faulty protection systems, untested interlocks that are faulty, or faulty design or construction. The fall of the Champlain Tower collapse in June of 2021 was probably due to unrepaired structural damage, or latent failures. They can also be personnel issues like overworked important employees, time pressure, or inadequate equipment. The COVID-19 epidemic has caused health care workers to experience great amounts of stress that could lead to being a latent condition.

The point of this model is that there are holes in the defenses all the time and if the conditions line up an accident can happen.

The accident causation models discussed earlier are considered Simple Linear or Complex Linear (Safeopedia 2017). That is that accidents are a culmination of a series of events or circumstances. There is a sequential interaction of events. An event occurs and his leads to the next event. Heinrich's Domino Theory (1931) is the classic example. As shown earlier, the sequence is broken by removing one of the events the disaster will be avoided.

Complex Linear presumes that accidents are the result of a combination of latent hazards and unsafe acts that continue to happen in a sequential way. The model considers a variety of factors that include the environmental as well as organizational effects. The application of the model enables the set-up of safety barriers and defenses along the timeline of the events (contributing factors). The Swiss cheese model is considered Complex Linear (Reason 1990, 1997).

Complex non-linear accidents are the results of a combination of mutually interacting variables occurring in real world environments (Safeopedia 2017). These models seek to understand the interactions through careful analysis. A systemic model focuses on interactions and functions of the system rather than just individual events. Accidents are regarded as emergent rather than resultant phenomena.

Hollangel and Örjan (2004) and Hollangel (2012) FRAM (Functional Resonance Accident Model) is an example of a complex non-linear accident causation model. This is a very complex model. Hollangel's idea is that in a system there are numerous interactions between components and systems. Instead of being a linear sequence of events, one activity/event can influence one or more of the other activities/events. A path can lead to success or failure depending on the variability in the activities/events. An accident can be caused by the variability in the system. The variability is a result of the variable and different conditions, group interactions, resources allocation, time available, control functions, and many more possibilities. Hollangel describes resonance of a system as a function of an activity/events' variability. This means that if its variability is unusually high then there could be consequences spreading dynamically to the other functions of the system through not necessarily identifiable couplings. FRAM enables a better understanding of a socio-technical system, by avoiding decomposing the system into smaller components and characteristics. The process itself forces questioning rather than finding straight clear answers, as it does not include the typical cause-effect models. The process is time consuming and requires the accident analyst to do "what-iffing" to develop the best model of an accident. Figures 1.8 and 1.9 show the basic unit of the FRAM model and an example of a possible connection.

Risk assessment techniques can also be used to develop accident sequence models as well. Please consult Ostrom and Wilhelmsen (2019) for further information.

We developed a risk model of aircraft inspection, factoring in the safety culture of the maintenance and inspection organization. The complete article is found at Ostrom and Wilhelmsen (2015). This work was performed as part of a NASA sponsored study on the risk factors associated with aircraft inspection. The basic inspection model is represented in Figure 1.10. DVI is detailed visual inspection and GVI is general visual inspection. As the names imply, a DVI is a much more detailed examination of an aircraft part and a GVI is more of a walk by examination of an aircraft.

Figure 1.8 Basic unit of the FRAM model. Source: Redrawn by Ostrom (2022).

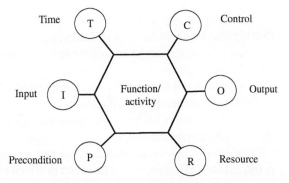

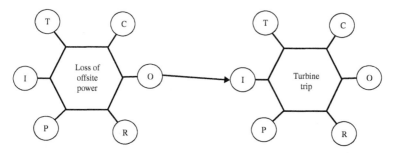

Figure 1.9 Example connection in a FRAM model. Source: Redrawn by Ostrom (2022).

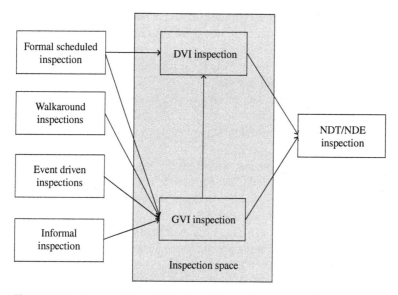

Figure 1.10 Representation of the aircraft inspection process. Source: Ostrom and Wilhelmsen (2015).

Depending on the results of a GVI or DVI a non-destructive examination (NDE) technique will be used to determine if an anomaly in an aircraft part is truly damage. Many factors contribute to the success of failure of the inspection process.

These factors reside in the physical condition of the inspector, conditions of the inspection event, nature of the anomalies, process parameters, and organizational system. These factors are normally called performance shaping factors (PSFs) in the risk assessment and human factors domains.

During the study we developed a risk framework of the inspection process. A risk framework is basically the context in which safety/risk is considered for a system and/or organization. Several different alternatives for depicting the risk

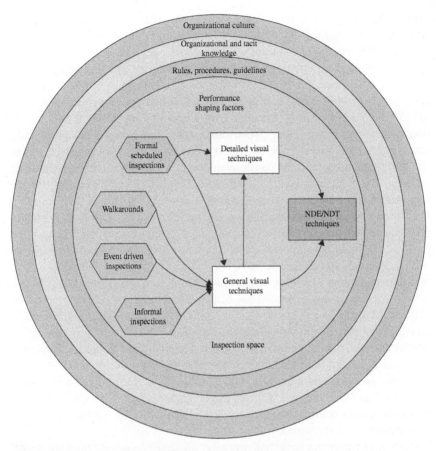

Figure 1.11 Inspection risk framework. Source: Ostrom and Wilhelmsen (2015).

framework for visual inspection were considered. Figure 1.11 shows our final depiction.

The overarching factor in this depiction is the organizational culture. In this framework, there are four generalized entry points into the visual inspection space. These include the following: formal scheduled inspections, walk-around inspections, event driven inspections, and informal inspections. As the graphic depicts, we found that the organizational factors were the overriding contributor to success or failure of the inspection process.

1.4 Safety and Common Sense

I recently was required to take safety training so I can have easier access to the Center for Advanced Energy Studies (CAES) building, which is absolutely a great policy. However, a term that was used was that certain aspects of safety were "Common Sense." I don't believe safety is common sense. In fact, I don't believe in the concept of common sense at all. As discussed in the Section 1.1 in the first few pages of this chapter, I brought up the fact that preparing a rabbit for eating if someone hasn't done it before, is an alien concept. It isn't common sense. As Mark Woodward says in his podcast, if he hands a loaded shotgun to his 13-year-old son without instruction, does his son have the common sense to operate it without instruction (Woodward 2018)? This goes for operating tools, using safety equipment properly, not to mention complex systems like flying airplanes, operating nuclear power plants, etc. I will always remember being called by the front gate security at the PPG chemical plant I worked at in the early 1980s and told a tank truck driver was delivering a load of chemicals and he refused to take an escape respirator. Anyone who worked or entered our plant was required to carry an escape respirator. So, I hiked down to the gate and told the driver I would have to accompany him into the plant if he wouldn't take an escape respirator. He told me he had a respirator and showed me a half-face piece cartridge respirator. However, it had dust purification cartridges. I told him it wasn't acceptable, and he needed one of ours. He refused and, so, I followed him to the point where he had to unload the chemicals. The driver got out of the truck, and I noticed the respirator was hanging upside down on his neck by the bottom strap. I asked him if he knew how to use it. He proceeded to put the respirator up to his face, with the chin cup over his nose and the nose cup over the chin. So, while the operators hooked up the truck to the storage tank, I stood by in the event there was a release, with an extra escape respirator, because the driver obviously had the common sense to know how to use his dust cartridge respirator correctly. Yes, that is sarcasm.

Safety is not common sense. All workers, whether 12-year-old mowing lawns, undergraduate students in chemistry labs, graduate students in nuclear labs, new skilled trade workers, office workers, and engineers must have training and education on how to work safely. This includes people in their homes standing on chairs to reach top shelves or to water plants. Remember, a 6-ft fall can kill a person.

Organizations must not rely on "common sense" to form the basis of their safety culture. They must develop their safety cultures with sound safety management.

1.5 Interviews with Safety Professionals

Five people/organizations were interviewed or supplied information on five questions concerning safety culture. The five questions were:

1. How do you define safety culture?
2. How does safety culture impact the operation of your facility/company/oversight function?

3. Describe an incident/accident that a negative safety culture impacted the outcome.
4. Describe an incident where a positive safety culture has prevented a catastrophe.
5. How does your organization assess safety culture?

The five professionals/organizations were:

- Ron Farris – Human Performance Specialist
- Shane Bush – Human Performance Specialist
- Lt Donald Olsen – US Navy
- Leif Åström – Former Wing Commander, Swedish Air Force
- United States Nuclear Regulatory Agency

The interviews were recorded using Zoom and the Dictate tool on Microsoft Word. Efforts were made to edit double words, vocalizations like um and ah from these interviews. However, the interviews for the most part are in the words and phrases of the interviewees. The information that was supplied by the other people/organizations are exactly as provided.

Ronald Farris, Consultant

Mr. Farris is a co-author of the book; Critical Steps: Managing What Must Go Right in High-Risk Operations, CRC Press, 2022, By Tony Muschara, Ron Farris, Jim Marinus.

1. **How do you define safety culture?**
   The collective behaviors of the organization, good or bad, that produce the work outcomes.
2. **How does safety culture impact the operation of your facility/company/ oversight function?**
   I am a consultant that works with numerous high-risk industry partners in support of improving organizational performance and safety improvement. For me culture is something that directly impacts these organizations' bottom line. It is always difficult to measure the current safety culture as the workforce is not static, that is, people come and go on a regular basis. Additionally, leaders are always changing from frontline supervisors clear up to the top leader. This constant state of change means that if you measure safety culture, what you get is a result that is not the current safety culture. In other words, it is a lagging indicator of the current state not the actual one.
   As for the impact on operations, I believe that collective behaviors directly impact work outcomes. My simple perspective based on my military and

---

(Continued)

(Continued)

civilian experience is simple. Leaders set standards and expectation, how these standards and expectations are enforced and reinforced by front-line leaders (foreman, crew-chiefs, supervisors, and frontline-managers), creates the culture of performance, and this culture creates the work outcomes.

3. **Describe an incident/accident that a negative safety culture impacted the outcome.**

As a consultant I will provide an example of a near death accident that was experienced by a petrochemical client. The safety culture at this refinery was one of accepting the unacceptable. The unacceptable was obvious in that there were numerous housekeeping issues such as tools and debris strewn about in the production areas, numerous tripping hazards in the facility, and there were obvious OSHA infractions such as missing or dangerous handrails. When I got a tour to go see the location of the accident, nobody seemed to notice the issues except me and the corporate safety director. The accident was straight forward. At the top of a platform that was about 10 ft off the ground an operator fell backwards after losing their balance, fell backwards, landing on the concrete floor. The individual sustained bruises and contusions but had come within inches of their head hitting a concrete support that had a sharp corner. The accident was a result of the individual having to use an inadequate fixed ladder that lacked either fixed siderails at the landing and or a cage.

4. **Describe an incident where a positive safety culture has prevented a catastrophe.**

This is a tough one, but I will use an example of what we teach organizations to do to proactively prevent incidents. When organizations are proactively seeking to prevent incidents with consistency, they have a good safety culture. As a consultant I want organization to consciously seek out landmines, report them, and resolve them. A landmine defined: A metaphor for an undetected hazardous condition in the workplace that is poised to trigger harm. A workplace condition that increases the potential for an uncontrolled transfer of energy, matter, or information with one action – an unexpected source of harm – unbeknownst to the performer. Includes compromised or missing defenses; also known as an "accident waiting to happen."

As a consultant I teach organizations to identify a variety of visual cues that something is not right such as landmines (latent conditions) and to respond by calling a STOP-Work and to seek technical help. A strong safety culture is one that instills "Mindfulness" and a healthy "sense of unease," that produce recognition and control of landmines. Formal prework discussion prime front-line workers to expect and detect landmines. Often what I find is that if these landmines are not promptly corrected, become workarounds.

I believe that latent error prevention that creates these latent conditions is not the objective of organizational leaders, rather it is their responsibility to identify and correct the latent conditions left behind. In his book, Human Error, Dr. James Reason says, that "it is neither efficient nor effective to prevent "latent errors," but to quickly find and detect the unsafe outcomes of these errors." "Latent conditions accumulate everywhere within an organization and pose an ongoing threat to safety."

5. **How does your organization assess safety culture?**

   As a consultant in a company that starts with assessing reliability, our assessment includes safety culture. We typically measure the following:

   - Culture surveys (what people think/believe)
   - Number of work-arounds present
   - Number of deferred maintenance items
   - Leadership field presence (top to frontline leaders)
   - Number of STOP Work used by workforce
   - Near Misses fully investigated (free lessons treated like an actual incident)
   - Formalized use of error detection/prevention techniques
   - Work Instruction adherence (Standard Operating Procedures and Maintenance Instructions)
   - Use of Defense-In-Depth approach to protect assets from built-in hazards
   - Presence of formal pre-work discussions (prejob briefings, huddles, etc.) prior to work activities

---

Shane Bush – Human Performance Professional

---

1. **How do you define safety culture?**

   In consulting I'd always been told that a safety culture is the accumulation of behaviors right. It's the accumulation of behaviors of the workers and the management, just basically anybody in the way they're exhibited.

   I still believe that to be true. I still believe that a safety culture is defined by what you see in what you hear, which takes me back to the definition of a behavior.

(Continued)

(Continued)

I also learned early on, through behavior-based safety and other theories that behaviors can quite often be described as anything you see or hear. So, if that's a behavior you want to know how your culture is doing, then you look at the behaviors being exhibited within that culture. That culture is the accumulation of all the behaviors. Now having said that, we have grown so much over the last several years at least in my experiences working with so many companies, big name companies that the culture has also got a lot of hidden parts of it. So, I think in addition to what you see and what you hear there's the unwritten rules as we sometimes describe it, or there's what I say but not what I do kind of messages or the opposite probably.

I think culture is something that we will continue to learn more and more about and define more and more. I think we tried to keep it simple.

I believe that it's just too difficult to keep the idea of culture simple, that there's too many attributes that have to come in for a culture to work.

2. **How does safety culture impact the operation of your facility/company/oversight function?**

There's the unwritten or the behind the doors talked about topics that drives a culture as well. So, again, I believe that culture still is described as what you see in what you experienced there, but I also believe now that there are many things that are very difficult to measure that can, if not drive a culture heavily, influence the culture.

I've been given permission to use some names and some examples but obviously I've signed documents where some stuff I can't share. From my experience and just for a reminder to you in the human performance consulting I do for the audience, it is obviously dealing with human error and human fallibility. Human error and human fallibility occur in many more areas that just safety.

Having said that, there's no doubt in all my years of what I currently am doing, and, for that matter, my whole career in safety has always been a topic that is universal.

It doesn't matter whether I'm going into Disney World, it doesn't matter whether I'm going to Cirque du Soleil, going for food, to do we government or non-government, safety is the number one topic.

As an example, with human performance, then they realize that it's much bigger than just safety it's a business initiative.

But safety is such a cornerstone part of any business that without a good safety culture. I don't care how good of a product you produce. I don't care how well you've done in improving it.

If safety is not at the competitive levels of either your peers, or what is expected. It will eventually affect your business period.

That is why so many companies in my opinion, spend so much time on safety is because, one, it's a building block that I to have as best as I can and then we learned so much and safety, because a lot of people realize it really needs to be a business initiative. I think the first relationship, we made with company components was quality, safety and quality have to be hand in hand, without safe you don't have good quality without good quality don't have good safety, W we spent years talking about that. And then it just basically moved into production and moved into I mean, even in six sigma and lean manufacturing and all these other product process improvement initiatives that are out there, they still come back and want to reevaluate. To have the improvements we've suggested hindered or caused any sort of harm to our safety or safety culture.

I don't see any day ever in the future in which safety is not a number one subject or a priority, which course when you, when you learn about safety with some of the behavior list they'll say, safety can't be a priority safety has to be a value.

Right. We've heard that a lot and practice priorities change and priorities, switch.

But values are always the same. One example and this is what I can talk about because it's open, and it's a well-known example is at the Idaho National Laboratory (INL), this has been years ago so this activity involves previous contractors. Many people are unaware that we have Three Mile Island (Nuclear Power Plant) (damaged) fuel at the INL, and we had to move it from one location to another. It involves taking it, I believe it was in in stored in water, to begin with but anyway it was in one facility, but it had to go to dry storage,

I know that it ended up at one of our facilities and dry storage.

So, we called it a campaign, and it was about a six-month campaign, the beginning to the end we planned around this we're going to do it in six months.

As a lab, we're ahead of schedule and under budget. Bonuses were being talked about etc. Corporate then came in and they looked at the safety statistics what they could look at the time which is your typical total recordable case rate, daily case rate, the stuff that quite often comes up.

And for this given organization that had again. Come in, ahead of schedule under budget, their safety statistics were almost twice in the negative way,

(Continued)

(Continued)

twice as bad as the regular company average, and the actual corporate said their words not mine, that we would consider this an overall failure.

Because even though you came in ahead of schedule and under budget, and we didn't really have any big safety incidents. There were numerous indicators that we easily could have had a major incident. This was my first impression that corporations really understood the importance of safety and then of course you get into the big events like you had with the Deepwater Horizon, where you have potential of management going to jail because of safety.

This takes it to a whole other level. Maybe not for the right reason, maybe they're doing it because they don't want to go to jail but at least they're doing it right.

I see safety will always be the foundation, no matter what approval processes, whether it's voluntary Protection Program (VPP), whether it's integrated safety management system, or for process improvements, it'll always be a foundation block that we absolutely need to continually improve and there's been some companies that have done wonderful in improving their safety cultures and seeing great benefits.

One other quick example I want to give you. When we implemented the VPP at the INL I was a part of the core team, my specific assignment was teaching management the different principles of VPP. One of the things that we did specifically talk about was how do we, if we're going to grow a culture at the INL, how are we going to get the subcontractors up to speed. How do we get the subcontractors to take on that same culture?

We actually got approval eventually that subcontractor could not even bid on a job at the INL unless they met certain criteria, they were having safety meetings they were talking about hazard analysis, they were reinforcing the use of VPP all the things we find routine today.

It was the solid companies that really had good programs that were extremely thankful because they were trying to compete with fly by night companies that were really skimming and cutting corners.

So, once again, safety came back to not only address safety, but now you had to be good performer in all areas in order to have the resources to compete with companies that are required to be at a certain level in certain performance criteria.

I thought that was a good example of how you can take safety and get a lot higher caliber subcontractor or workers or whatever, by what you reward and what you don't reward.

3. **Describe an incident/accident that a negative safety culture impacted the outcome.**

   One that I actually, and again I have to be careful what I say here but I was working with a company in Montana I can't tell you the company's name, but involves a bunch of linemen.

   And these linemen, for whatever reason, they had been picked up by a company to do a documentary on them.

   And the documentary meant that they were followed for a couple of weeks with cameras and lights and all sorts of stuff.

   So my involvement obviously came after the incident or the accident, but to kind of set it up, they had a call out one evening in which they had to go out and replace what's called an H structure, which basically means two poles one cross member it looks like an H, and they videoed it from numerous corners like I mentioned, it was at night.

   They had floodlights on and they were replacing a rotted power pole. A worker grabbed it with a great big grappler, a great big tool that can hold it. They disassembled all the lines and everything on it. Then that person takes that pole and lays it down. Grabs the new poll and he or she holds it in place while they fasten the lines and the cross member, etc.

   SA little bit of context here is important, you're outside it's blowing. It wasn't deep winter, but it was cold weather, so you're not going to verbally be able to hear each other so the workers usually communicate with sign language. They have sign language, the universal with that field or that particular industry. But what the person is supposed to do is take that pole hold it, and then someone on the ground chokes it off with a chain in a smaller little hand driven crane that they can actually take the weight off of it. That person up there can loosen their grip because they really can't go up and down very easy, it's too big of machinery, but they loosen their grip and then the person down below, actually lowers the pole down into the hole and then they do all their work on the ground.

   Over time there is system drift or process drift. They quit confirming with sign language what they were doing because they can visually see each other and they just figured; well I can see that they're doing so why do I need them to confirm something I can visually see. So, on this day the person at the bottom chokes it off and walks away and the person up top assumes good to go.

   Unbeknown to him, the person at the bottom hadn't completely choked it off. I think they call it a turn buckle, that you put through the chain and then you could pull it.

(Continued)

(Continued)

As he was walking away when the person up top loosens their grip the pole completely went out of their control and fell directly on the top of the employee and killed him instantaneous.

By the way, another side note here is when I interviewed the CEO of the company; he talked about the difficulty of visiting a wife and two kids and explaining why their father is not going to go home because of something as simple as communication.

I have found is the absolute number one source of errors and mistakes that result to unwanted outcomes, especially safety communicating it might be verbal nonverbal and one thing we have taught since this incident was communicate.

One thing we have taught since this incident was to communicate your intent and what we call critical risk important steps. Communicate your intent. In other words, if either one of them had communicated they wouldn't be dead

The person up on the pole got their attention says I'm going to loosen the grip. There's no doubt the guy on the ground would have said hold the pole (Shane crossing his arms over his chest) which is a universal sign for stop what you're doing. A worker might indicate that he is getting ready to leave or not done, whatever that language is. It's so sad that we have such significant catastrophic outcomes, where a simple tool, if they continue to use it, could have prevented it and to be home safely.

It is all about your willingness of being diligent because we use it for a while, whatever it is we get a new tool, we get a new process, we use it for a while, we kind of get a little bit lazy. Honestly, the one other thing that I have found is, I am totally convinced, that no matter how much training you give people, no matter how much you talk about defenses or controls or hazards, everybody does real time hazard assessment in your head. You know the rules you know the requirements, but you will look at something, and you'll immediately start running rationalizations it in your head as to why you do or don't need to do something this time, or why you can do something this time that you don't normally do. This real time rationalization of hazard control I think is a huge thing that needs to still be dealt with.

What a sad story. What a sad story that could have prevented with one small communication tool.

There's no such thing as common sense. There's a book out there that says that there's no such thing as common sense, common sense to you is absolutely going to be different or interpret different than me. That's your point is that it is such a personal thing that it does not exist. I totally agree 100% it doesn't exist. And by the way, there's another book out there that I can't remember the title of, but it talks about there's no such thing as a routine task. There's no such thing. All you have to do is change how much sleep the worker got from one night to the next and the task is different, or you got to do is change the fact that you and your spouse had an argument. The task is different. So, there's no such thing as a routine task and there's no such thing as common sense. And the more we understand that the more we can deal with this real time hazard assessment that we tend to do as human beings.

So, again, you've kind of talked about this but describe an incident where a positive safety culture of preventative out a catastrophe. I'll let me give you a really good one, and this one again, I can talk about.

So, as again you may be aware, I am, or was I was getting ready to leave my current job, but I was asked four years ago to get the Human Performance Improvement (HPI) program back up and, and we're calling it a reinvigoration of HPI at the INL.

There are three categories that we're always looking at simultaneously always and its safety mission, interruption, and physical damage.

So, if there's anything that an error or mistake or misinterpretation, if it happened could result in a safety incident.

Mission interruption which is kind of a catch all phrase involving reputation as you reputation of the lab and or physical damage to something you hold a value, we're going to concentrate on those steps.

So having said that, we have a biofuels test apparatus where they take these biofuel experiments and the containment that they put them in. The lid weighs, I have to guess about a ton.

They have a lifting device that looks like a forklift but it's a tele-handler if you're familiar with the tale-handler. It not only can go up and down, but it can go in and out and it can even tilt.

They were going in and using this tele-handler. A little bit of context here is important when they actually installed the tele-handler, they had to install it high above the deck because of all the equipment in the way. So, the straps are quite long. That is going to play into this. So you got straps coming down X feet, they hooked to the lid. Every time the lid is either lifted, or lowered, workers have to be out 25-ft radius away.

This has been the practice for seven or eight years.

One day, this was about two years ago, they're doing the job.

(Continued)

(Continued)

What they were doing in this case as they lifted up and they had replaced the experiment successfully. Now they're ready to lower it back down, which is about a 3-ft drop is a better way to say it.

Now, all the workers go out 25 ft.

As this thing is lowering, where do you think all the workers' attention is on this lid, on the tele-handler, on the experiment, where are we looking? I'm looking at the experiment. Right, so nobody's paying attention to anything but the experiment, which is how it has always been. And as they're watching it, everything is looking normal because it's coming down; unbeknown to them the tele-handler has two levers side by side; they are co-located and look the same. One lifts the tines up and down; the other level tilts the tines. The operator accidentally grabbed the wrong one and started tilting the tines, which gave us the exact same immediate action it lowered it, but it tilted it to such a level that the whole lid slid off of this tines and boom, it came crashing down.

Now, we actually had a huge safety celebration that day. Management actually had a safety celebration that day because for seven years, it finally paid off to have workers stand back 25 ft.

But remember our second category mission interruption. Did we celebrate mission interruption? No, because now we were behind schedule and had equipment damage. So, two of our three things, but from safety was concerned, we celebrated safety.

One thing that we've started to do is you really need to look at the different components of a job or an activity and separate safety from all the other stuff.

And if the other stuff doesn't go well, address it accordingly. But if the safety components go well, you need to celebrate that so you still get the continued behavior you want.

When we talk about resiliency and again what I'm about to share with you I presented a national conference and now I've got people using it everywhere with a lot of really good feedback on how it has helped him in just about everything and anything (Boyce et al. 2020)

They've been doing related to looking at the potential of an unwanted outcome mind you related to human error and human fallibility, but it's just continuing to cross into the organizational performance etc.

So basically, this is how the resiliency tool works (Figure 1.12). If you take any activity, doesn't matter what the activity is and the pictures are intended to, to help with examples, in this case, obviously a railroad crossing.

BushCo HPI.com

Failing safely
"Building a capacity for resiliency"

Increasing resiliency ↑

5 **integrates permanent barriers** intended to permanently prevent error and/or the consequences of error for a specific critical step.

4 **integrates temporary barrier(s)** intended to prevent specific errors and/or the consequence of error for a specific critical step.

3 **integrates** safety **tools** such as: sensory warning devices, visual delineated boundaries, or other specialized tools that cue individuals they are in a hazardous area or zone.

2 **provides personnel with job aids** such as checklist(s), labels, signs, procedures (?) that communicates the behavior expectations at the critical step.

1 **depends on people** not making errors, regardless of knowledge, skills, and ability, to perform the critical step error free (e.g. self-checks, memory requirements, Simple/Routine Communications).

Improved capacity to manage human error at critical/risk important steps →

Figure 1.12 Concept of resiliency. Source: Shane Bush.

But if you look at the pictures what they're intended to do is give you an idea on how resilient you are.

And again, this could be safety so if you're in a number one it depends on people not making errors regardless of knowledge, skills, and abilities. Low resiliency because we know people now remember we're talking critical and risk important steps, not everything or you're going to go broke. But number two if you provide them at least with a job aid or a checklist or label or sign. You've increased the resiliency and then three safety tools for temporary bound us five you engineer that now is this related to the, to the hierarchy of controls. Yes, but it's specific to human error and human fallibility. What we have found with this is people are using it.

You may not be aware but we had a job where we actually put out a request for proposal, we granted it to the subcontractor, they came in, they built the building, the prime contractor wrote them a check, and through that whole process we didn't provide the proper details. If you want to piss off management spend millions and not get what you paid for.

Now while that wouldn't fall into a safety thing, the scale absolutely applied, because it was mission interruption.

(Continued)

(Continued)

Within the last two years we sent out, at least on one occasion, we made an offer to a candidate, who was the wrong candidate because of all the corporate crap and interrupted our process and everything. The wrong candidate accepted it. I don't know how far along they got but it was far enough along of the person selling their house, to quit their job that we're in the middle of a lawsuit.

So, this resilience scaling has really turned out to be a really cool thing.

The other thing that I think is important and this is something that we've really been pushing a lot.

I work with Todd Conklin and have a good explanation of our resilience process. Listen what he says and this is basically our premise right now every time we go out. This is exactly how we approach safety and the private world.

So having said that, do you see Todd on there yet. Yes. Okay. Let's listen to this clip. Just when I started here. Tell me if you can hear him.

"So really about automobile safety in the 1950s. Back then, the automotive industry said these words, asking drivers to not wreck is not very effective.

It feels like the right thing to say, but it doesn't reduce the number of vehicle accidents.

Our job has to change. And so what they said is, let's assume everybody driving in a car has a chance that they're going to wreck. Because when you deal with uncertain outcomes. You've got to use certain terms. And so what they said is, let's not manage the absence or presence of the failure.

Let's actually manage the cars, ability to handle the failure.

And the idea became that we can actually make cars safer and safer and safer. Now what's interesting is if Industrial Safety were in charge of vehicle safety.

I don't think they could make the claim of zero fatalities.

Because if Industrial Safety were in charge of vehicle safety here's what it would look like in your company, we'd have a program where before you got to drive, you have to take the mandatory don't wreck the car training. And then once a year, you'd have to take the don't wreck refresher class, but that's on the web. So it's pretty easy to take. And then we tell everybody to drive on not wrecking louder and slower. What part of not wrecking did you not hear the last time we did this.

In fact, I'm going to suggest, our target for safety has to change.
I think this is our new definition.
Safety is not the absence of accidents.

This is the theme we use all the time. Now, when we look at a process, we don't look at it as you're well aware for years and years when we go into new companies how healthy is your safety program oh good How do you know, well look at our total recordable rate look at our day with a trade look at our close calls or near misses or whatever and it's like, that tells me nothing and the question is, what did you get that by luck or did you get that because you're good?

What we do now is we go into any system or process help them identify their critical risk important steps, and then help them put the appropriate level of defense in place to catch the air but more importantly reduce the consequences of error at those steps, whether its safety, whether its mission interruption, or whether it's physical damage. I'm telling you, it's amazing because, and then they have to put it on a scaling, they actually have to go in identified the steps, given a scaling of where we're currently at whatever that is. Our guidance and I can even send you the procedure we have because it's only five pages.

The guidance says, take these critical and risk important steps and get the resiliency to as high as reasonably achievable. So, as you probably picked up on we stole the ALARA (As Low As Reasonably Achievable) concept right to as high as reasonably achievable. We just flipped it, and that way they don't feel like, "Oh my gosh, there's no way I can get every step to a five." We're not asking you to just get it to as high as reasonably achievable.

This graph that I showed a second ago is the organization's accountability to get safety to a level, and then the accountability goes to the worker. We've got it in there and by the way we've got a critical step here, and it's relying on pure check to be successful,

It's relying on procedure adherence to be successful it's relying on an independent verification to be successful. So, you've got organizational accountability; get it the best you can.

Then the individual accountability which we absolutely felt was critical in the bar for company leadership to buy into this, because it didn't, we didn't want to make it look like we're handing all the accountability off to the company in this endeavor. We've got to keep that individual accountability, but it comes after we've developed processes to the best we can.

I in fact have my students fill out the exam for getting their certificate for the class I teach. So, I do give them an option go watch the last, what's it

(Continued)

(Continued)

called Deepwater Horizon. I tell the students go watch it and identify the precursors the critical steps, the performance modes, etc. and write me a report and I'll let you do that if you want you can enjoy a movie.

It's amazing. It's, I mean I've had books sent to me by students.

Every time I hear a story about wannabe rescuers dying in an emergency event, it's like it usually takes three rescuers dying before the realization of a hazardous situation. One goes down, a second person doesn't know what's going on, so they go down and the third by the third person they're kind of thinking, okay there's a pattern here, what is it?

Whether it's confined space problem or whatever, right, or steam. With that seems, by the way, that sounds so familiar to our TES (Thermal Energy Storage) incident, we just had. Are you familiar with our TES incident?

TES is the name of the experiment. It is where they were trying to vent off what they thought was water, but it was hot oil.

The reason that that we came up with that mission interruption term was because there were so many people at my workplace trying to exempt themselves from human performance, you know, it's people who turn valves or people will flip switches and when we start talking about well what about people who get a job offer when they shouldn't worry about contracts you get issued when it shouldn't or what about and QA standards that are met. Again, it usually comes back that human error was a big role. So, I feel so fortunately that again my internal sponsor is very supportive. I must give him credit. He has always been supportive to me.

Lt Donald Olsen – US Navy

1. **How do you define safety culture?**

   Having a culture meaning something, where you're from top to bottom kind of unified on the idea, when it comes to safety for that it goes beyond just having training shows to some extent that the community is holes invested in safety, but it's kind of limited in getting that full investment for everyone, because we have training for so many things. I think kind of dulls the impact of training, but when you hold people accountable, and show that there is that accountability especially at higher levels, that's where we really start to see a lot more investment from people all throughout the ranks oftentimes. We end up holding the lowest guy accountable and that ends up with people, saying, obviously, the community doesn't really care that much about it, because they were trained by so and so or so and so was in charge when this happened and yet they got a slap on the wrist, whereas this other guy got fired. I've seen good and bad safety culture, and as far as good safety culture goes, it has been best, where that accountability was equal throughout all the ranks.

2. **How does safety culture impact the operation of your facility/company/ oversight function?**

So, for the new clear community in the Navy, we have these watch standing principles. I have these written down, and I don't mess them up, but we've got integrity level of knowledge formality, questioning attitude, procedural compliance, and forceful backup. Each one of these ends up really coming into play, when it comes to having a positive impact on operations through our safety culture for integrity. Having that encouragement of people being open and honest about what happened really has a huge positive impact on safety culture and our ability to operate and correct our mistakes. We don't make repeat errors, whenever a ship has any kind of mishap that incident will end up being very thoroughly investigated. Punitive measures occasionally come into play, but that's usually only, if someone was blatantly negligent or did something maliciously, because we want everyone to be very willing to be open and honest. We don't want people afraid of negative impacts on their career just because they were saying OK, if I did this thing the wrong way, it ended up causing this thing to happen and now this thing is broken, or someone got hurt. We want that discourse to be able to happen uninhibited. So, the repercussions are usually minimal, whenever it comes to someone, who's being open and honest for what they've done. So that what went wrong is as soon as that investigation is done that information is promulgated throughout the entire fleet. This helps not just a single ship being one part of the fleet; it helps not just them to avoid this mistake again, it also helps every other ship that has similar symptoms to not make this the same exist for level of knowledge on that. We really expect even at the lower levels, especially at the lower levels, we want them to be subject matter experts (SMEs), if they're knowledgeable about the system. This has a huge positive impact on their ability to be safe around the system. The more they know about the risks involved and the more they know about how to use the system properly, the less likely we are to run into safety concerns, whenever you're talking about human reliability analysis, you often end up considering the level of expertise of the personnel involved. I've seen that more experienced personnel

(Continued)

(Continued)

are far safer in operations and people who have that knowledge, then formality and procedural compliance kind of come into play together for this. Having that well written procedure and then going through the formal process of doing the same way every single time really helps you to promote safety. If it worked safely this one time and you, did it this one way, it should work safely doing it every single time, if you stick to what has been done and what is written for you to do, and then with that formality having a formal process not just, when it comes to procedural compliance, but also whenever it comes to things like watch turnover. You want to make sure that you have a set process where the people who are going to be taking the watch being the people, who are going to be on their shift essentially. It's working in shifts, the people who are coming on shift will get full scope of what's going on before they end up taking control of it and that helps set some continuity throughout, and as long as everyone does everything the same way. It makes it easy for someone to step in and then for that questioning attitude. I don't really promote this as much, as it's promoted in the Navy, but having a questioning attitude and being willing to ask the question and actually say, is this safe or is this how we are supposed to do it, that is very highly encouraged and when you look at a lot of major problems that have happened in engineering and in industry over history, a lot of these problems could have been mitigated, if someone had the courage to actually speak up and ask a question. Part of this requires the culture to not have punitive measures for those, who are willing to ask the hard questions and then that forceful backup this concept is where someone is willing to say to step in and say you are not doing this correctly. This is how it should be done now. This often happens when you got supervisors who are talking to subordinates and they're saying, "hey you're not doing this correctly," and that's common place in a lot of workplaces, but what the Navy does an excellent job with is empowering subordinates to be able to do this now. It's still within the military structure. They're still doing it with tact and doing it with a lot of respect in it, but they are the subject the enlisted the lower ranking people are the SMEs meaning, they're the ones who in their area. They should know the most at least from a technical standpoint. Now the officers and the people managing them won't necessarily know quite as much about that they know. Where they are more like a Jack of all trades where they know a little bit about everything in the plant and how it comes together. Let me play empowering those lower ranking people to be able to say, "excuse me Sir, excuse me ma'am." This is what the book says on this. We're not doing this in accordance with what the book says. I can't comply with that order because as a matter of fact, the book was signed off by and Admiral, who's far higher ranking than the officer who's ordering them to do something. It's not like they're disobeying a lawful order or anything like that, but having that empowerment to give subordinates the ability to raise their hand and say "no we have to do it this way because this is how it's written" is something that has always really helped me and it really helps I think maybe as a whole in making sure that we are as safe as we can be in our operations.

3. **Describe an incident where a positive safety culture has prevented a catastrophe.**

   I can't go into specifics, but I can talk about just generally, what we do in the Navy and how it regularly ends up preventing catastrophe, and this feeds back into that a safety culture aspect. Even at the very lowest level, let's say you've got a brand-new junior sailor. They've just come into the Navy and they're doing their first maintenance item, so they are instructed to do something, let's say as simple as replacing a light bulb and believe it or not yes, we have procedures for replacing light bulbs. They have this maintenance requirement card, or an MRC and they're supposed to go through step by step and at the very beginning of every single MRC no matter how simple it is it's going to have some safety precautions listed up at the top and a part of that is always going to regardless of what it is it will always be referring to some form of the NSTM or Navy service technical manual. You'll have to check me on that one, but it's going to be referring to the safety portions of this and whatever safety portions are applicable to it, but it always at least includes general safety, which has everything from holding onto hand rails, while going down a ladder well which is steep stairwell essentially, and believe it or not a lot of people regularly go down ladder wells without holding on to the hand rails and it's just not worth it, because ships can move unpredictably. If a wave pops up and hits you suddenly, you don't see what's happening you could end up falling or getting hurt. God forbid crashing into something suddenly and, if you're not holding onto something you could get hurt. I've known of several people who have fallen and gotten hurt on ladder wells, but it will also include electrical safety portions and part of this. They do their maintenance, and they replace the light bulb OK. That's a good start and they, hopefully, followed the full procedure and had all their safety equipment together etc. Now we have what's called a spot check, and that's, where a supervisor is going to come and check, that they know how to do the maintenance properly.

(Continued)

(Continued)

There are several different forms in which this can be done sometimes, it's historical, and you'll have to follow along behind them and check too. They will walk you through what they did in the past, and you'll just see that they knew how to do it, and then sometimes you can actually just watch them actively doing the maintenance, but in almost every single one of these, we as supervisors are encouraged to ask them about some of the specifics, about the safety portions of the NSTM (please define), that they used and now the expectation isn't that they had to read the entire chapter of the NSTM every single time. It's good to make sure that they, at some point, read it, and that they are familiar with what's needed from that, and then whenever it comes to things that are more technical in nature like electrical safety, they better have had that open and actually gone through all of the actual safety precautions, because those are those are steps in air required procedure, but for general safety, you're wanting to make sure, they have a good idea of what's happened now a little bit more on electrical safety; a lot of the incidences that you hear about in the public world and in the military often end up being related to electrical shock, and this can be catastrophic both for the people involved, but it can also damage equipment which on a warfighting platform. That's not good, especially, if you're out to sea and you need to be able to make it back home, so going through all those actual initial voltage verifications and every bit of the checks is extremely important and that is why we very regularly end up having additional supervisors, or will, if for everything on, honestly, we end up having additional supervisors. There is as someone, who's not hands on in it, and they can check and make sure that everything is being done in accordance with the safety requirements.

4. **How does your organization assess safety culture?**

   For assessments, we have a lot of different ways that we do this. I touched a little bit on kind of the lowest level on the last question, where we have the spot checks for maintenance stuff and that's part of the self-policing, that ships are supposed to be able to accomplish, and in order, to make sure, that ships can check themselves doing their own self assessments. We have outside entities come in and investigate them, but I'll touch a little bit more on ship's own self assessments first, so in the new clear community on top of having there are spot checks. We also have audit and surveillance programs. There is a whole lot of reviewing paperwork making sure paperwork is all up to date. Believe it or not that is actually a huge impact on safety for the nuclear side, because if your paperwork is not correct sometimes, that can be, or if there are errors or not necessarily errors. If there are things that are noticed whenever you reviewing this old historical paperwork, you can end up catching things that weren't originally caught by the operators and you can find problems with the ship before it ends up becoming a bigger problem, and obviously structural integrity, and chemistry controls, and radiological controls. These are all things that can end up causing huge incidences and in problems. You could lose the public's trust in our ability to operate nuclear reactors, if you end up with some of these more severe problems occurring, but we've got these surveillance programs as well where this may not be a the MRC card type maintenance and not the standard maintenance kind of things, but maybe something a little more complex or an operation you could be observing an operation like the startup of a nuclear reactor or something like that. These are going to be experienced personnel to know a lot about that area and then they're going to come in as an observer. They can step in if they need to for the sake of safety, but they're just there for assessment purposes and that's within ships force, and then, if you're working with the shipyard, they're going to have their own civilian entities. They're going to do very similar things for their own personnel and for the ships force, because you're going to have them both working together quite a bit, but then at a higher level, you've got the type commanders. So, a command is going to be underneath the overarching fleet. Let's say the Pacific fleet would have several type commanders, he would have SURFPAC and AIRPAC audit commanders, who were in charge of the surface fleet on the Pacific side or the aviation side of it, which includes aircraft carriers on the Pacific side. These leaders are both administratively and operationally in charge of the individual units, so part of that includes their need to perform assessments, and they've got several entities that can help do this, but you've got like mobile training teams in the float training group. They are two of the biggest examples of assessment teams that will actually go out from ship to ship, no matter where they are in the world they will even fly out to ships on deployment, and then they will do these multi-day sometimes over a week long assessments, where they're doing extremely in depth review of all the administrative stuff, but also checking for our ability to perform operations, and a huge part of what they check is can we notice the same problems. They notice and then we notice a problem, or if they notice a problem first, and we don't notice it until they bring it up how do we correct that and it's looking for the ship's ability to do that self-releasing. I was talking about an issue that's been noted, how quickly

(Continued)

(Continued)

is that issue communicated and where is that communicated all the way up to the appropriate level, or did it die ahead the at a lower level. It should have this secret and it did end up being held at the divisional level, or did it make it all the way up to the department head and who can then choose whether, or not it's worth the captain's time to know about it and then how did they fix it, how quickly did they fix it, and did they follow procedure, when they fixed it. It just gives them even more that they can observe as you're trying to fix these issues, and a lot of times these end up coming down to operational issues. Someone messes up procedurally and they catch it, if they're looking for obvious. You can't go back in time and fix that, but they're looking for your ability to then promulgate that information out to the relevant people who could also make that same mistake, and how are you doing. We've noted this issue three times in the last couple of weeks, obviously, this is a little more than just on an individual level that people are making mistakes. We're going to train on it as a whole or who knows, maybe, it is just a one-off thing, and just that one individual needs to do a little bit of extra, when it comes to their qualifications or improving their level of knowledge. We call those upgrades, but so sometimes it's as simple as an individual made a mistake because we are humans, and we will make mistakes and that individual just needs to recognize that and then do a little bit of learning on it too. They themselves can avoid repeating that error, then at an even higher level, we have a bigger entity that can end up doing investigations. Like for the nuclear side naval reactors will come out and do big investigations are not necessarily assessments. They'll do a big assessment on ships periodically, and every single ship ends up with an assessment on a regular basis, and these are very much outside entities. If they end up noticing, the ship is not safe to operate, the type of commander also gets in trouble. That encourages the type of commanders to really thoroughly investigate on that lower level, and do very thorough assessments, then that way when naval reactor shows up, they are noticing this ship has significant problems that haven't been addressed by the mobile training teams, and they're making sure that the ship is in fact going to do everything, they need to do to maintain that public trust that we need in order to continue to operate many nuclear reactors right off of some of our busiest ports. It's very important to our national safety and we need to maintain that trust.

Leif Åström – Former Swedish Air Force Wing Commander and Aviation Accident Investigator

The following was derived from a slide presentation provided by Mr. Åström and not an interview (Åström 2019).

Mr. Åström's experiences included:

Swedish Air Force
- **Squadron Commander attack at Air Force Base (AFB) F6** (10 years)
  - "Supplier" in the Command Chain
- **Wing Commander Flying at AFB F6**
  - Operations Commander in war
  - "Customer" in the Command Chain
- **Also worked with:**
  - System Safety Analysis Gripen (Saab)
  - Mid-Life Update of AJ 37/S 37 Viggen ("the AJS-mod")
  - Development of Command/Control/Mission Support System
  - Electronic Warfare system (VMS I) for JAS 39 Gripen
  - Responsible at HQ for Flight Safety JAS 39 Gripen
- **Chef FLSC** *(Flygvapnets LuftstridsSimuleringsCentrum)*
  - Combat Training of pilots
  - Tactics Development
  - Weapons and System studies

Private Consultant
- Air Power, EW
- Accident Investigations

Sweden is a relatively small country. The population was approximately 7.8 million people in the 1960s. The current population is 10.4 million. However, it is very technologically and industrially advanced. Before 1990 Sweden had the fourth largest air force in the world. In the 1960s the Swedish Air Force (SAF) had a problem. They lost 215 aircraft and 97 crew members to accidents. This loss was comparable to combat losses in a war.

The SAF knew something needed to be done. The quote provided says: "We cannot continue to crash ourselves away in peacetime, we must do something!"

The SAF blamed the pilots and crews involved in the accidents and it was said the pilots didn't have the "right stuff."

(Continued)

(Continued)

The SAF conducted a thorough analysis and in the late 1960s they introduced:

- Simulators
- A modern and readily accessible system for submission of Occurrence Reports

The SAF introduced the following changes:

- 1970s:
  - Better Pilot Selection
    - Knowledge tests
    - Psychological tests
    - Medical tests
    - Ability tests
- 1980s:
  - Leadership training:
    - 2-ship and 4-ship leader
    - Squadron commander (including Competence Evaluation)
    - Wing commander
    - Flight training focused on training rather than selection
- 1990s:
  - Simulators
    - Realistic training of dangerous procedures
    - Instructors with dedicated training
- 2000s:
  - Improved simulators
    - Outside world visual system
    - Realistic cockpit motion
    - Networked simulators
  - Dynamic simulator – Fighter jets can operate up to 9-G forces and pilots have to be able to withstand that level of G-force. Figure 1.13 shows the G-force simulator and the results of the testing on the human body.

Figure 1.13 G-force simulator. Source:
(i) Courtesy of Sven E Hammarberg.

The SAF developed a "toolbox" to combat the high incidence of mishaps. Figure 1.14 shows the reduction in incidence over time as the changes were implemented.

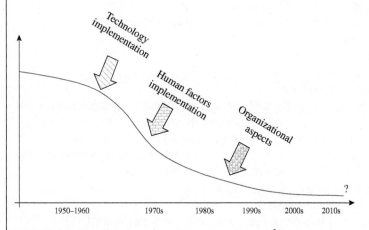

Figure 1.14   Swedish air force incident trend. Source: Åström (2022).

The SAF toolbox contained:

- Technology
  - Reliability of systems
  - System safety analysis of systems
  - Man–machine interaction improvements

(Continued)

(Continued)

- Human Factors
  Human capabilities and limitations
  - Selection of pilots
  - More focused and better training
  - Better planning and preparations
  - Making use of Lessons Learned
- Organizational aspects
  - Balance tasks – Resources
  - Safety culture of the fighter squadrons
- Continued monitoring
  - Medical evaluations
    - Physician examinations
    - Fitness and circulation testing of pilots
    - G-force tolerance testing over the career of the pilots
  - Knowledge evaluations
  - Ability evaluations
    - Simulator evaluations
    - Live flight exercises
  - Monitoring should be taken seriously because two recent commercial airline events occurred because the pilots possible had mental disorders. These are the Germanwings Flight 9525 in which Andreas Lubitz, the young co-pilot, deliberately crashed a Germanwings airliner into the French Alps on 24 March 2015, killing himself and 149 other people, started flying as a teenager. This was called "suicide by pilot (BBC 2017)."
- Traditional risk management
  - The focus of traditional risk management was on (Figure 1.15)
    - Consequences, direct causes
    - Non-safe actions by operational personnel
    - To not "function safe" is punished
    - Takes care only of identified safety problems
    - *WHAT? WHEN? WHO?* But not always *HOW? WHY?*
- Swedish Air Force took the attitude that
  - "If we know about incidents, we can avoid accidents"
    - Let us monitor and correct less serious occurrences
    - In that way we can discover and correct serious problems
    - Occurrence management was implemented that reports were easier to enter and information was easier to access.

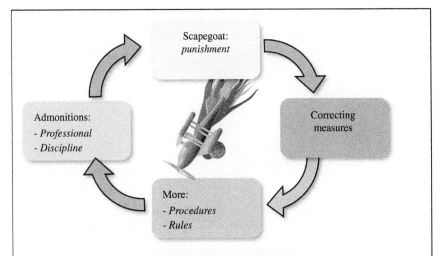

Figure 1.15 Traditional risk management. Source: Redrawn by Ostrom (2022).

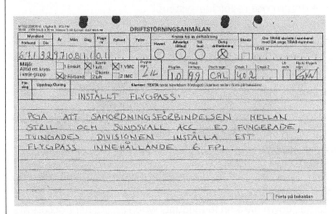

Figure 1.16 Example occurrence report form. Source: Åström (2022).

- Figure 1.16 is an example of an occurrence report. Note it is in Swedish.
- The SAF established a system for Occurrence Reports
  - In service since the late 1960s
  - An open and honest system – No punishment
  - Methodically structured, computerized and continuously developed
  - Data base today, comprising more than 100,000 reports
  - About 4000 reports are submitted annually from crew members, technicians, fighter controllers, and others

(Continued)

(Continued)

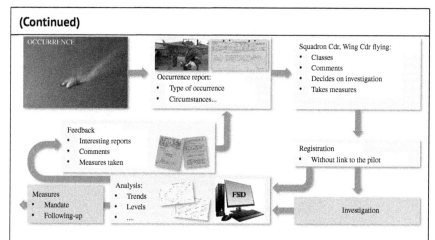

Figure 1.17 Occurrence report system. Source: Åström (2022).

- Figure 1.17 is the occurrence reporting system.
- Open, honest reporting prerequisites:
  - Non-Blame Culture!
- The system must not be used to reprimand involved individuals.

Confidence in the System

- Is built slowly
- Prerequisites that anonymity is possible
- Can be ruined in no time by one person's mistake

No-Blame Culture: Many organizations contend to have no-blame cultures. However, examples of how an organization that claims to have a no-blame culture but then comes down hard on individuals when an incident occurs are not uncommon.

One example is when a fighter experienced a fuel loss during flight:

- While flying suddenly a large quantity of fuel was missing
- The pilot aborted and wrote an Occurrence Report (honest!)
  - "I checked the fuel before start but could not swear on it in this case."
- The AFB Commander: Reprimand the pilot for misconduct!

A dialysis case:

- A nurse unintentionally shut down fluid level alarms and the patient died.
- Shortcomings in equipment design and the equipment was scrapped soon thereafter.
- Court of Appeal: Blamed nurse for other persons death.

– The nurse was particularly blamed, as she had earlier remarked on the short-comings.
– Could not happen with an open reporting culture.

Occurrence Report

– The pilot/operator is responsible to report deviations:
  • Personal mistakes (errors, breaking clearances, etc.) or potential mistakes
  • Events that have affected flying negatively (technical problems, faulty information, organizational circumstances, etc.)
– To not write an Occurrence Report is
  • An active distancing from the Occurrence Reporting System – Not acceptable!
  • Highly unusual
– Writing an Occurrence Report gives
  • No protection when rules have deliberately been broken, for instance the pilot of the C130 Hercules was banned from flying after breaking clearance rules (Figure 1.18).

Figure 1.18 Pilot flies C130 too low in violation of flight rules. Source: Åström (2022).

  • Occurrence Reporting must
    – Be quick and simple
    – Provide analyzable info
      • Showing trends, level alarms, etc.
      • Fundamental for actions

(Continued)

(Continued)

- • Base for feedback on actions and risks
  - − Without actions and feedback the system dies!
- • Examples of info in Occurrence Reports
  - − Occurrence category
  - − Occurrence type
  - − Flight phase
  - − Cause of occurrence (if obvious):
    - • Function
    - • Pilot error
- • Analyzable statistics requires
  - − Thought-out report codes
  - − Codes being stable over time
- • Analysis can for example provide
  - − Trends
  - − Level alarms
- • Examples:
  - − Equipment failures
  - − Pilot error
  - − Broken clearances
- • Feedback to operators
  - − Essential for maintained willingness to report!
  - − Readily accessible for all concerned personnel!
  - − While the occurrences still attract interest...
- • Example: Fly Day (FlygDags)
  - − Every month
  - − Occurrences of the month of special interest
    - • Comments from Squadron Cdr/Wing Cdr Flying
    - • Comments from HQ Flight Safety Section
    - • Measures taken when problems/failures are found
- • Examples of failures:
  - − False Ground Collision Warning on the JAS39 A/B aircraft. Many reports were given on this problem.
    - • Pilots' belief was: "By now they must have got it..."
    - • Flight Safety's belief was: "Seems like the problems are solved..."
    - • Consequence:
      - • Contributed to a belly landing with a JAS39 C Gripen
- • The seriousness in an occurrence must never be played down, neither in the report or in the following management.

  – Personal mistakes or potential mistakes
- Incident 2000-01-25 with a Swedish
  C-130 Hercules at Kebnekaise

   There was a near collision to the same mountain, but the report was played down.

- In 2012 a Norwegian C130J crashed at the same location, with five killed. Figure 1.19 is a photograph of the crash site.

Figure 1.19 Crash site of Norwegian C130J. Source: Åström (2022).

2012-03-15: Norwegian C-130J, 5 killed

   One of the tools for solving aircraft related issues was the Gripen's Chief Engineer's Problems Resolution List, called the CI PAL. The engineering group met once a month to discuss and find solutions for aircraft engineering issues. At first, the meetings were not successful because the issues were not taken seriously and detested by industry. However, with persistence the program was implemented as was found to be one of the prerequisites for the Gripen export success. The CI PAL was mentioned as the base for Saab's present quality system. Figure 1.20 shows a Gripen fighter being flown by Wing Commander Åström.

(Continued)

(Continued)

Crisis Management

– The SAF's current policy is to eliminate before they result in a crisis.
• Economic implications are

> If you think safety is expensive...
> ...try an accident!

• Direct costs for an accident:
  » Damages (personnel, equipment)
  » Technical, operational costs
  » Disruptions in processes, production, etc.
• Indirect costs for an accident:
  » Immaterial costs: Reputation, brand, lost orders, etc.
  » Losses: Lost revenues, machine time, work time, investigation time, etc.
  » Sanctions: Lawsuits, fines

Figure 1.20 Gripen aircraft flown by wing commander Åström. Source: Åström (2022).

Figure 1.21 shows a graph that correlates cost with risk and losses. The graph shows that there is an optimal point. This point is called ALARP or As Low as Reasonably Practicable.

Example SAF Accident
On April 19, 2007 a JAS39 C Gripen at Vidsel airbase crashed. Before landing the pilot was experiencing high G-loading when turning to the downwind leg before landing. During reduction of the G-load the canopy exploded, and the pilot was ejected. Figure 1.22 shows the crashed aircraft.

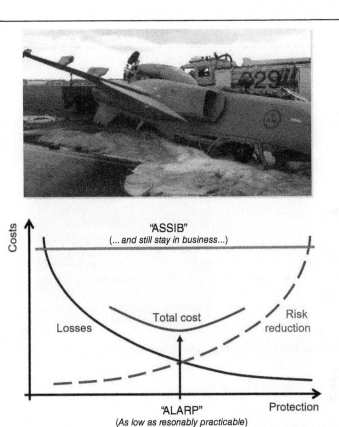

Figure 1.21 Risk/reward graph. Source: Åström (2022).

The maintenance/flight data recorder had to crash protection and the memory cards were torn away. However, the data was retrieved.

The pilot said he had his hand on the throttle and the stick. The flight data recorder showed that the aircraft functioned flawlessly, no fault alarm before ejection and the throttle and Stick movements gave credibility to the pilot's statement. Examination of the ejection seat showed no signs of malfunctions or discrepancies. However, the ejection mechanism was triggered.

The main question from the initial accident investigation was how could the ejection mechanism be activated during flight without the handle being touched by the hands?

Tests were carried out in the dynamic simulator with the handle from the crashed aircraft and two other designs of the ejection seat handle (Figure 1.22).

(Continued)

(Continued)

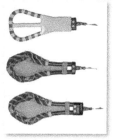

C/D version

A/B version

Extended
A/B version

Figure 1.22 Ejection seat handle testing. Source: Åström (2022).

Tests in Dynamic simulator found that

- Four pilots of four were ejected
- Retaining forces were measured on 15 mechanisms:
  o Initial force (ball lock for the handle)
  o Firing force (wedge lock for firing pins)

Further examination found that

- There were uncocking problems frequently reported for more than 10 years.
- In systems report and risk analysis: Pilot's thighs can affect the handle.
- No detailed analysis concerning unintentional ejection.
- Documentation – Forces on the ejection mechanism.
- There was no requirement for lowest firing forces.

However, an old requirement documents at FMV found that the Initial force should be 150 N and the firing force should be 220–290 N.

From a delivery of 70 ejection seats, 54 seats were below specified initial force, 48 seats were below lowest firing force. There had been a change in personnel and only the highest firing force was checked.

A description in the pilot's manual (AOM) said: "Push the handle back to its parking position." Due to this AOM-text, the pilots perceived the problem with ejection seat as normal and acceptable. Uncocking problems therefore often resulted in no Occurrence/Technical Reports.

The following were the modification of the ejection mechanism:

- 2000-11-01: Decision on modification, was included in Mod package C
- 2005-04-13: Mod package C was stopped for economic reasons
- 2006-08-16: The measures were included in a Technical Order (TO)

The case was therefore removed from Problem List Gripen JAS39.

In the new TO, modification of the ejection mechanism was missing.

The accident was caused by shortcomings in the Quality Assurance of the Flight Safety Process within and between Saab, FMV, and FM.

These shortcomings resulted in activation of the aircraft's rescue system after repeated high G-loads due to

- The shape of the handle
- The surface friction of the handle
- The low holding forces of the mechanism

A sobering thought that the Gripen had participated in numerous international air displays, involving high G-loads and it is lucky the ejection seat was not activated.

Nuclear Regulatory Commission (NRC) Response to Questions from Lee Ostrom

Please note that the references for this section are at the end of the section and are listed independently from the rest of the chapter.

1. **How do you define safety culture?**

   The NRC's Safety Culture Policy Statement (SCPS) sets forth the Commission's expectation that individuals and organizations establish and maintain a positive safety culture commensurate with the safety and security significance of their activities and the nature and complexity of their organizations and functions. The SCPS is not a regulation. It applies to all licensees; certificate holders; permit holders; authorization holders; holders of quality assurance program approvals; vendors and suppliers of safety-related components; and applicants for a license, certificate, permit, authorization, or quality assurance program approval, subject to NRC authority. In addition, the Commission encourages the Agreement States (States that assume regulatory authority over their own use of certain

(Continued)

(Continued)

nuclear materials), their licensees, and other organizations interested in nuclear safety to support the development and maintenance of a positive safety culture within their regulated communities. The NRC's Web site contains more information on Agreement States.[1]

The SCPS addresses both safety and security. Organizations should ensure that personnel in the safety and security sectors appreciate the importance of each and should emphasize the need for integration and balance to achieve both safety and security in their activities. Safety and security activities are closely intertwined. Although many safety and security activities complement each other, instances may occur in which safety and security interests create competing goals. It is important that consideration of these activities be integrated so that one activity does not diminish or adversely affect the other; therefore, mechanisms to identify and resolve these differences should be established. A safety culture that accomplishes this would include all nuclear safety and security issues associated with NRC-regulated activities.

The SCPS defines nuclear safety culture as the core values and behavior resulting from a collective commitment by leaders and individuals to emphasize safety over competing goals to ensure protection of people and the environment.

The SCPS includes nine traits that further define a positive safety culture. These traits describe patterns of thinking, feeling, and behaving that emphasize safety, particularly in goal conflict situations, such as when safety goals conflict with production, schedule, or cost goals. The traits listed in Table 1.1 are not all-inclusive. Some organizations may find that one or more of the traits are particularly relevant to their activities. In addition, the SCPS may not include traits that are important in a positive safety culture. More information on the SCPS appears on the NRC's Web site.[2]

2. **How does safety culture impact the operation of your facility/company/ oversight function?**

The NRC recognizes that it is important for all organizations performing or overseeing regulated activities to establish and maintain a positive safety culture. The NRC's approach to safety culture is based on the premise that licensees bear the primary responsibility for safety. The NRC provides oversight of safety culture through expectations detailed in policy statements and the oversight processes.

Table 1.1 Safety culture policy statement traits.

Leadership safety values and actions	Problem identification and resolution	Personal accountability
Leaders demonstrate a commitment to safety in their decisions and behaviors	Issues potentially impacting safety are promptly identified, fully evaluated, and promptly addressed and corrected commensurate with their significance	All individuals take personal responsibility for safety
Work processes	**Continuous learning**	**Environment for raising concerns**
The process of planning and controlling work activities is implemented so that safety is maintained	Opportunities to learn about ways to ensure safety are sought out and implemented	A safety conscious work environment is maintained where personnel feel free to raise safety concerns without fear of retaliation, intimidation, harassment, or discrimination
Effective safety communications	**Respectful work environment**	**Questioning attitude**
Communications maintain a focus on safety	Trust and respect permeate the organization	Individuals avoid complacency and continually challenge existing conditions and activities to identify discrepancies that might result in error or inappropriate action

Beginning in 1989, the NRC published the first of three policy statements about safety culture at nuclear power plants. The first policy statement describes the Commission's expectations for the conduct of operations in control rooms. The NRC published the second policy statement in 1996 to establish the Commission's expectation for maintaining a safety conscious work environment (SCWE) in which workers are able to raise nuclear safety concerns without fear of retaliation. Finally, the NRC published the third policy statement, the SCPS, in 2011 to establish the Commission's expectations for licensees to maintain a strong safety culture.

The NRC staff developed outreach materials that can be used to educate stakeholders about safety culture and the NRC's SCPS. These materials include brochures, posters, "Safety Culture Case Studies," "Safety Culture Trait Talk," "Safety Culture Journey," and "Educational Resource." As part of

(Continued)

(Continued)

ongoing outreach activities, the NRC staff makes presentations at meetings and conferences in a variety of industry forums to share information with stakeholders about the SCPS and participates in international and interagency efforts on safety culture. The NRC Web site provides all of the NRC outreach and educational materials and the SCPS.[3].

Safety Culture is included in the NRC's oversight processes, the Allegation Program and the Enforcement Program. The NRC provides training to inspectors and other experienced staff to become qualified as safety culture assessors for safety culture assessments under Inspection Procedure (IP) 95 003, "Supplemental Inspection for Repetitive Degraded Cornerstones, multiple Degraded Cornerstones, Multiple Yellow Inputs, or One Red Input." The Reactor Oversight Process (ROP) is the NRC's program for assessing the performance of operating commercial nuclear power reactors, which includes guidance and procedures for inspecting and assessing aspects of licensees' safety culture. In addition to the oversight processes, the NRC's Allegation Program evaluates and responds to safety culture concerns raised by workers at the agency's licensed facilities and by the general public. Furthermore, the Allegation Program and Enforcement Program address safety culture through the use of chilling effect letters (CELs) and confirmatory orders (COs). (*See question 5 for more information on NRC assessment and oversight of safety culture*).

3. **Describe an incident/accident that a negative safety culture impacted the outcome.**

   The NRC has developed Safety Culture Case Studies to provide real-life events where review of the circumstances surrounding the event and the results of the investigations found clear examples of the role that safety culture played in contributing to or in lessening the causes and consequences of the event.

   These case studies are learning tools. Those responsible for regulating or using radioactive material in a safe and secure manner should not become complacent and should be open to learning from the mistakes and the problems others have faced in an effort to prevent recurrences.

   The NRC has also developed a Safety Culture Case Study User Guide to help individuals and organizations use the various case studies more effectively, providing them with a better understanding of why a strong safety culture and safety-first focus are critically important. It is recommended that you review the User Guide prior to reviewing the case studies.[4]

- *June 2009 Collision of Two Washington Metropolitan Area Transit Authority Metrorail Trains Near Fort Totten Station, Washington, DC*: The case study will enable the staff to identify and learn from the findings made by the National Transportation Safety Board (NTSB) about the collision of two Washington Metropolitan Area Transit Authority (WMATA) Metrorail trains in June 2009. The collision resulted in the loss of nine lives and multiple passenger injuries. The NTSB report points out that WMATA failed to implement many significant attributes of a sound safety program. Many of these attributes parallel the positive safety culture traits the NRC has incorporated into its Safety Culture Policy Statement.[5]
- **Partial Collapse of the Willow Island Cooling Tower**: This case study describes the safety culture shortcomings that led directly to the partial collapse of a natural draft, hyperbolic, reinforced concrete cooling tower at Willow Island, WV, on April 27, 1978. The cooling tower was one of two towers being constructed by Research-Cottrell, Inc., for the Pleasants Power Station, a coal-fired electric power station owned by Allegheny Energy Supply Company. The resulting deaths of 51 construction workers represent one of the costliest accidents, in human terms, of any in the US construction industry's history.[6]
- **April 2010 Upper Big Branch Mine Explosion – 29 Lives Lost**: This case study provides the regulated community with the findings of West Virginia Governor Manchin's appointed independent investigation panel and the results of an investigation conducted by the US Department of Labor's Mine Safety and Health Administration (MSHA). Many of these findings contrast starkly with the positive safety culture traits that the NRC has incorporated into its safety culture policy statement.[7]

4. **Describe an incident where a positive safety culture has prevented a catastrophe.**
   - **US Airways Flight 1549: Forced Landing on Hudson River**: This case study is intended to provide a useful tool for the US Nuclear Regulatory Commission (NRC) staff as it interacts with their stakeholders. It will also enable the regulated community to identify with and learn from the findings made by the National Transportation Safety Board (NTSB) regarding the forced landing on the Hudson River by US Airways Flight 1549. The positive safety culture traits that contributed to the successful outcome of this event parallel those traits the NRC has incorporated into its SCPS.[8]

(Continued)

(Continued)

5. **How does your organization assess safety culture?**

Safety Culture Assessor Training

Qualification as a safety culture assessor requires the completion of a variety of activities, each of which is designed to help gather information or practice a skill that may be important during safety culture assessments. When qualified, the safety culture assessor will have demonstrated the following competencies:

- understanding of the regulatory processes used to achieve the NRC's regulatory objectives
- mastery of the techniques and skills needed to collect, analyze, and integrate information using a safety culture focus to develop a supportable regulatory conclusion
- the personal and interpersonal skills needed to carry out assigned regulatory activities, either individually or as part of a team

Additional information on IMC 1245, "Qualification Program for New and Operating Reactor Programs," Appendix C12, "Safety Culture Assessor Training and Qualification Journal," dated February 1, 2016, appears on the NRC's Web site.[9]

Safety Culture Common Language

Before work began on the 2011 SCPS, the nuclear power industry approached the NRC about developing a shared set of terms to describe safety culture. With insights gained during the development of the SCPS, the Office of Nuclear Reactor Regulation, along with the Institute of Nuclear Power Operations and Nuclear Energy Institute, hosted a series of public workshops beginning in December 2011 to discuss the idea of a safety culture common language. The intent of this initiative, as requested by the industry, was to align terminology between the NRC's inspection and assessment processes within the ROP and the industry's assessment process. This initiative was within the Commission-directed framework for enhancing the ROP treatment of cross-cutting areas to more fully address safety culture.

NUREG-2165 documents the outcomes of the public workshops to develop a common language to describe safety culture in the nuclear industry. These workshops included panelists from the NRC, the nuclear power industry, and the public. NUREG-2165 outlines a suggested common language for classifying and grouping traits and attributes of a healthy nuclear safety culture. The common language initiative identified 10 traits of a healthy safety culture (the 9 traits from the SCPS plus a 10th trait, decision-making), 40 aspects included under those traits, and numerous examples for each aspect. These common language traits and aspects have been incorporated under the three cross-cutting areas of the ROP. NUREG-2165 is available on the NRC's Web site.[10]

Regulatory Oversight

Oversight Processes

Reactor Oversight Process: The NRC's approach to safety culture is based on the premise that licensees bear the primary responsibility for safety. The ROP is the NRC's program for assessing the performance of operating commercial nuclear power reactors. The ROP uses inputs from performance indicators and inspection findings to develop conclusions about a licensee's safety performance. Performance is evaluated systematically and continuously through planned inspections and mid-year and end-of-year assessment meetings.

The ROP framework contains three key areas and seven cornerstones, as shown in Figure 1.23. Each cornerstone has corresponding performance indicators and inspection procedures to assess licensee performance. Safety culture is considered within three cross-cutting areas: (i) human performance, (ii) problem identification and resolution, and (iii) SCWE.

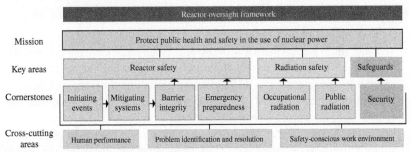

Figure 1.23 Reactor oversight framework. Source: NRC 2022/United States Nuclear Regulatory Commission/Public Domain.

Based on the NRC's assessment of safety performance, licensees are assigned to a column in the ROP Action Matrix, and that placement in the ROP Action Matrix determines the level of NRC oversight for that particular licensee. The NRC's approach to safety culture assessment is a graded process (see Figure 1.24). The extent and complexity of a safety culture assessment is generally based on a licensee's placement in the ROP Action Matrix. The scope and complexity increases with increased oversight, and the focus of the assessment may be tailored based on the

(Continued)

(Continued)

original performance deficiency. An assessment may focus more heavily on one part of the plant or on one area of safety culture, such as the SCWE.

Figure 1.24 ROP action matrix. Source: 2022/United States Nuclear Regulatory Commission/Public Domain.

The NRC's ROP Action Matrix with the four columns of increasing oversight is based on performance indicators and inspection findings. Licensees in column 1 are subject to the NRC's baseline inspection program. As licensees move to columns 2, 3, or 4, they are subject to additional oversight in the form of supplemental inspections.

Qualified NRC safety culture assessors evaluate the licensee's third-party safety culture assessment and then determine the scope of the NRC's assessment based on that evaluation. The NRC assessors conduct the assessment on site and identify and document safety culture themes in the inspection report. The assessors also review the licensee's planned and completed corrective actions to evaluate whether they address the identified safety culture themes and whether the licensee needs to develop follow-up actions to deal with any remaining concerns. The NRC Web site contains a detailed description of the ROP[11] and NUREG-1649, "Reactor Oversight Process," Revision 6, issued July 2016.[12]

Construction Reactor Oversight Process: The Office of New Reactors (NRO) staff revised the cROP based on the ROP assessment program methodology, including the use of safety culture traits and cross-cutting issues, and completed a pilot of the revised cROP in December 2012.

Based on the results of the pilot program, NRO revised the construction oversight process, including IMC 0613, "Power Reactor Construction Inspection Reports," dated February 9, 2017, and IMC 2505, "Periodic Assessment of Construction Inspection Program Results," dated January 6, 2017, to provide guidance in assessing the safety culture of a construction site. NRO revised IMC 0613 to include a list of cross-cutting aspects that can be assigned to inspection findings. Assigned cross-cutting aspects are generally associated with the root causes of performance deficiencies and are evaluated to identify cross-cutting themes that are assessed as outlined in IMC 2505. IMC 2505 also references the supplemental inspection procedures that are used when safety performance declines at a construction site. These procedures provide NRC inspectors with guidance on how to assess the safety culture at a construction site with escalating levels of efforts commensurate with the significance of a site's performance decline. The supplemental inspection procedures also give NRC inspectors the tools to communicate safety culture issues to stakeholders. The NRC's Web site contains IMC 0613 and IMC 2505[13] and additional information on the cROP.[14]

Allegation and Enforcement Programs

In addition to violations of NRC regulations and potential wrongdoing, the NRC's Allegation Program and Enforcement Program address discrimination against licensee employees for raising allegations related to nuclear safety concerns and the potential resulting chilling effect on the employee or coworkers. An allegation is a declaration, statement, or assertion of impropriety or inadequacy associated with NRC-regulated activities, the validity of which has not been established. Workers in the nuclear industry or members of the general public have the option of reporting nuclear safety concerns directly to the NRC.

Safety Conscious Work Environment: The Commission describes a SCWE as a work environment in which employees are encouraged to raise safety concerns and concerns are promptly reviewed, given the proper priority based on their potential safety significance, and appropriately resolved with timely feedback to the originator of the concerns and to other employees as appropriate. Fostering an environment for raising concerns continues to be an important attribute of a positive nuclear safety culture and is incorporated as one of the traits of a positive safety culture in the NRC's SCPS as "Environment for Raising Concerns." The NRC's Web site contains additional information on SCWE.[15]

(Continued)

(Continued)

The NRC places a high value on nuclear industry employees being free to raise potential safety concerns to both licensee management and the NRC regardless of the merits of the concern. Unlawful adverse actions taken against an employee for raising safety concerns or perceptions that such actions have been taken may create a "chilling effect" on the employee or other workers who may wish to raise concerns (i.e. the employees may not feel that they are free to raise concerns without fear of retaliation). When the chilling effect is not isolated (e.g. multiple individuals, functional groups, shift crews, or levels of workers within the organization are affected), the NRC refers to the situation as a "chilled work environment."

If the NRC suspects that an organization has created a chilled work environment, the agency may ask the licensee for more information, or the NRC will investigate through follow-up inspections. If the NRC is concerned about the licensee's awareness of, or efforts to address a known chilled work environment, the agency may issue a CEL. A CEL is a public way for the NRC to communicate with the licensee, the public, and the licensee's employees. The intent of such action is, in part, to prompt the licensee to act to mitigate the chilling effect that the discriminatory act or other event has caused. The NRC's Allegation Program includes guidance on the agency's SCWE policy and issuance of CELs and is available on the NRC's Web site.[16]

The NRC's Enforcement Policy ensures, through appropriate enforcement action against a licensee or licensee contractor (and, when warranted, against the individual personally responsible for the act of discrimination), that adverse employment actions taken against licensee or contractor employees for raising safety concerns do not have a chilling effect on the individual or others who may wish to report safety concerns. The NRC vigorously pursues actions against licensees or licensee contractors who discriminate against their employees for raising nuclear safety concerns. The NRC's Enforcement Program includes information on sanctions for discrimination against employees who raise safety concerns and is available on the NRC's Web site.[17]

Safety Culture Corrective Actions: Through the identification of cross-cutting issues, safety culture assessments in supplemental inspections, or findings of discrimination or chilling effect, the NRC typically documents the concerns publicly, and the licensee responds to the concerns with planned corrective actions. The NRC may also use its Alternative Dispute Resolution Program (ADR) to resolve discrimination and wrongdoing cases or other specific cases subject to enforcement action through mediation rather than through the NRC's traditional enforcement processes. The ADR Program comprises two entirely different subprograms: (i) a pre-investigation ADR subprogram and (ii) an enforcement ADR subprogram. The first subprogram, commonly referred to as "early ADR," is offered before the initiation of an investigation by the NRC's Office of Investigations (OI). Early ADR is available to allegers and their employers for resolving allegations of discrimination only. It is an opportunity for the parties to resolve their concerns early in the process. The second subprogram, formerly referred to as "post-investigation ADR," is offered after the completion of an OI investigation. Enforcement ADR is available to licensees (including contractors and employees) and the NRC for resolving wrongdoing or discrimination cases in which the NRC has concluded that enforcement may be warranted. Enforcement ADR is also available for escalated non-willful (traditional) enforcement cases with the potential for civil penalties (not including violations associated with findings assessed through the ROP). The enforcement ADR subprogram documents agreements between the NRC and the licensee on the licensee's planned actions, which then becomes the basis for a CO. The CO is legally binding. The NRC's Web site contains additional information on the Enforcement Program[18] and the ADR Program.[19,20]

References for the NRC Section

1 NRC (2022). Agreement state program. http://www.nrc.gov/about-nrc/state-tribal/agreement-states.html (accessed February 2022).

2 NRC (2022). Safety culture policy statement. http://www.nrc.gov/about-nrc/safety-culture/sc-policy-statement.html (accessed February 2022).

3 NRC (2022). Outreach and education materials. http://www.nrc.gov/about-nrc/safety-culture/sc-outreach-edu-materials.html (accessed February 2022).

4 NRC (2022). Safety culture case study user guide. https://www.nrc.gov/docs/ML1519/ML15196A440.pdf (accessed February 2022).

5 NRC, Safety Culture Communicator (2011). US Airways Flight 1549: Forced landing on Hudson river. https://www.nrc.gov/docs/ML1122/ML11228A218.pdf (accessed February 2022).

6 NRC, Safety Culture Communicator (2011). June 2009 Collision of Two Washington Metropolitan Area Transit Authority Metrorail Trains Near Fort Totten Station, Washington, DC. https://www.nrc.gov/docs/ML1115/ML11159A220.pdf (accessed February 2022).

7 NRC, Safety Culture Communicator (2011). Partial collapse of the Willow Island cooling tower. https://www.nrc.gov/docs/ML1126/ML11264A019.pdf (accessed February 2022).

(Continued)

(Continued)

8 NRC, Safety Culture Communicator (2012). April 2010 Upper big branch mine explosion—29 Lives Lost. https://www.nrc.gov/docs/ML1206/ML12069A003 .pdf (accessed February 2022).

9 NRC, Inspection Manual Chapter (IMC) 1245 (2011). Qualification program for new and operating reactor programs, Appendix C12. Safety Culture Assessor Training and Qualification Journal, ADAMS Accession No. ML16020A397. https://adamsxt.nrc.gov/navigator/#:~:text=Accession %20Number-,ML16020A397,-ML16020A397 (accessed February 2022).

10 NRC, NUREG-2165. Safety culture common language. http://www.nrc.gov/ reading-rm/doc-collections/nuregs/staff/sr2165 (accessed February 2022).

11 NRC. ROP framework. http://www.nrc.gov/reactors/operating/oversight/rop-description.html (accessed February 2022).

12 NRC, NUREG-1649 (2022). Reactor oversight process. Revision 6. http://www .nrc.gov/reading-rm/doc-collections/nuregs/staff/sr1649 (accessed February 2022).

13 NRC, IMC 0613 (2022). Power reactor construction inspection reports and IMC 2505, Periodic assessment of construction inspection program results. http://www.nrc.gov/reading-rm/doc-collections/insp-manual/manual-chapter/index.html (accessed February 2022).

14 NRC (2022). Construction reactor oversight process (cROP). http://www.nrc .gov/reactors/new-reactors/oversight/crop.html (accessed February 2022).

15 NRC (2022). Safety conscious work environment. http://www.nrc.gov/about-nrc/safety-culture/scwe.html (accessed February 2022).

16 NRC (2022). Safety conscious work environment policy guidance. http://www .nrc.gov/about-nrc/regulatory/allegations/scwe-mainpage.html (accessed February 2022).

17 NRC (2022). Sanctions for discrimination against employees who raise safety concerns. http://www.nrc.gov/about-nrc/regulatory/enforcement/sanctions .html (accessed February 2022).

18 NRC (2022). Enforcement program overview. http://www.nrc.gov/about-nrc/ regulatory/enforcement/program-overview.html (accessed February 2022).

19 NRC (2022). Pre-investigation ADR. http://www.nrc.gov/about-nrc/ regulatory/enforcement/adr/pre-investigation.html (accessed February 2022).

20 NRC (2022). Enforcement ADR. http://www.nrc.gov/about-nrc/regulatory/ enforcement/adr/post-investigation.html (accessed February 2022).

1.6   Chapter Summary

This chapter covered a great amount of information on early safety professionals, culture, and safety culture and the problems that can arise when there isn't a positive safety culture in an organization. There is always a safety culture, whether it is positive or negative. Accidents can occur in positive safety cultures but are more likely to occur in negative safety cultures. I wish to thank Ron Farris, Shane Bush, Lt Olsen, Wing Commander Åström, and the NRC for their contributions to this chapter.

References

Åström, L. (2019). *Safety Culture & Crisis Management (PowerPoint Slides)*. Sweden, AB: GKN Aerospace.

BBC (2017). Germanwings crash: who was co-pilot Andreas Lubitz? https://www.bbc.com/news/world-europe-32072220 (accessed February 2022).

Bird, F.E. and Germain, G.L. (1996). *Practical Loss Control Leadership*. Loganville, GA: Det Norske Verita.

Charles Rivers Editor, *Dagestan: The History and Legacy of Russia's Most Ethnically Diverse Republic* 2019.

Franco, G. and Fusetti, L. (2004). Bernardino Ramazzini's early observations of the link between musculoskeletal disorders and ergonomic factors. *Applied Ergonomics* 35 (1): 67–70. https://doi.org/10.1016/j.apergo.2003.08.001.

Heinrich, H.W. (1931). *Industrial Accident Prevention: A Scientific Approach*. New York, NY: McGraw Hill Book Company [Google Scholar].

Heinrich, H.W., Peterson, D., and Roos, N. (1980). *Industrial Accident Prevention*, 5e. New York, NY: Mcgraw Hill.

Hollnagel, E. (2012). *FRAM, The Functional Resonance Analysis Method: Modelling Complex Socio-Technical Systems*. Ashgate Publishing, Ltd.

Hollnagel, E. and Örjan, G. (2004). The functional resonance accident model. *Proceedings of Cognitive System Engineering in Process Plant*, Tohoku University, Sendai, Japan.

David D Boyce, Shane Bush, Danielle E Lower, Tyson Todd Allen, John K Epperson. (2020) *INL Human Performance Improvement Guide, GDE-863* (No. INL/MIS-20-57536-Rev000). Idaho National Lab.(INL), Idaho Falls, ID (United States).

Ostrom, L. and Wilhelmsen, C. (2015). *Factoring in Safety Culture into a Risk Framework*. Vienna, Austria: Presented at the International Atomic Energy Agency.

Ostrom, L.T. and Wilhelmsen, C. (2019). *Risk Assessment: Tools, Techniques, and Their Applications*, 2e. Wiley.

Ostrom, L.T., Wilhelmsen, C.A., and Kaplan, K. (1993). Assessing safety culture. *Journal of Nuclear Safety* 34 (2).

Pidgeon, N. (1991). Safety culture and risk management in organizations, business. *Journal of Cross-Cultural Psychology* 22: 129–140.

Reason, J. (1990). The contribution of latent human failures to the breakdown of complex systems. *Philosophical Transactions of the Royal Society (London), series B.* 327: 475–484.

Reason, J. (1997). *Managing the Risks of Organizational Accidents*. Aldershot, UK: Ashgate Publishing Limited.

Safeopedia (2017). What does accident causation model mean?. https://www.safeopedia.com/definition/617/accident-causation-model-occupational-health-and-safety#:~:text=Simple%20linear%20models%20(Heinrich%2C%201931, accident%2C%20injury%2C%20etc. (accessed February 2022).

Woodward, M. (2018). Safety Myths Debunked! Safety Isn't Common Sense. https://previsorinsurance.com/blog/safety-myths-debunked/ (accessed February 2022).

2

History of Safety Culture

This chapter provides an overview of how safety culture has evolved over the years. This overview looks beyond safety in the workplace and shows how safety and health issues were viewed over the centuries.

2.1 Life Expectancy and Safety

We all complain about how hard life is at times. However, it is nothing like how life was in past centuries, or even past decades. A simple accident in 1900 could wind up killing a person because of a bad infection. There was also a lack of very basic safety protective equipment, like safety glasses and hearing protection. In general, safety in the workplace and at home has shifted dramatically over the centuries. Sharon Basaraba (2020) summarizes how life expectancy has changed over the centuries. 30,000 years ago, life expectancy may have exceeded 30 years of age for the first time. In Greek and Roman times life expectancy was between 20 and 35 years of age and by the 1500s it hovered between 30 and 40 years of age. This doesn't mean that everyone had died by 40, but that 30–40 was the average life expectancy. So, many infants died at or close to birth and some people lived to be 70. People died due to disease, poor nutrition/famine, and accidents. As Ms. Basaraba writes, if a person lived past 15, they could have a long life. In the 1900s life expectancy increased to approximately 75 years of age.

The most current life expectancy data from the United Nations projections show modest increases in life expectancy (United Nations 2021):

- The current life expectancy for United States in 2022 is 79.05 years, a 0.08% increase from 2021.
- The life expectancy for United States in 2021 was 78.99 years, a 0.08% increase from 2020.

Impact of Societal Norms on Safety, Health, and the Environment: Case Studies in Society and Safety Culture, First Edition. Lee T. Ostrom.

- The life expectancy for United States in 2020 was 78.93 years, a 0.08% increase from 2019.
- The life expectancy for United States in 2019 was 78.87 years, a 0.08% increase from 2018.

When my children were much younger in the 1990s, we played the game Oregon Trail quite a bit (Softkey Multimedia 1995). In the game an avatar family crosses the great plains, on their way to Oregon's Willamette Valley. This actual migration occurred in the mid- to later-1800s (Oregon 2021). During the game the family would experience the simulated events that the actual migrants did. Almost every time we played one or more of the avatar family members died from a disease or from a broken leg. The game, of course, is a simulation. In real life the situations were much worse. One side of my great grandparents traveled the Oregon Trail in the 1880s. My great grandmother got sick near Boise, ID and my great grandfather had to find temporary homes for their six kids, until she was well enough to continue their way. When she got well, my great grandfather had to round up all the kids and some of the families had moved on to other locations. They eventually made it to what is now central Washington, where they homesteaded and had six or seven more children. It is not known if other children died during childbirth.

Life was much different before the introduction of modern antibiotics (Healthy Children 2021):

- Before antibiotics, 90% of children with bacterial meningitis died. Among those children who lived, most had severe and lasting disabilities, from deafness to mental retardation.
- Strep throat was at times a fatal disease, and ear infections sometimes spread from the ear to the brain, causing severe problems.
- Other serious infections, from tuberculosis to pneumonia to whooping cough, were caused by aggressive bacteria that reproduced with extraordinary speed and led to serious illness and sometimes death.

Before the discovery of antibiotics, physicians and pharmacists relied on drugs like sulfa drugs, based on the compound sulfonamide. This class of drugs was introduced by Gerhard Domagk in 1935 (Science History Institute 2021). The first drug introduced was Prontosil and was used to treat staphylococcal (Staph) and streptococcal (Strep). Prior to the introduction of sulfa drugs diseases like syphilis were treated with chemicals like mercury compounds. In fact, mercury was used to treat this heinous disease from the sixteenth to the early twentieth century (Oregon Health & Science University [OHSU] 2016). Mercury had devastating side effects as well, mercury poisoning and disfigurement. However, syphilis was worse. Diseases on the surface of the skin were controlled using iodine/iodine compounds (Wounds International 2011) and Mercurochrome (Northeast Waste

Management Officials' Association 2021). Iodine was first used in the 1800s and is still used to this day as a topical antiseptic. The current form is an Iodophor, which is less toxic than elemental Iodine. Mercurochrome is an organo-mercury compound in aqueous solution, used to prevent infection in minor wounds. The use of Mercurochrome is now banned in the United States because of the use of Mercury.

The OHSU article also discusses the relationship between military personnel and venereal disease. An outbreak of syphilis was a first during a military campaign in Europe. Military medicine was challenged by the effect of sexually transmitted disease on combat readiness, and thus national security. Note that it was readiness and not necessarily the soldiers' and sailors' health the military was worried about.

The US military began distributing prophylactics and informational materials during World War I to help at least control venereal disease. However, these actions were positive and matched well with the Progressive-era priorities for public health, which emphasized education and infectious disease prevention.

Venereal disease was not a new problem in World War I. In the US Civil War, there were 183,000 cases of venereal disease documented by the Surgeon General of the United States Army (Case Western University 2021). The Army also court martialed over 100,000 soldiers for sexual misconduct during the war. The most interesting idea the Army did was to regulate prostitution in Nashville and Memphis, TN from 1863 to 1865. The Case Western article has a quote from one soldier who had written his wife that, "you would think that there was not a married man in the regiment."

Accidental death was quite common in past centuries as well. Of course, as stated earlier, a simple accident that caused a lesion could result in death because of a bacterial infection. Gangrene was a very common type of infection where the only treatment in past centuries was the amputation of the affect limb. Clostridium perfringens is one of the bacterial species that causes Gangrene (Mayo Clinic 2021). It is an anaerobic bacterium meaning it lives without air, so in deeper tissue.

The types of accidents people had in past centuries were quite similar to the types of accidents today. The difference is the technology of the time. Ricky Longfellow (2017) wrote an article on the US Department of Transportation Highway History website that discusses stagecoach accidents. It reads:

> Although the pioneers didn't have to worry about that rare freeway pile-up or train derailment, they did have their share of public transportation accidents. Many of these involved the convenience of stagecoach travel.
>
> A history of stagecoach accidents recorded as early as the 1700s tells of wheels falling off, horses out of control, and brakes failing. Crudely built bridge floors either fell through or shifted causing the stage to overturn.

Luckily, there were not many fatalities to people, but the horses pulling these conveyances would occasionally suffer injuries or death.

Today's small bridges over creeks and narrow rivers are sturdy and safe. It takes only a few moments to cross them and many times we do so without even knowing, as the roads and bridges are of similar material. However, early plank and log bridges were a hazard to stagecoach travelers. The planks or logs would move apart or bunch up; either way it caused a tremendous problem for vehicles and horses alike. In the early 1800s, in North Carolina, the front wheels of a coach got stuck in the planks causing the horses to bolt, which pulled the planks along with the coach swiftly across the bridge. The passengers were not injured, but one horse was killed.

Sometimes negligence would be the cause of an accident. In the 1820s, the stagecoach companies were in competition. One incident in New York tells of a driver who sped up and tried to pass the competition going uphill. Consequently, he upset the coach with the passengers inside. Again, no one was killed but injuries were reported.

Lastly, "driving under the influence" appears to be a current day offense when it is actually as old as the beverage consumed itself. An accident in the early 1800s reveals one inebriated driver who gathered up enough speed to overturn the coach, seriously injuring some of the passengers.

Railroad accidents were also very frequent.

Several websites contain detailed information about past railroad accidents and train wrecks. The following are just two examples:

Northern Pacific Rotary Accident February 11, 1903 (Provided by permission from Steve Shook) (Figure 2.1).

A remarkable railway accident, which resulted from a snow slide, occurred on February 10. The structure wrecked is the celebrated S-bridge, located on the Northern Pacific railway and crossing a small stream known as Willow creek, near Mullan, Idaho. This "bridge" is a single-track timber trestle some 300 ft long and built on a reverse or ogee curve, the sinuous plan of the structure being necessitated by the loop that the railway makes at this point in its climb over the bitter Root summit. At the time of the accident the snow in the bottom of the gorge buried the trestle bents some 30 ft deep and had completely blocked the road ahead of the passenger train, which, with two extra locomotives and a broken-down rotary snowplow, was retreating from an attack of the snow-blocked road to the shelter of the nearest station in the rear.

The train consisted of a rotary snowplow with its pushing engine; the passenger engine followed by two passenger cars; a freight train caboose, and a helping engine at the rear end, and was standing with the rear engine, the caboose, and the rearmost passenger car well over the trestle when the accident occurred. In the caboose were the crew of the rear engine and some others of the train crew, making eight persons in all, and in the rear passenger car there were some eight

Figure 2.1 Northern Pacific Rotary Accident February 11, 1903. Source: Courtesy of Steve Shook.

or nine persons. The accident was caused by a snow slide that started far up the mountain side and sweeping down the narrow gorge tore out the trestle structure for a short length directly under the standing train, the caboose and rear locomotive of which plunged into the gap and were almost completely buried in the deep snow. The rear passenger car hung with one end over the gap and tipping down at a considerable angle but was prevented from falling by the coupling to the car ahead. Of the passengers in the caboose, one was rendered unconscious by the fall of that car, but soon recovered, and the others were practically uninjured. The deep bed of snow into which the car fell had cushioned the fall and preserved the lives of those who underwent the 80-ft drop. The same agent also preserved the life of a frightened passenger in the car ahead who leaped from a window into the gorge below with no more serious consequence than a ducking in the soft snow. The other passengers in the rear car climbed into the car ahead with no injuries more serious that fright at their narrow escape.

2.2 Consumer Items and Toys

There are so many examples of consumer items and toys with safety and health issues it would be several book volumes to contain them all. The following are just a few.

2.2.1 Vintage Toys and Other Items

I have always been interested in rocks and minerals. However, it has mostly been a hobby and not my profession. When I was maybe a third grader, around 1964 or 1965, my parents bought me a rock/mineral sample kit. I can still remember opening the box and identifying all the samples, but we couldn't find the asbestos sample. We dug around in the box for a long time and then reread the directions. The asbestos sample was the copious amount of packing in the box. So, the company sent a known hazardous material as packing in a science kit. It will never be known if children or parents experienced any problems from this.

Other vintage toys contain many heavy metals. An article by the National Environmental Health Association (2015) discusses the amount and types of heavy metals found in vintage toys, including Barbie dolls, Fisher Price Little People figurines, and My Little Pony dolls and accessories.

My dad played with what were called "Tin" soldiers. I have a couple of these metal soldiers. I used my research Olympus X-ray fluorescence (XRF) Spectrometer to analyze the metals in several of my old toys and other possessions I have be given by various family members over the years. Figure 2.2 shows the items I scanned, and Figure 2.3 shows the XRF in use. Figure 2.4 shows the results on the XRF display from the "Tin" soldier. The toy soldiers were from the late 1920s or early 1930s. The Native-American coin bank was given to me when I was a pre-teen. I am not sure when it was made. The horse is Dala Horse from Sweden

Figure 2.2 Vintage items that were scanned. Source: Ostrom (2022).

Figure 2.3 Native American coin bank being scanner. Source: Ostrom (2022).

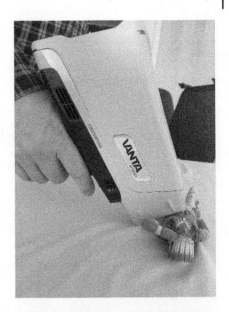

Figure 2.4 Scan results for "tin" soldier. Source: Ostrom (2022).

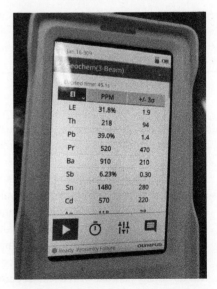

from 1961. The toy car is from 1968 or 1969. The GI Joe is from the mid-1960s. The Swedish copper pan is from the 1800s. The Table 2.1 lists the metals the XRF detected in these items. Note that the values for the metals should be viewed as a good approximation and not the absolute value. Also, the values will vary, depending on where on the item it is scanned. This is due to variations in

Table 2.1 X-ray spectrometer analysis results of scanning items.

Native American Coin Bank	1961 Dala Horse	1968/1969 Toy Car	Early 1960s GI Joe	Late 1920s/ Early 1030s Toy Soldier	1800s Copper Pan
Iron 46.00%	Light elements 94.11%	Zinc 80.3%	Light elements 98.22%	Lead 33.90%	Copper 96.45%
Light elements 18.92%	Lead 1.34%	Titanium 14.55%	Titanium 0.958%	Light elements 31.1%	Potassium 0.763%
Titanium 17.64%	Sulfur 1.28%	Aluminum 2.2%	Aluminum 0.238%	Sulfur 13.83%	Lead 0.647%
Sulfur 6.51%	Silicon 0.566%	Magnesium 1.11%	Silicon 0.198%	Calcium 5.50%	Thorium 0.464%
Silicon 2.86%	Magnesium 0.440%	Silicon 0.768%	Sulfur 0.152%	Antimony 5.25%	Sulfur 0.445%
Lead 2.10%	Barium 0.429%	Phosphorus 0.461%	Potassium 0.064%	Titanium 2.87%	Calcium 0.415%
Magnesium 0.990%	Calcium 0.403%	Molybdenum 0.199%	Calcium 0.053%	Arsenic 1.81%	Silver 0.215%
Cobalt 0.971%	Chromium 0.400%	Copper 0.095%	Iron 0.033%	Chromium 1.47%	Aluminum 0.189%

paint and other materials. Light elements for the XRF are, in descending atomic weight, fluorine, oxygen, nitrogen, carbon, boron, beryllium, lithium, helium, and hydrogen. The results for the scan are the top eight elements that the XRF detected.

This simple analysis shows there is no significant amount of tin (chemical symbol Sn) in the "Tin" soldier. However, there is a great amount of lead. It might be good to only display many of the vintage items and not use as toys for children or for cooking. The vintage GI Joe contains no large amounts of heavy metals, and the titanium is more than likely a component of the pigment, the same for the small car. Lead in pigments was quite common before the 1970s. The Dala Horse was obviously painted with a leaded paint and should only be a display item. The US Consumer Product Safety Commission (CPSC) (2018) has mandated several recalls of lead containing items over the years that have come from countries like China. The CPSC is also on the lookout for items like children's clothing that is not flame retardant and other safety issues with consumer items. These issues are becoming less over the years, but watchdog groups, like the CPSC, are helping to ensure consumers are protected from dangerous good.

2.3 Flawed Cars

Ralph Nader was a crusader for public safety and one of the products he sought to have removed from causing harm to the public was the Chevrolet Corvair. His book "Unsafe at any Speed" (1965) brought to life the unsafe nature of that vehicle and led to more scrutiny of unsafe products. The major issues with the Corvair were its unusual rear-engine lay-out and the car's suspension. That design led to unstable emergency handling, according to Nader. There was much discussion as to whether this design made the car much more prone to roll over (Sorokanvich 2017).

2.4 Ford Pinto

The Ford Pinto was manufactured from 1971 to 1976 and its Mercury brand cousin, the Bobcat, was manufactured from 1975 to 1976. Several accidents had been reported where a Pinto was struck from behind and the gas tank exploded (The Center for Highway Safety 2021). According to this article, In April, 1974, the Center for Auto Safety petitioned the National Highway Traffic Safety Administration to recall Ford Pintos due to defects in the design of the strap on gas tank which made it susceptible to leakage and fire in low to moderate speed collisions. The Center's petition was based upon reports from attorneys of three deaths and four serious injuries in such accidents. This petition was not addressed by the National Highway Traffic Safety Administration (NHTSA) until 1977. The NHTSA stated in their 1978 report that:

> A formal defect investigation case was initiated on September 13, 1977, based upon allegations that the design and location of the fuel tank in the Ford Pinto make it highly susceptible to damage on rear impact at low to moderate closing speeds. On August 10, 1977, a press conference was held in Washington D.C., to announce the release of an article entitled, "Pinto Madness," which was published in the September/October of Mother Jones magazine. The article made several allegations concerning the safety of the Pinto fuel tank. The most significant of these charges as related to the National Highway Traffic Safety Administration's (NHTSA) defect investigation are as follows:
>
> 1. That the Pinto fuel tank is designed and located so that in rear-impact collisions at low to moderate speeds, it is displaced forward until it impacts the differential housing on the rear axle, resulting in tank cuts and/or puncture. The leakage of gasoline thus presents a significant fire hazard.

2. That the Ford Motor Company had knowledge of this "defect" during the developmental phase of the Pinto through its own test programs but concluded that it was more cost effective to produce the vehicle without modifications which would have corrected the problem but added to the production cost.

An investigation was conducted, and the Pinto and Bobcat were recalled in September 1978. The Center for Auto Safety article (2021) stated that the recall entailed the following:

The modifications included a longer fuel filler neck and a better clamp to keep it securely in the fuel tank, a better gas cap in some models, and placement of a plastic shield between the front of the fuel tank and the differential to protect the tank from the nuts and bolts on the differential and another along the right corner of the tank to protect it from the right rear shock absorber. Recall notices were mailed in September 1978 and parts were to be at all dealers by September 15, 1978. However, between June 9, 1978, and the date when parts were available to repair the estimated 2.2 million vehicles, six people died in Pinto fires after a rear impact.

Faulty consumer products continue to be manufactured and sold. For instance, a CPSC report from 2021 provided the injuries from off-road vehicles with more than two wheels (CPSC 2021).

2.5 Off-Highway-Vehicle-Related Fatalities Reported

- As of September 2021, the year 2018 is the most recent year of reporting for fatalities that CPSC considers complete. CPSC staff is aware of 2211 deaths associated with Off-Highway Vehicles (OHVs) that resulted from 2156 incidents during the three-year period from 2016 through 2018.
- Of the OHVs involved in those 2211 reported deaths, CPSC staff classifies 1591 as All-Terrain Vehicles (ATVs), 506 as Recreational Off-Highway Vehicles (ROVs), and 47 as Utility-Terrain Vehicles (UTVs). For the remaining 67 deaths, CPSC staff does not know the vehicle classification, but staff concludes that the vehicle is either an ROV or UTV.
- CPSC staff divided these 2211 reported deaths across various age groups: under 12 years (6%), 12–15 (7%), 16–24 (15%), 25–34 (15%), 35–44 (13%), 45–54 (15%), and 55+ (29%). Children under 12 represent about half (48%) of the fatalities among the combined under-16 age group.

- CPSC staff observed that OHV overturns and/or collisions (e.g. with other vehicles or stationary objects, such as trees) were the most common fatality hazards.
- Off-Highway-Vehicle-Related Emergency Department-Treated Injury Estimates
- Over the full five-year period from 2016 through 2020, CPSC staff estimates that there were 526,900 emergency department-treated injuries associated with OHVs (ATVs, ROVs, and/or UTVs) in the United States. This corresponds to an estimated annual average of 105,400 emergency department-treated injuries over the period.
- Although these estimated injuries do not trend in a single direction over the period, there is a statistically significant decrease estimated from 115,500 in 2016, to 95,000 in 2018, followed by a significant increase to an estimated 112,300 in 2020.
- CPSC staff divided injuries during the 2016 through 2020 period across various age groups: under 12 years (13%), 12–15 (13%), 16–24 (23%), 25–34 (20%), 35–44 (13%), 45–54 (9%), and 55+ (8%). This distribution of estimated injuries appears to be more heavily weighted towards younger ages than the distribution of reported fatalities.
- In the most recent year 2020 estimated OHV-related emergency department-treated injuries for all ages, CPSC staff found that:
- The most common diagnoses were fractures (30%) and contusions/abrasions (18%).
- The affected body parts were primarily: the head and neck (30%), the arm (shoulders to fingertips, 30%), the torso (20%), and the leg (20%).
- Victims were more frequently identified as male (68%) than female (32%).
- Most were treated and released (78%) or hospitalized (19%).
- Hospitalizations (meaning cases treated and admitted or transferred to another hospital) were found significantly increased in the year 2020, compared with the 4 prior years.

The difference between the 1970s and now is that flawed consumer items are identified sooner and then recalled much sooner.

The Takata airbag recall is an example of a more recent defective product recall. 67 million Takata airbags have been recalled (NHTSA 2021). In fact, my pickup was added to the list recently and I am having it replaced soon.

2.6 Work Relationships

Over the centuries work relationships have changed drastically. Prior to the industrial revolution most workplaces were small shops or work being performed in people's houses or barns. Medievalists.net (2021) list 20 common jobs from the Middle Ages. These were:

- Butcher
- Baker
- Stonemason
- Weaver
- Mason
- Winemaker
- Farmer
- Watchman
- Cobbler/Shoemaker
- Wheelwright
- Roofer
- Locksmith
- Tanner
- Tax Collector
- Belt Maker
- Grocer/merchant
- Armorer
- Carpenter
- Cook
- Blacksmith

One not listed here is the Fuller. What the fuller did was take raw wool clothe and make it more wearable (Deron 2015). Wool fullers basically "fulled" whatever cloth they were working with using their feet and stale urine. The ammonium salts that are contained in the stale urine help soften and cleanse the cloth. It also helps brightened white clothing as well. Urine was even taxed because of how often it was used for fulling purposes. People would go door to door and collect (buy) urine to use in the fulling process.

At this time the skilled crafts utilized the master, journeyman, and apprentice system (Perrin 2017; Wilson-Slack 2018). This was called the "Guild System." An apprentice was a novice or learner of a trade like mason or blacksmith. This person, most likely a male, was "apprenticed" to a journeyman at the ripe old age of 10–15 years of age. This apprenticeship lasted for three to seven years, depending on the trade. The apprentice was more than likely not paid but was given a place to live and food by the journeyman. Once the apprentice finished his training, he became a journeyman himself.

A journeyman was still under the supervision of his master. He could be paid for his labor and worked independent of the master. The journeyman could visit other masters and/or provide services to customers. The apprentice was nearly always at the side of the master. The journeyman was skilled enough to be hired and was ready to leave the shop of his master. The journeyman was, therefore, a competent, job-worthy craftsman who was no longer a beginner or apprentice.

It would take a journeyman many years to then rise to or qualify as a master. A journeyman would need to develop a "masterpiece" that would be assessed by the entire guild to determine if he should attain the level of a master.

The master craftsman would then be permitted by the guild to take on apprentices of his own and perpetuate the process of training others in the art. The master was a mentor or one who guided and coached not only the hands of his apprentice but also his mind.

In this model, it was important that a master's journeyman and apprentices were safe. If an apprentice or journeyman were hurt, it would affect the master's or journeyman's income. The higher-ranking craftsman would train the lower ranking tradesmen on how to make tools and how to perform tasks safely.

The journeyman and apprentice system lives on in some current trades like electricians and plumbers. In fact, in the current job climate there is a real push in many industries to increase the number of apprentices. Next Step Idaho (Idaho Department of Education 2021) is a program to have more young people become apprentices in skilled craft careers.

So, why and when did the guild system really fall apart?

In an article called "The Natural Tie Between Master and Apprentice has been Rent Asunder: An Old Apprentice Laments Changes in the Workplace, 1826" the author provides a basis as to why the system failed in the late 1700s and early 1800s (History Matters 2021). It reads:

> The urban workplace changed dramatically in the early decades of the nineteenth century. The American Revolution, with its rampant egalitarianism, dissolved much of the paternalistic control once wielded by fathers, masters and other authority figures, as the anonymous author "Old Apprentice" made clear in his set of three letters to the New York Observer in 1826. But significant blame for this erosion rested with the manufacturers themselves. Eager to seize upon new markets with expanded production, they divided up tasks to produce cheaper clothing or shoes. Semiskilled and unskilled women and children performed this labor rather than apprentices or other workingmen of the traditional artisanal system. These changes also dissolved the traditional residential patterns, pushing working men out into the housing market. A loss of reciprocity and responsibility occurred on both sides.

One of the letters that was written by the old apprentice and published in the New York Observer October 14, 1826, and shown in the History Matters article (2021) reads:

> ... The tendency of our laws, which give masters no control over their apprentices, or the manner in which these laws are enforced or abused, by

affording to unruly apprentices inducements to complain of, and to mortify and perplex their masters, has induced the solution on the part of the most respectable master mechanics, not to take apprentices at all. It is a fact well known to many, that there are great numbers of poor and friendless boys in our streets, who are yet honest, and desirous to work, but who, in consequence of this state of disorganization, are unable to obtain the knowledge of a trade. These may be seen wandering about our docks, or lounging at the Intelligence Office, until their scanty funds are exhausted, when to avoid starvation, they resort to pilfering, and are at length brought to the House of Refuge or in some other place of confinement less favorable to the reformation of character.

I had occasion, recently, to look at the form of an indenture; and presuming that I should find several in the first shop I entered, I stepped into my tailors, when to my surprise I found that, although one of the largest establishments in our city, the proprietors had not a single apprentice. I then went to several other tailors, jewelers, watch-makers, printers, and boot-makers before I could succeed. In enquiring the cause, I found that they all agreed (with a single exception), in the reasons here assigned, viz. the insufficiency of the existing laws to compel an apprentice to do his duty, and the power given to an obstinate and exasperated boy, in case of even moderate punishment, to drag his masters before a court, exposing him to the degradation of unmerited punishment, or at least subjecting him to expense, loss of time and the mortifying sneers of the rest of the boys who thus learn that they may pursue the same course with impunity.

That apprentices should be protected from the abuse of capricious or tyrannical maters, is in accordance with justice and the spirit of our government; but that a master should not have the power of compelling to the performance of his duty, an obstinate boy, who has been of little or no use in the early part of his apprenticeship, and who, when he has half learned his trade, will court the breaking of his indentures that he may be employed elsewhere as a journeyman, is a monstrous extension of Mr. Jefferson's "free and equal" declaration. It is not only subversive of good order and government between masters and apprentices, but it is subversive also of these principles, in the exercise of which in their purity, we have the surest, indeed the only guarantee of the continuance of our boasted independence....

The factory system really came about with the concept of interchangeable parts. Eli Whitney is credited with this idea, but it was being used in other industries (Eli Whitney Museum and Workshop 2021). Factories, instead of small workshops, could be developed once products could be produced using standardized parts. Prior to the idea of interchangeable parts muskets, for instance, were made by skilled craftsman and were unique, even if they followed a standard design. Once interchangeable parts came along, anyone with less skills could assemble and repair muskets or other firearms.

The next big step in the development of the factory system was Frederick Taylor's concept of Scientific Management (1929). His work and theory had a profound impact on work in general and on safety.

Taylor's four principles are as follows:

1. Replace working by "rule of thumb," or simple habit and common sense, and instead use the scientific method to study work and determine the most efficient way to perform specific tasks.
2. Rather than simply assign workers to just any job, match workers to their jobs based on capability and motivation and train them to work at maximum efficiency.
3. Monitor worker performance and provide instructions and supervision to ensure that they're using the most efficient ways of working.
4. Allocate the work between managers and workers so that the managers spend their time planning and training, allowing the workers to perform their tasks efficiently.

I have used the example of Frederick Taylor to demonstrate some of the reasons ergonomic issues developed. Using his philosophy of standardization and assigning workers with certain capabilities to certain jobs is counter to how ergonomists develop tasks and workplaces. According to Grandjean (1988), we should fit the tasks to the workers and not the other way around. Standardizing workplaces cause most all workers to not be at their optimal work heights and the job demands to exceed worker capabilities.

By the mid-twentieth century workplaces in the United States were going the wrong way regarding safety and health. In 1939 the number of occupational fatalities in the United States was at least 16,600 (BLS 2021; New York Times 1940).

2.7 Food

Food and culture are very interesting topics. Up until modern times even people in developed countries didn't have enough food. Roser and Ritchie (2022) discuss the current state of food insecurity in the world. They summarized the data, and it is presented as follows:

- 8.8% of the world's population are undernourished – this means they have a caloric intake below minimum energy requirement.
- 663 million people globally are undernourished.

- 22% of children younger than five are "stunted" – they are significantly shorter than the average for their age, as a consequence of poor nutrition or repeated infection.
- 9% of the world population – around 697 million people – are severely food insecure.
- One-in-four people globally – 1.9 billion – are moderately or severely food insecure.

Food quality is another big issue. In the United States there are food recalls quite regularly.

Upton Sinclair wrote The Jungle in 1906 about the problems in the meatpacking industry. Originally, the book was to illustrate the disparities in society and how immigrants were being exploited, but his work brought attention to the poor quality of meat and meat products (Sinclair 1906). The Federal Food and Drugs Act, better known as the Wiley Act, was signed by President Theodore Roosevelt on June 30, 1906 (Food and Drug Administration 2021a). This act followed approximately 100 laws passed since 1879 on food and drug quality.

An example of the types of recalls that are still occurring in the United States is shown next (United States Department of Agriculture 2022):

> WASHINGTON, Sept. 13, 2021 – SAS Foods Enterprises Inc., an Elk Grove Village, Ill., establishment, is recalling approximately 3,768 pounds of beef and chicken empanada products that were produced without the benefit of federal inspection and bearing a label with a false USDA mark of inspection, the U.S. Department of Agriculture's Food Safety and Inspection Service (FSIS) announced today.
>
> The frozen, fully cooked beef and chicken empanada items were produced on various dates from Jan. 1, 2020, through Sept. 11, 2021. The following products are subject to recall:
>
> 1-lb. zip-lock bags or clear, plastic containers with "SAS Food EMPANADAS DE POLLO CHICKEN PATTIES."
>
> 1-lb. zip-lock bags or clear, plastic containers with "SAS Food EMPANADAS DE CARNE BEEF PATTIES."
>
> The products subject to recall bear establishment number "EST. 38548" inside the USDA mark of inspection; however, the recalling company has no affiliation with Establishment 38548. These items were shipped to retail consignees in Illinois, Indiana, Minnesota, Ohio, and Wisconsin.
>
> The problem was discovered after FSIS received an anonymous tip and initiated an investigation.
>
> There have been no confirmed reports of adverse reactions due to consumption of these products. Anyone concerned about a reaction should contact a healthcare provider.

FSIS is concerned that some product may be frozen and in consumers' freezers. Consumers who have purchased these products are urged not to consume them. These products should be thrown away or returned to the place of purchase.

Food quality around the world is even worse in some countries. My wife, who is originally from Ukraine, talks about the poor quality of sausage and not knowing what the actual ingredients are. The United States imports roughly US$ 4.6 billion worth of food from China (Minnesota Department of Agriculture 2017).

ERS researchers analyzed FDA refusals of food import shipments originating from China by type of violation (United States Department of Agriculture 2009). Here, the term violation refers to products that appear to violate one or more of the laws enforced by FDA, such as those dealing with adulterated or misbranded products. FDA refusals of food shipments from China peaked in early 2007, just before a series of highly publicized incidents. In 2007, FDA issued import alerts for wheat gluten, rice protein products, and five kinds of farm-raised fish and shrimp from China. Customs statistics show that shrimp imports from China slowed after the FDA alert was issued. FDA refusals of Chinese food shipments reflect the mix of products imported: fish and shellfish, fruit, and vegetable products account for most refusals. Most Chinese food imports are processed to some degree, and the most common problems cited by FDA, "filth," unsafe additives, inadequate labeling, and lack of proper manufacturer registrations, are typically introduced during food processing and handling. Another of the most common problems, potentially harmful veterinary drug residues in farm-raised fish and shrimp, is introduced at the farm. FDA cites harmful pesticide residues and pathogens in Chinese food shipments less frequently.

In many of these food safety cultural issues the population accepts the poor-quality food because in many cases the poor-quality food is cheaper. For instance, food processed in China is cheaper than locally processed food (A&A 2021). Think about how that food traveled. Apples grown in the United States might get imported into China, processed there, and then shipped back to the United States. Even with all this shipping the juice from China can be cheaper that juice processed locally. How much of the nutritional quality of the juice is of question? Honey from China is a good example of how a nutritional product goes from beneficial to only a sweet, sticky syrup (Ungoed-Thomas and Leake 2021). Five years ago, Mexican farmers were paid 47 pesos (£1.73) per kilogram for their organic honey by a local fair-trade co-operative, but the price has dropped to just 35 pesos per kilogram. Many of Mexico's estimated 42,000 beekeepers were giving up their hives because they can't compete with the price of honey from China. However, as Ungoed-Thomas and Leake discuss, the honey that comes from China is not 100% honey. It is a blend of syrup and some honey. In addition, some of the most nutritional parts of the honey portion of the Chinese have been removed (Olmsted 2016). The reason for this is the price of the honey from China is much less.

The bottom line is that cheaper food is often, the driving force for consumers, rather than more costly nutritionally sound products. Though, with the level of food insecurity in the world, cheaper alternatives are necessary. However, these cheaper alternatives should be nutritious.

2.7.1 Food Trends and Culture

Food trends come and go. Some of these trends are beneficial health wise and some, not so much. The following paragraphs present a few of the trends from history.

2.7.1.1 The Tomato

The tomato is the world's most popular fruit. Annual production is approximately 60 million tons. However, in 1595 it was thought to be poisonous and European society only grew it as an ornamental oddity. Cortez discovered tomatoes growing in Montezuma's gardens in 1519 and brought back seeds to Spain. The Italians were the first to begin to cultivate tomatoes and to use them in food. How they became accepted was not a linear path. It is speculated that nobles felt they were an aphrodisiac, while the masses saw animals that ate them and did not die. In the United States, Robert Gibbon Johnson ate one on a New Jersey courthouse steps and did not die. This convinced skeptical Americans the tomatoes were not toxic (Veggie Cage 2022).

2.7.1.2 Fad Diets

There are so many fad diets that have been in the United States culture over the years that several books could be written. In fact, there are several books on fad diets. Promoters of these diets might take offense on me calling them "Fad." However, some of them come and then go when the people on the diets determine they cannot follow the diets' rules for very long. To avoid any backlash, I will only discuss a couple from the past.

One of these diets is the Grapefruit Diet. A recent article in Cleveland Clinic Health Essentials (CCHE) discusses this fad diet that has been around since the 1930s (Cleveland Clinic 2021).

The Grapefruit fad diet was debuted in the 1930s amidst claims that grapefruit contains fat-burning enzymes that melt off pounds in the form of fat. Almost a 100 years later the diet still draws attention and has followers.

The diet consists of eating grapefruit at every meal. The CCHE article says that the short-term diet's main selling point is that it can help you lose 10 pounds during a 10- to 12-day detox. After that, the primary focus of the diet is cutting way back on calories. Some plans suggest limiting yourself to as little as 800 calories in a day. That's less than half of what is recommended for a healthy diet. Obviously, the calorie reduction has a large component of the weight loss regime.

The classic grapefruit diet includes eating foods rich in protein and high in fat and cholesterol. The diet also specifies eating fewer sugars and carbohydrates.

The CCHE article states that a sample meal on the grapefruit diet might include:

- Half a grapefruit or 8 oz 100% grapefruit juice without adding sugar.
- Salad or a red or green vegetable cooked in butter or spices.
- Meat or fish cooked in any manner.
- One cup of coffee or tea without cream or sugar.

Grapefruit is roughly 88% water, so eating it with meals tends to make you feel full faster.

As stated earlier, the grapefruit diet's quick weight loss claims, that's more a byproduct of severely slashing calories. "You lose weight on the plan because you're not eating as much," says Lauren Sullivan, RD, who is quoted in the article.

As stated in the article and collaborated in other sources, grapefruit does impact the efficacy of certain medications. The FDA provides a list of medications that are impacted by grapefruit. These are:

- Some statin drugs to lower cholesterol, such as Zocor (simvastatin) and Lipitor (atorvastatin).
- Some drugs that treat high blood pressure, such as Procardia and Adalat CC (both nifedipine).
- Some organ-transplant rejection drugs, such as Neoral and Sandimmune capsule or oral solution (both cyclosporine).
- Some anti-anxiety drugs, such as BuSpar (buspirone).
- Some corticosteroids that treat Crohn's disease or ulcerative colitis, such as Entocort EC and Uceris tablet (both budesonide).
- Some drugs that treat abnormal heart rhythms, such as Pacerone and Cordarone tablet (both amiodarone).
- Some antihistamines, such as Allegra (fexofenadine).

Grapefruit juice does not affect all the drugs in the categories earlier. The severity of the interaction can be different depending on the person, the drug, and the amount of grapefruit juice you drink. Talk to your health care provider or pharmacist, and read any information provided with your prescription or non-prescription (OTC) drug to find out:

The US Food and Drug Administration (FDA) requires warnings on some medications related to their use with grapefruit or grapefruit juice (FDA 2021b).

The CCHE article says that don't let the dubious claims about the grapefruit diet shape your opinion on the tropical fruit. Overall, grapefruit offers an impressive nutritional resume. Highlights include:

- Serving as an excellent source of Vitamin C and Vitamin A.
- Packing a high fiber content that can boost heart health.
- Cancer-fighting antioxidant properties.
- A low score on the glycemic index, meaning it doesn't cause blood sugar levels to spike.
- On top of that, grapefruit packs a lot of taste without a lot of calories.

One new version of diet entails taking grapefruit extract and not eating the fruit.

2.8 Genetically Modified Organisms (GMO) Foods

There are many sides to Genetically Modified Organisms (GMO) foods. Many people are strongly against the use of GMO foods and many people don't care. Food producers are for GMO foods for economic reasons. This brief section does not advocate for GMO foods or condemns them. The following just presents some of the facts surrounding the foods. It is important to note that humans have been manipulating the genetics of animals, vegetables, and fruits for probably 10,000 years (Wetherell 2019). Wheat, cattle, apples, and pigs are all examples of food items that have been bred to where they are today. In fact, all the dog breed characteristics that are expressed in the hundreds of certified dog breeds were contained in the genetic material of the ancient dogs (Black 2021; Callaway 2020; Bergstrom et al. 2020).

The difference with GMO organisms is that the new genetic material that is used to make changes to the plants or animals is placed in the genome artificially. This can be done by gene splicing or by using a bacterium or virus, along with other similar methods (The Royal Society 2021). Table 2.2 shows some of the food items that are currently GMO approved by the FDA.

Phillips (2008) article presents the potential reasons why GMO foods might be problematic. These reasons are:

- Changes to the organism's metabolism, growth rate, and/or response to external environmental factors. These consequences influence not only the GMO itself, but also the natural environment.

Table 2.2 Examples of GMOs resulting from agricultural biotechnology.

Genetically conferred trait	Example organism	Genetic change
Approved commercial products		
Herbicide tolerance	Soybean	Glyphosate herbicide (Roundup) tolerance conferred by expression of a glyphosate-tolerant form of the plant enzyme 5-enolpyruvylshikimate-3-phosphate synthase (EPSPS) isolated from the soil bacterium *Agrobacterium tumefaciens*, strain CP4
Insect resistance	Corn	Resistance to insect pests, specifically the European corn borer, through expression of the insecticidal protein Cry1Ab from *Bacillus thuringiensis*
Altered fatty acid composition	Canola	High laurate levels achieved by inserting the gene for ACP thioesterase from the California bay tree *Umbellularia californica*
Virus resistance	Plum	Resistance to plum pox virus conferred by insertion of a coat protein (CP) gene from the virus

Source: Phillips (2008).

- Potential health risks to humans include the possibility of exposure to new allergens in genetically modified foods, as well as the transfer of antibiotic-resistant genes to gut flora.
- Horizontal gene transfer of pesticide, herbicide, or antibiotic resistance to non-GMO organisms. This could potentially spread disease among both plants and animals.
- Expression of unintended genes that might produce potentially toxic proteins that could affect other animals and plants.

The benefits of GMO foods are (National Library of Medicine [NLOM] 2021):

- More nutritious food
- Tastier food
- Disease- and drought-resistant plants that require fewer environmental resources (such as water and fertilizer)
- Less use of pesticides
- Increased supply of food with reduced cost and longer shelf life
- Faster growing plants and animals
- Food with more desirable traits, such as potatoes that produce less of a cancer-causing substance when fried
- Medicinal foods that could be used as vaccines or other medicines

The NLOM article goes on to say that the concerns with GMO foods so far are unfounded. None of the GMO foods used today have caused any of these problems. The US FDA assesses all GE foods to make sure they are safe before allowing them to be sold. In addition to the FDA, the US Environmental Protection Agency (EPA) and the USDA regulate bioengineered plants and animals. They assess the safety of GE foods to humans, animals, plants, and the environment.

2.8.1 Messenger Ribonucleic Acid (mRNA) Vaccines

Since we are talking about modifying genetics, I felt it was a good place to briefly present how mRNA vaccines work. The following is from the CDC website (CDC 2022):

> To trigger an immune response, many vaccines put a weakened or inactivated germ into our bodies. Not mRNA vaccines. Instead, mRNA vaccines use mRNA created in a laboratory to teach our cells how to make a protein—or even just a piece of a protein—that triggers an immune response inside our bodies. That immune response, which produces antibodies, is what protects us from getting infected if the real virus enters our bodies.
>
> COVID vaccine
>
> First, COVID-19 mRNA vaccines are given in the upper arm muscle. The mRNA will enter the muscle cells and instruct the cells' machinery to produce a harmless piece of what is called the spike protein. The spike protein is found on the surface of the virus that causes COVID-19. After the protein piece is made, our cells break down the mRNA and remove it.
>
> Next, our cells display the spike protein piece on their surface. Our immune system recognizes that the protein doesn't belong there. This triggers our immune system to produce antibodies and activate other immune cells to fight off what it thinks is an infection. This is what your body might do to fight off the infection if you got sick with COVID-19.
>
> At the end of the process, our bodies have learned how to protect against future infection from the virus that causes COVID-19. The benefit of COVID-19 mRNA vaccines, like all vaccines, is that those vaccinated gain this protection without ever having to risk the potentially serious consequences of getting sick with COVID-19. Any temporary discomfort experienced after getting the vaccine is a natural part of the process and an indication that the vaccine is working.

2.9   Traffic Safety

This is a huge topic and will not be tackled in depth in this book. However, the following are some key points (NHTSA 2021):

- Traffic related fatalities have been going down over the years (see Table 2.3).
- Drunk driving continues to be one of the largest contributors to traffic fatalities.
- Not wearing seatbelts is a major contributor to not surviving a potentially fatal accident.
- Distracted driving is another significant contributor to traffic accidents.
- The incorporation of automatic airbags in vehicles has contributed to the survivability of potentially fatal accident.

We can also say that not realizing what the consequences of a car accident could have and poor judgement continues to be a major cause of accidents as well. Those that have not been in a serious accident do not realize the liability they have if they are the cause of the accident. A 4000-to-7000-lb (1800-to-3181-kg) vehicle, traveling at speeds above 30 miles per hour (mph) (50 kph) has tremendous kinetic energy. Newer cars offer better protection, but not enough to always protect the occupants.

We live in snow country and every year after the first couple snow falls there are a large number of accidents because drivers continue to drive like the roads are dry. We were recently in Las Vegas and were driving in the HOV lane on I-15, heading north. The HOV lane is actually two lanes near the intersection of US 95. A pickup went into the left lane and wanted to pass, except the left HOV lane goes left to Hwy 95. He saw his mistake and instead of slowing down, he sped up and bounced over the curbing at the intersection and went into our lane, barely missing the front of our car. Obviously, his judgement was flawed.

Cell phone usage has become the major source of distraction over the past couple years. NHTSA (2021) states: "Texting is the most alarming distraction. Sending or reading a text takes your eyes off the road for five seconds. At 55 mph, that's like driving the length of an entire football field with your eyes closed. You cannot drive safely unless the task of driving has your full attention. Any non-driving activity you engage in is a potential distraction and increases your risk of crashing. Using a cell phone while driving creates enormous potential for deaths and injuries on US roads. In 2019, 3142 people were killed in motor vehicle crashes involving distracted drivers."

Table 2.3 Motor vehicle traffic fatalities and fatality rates 1899–2019.

Motor vehicle traffic fatalities and fatality rates, 1899–2019

Year	Total fatalities	Vehicle miles traveled (millions)	Fatality rate per 100 million VMT	Year	Total fatalities	Vehicle miles traveled (millions)	Fatality rate per 100 million VMT	Year	Total fatalities	Vehicle miles traveled (millions)	Fatality rate per 100 million VMT
1899	26	—	—	1940	32,914	302,188	10.89	1981	49,301	1,555,308	3.17
1900	36	—	—	1941	38,142	333,612	11.43	1982	43,945	1,595,010	2.76
1901	54	—	—	1942	27,007	268,224	10.07	1983	42,589	1,652,788	2.58
1902	79	—	—	1943	22,727	208,192	10.92	1984	44,257	1,720,269	2.57
1903	117	—	—	1944	23,165	212,713	10.89	1985	43,825	1,774,826	2.47
1904	172	—	—	1945	26,785	250,173	10.71	1986	46,087	1,834,872	2.51
1905	252	—	—	1946	31,874	340,880	9.35	1987	46,390	1,921,204	2.41
1906	338	—	—	1947	31,193	370,894	8.41	1988	47,087	2,025,962	2.32
1907	581	—	—	1948	30,775	397,957	7.73	1989	45,582	2,096,487	2.17
1908	751	—	—	1949	30,246	424,461	7.13	1990	44,599	2,144,362	2.08
1909	1,174	—	—	1950	33,186	458,246	7.24	1991	41,508	2,172,050	1.91
1910	1,599	—	—	1951	35,309	491,093	7.19	1992	39,250	2,247,151	1.75
1911	2,043	—	—	1952	36,088	513,581	7.03	1993	40,150	2,296,378	1.75
1912	2,968	—	—	1953	36,190	544,433	6.65	1994	40,716	2,357,588	1.73
1913	4,079	—	—	1954	33,890	561,963	6.03	1995	41,817	2,422,823	1.73
1914	4,468	—	—	1955	36,688	605,646	6.06	1996	42,065	2,484,080	1.69
1915	6,779	—	—	1956	37,965	627,843	6.05	1997	42,013	2,552,233	1.65
1916	7,766	—	—	1957	36,932	647,004	5.71	1998	41,501	2,628,148	1.58
1917	9,630	—	—	1958	35,331	664,653	5.32	1999	41,717	2,690,241	1.55
1918	10,390	—	—	1959	36,223	700,480	5.17	2000	41,945	2,746,925	1.53
1919	10,896	—	—	1960	36,399	718,762	5.06	2001	42,196	2,795,610	1.51
1920	12,155	—	—	1961	36,285	737,421	4.92	2002	43,005	2,855,508	1.51

Year	Fatalities	VMT	Rate		Year	Fatalities	VMT	Rate		Year	Fatalities	VMT	Rate
1921	13,253	55,027	24.08		1962	38,980	766,734	5.08		2003	42,884	2,890,221	1.48
1922	14,859	67,697	21.95		1963	41,723	805,249	5.18		2004	42,836	2,964,788	1.44
1923	17,870	84,995	21.02		1964	45,645	846,298	5.39		2005	43,510	2,989,430	1.46
1924	18,400	104,838	17.55		1965	47,089	887,812	5.30		2006	42,708	3,014,371	1.42
1925	20,771	122,346	16.98		1966	50,894	925,899	5.50		2007	41,259	3,031,124	1.36
1926	22,194	140,735	15.77		1967	50,724	964,005	5.26		2008	37,423	2,976,528	1.26
1927	24,470	158,453	15.44		1968	52,725	1,015,869	5.19		2009	33,883	2,956,764	1.15
1928	26,557	172,856	15.36		1969	53,543	1,061,791	5.04		2010	32,999	2,967,266	1.11
1929	29,592	197,720	14.97		1970	52,627	1,109,724	4.74		2011	32,479	2,945,194	1.10
1930	31,204	206,320	15.12		1971	52,542	1,178,811	4.46		2012	33,782	2,963,497	1.14
1931	31,963	216,151	14.79		1972	54,589	1,259,786	4.33		2013	32,893	2,982,941	1.10
1932	27,979	200,517	13.95		1973	54,052	1,313,110	4.12		2014	32,744	3,020,377	1.08
1933	29,746	200,642	14.83		1974	45,196	1,280,544	3.53		2015	35,484	3,089,841	1.15
1934	34,240	215,563	15.88		1975	44,525	1,327,664	3.35		2016	37,806	3,173,815	1.19
1935	34,494	228,568	15.09		1976	45,523	1,402,380	3.25		2017	37,473	3,210,248	1.17
1936	36,126	252,128	14.33		1977	47,878	1,467,027	3.26		2018	36,835	3,240,327	1.14
1937	37,819	270,110	14.00		1978	50,331	1,544,704	3.26		2019	36,096	3,261,772	1.11
1938	31,083	271,177	11.46		1979	51,093	1,529,133	3.34					
1939	30,895	285,402	10.83		1980	51,091	1,527,295	3.35					

Total Traffic Fatalities (1899–2019): 3,830,591

Note: A traffic fatality is defined as a death that occurs within 30 days after a traffic crash.

Sources: **Traffic fatalities, 1899–1974:** National Center for Health Statistics, *HEW and State Accident Summaries* (adjusted to 30-Day Traffic Deaths by National Highway Traffic Safety Administration (NHTSA)); **1975–2019:** NHTSA, Fatality Analysis Reporting System (FARS) 1975–2018 Final and 2019 Annual Report File.

Vehicle Miles Traveled (VMT) – Federal Highway Administration (FHWA) – Not Available for Years 1899 – 1920.

Traffic Safety Facts Annual Report, May 2021

2.10 Public Acceptance of Seatbelts and Masks for Protection from Respiratory Disease

NHTSA reports that "One of the safest choices drivers and passengers can make is to buckle up. Many Americans understand the lifesaving value of the seat belt – the national use rate was at 90.3% in 2020. Seat belt use in passenger vehicles saved an estimated 14,955 lives in 2017. Understand the potentially fatal consequences of not wearing a seat belt and learn what you can do to make sure you and your family are properly buckled up every time."

Table 2.4 compares the number of fatalities with and without constraints from 2000 to 2019. This table clearly shows that survivability is much better with vehicle

Table 2.4 Comparison of the number of fatalities with and without constraints from 2000 to 2019.

Injury severity/ year	Restraint use							
	Restrained		Unrestrained		Unknown		Total	
	Number	Percent	Number	Percent	Number	Percent	Number	Percent
2000	11,787	36.6	17,810	55.3	2,628	8.2	32,225	100.0
2001	11,946	37.3	17,517	54.7	2,580	8.1	32,043	100.0
2002	12,532	38.2	17,798	54.2	2,513	7.7	32,843	100.0
2003	12,967	40.2	16,764	51.9	2,540	7.9	32,271	100.0
2004	13,250	41.6	16,432	51.6	2,184	6.9	31,866	100.0
2005	13,063	41.4	16,248	51.5	2,238	7.1	31,549	100.0
2006	12,710	41.4	15,635	51.0	2,341	7.6	30,686	100.0
2007	12,322	42.4	14,446	49.7	2,304	7.9	29,072	100.0
2008	10,691	42.0	12,925	50.8	1,846	7.3	25,462	100.0
2009	10,190	43.5	11,545	49.2	1,712	7.3	23,447	100.0
2010	9,969	44.8	10,590	47.5	1,714	7.7	22,273	100.0
2011	9,471	44.4	10,215	47.9	1,630	7.6	21,316	100.0
2012	9,746	44.7	10,370	47.6	1,663	7.6	21,779	100.0
2013	9,840	46.4	9,622	45.3	1,761	8.3	21,223	100.0
2014	9,961	47.3	9,410	44.7	1,679	8.0	21,050	100.0
2015	10,763	47.5	9,975	44.1	1,903	8.4	22,641	100.0
2016	11,343	47.7	10,463	44.0	1,981	8.3	23,787	100.0
2017	11,488	48.5	10,116	42.8	2,059	8.7	23,663	100.0
2018	11,055	48.4	9,845	43.1	1,945	8.5	22,845	100.0
2019	10,815	48.7	9,466	42.6	1,934	8.7	22,215	100.0

restraints, than without. My dad was a truck driver before 1985 and swore that getting thrown from a truck in a crash was much better than being restrained by seat belts. I was never sure what changed his mind. However, after he retired, he always wore seatbelts.

The excuses people used and still use to not wear seatbelts are (State of Delaware 2007):

(1) I am afraid of getting stuck in a crashed car – If you are not buckled up when the wreck occurs, you are more likely to be killed or knocked unconscious, and therefore will be unable to get out of the car at all. When you are buckled up, you are more likely to remain in place, in control of the vehicle and conscious to make smart decisions.

(2) It irritates the skin on my neck or chest – most newer vehicles have adjustable shoulder height positioners that allow occupants to move the shoulder belt height up or down for a more comfortable fit; for older vehicles, occupants may consider wearing clothes with a higher neck to provide some extra padding if this is a big concern for you

(3) It makes me feel restrained – that's the function of a seat belt; in a crash it keeps you in your seat so you're unable to be thrown around or out of the vehicle where you're four times more likely to be killed than if you remain the car. All driver's side seat belts allow free movement of the occupant until a crash occurs (or in some instances until you jam on your breaks!)

(4) I am too big to wear a seat belt, it doesn't fit – For some individuals, purchasing a seat belt extender may work to solve this issue.

(5) I can't look over my shoulder before turns – Yes you can. A seat belt doesn't restrain your head, it restrains your chest.

(6) I forgot – This is a common excuse, and is why most vehicles have annoying seat belt reminder systems that beep every minute or so when the vehicle senses that the restraint system isn't in use.

(7) Nobody tells me what to do in my car – Every state has a variety of traffic laws that mandate what people can or cannot do. For instance, it is illegal to drink and drive, it is illegal to speed, and it is illegal to drive or ride without a seat belt.

(8) I have an air bag, I don't need a seat belt – Air bags are Supplemental Restraint Systems, meaning they are designed to work in conjunction with seat belts not as a restraint system alone. Air bags are not soft cushy pillows. They deploy at approximately 250 mi an hour (the blink of an eye), and begin to deflate immediately after deployment. If you are not restrained by your seat belt, you will go into the air bag, and since it starts to deflate immediately, you will quickly go into the steering column or through the windshield.

(9) I can't wear a seat belt because I can't feed my baby with it on – If you're driving, your eyes should be focused forward. If you're trying to feed your baby in the backseat, you can't possibly be focusing your attention on the road and you are risking both of your lives. If you're a passenger and need to feed

your child a bottle, you should sit in the back seat with the child, and both of you should be properly restrained. Nursing mothers should never try to feed their child while the vehicle is moving. You never know when someone may hit your vehicle and the laws of physics will make it impossible for you to hold onto your child in a crash. Please pull over to a safe location to nurse your child.

(10) I have a medical condition; I can't wear it – This can be a valid excuse but only if a doctor provides you with a written medical note. If this is the case, make sure to carry it in your purse or wallet so that the doctor's instructions remain with you if you are a passenger in someone else's vehicle.

In the early 1980s when seatbelt laws were first introduced there was huge pushback, probably in proportion to the pushback on the use of masks now to help control the COVID-19 virus. David Roos (2020) writes in an article on History.com that when seat belt laws were first introduced that there was a great backlash. In his article he states that "When David Hollister introduced a seat belt bill in Michigan in the early 1980s that levied a fine for not buckling up, the state representative received hate mail comparing him to Hitler. At the time, only 14 percent of Americans regularly wore seat belts, even though the federal government required lap and shoulder belts in all new cars starting in 1968."

Does this sound familiar? At this writing the pandemic is still in high gear and there have been riots in the United States and other countries over mask usage and other COVID-19 control measures. A recall election was held for the Governor of California, Gavin Newsome, over masks and other control measures (Blood and Ronayne 2021).

Opponents to the mandating of seatbelt use, wearing masks, and smoking bans claim that this impinges on their personal freedoms. In a Washington Post article from 2020 Rozsa et al. (2020) state that this in an All-American story. The authors state "Mask-wearing for some people is an identifier of broader beliefs and political leanings. Like so many issues rooted in science and medicine, the pandemic is now fully entangled with ideological tribalism. This has played out before: helmets for motorcyclists, seat belts in cars, smoking bans in restaurants. All of those measures provoked battles over personal liberty."

However, what is misunderstood is that if someone smokes and develops Chronic Obstructive Pulmonary Disease (COPD), like my dad and my father-law did, it also impacts the family and creates a burden on the medical community. An accident that causes the death of someone not wearing a seatbelt impacts many people, Emergency Medical personnel, fire fighters, the family, and others. It also increases insurance rates for all drivers.

Not wearing masks is analogous. It can lead to others becoming ill and may make the pandemic last longer. Personal freedoms are only freedoms if they don't impinge on others' freedoms.

NHTSA lists the facts of seatbelt use:

1. Buckling up is the single most effective thing you can do to protect yourself in a crash
   Seat belts are the best defense against impaired, aggressive, and distracted drivers. Being buckled up during a crash helps keep you safe and secure inside your vehicle; being completely ejected from a vehicle is almost always deadly.
2. Air bags are designed to work with seat belts, not replace them
   If you don't wear your seat belt, you could be thrown into a rapidly opening frontal air bag. Such force could injure or even kill you. Learn about air bag safety.
3. Guidelines to buckle up safely
   The lap belt and shoulder belt are secured across the pelvis and rib cage, which are better able to withstand crash forces than other parts of your body.
   Place the shoulder belt across the middle of your chest and away from your neck.
   The lap belt rests across your hips, not your stomach.
   NEVER put the shoulder belt behind your back or under an arm.
4. Fit matters
   Before you buy a new car, check to see that its seat belts are a good fit for you.
   Ask your dealer about seat belt adjusters, which can help you get the best fit.
   If you need a roomier belt, contact your vehicle manufacturer to obtain seat belt extenders.
   If you drive an older or classic car with lap belts only, check with your vehicle manufacturer about how to retrofit your car with today's safer lap/shoulder belts.
5. Seat belt safety for children and pregnant women.
   Find out when your child is ready to use an adult seat belt and learn about seat belt safety when you're pregnant.
   If You're Pregnant: Seat Belt Recommendations for Drivers and Passengers
   If you're pregnant, make sure you know how to position your seat and wear a seat belt to maximize your safety and the safety of your unborn child. Read our recommendations in the following text or view the instructional diagram version of our seat belt recommendations for pregnant drivers and passengers (PDF 497 KB).
   I'm Pregnant. Should I Wear a Seat Belt?
   YES – doctors recommend it. Buckling up through all stages of your pregnancy is the single most effective action you can take to protect yourself and your unborn child in a crash.
   NEVER drive or ride in a car without buckling up first!

6. What's the right way to wear my seat belt?

   The shoulder belt away from your neck (but not off your shoulder) and across your chest (between your breasts), making sure to remove any slack from your seat belt with the lap belt secured below your belly so that it fits snugly across your hips and pelvic bone.

   NEVER place the shoulder belt under your arm or behind your back.

   NEVER place lap belt over or on top of your belly.

7. Should I adjust my seat?

   YES – Adjust to a comfortable, upright position.

   Keep as much distance as possible between your belly and the steering wheel.

   Comfortably reach the steering wheel and pedals.

   To minimize the gap between your shoulder and the seat belt, avoid reclining your seat more than necessary.

   Avoid letting your belly touch the steering wheel.

8. What if my car or truck has air bags?

   You still need to wear your seat belt properly.

   Air bags are designed to work with seat belts, not replace them.

   Without a seat belt, you could crash into the vehicle interior, other passengers, or be ejected from the vehicle.

   My Car Has an ON–OFF Air Bag Disabling Switch. Should I turn it off?

   NO – Doctors recommend that pregnant women wear seat belts and leave air bags turned on. Seat belts and air bags work together to provide the best protection for you and your unborn child.

9. What should I do if I am involved in a crash?

   Seek immediate medical attention, even if you think you are not injured, regardless of whether you're the driver or passenger

2.11 Radiation Hazards and Safety

My first real exposure to radiation safety was while I was working at the PPG chemical plant in LaPorte, TX. One of the storage tanks had a Kay-Ray level gauge. The device had a Cesium-137 source that emitted Beta Particles. These will be explained next. One day the Nuclear Regulatory Commission (NRC) came in and asked, "Who is your Radiation Safety Officer (RSO)?" The plant manager said, "Why Lee is." I happened to be out of the plant that day, which was lucky because I wasn't an RSO. However, after that day I was. I, of course, knew the basics of radiation from physics and chemistry classes, but I was by no means an expert. At the Idaho National Engineering Laboratory (INEL) (now Idaho National Laboratory (INL), I became much more educated on radiation and the hazardous effects.

The next case study discusses the exposure to radium by mostly women who painted clock/watch faces with radium paint. The exact nature of their exposure will be discussed next. In Chapter 8 is devoted to accidents and incidents involving radioactive materials, also called radioisotopes.

The copper pan from Sweden has 0.464% thorium. Every isotope of thorium is radioactive. However, the half-lives vary greatly. I used my Geiger counter to see how radioactive the pan was. Yes, the Geiger counter does detect some activity, but very small.

The following is a high-level overview of radiation hazards and safety. The following discussion provides background on radiation that will explain why radiation is hazardous. This information was excerpted from the NRC website called "Radiation Basics."

2.11.1 Radiation

Radiation is energy given off by matter in the form of rays or high-speed particles. All matter is composed of atoms. Atoms are made up of various parts; the nucleus contains minute particles called protons and neutrons, and the atom's outer shell contains other particles called electrons. The nucleus carries a positive electrical charge, while the electrons carry a negative electrical charge. These forces within the atom work toward a strong, stable balance by getting rid of excess atomic energy (radioactivity). In that process, unstable nuclei may emit a quantity of energy, and this spontaneous emission is what we call radiation.

Matter gives off energy (radiation) in two basic physical forms. One form of radiation is pure energy with no weight. This form of radiation – known as electromagnetic radiation – is like vibrating or pulsating rays or "waves" of electrical and magnetic energy. There are two types of wave radiation; non-ionizing like infrared, microwave, ultra-violet light and visible light, and ionizing radiation like X-ray and gamma rays. Wave radiation can be harmful. For instance, too much exposure to ultra-violet radiation can lead to melanoma. X-ray radiation and gamma radiation can be a cause to many types of cancers.

The other form of radiation, known as particle radiation, is fast-moving particles that have both energy and mass (weight). This less-familiar form of radiation includes alpha particles, beta particles, and neutrons.

Through radioactive decay radioisotopes lose their radioactivity over time. This gradual loss of radioactivity is measured in half-lives. Essentially, a half-life of a radioactive material is the time it takes one-half of the atoms of a radioisotope to decay by emitting radiation.

Radioactive isotopes are natural in the environment. For instance, in Southeast Idaho there is great amount of Radon. Radon is a gas and is the product of decay of Uranium into Radium. One isotope of Radon-222 only has a half-life of

3.8 days. Radon is a hazardous material and is found in many unventilated base-ments. Radon can lead to lung cancer. Potassium 40 is another radioactive element that is quite common. If one holds a Geiger counter up to one's thyroid (neck area) there is usually a good indication of radiation decay. This is due to possibly natural radioactive Iodine.

What heavy radioactive elements, like uranium, were created in the first place is theorized as coming from either stars that have exploded in the form of a super nova or from the collision of two neutron stars (World Nuclear Association 2021). One of the reasons the core of the earth is heated is from the decay of natural uranium.

In some elements, the nucleus can split because of absorbing an additional neutron, through a process called nuclear fission. Such elements are called fissile materials. One particularly notable fissile material is uranium-235. This is the isotope that is used as fuel in commercial nuclear power plants.

Only certain radioactive elements fission. These elements are called fissile mate-rials. When a radioactive element nucleus fission, it causes three important events that result in the release of energy. Specifically, these events are the release of radi-ation, release of neutrons (usually two or three), and formation of two new nuclei (fission products). These fission products are a combination of other radioactive or non-radioactive elements.

The particles released by radioactive decay or fission include:

- Alpha particles
- Beta particles
- Neutrons

Alpha particles are charged particles, which are emitted from naturally occur-ring materials (such as uranium, thorium, and radium) and man-made elements (such as plutonium and americium). These alpha emitters are primarily used (in very small amounts) in items such as smoke detectors.

In general, alpha particles have a very limited ability to penetrate other materi-als. In other words, these particles of ionizing radiation can be blocked by a sheet of paper, skin, or even a few inches of air. Nonetheless, materials that emit alpha particles are potentially dangerous if they are inhaled or swallowed, but exter-nal exposure generally does not pose a danger. Alpha particles are the same as a Helium atom nucleus and are comprised of two neutrons and two protons. They are relatively large as compared with the other particles.

Beta particles, which are like electrons, are emitted from naturally occurring materials (such as strontium-90). Such beta emitters are used in medical applica-tions, such as treating eye disease.

In general, beta particles are lighter than alpha particles, and they generally have a greater ability to penetrate other materials. As a result, these particles can travel a few feet in the air, and can penetrate skin. Nonetheless, a thin sheet of metal or plastic or a block of wood can stop beta particles.

Neutrons are high-speed nuclear particles that have an exceptional ability to penetrate other materials. Neutrons are the only form of radiation that can make objects radioactive. This process, called neutron activation, produces many of the radioactive sources that are used in medical, academic, and industrial applications (including oil exploration).

Because of their exceptional ability to penetrate other materials, neutrons can travel great distances in air and require very thick hydrogen-containing materials (such as concrete or water) to block them. Fortunately, however, neutron radiation primarily occurs inside a nuclear reactor, where many feet of water provide effective shielding.

Under the right isotope mass and geometric conditions, fissile materials can have a criticality incident. That is a runaway fission of the isotope that doesn't cause an explosion. Instead, it is a blue flash of neutrons. This blue flash or "blue glow" can also be attributed to Cherenkov radiation. This phenomenon is observed in if either water is involved in the system or when the blue flash is experienced by the human eye (Clayton et al. 2009). For those individuals who are close to the criticality event it can be lethal. The observer is bombarded with billions of neutrons. This section is not dedicated to nuclear accidents. Several criticality incidents will be described in Chapter 8.

Gamma rays and X-rays consist of high-energy waves that can travel great distances at the speed of light and generally have a great ability to penetrate other materials. For that reason, gamma rays (such as from cobalt-60) are often used in medical applications to treat cancer and sterilize medical instruments. Similarly, X-rays are typically used to provide static images of body parts (such as teeth and bones) and are also used in industry to find defects in welds.

Despite their ability to penetrate other materials, in general, neither gamma rays nor X-rays can make anything radioactive. Several feet of concrete or a few inches of dense material (such as lead) can block these types of radiation.

2.11.2 Measuring Radiation (CDC 2021)

Most scientists in the international community measure radiation using the System Internationale (SI), a uniform system of weights and measures that evolved from the metric system. In the United States, however, the conventional system of measurement is still widely used.

Different units of measure are used depending on what aspect of radiation is being measured. For example, the amount of radiation being given off, or emitted, by a radioactive material is measured using the conventional unit curie (Ci), named for the famed scientist Marie Curie, or the SI unit becquerel (Bq).

The radiation dose absorbed by a person (that is, the amount of energy deposited in human tissue by radiation) is measured using the conventional unit rad or the SI unit gray (Gy). The biological risk of exposure to radiation is measured using the conventional unit rem or the SI unit sievert (Sv) (CDC 2021).

When the amount of radiation being emitted or given off is discussed, the unit of measure used is the conventional unit Ci or the SI unit Bq.

A radioactive atom gives off or emits radioactivity because the nucleus has too many particles, too much energy, or too much mass to be stable. The nucleus breaks down, or disintegrates, in an attempt to reach a nonradioactive (stable) state. As the nucleus disintegrates, energy is released in the form of radiation.

The Ci or Bq is used to express the number of disintegrations of radioactive atoms in a radioactive material over a period of time. For example, one Ci is equal to 37 billion (37 X 109) disintegrations per second. The Ci is being replaced by the Bq. Since one Bq is equal to one disintegration per second, one Ci is equal to 37 billion (37 X 109) Bq.

Ci or Bq may be used to refer to the amount of radioactive materials released into the environment. For example, during the Chernobyl power plant accident that took place in the former Soviet Union, an estimated total of 81 million Ci of radioactive cesium (a type of radioactive material) was released.

When a person is exposed to radiation, energy is deposited in the tissues of the body. The amount of energy deposited per unit of weight of human tissue is called the absorbed dose. Absorbed dose is measured using the conventional rad or the SI Gy.

The rad, which stands for radiation absorbed dose, was the conventional unit of measurement, but it has been replaced by the Gy. One Gy is equal to 100 rad.

People are exposed to radiation daily from different sources, such as naturally occurring radioactive materials in the soil and cosmic rays from outer space (of which we receive more when we fly in an airplane). Some common ways that people are exposed to radiation and the associated doses are shown as follows:

Source of exposure Dose in rem Dose in sievert (Sv)

- Exposure to cosmic rays during a roundtrip airplane flight from New York to Los Angeles 3 mrem 0.03 mSv
- One dental X-ray 5 mrem 0.05 mSv
- One chest X-ray 10 mrem 0.1 mSv
- One mammogram 70 mrem 0.7 mSv
- One year of exposure to natural radiation (from soil, cosmic rays, etc.) 300 mrem 3 mSv

2.11.3 Health Effects of Radiation (EPA 2021)

Ionizing radiation has sufficient energy to affect the atoms in living cells and thereby damage their genetic material (DNA). Fortunately, the cells in our bodies are extremely efficient at repairing this damage. However, if the damage is not repaired correctly, a cell may die or eventually become cancerous.

Exposure to very high levels of radiation, such as being close to an atomic blast, can cause acute health effects such as skin burns and acute radiation syndrome ("radiation sickness"). It can also result in long-term health effects such as cancer and cardiovascular disease. Exposure to low levels of radiation encountered in the environment does not cause immediate health effects but is a minor contributor to our overall cancer risk.

A very high level of radiation exposure delivered over a short period of time can cause symptoms such as nausea and vomiting within hours and can sometimes result in death over the following days or weeks. This is known as acute radiation syndrome, commonly known as "radiation sickness."

It takes a very high radiation exposure to cause acute radiation syndrome – more than 0.75 gray (75 rad) in a short time span (minutes to hours). This level of radiation would be like getting the radiation from 18,000 chest X-rays distributed over your entire body in this short period. Acute radiation syndrome is rare and comes from extreme events like a nuclear explosion or accidental handling or rupture of a highly radioactive source.

Exposure to low levels of radiation does not cause immediate health effects but can cause a small increase in the risk of cancer over a lifetime. There are studies that keep track of groups of people who have been exposed to radiation, including atomic bomb survivors and radiation industry workers. These studies show that radiation exposure increases the chance of getting cancer, and the risk increases as the dose increases: the higher the dose, the greater the risk. Conversely, cancer risk from radiation exposure declines as the dose falls: the lower the dose, the lower the risk.

Radiation doses are commonly expressed in millisieverts (international units) or rem (US units). A dose can be determined from a one-time radiation exposure or from accumulated exposures over time. About 99 percent of individuals would not get cancer because of a one-time uniform whole-body exposure of 100 millisieverts (10 rem) or lower. At this dose, it would be extremely difficult to identify an excess in cancers caused by radiation when about 40% of men and women in the United States will be diagnosed with cancer at some point during their lifetime.

Risks that are low for an individual could still result in unacceptable numbers of additional cancers in a large population over time. For example, in a population of one million people, an average one-percent increase in lifetime cancer risk for individuals could result in 10,000 additional cancers. The EPA sets regulatory limits and recommends emergency response guidelines well below 100 millisieverts (10 rem) to protect the US population, including sensitive groups such as children, from increased cancer risks from accumulated radiation dose over a lifetime.

Understanding the type of radiation received, the way a person is exposed (external vs. internal), and for how long a person is exposed are all important in estimating health effects.

The risk from exposure to a particular radionuclide depends on:

- The energy of the radiation it emits.
- The type of radiation (alpha, beta, gamma, X-rays).
- Its activity (how often it emits radiation).
- Whether exposure is external or internal.
- External exposure is when the radioactive source is outside of your body. X-rays and gamma rays can pass through your body, depositing energy as they go.
- Internal exposure is when radioactive material gets inside the body by eating, drinking, breathing, or injection (from certain medical procedures). Radionuclides may pose a serious health threat if significant quantities are inhaled or ingested.
- The rate at which the body metabolizes and eliminates the radionuclide following ingestion or inhalation.
- Where the radionuclide concentrates in the body and how long it stays there.

Children and fetuses are especially sensitive to radiation exposure. The cells in children and fetuses divide rapidly, providing more opportunity for radiation to disrupt the process and cause cell damage. EPA considers differences in sensitivity due to age and sex when revising radiation protection standards.

Many of the radioactive isotopes are also heavy metals. Many also have affinity for accumulating in human bones. Two examples are Plutonium and Radium. Once in the bones the radioactive materials continue to release particles and waves that affect the surrounding tissue. One particularly hazardous radioactive material is Polonium 210 (Biggers, 2017). Its physical half-life is 140 days. Its biological half-life is 40 days, so it takes 40 days for biological processes to eliminate half of the Polonium-210 in the body. One gram of Polonium 210 is estimated to be able to kill 50 million people and make another 50 million ill. However, even though the material is natural, it is very rare because the materials half-life is so short. Alexander Valterovich Litvinenko, who had been a Soviet Union intelligence officer, was poisoned by Polonium 210 in 2006, around November 1st and died on November 23rd. The latest theory was he was poisoned by Russia's Federal Protective Service (FSB) (Newman 2021).

2.11.4 Uses of Radiation (NRC 2020)

Although scientists have only known about radiation since the 1890s, they have developed a wide variety of uses for this natural force. Today, to benefit humankind, radiation is used in medicine, academics, and industry, as well as for generating electricity. In addition, radiation has useful applications in such areas as agriculture, archaeology (carbon dating), space exploration, law enforcement, geology (including mining), and many others. The following uses are described as follows:

- Medical uses
- Academic and scientific applications
- Industrial uses
- Nuclear power plants

2.11.5 Medical Uses

Hospitals, doctors, and dentists use a variety of nuclear materials and procedures to diagnose, monitor, and treat a wide assortment of metabolic processes and medical conditions in humans. In fact, diagnostic X-rays or radiation therapy have been administered to about 7 out of every 10 Americans. As a result, medical procedures using radiation have saved thousands of lives through the detection and treatment of conditions ranging from hyperthyroidism to bone cancer.

The most common of these medical procedures involve the use of X-rays – a type of radiation that can pass through our skin. When X-rayed, our bones and other structures cast shadows because they are denser than our skin, and those shadows can be detected on photographic film. The effect is like placing a pencil behind a piece of paper and holding the pencil and paper in front of a light. The shadow of the pencil is revealed because most light has enough energy to pass through the paper, but the denser pencil stops all the light. The difference is that X-rays are invisible, so we need photographic film to "see" them for us. This allows doctors and dentists to spot broken bones and dental problems.

X-rays and other forms of radiation also have a variety of therapeutic uses. When used in this way, they are most often intended to kill cancerous tissue, reduce the size of a tumor, or reduce pain. For example, radioactive iodine (specifically iodine-131) is frequently used to treat thyroid cancer, a disease that strikes about 11,000 Americans every year.

X-ray machines have also been connected to computers in machines called computerized axial tomography (CAT) or computed tomography (CT) scanners. These instruments provide doctors with color images that show the shapes and details of internal organs. This helps physicians locate and identify tumors, size anomalies, or other physiological or functional organ problems.

In addition, hospitals and radiology centers perform approximately 10 million nuclear medicine procedures in the United States each year. In such procedures, doctors administer slightly radioactive substances to patients, which are attracted to certain internal organs such as the pancreas, kidney, thyroid, liver, or brain, to diagnose clinical conditions.

2.11.6 Academic and Scientific Applications

Universities, colleges, high schools, and other academic and scientific institutions use nuclear materials in course work, laboratory demonstrations, experimental research, and a variety of health physics applications. For example, just as doctors can label substances inside people's bodies, scientists can label substances that pass through plants, animals, or our world. This allows researchers to study such things as the paths that different types of air and water pollution take through the environment. Similarly, radiation has helped us learn more about the types of soil that different plants need to grow, the sizes of newly discovered oil fields, and the tracks of ocean currents. In addition, researchers use low-energy radioactive sources in gas chromatography to identify the components of petroleum products, smog and cigarette smoke, and even complex proteins and enzymes used in medical research.

Archaeologists also use radioactive substances to determine the ages of fossils and other objects through a process called carbon dating. For example, in the upper levels of our atmosphere, cosmic rays strike nitrogen atoms and form a naturally radioactive isotope called carbon-14. Carbon is found in all living things, and a small percentage of this is carbon-14. When a plant or animal dies, it no longer takes in new carbon and the carbon-14 that it accumulated throughout its life begins the process of radioactive decay. As a result, after a few years, an old object has a lower percent of radioactivity than a newer object. By measuring this difference, archaeologists can determine the object's approximate age.

As noted earlier, Chapter 8 presents many nuclear accidents, including nuclear power plant and medical misadministration events of radioisotopes.

2.11.7 Industrial Uses

We could talk all day about the many and varied uses of radiation in industry and not complete the list, but a few examples illustrate the point. In irradiation, for instance, foods, medical equipment, and other substances are exposed to certain types of radiation (such as X-rays) to kill germs without harming the substance that is being disinfected and without making the food radioactive. When treated in this manner, foods take much longer to spoil, and medical equipment (such as bandages, hypodermic syringes, and surgical instruments) are sterilized without being exposed to toxic chemicals or extreme heat. As a result, where we now use chlorine, a chemical that is toxic and difficult-to-handle, we may someday use radiation to disinfect our drinking water and kill the germs in our sewage. In fact, ultraviolet light (a form of radiation) is already used to disinfect drinking water in some homes.

Similarly, radiation is used to help remove toxic pollutants, such as exhaust gases from coal-fired power stations and industry. For example, electron beam radiation can remove dangerous sulfur dioxides and nitrogen oxides from our environment. Closer to home, many of the fabrics used to make our clothing have been irradiated (treated with radiation) before being exposed to a soil-releasing or wrinkle-resistant chemical. This treatment makes the chemicals bind to the fabric, to keep our clothing fresh and wrinkle-free all day, yet our clothing does not become radioactive. Similarly, nonstick cookware is treated with gamma rays to keep food from sticking to the metal surface.

The agricultural industry makes use of radiation to improve food production and packaging. Plant seeds, for example, have been exposed to radiation to bring about new and better types of plants. Besides making plants stronger, radiation can be used to control insect populations, thereby decreasing the use of dangerous pesticides. Radioactive material is also used in gauges that measure the thickness of eggshells to screen out thin, breakable eggs before they are packaged in egg cartons. In addition, many of our foods are packaged in polyethylene shrink wrap that has been irradiated so that it can be heated above its usual melting point and wrapped around the foods to provide an airtight protective covering.

All around us, we see reflective signs that have been treated with radioactive tritium and phosphorescent paint. Ionizing smoke detectors, using a tiny bit of americium-241, keep watch while we sleep. Gauges containing radioisotopes measure the amount of air whipped into our ice cream, while others prevent spillover as our soda bottles are carefully filled at the factory.

Engineers also use gauges containing radioactive substances to measure the thickness of paper products, fluid levels in oil and chemical tanks, and the moisture and density of soils and material at construction sites. They also use an X-ray process, called radiography, to find otherwise imperceptible defects in metallic castings and welds. Radiography is also used to check the flow of oil in sealed engines and the rate and way that various materials wear out. Well-logging devices use a radioactive source and detection equipment to identify and record formations deep within a bore hole (or well) for oil, gas, mineral, groundwater, or geological exploration. Radioactive materials also power our dreams of outer space, as they fuel our spacecraft and supply electricity to satellites that are sent on missions to the outermost regions of our solar system.

2.11.8 Nuclear Power Plants

Electricity produced by nuclear fission, splitting the atom, is one of the greatest uses of radiation. As our country becomes a nation of electricity users, we need a reliable, abundant, clean, and affordable source of electricity. We depend on it to give us light, to help us groom and feed ourselves, to keep our homes and businesses running, and to power the many machines we use. As a result, we use about one-third of our energy resources to produce electricity.

The purpose of a nuclear power plant is to boil water to produce steam to power a generator to produce electricity. While nuclear power plants have many similarities to other types of plants that generate electricity, there are some significant differences. Except for solar, wind, and hydroelectric plants, power plants (including those that use nuclear fission) boil water to produce steam that spins the propeller-like blades of a turbine that turns the shaft of a generator. Inside the generator, coils of wire and magnetic fields interact to create electricity. In these plants, the energy needed to boil water into steam is produced either by burning coal, oil, or gas (fossil fuels) in a furnace, or by splitting atoms of uranium in a nuclear power plant. Nothing is burned or exploded in a nuclear power plant. Rather, the uranium fuel generates heat through a process called fission.

Nuclear power plants are fueled by uranium, which emits radioactive substances. Most of these substances are trapped in uranium fuel pellets or in sealed metal fuel rods. However, small amounts of these radioactive substances (mostly gases) become mixed with the water that is used to cool the reactor. Other impurities in the water are also made radioactive as they pass through the reactor. The water that passes through a reactor is processed and filtered to remove these radioactive impurities before being returned to the environment. Nonetheless, minute quantities of radioactive gases and liquids are ultimately released to the environment under controlled and monitored conditions.

The USNRC has established limits for the release of radioactivity from nuclear power plants. Although the effects of very low levels of radiation are difficult to detect, the NRC's limits assume that the public's exposure to man-made sources of radiation should be only a small fraction of the exposure that people receive from natural background sources.

Experience has shown that, during normal operations, nuclear power plants typically release only a small fraction of the radiation allowed by the NRC's established limits. In fact, a person who spends a full year at the boundary of a nuclear power plant site would receive an additional radiation exposure of less than 1% of the radiation that everyone receives from natural background sources. This additional exposure, totaling about 1 millirem (a unit used in measuring radiation absorption and its effects), has not been shown to cause any harm to human beings.

2.11.9   Misuse of Radiation (EPA 2021)

Along with furniture, clothing, jewelry, dishes, and other treasures sold at thrift stores and antique shops, you might find some items that contain radioactive material. Some antiques were made and sold before scientists fully understood the health effects of radiation. Certain radioactive materials were used in antiques because of their unique color. For example:

Clocks, watches, and dials that glow-in-the-dark without the use of a battery may contain radium or tritium.

Ceramics made until the 1970s may have glazes colored with radionuclides.

Vaseline glass, or canary glass, contains a small amount of uranium. This gives the glass its yellow-green color. It also makes the glass glow bright green under a black light.

Cloisonné jewelry gets some of its yellow, orange, and off-white colors from small amounts of uranium in the glaze.

Radioactive antiques can continue to emit very low levels of radiation for thousands of years, if not longer. The amount of radiation these items emit is small. However, it can register on a hand-held Geiger counter if the object is close enough to the monitor.

Another interesting use of radiation, in this case, X-rays, involved using a fluoroscope to X-ray feet for sizing shoes (Museum of Radiation and Radioactivity 2021). The device used a 50 kV X-ray tube and began use in World War I and continued until the 1950s when a number of organizations issued warnings and the State of Pennsylvania banned the use of the devices in 1957. Figure 2.5 shows one of these devices.

2.11.10   Radium Dial Painters

The plight of the Radium Watch dial painters was recently portrayed in a movie called "Radium Girls." Radium 226 (^{226}Rn) was used in combination with a luminescence material, most likely Zinc Sulfide (ZnS), by the workers to paint the dials (Carr 2014).

Glow-in-the-dark paint is now made without radioactive material, but in the early 1900s radioactive materials were used to make paint that glowed (EPA 2021). Radium is one type of radioactive material that could be found in antiques. When radium was discovered in the early 1900s, people were fascinated by its mysterious glow and it was added to many everyday products, including paints. These paints were used on the dials of clocks and watches to make them glow-in-the-dark. This glow-in-the-dark paint was also used on airplane dials and gauges, which allowed people to read clocks, gauges, and dials at night with no other light.

Figure 2.5 Shoe fluoroscope. Source: Courtesy of The Grand Rapids Public Museum.

During the World War II, radium dials and gauges allowed pilots to fly at night without cockpit lights. This helped the pilots avoid being seen by enemy soldiers.

Radium is highly radioactive. It emits alpha, beta, and gamma radiation. If it is inhaled or swallowed, radium is dangerous because there is no shielding inside the body. If radium is ingested or inhaled, the radiation emitted by the radionuclide can interact with cells and damage them. During the production of radium dials, many workers who painted clock or instrument dials with radium developed cancer. To create fine tips on their paint brushes for small surfaces, many radium dial painters licked the bristles of their paintbrushes. In doing this, they often swallowed some of the radioactive paint. In the body, radium acts like calcium, so the radium that workers ingested was deposited into their bones. Many of these workers developed bone cancer, usually in their jaws. Eventually, scientists and medical professionals realized that these workers' illnesses were being caused by internal contamination from the radium they ingested. By the 1970s, radium was no longer used on watch and clock dials. The Radium dials also exposed the user to harmful radiation.

2.11.11 Safety Culture Issues

Why would a company purposely expose their employees to a hazardous substance and continue to do so even after the hazardous effects were becoming known? This is one of the questions that will be explored throughout this book. Profit is, of course, one of the motives. Readily availability of cheap labor is another.

2.12 The Occupational Safety and Health Administration (OSHA)

Earlier in this chapter the number of fatal accidents for 1939 was shown to be approximately 16,000. In 2020 that number was less than 5000 (BLS 2021). One of the reasons for this was the creation of the Occupational Safety and Health Administration (OSHA). The creation of OSHA had a profound effect on safety and health. Figure 2.6 shows how the accident rate has been reduced since 1972 (BLS 2021).

The purpose of the OSHA of 1970 (OSHA 2022b): "To assure safe and healthful working conditions for working men and women; by authorizing enforcement of the standards developed under the Act; by assisting and encouraging the States in their efforts to assure safe and healthful working conditions; by providing for

Chart 1. Incidence rates of nonfatal occupational injuries and illnesses, private industry,
12
10
8
6
4
2
1972
1977
1982
1987
1992
1997
2002
2007
2012
2017
Year

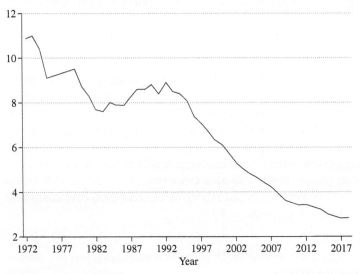

Figure 2.6 Nonfatal accident rate, 1972 to 2018. Source: BLS (2021).

research, information, education, and training in the field of occupational safety and health...."

The OSH Act gives workers the right to safe and healthful working conditions. It is the duty of employers to provide workplaces that are free of known dangers that could harm their employees. This law also gives workers important rights to participate in activities to ensure their protection from job hazards. Workers have the right to:

- File a confidential complaint with OSHA to have their workplace inspected.
- Receive information and training about hazards, methods to prevent harm, and the OSHA standards that apply to their workplace. The training must be done in a language and vocabulary workers can understand.
- Review records of work-related injuries and illnesses that occur in their workplace.
- Receive copies of the results from tests and monitoring done to find and measure hazards in the workplace.
- Get copies of their workplace medical records.
- Participate in an OSHA inspection and speak in private with the inspector.
- File a complaint with OSHA if they have been retaliated against by their employer as the result of requesting an inspection or using any of their other rights under the OSH Act.
- File a complaint if punished or retaliated against for acting as a "whistleblower" under the additional 21 federal statutes for which OSHA has jurisdiction.
- A job must be safe or it cannot be called a good job. OSHA strives to make sure that every worker in the nation goes home unharmed at the end of the workday, the most important right of all.

Employers have the responsibility to provide a safe workplace. Employers MUST provide their employees with a workplace that does not have serious hazards and must follow all OSHA safety and health standards. Employers must find and correct safety and health problems. OSHA further requires that employers must try to eliminate or reduce hazards first by making feasible changes in working conditions – switching to safer chemicals, enclosing processes to trap harmful fumes, or using ventilation systems to clean the air are examples of effective ways to get rid of or minimize risks – rather than just relying on personal protective equipment such as masks, gloves, or earplugs.

Employers MUST also:

- Prominently display the official OSHA poster that describes rights and responsibilities under the OSH Act.
- Inform workers about hazards through training, labels, alarms, color-coded systems, chemical information sheets, and other methods.

- Train workers in a language and vocabulary they can understand.
- Keep accurate records of work-related injuries and illnesses.
- Perform tests in the workplace, such as air sampling, required by some OSHA standards.
- Provide hearing exams or other medical tests required by OSHA standards.
- Post OSHA citations and injury and illness data where workers can see them.
- Notify OSHA within eight hours of a workplace fatality or within 24 hours of any work-related inpatient hospitalization, amputation, or loss of an eye.
- Not retaliate against workers for using their rights under the law, including their right to report a work- related injury or illness.

2.12.1 Who Does OSHA Cover

2.12.1.1 Private Sector Workers

Most employees in the nation come under OSHA's jurisdiction. OSHA covers most private sector employers and employees in all 50 states, the District of Columbia, and other US jurisdictions either directly through Federal OSHA or through an OSHA-approved state plan. State-run health and safety plans must be at least as effective as the Federal OSHA program.

2.12.1.2 State and Local Government Workers

Employees who work for state and local governments are not covered by Federal OSHA, but have OSH Act protections if they work in those states that have an OSHA-approved state plan. The following 22 states or territories have OSHA-approved programs:

Alaska	Arizona	California
Hawaii	Indiana	Iowa
Kentucky	Maryland	Michigan
Minnesota	Nevada	New Mexico
North Carolina	Oregon	South Carolina
Tennessee	Utah	Vermont
Virginia	Washington	Wyoming
Puerto Rico		

Five additional states and one US territory have OSHA-approved plans that cover public sector workers only:

Connecticut	Illinois	Maine
New Jersey	New York	Virgin Islands

Private sector workers in these five states and the Virgin Islands are covered by Federal OSHA.

2.12.1.3 Federal Government Workers

Federal agencies must have a safety and health program that meets the same standards as private employers. Although OSHA does not fine federal agencies, it does monitor federal agencies and responds to workers' complaints. The United States Postal Service (USPS) is covered by OSHA.

2.12.1.4 Not Covered Under the OSHA Act

- Self-employed;
- Immediate family members of farm employers; and
- Workplace hazards regulated by another federal agency (for example, the Mine Safety and Health Administration, the Department of Energy, or the Coast Guard).

OSHA is not without controversy. Most recently, OSHA was to administer a mandatory COVID-19 vaccine program for businesses over 100 employees. This mandate was struck down by the United States Supreme Court. On January 13, 2022 OSHA provided the following statement (OSHA 2022a):

US Secretary of Labor Marty Walsh issued the following statement on the Supreme Court ruling on the department's OSHA's emergency temporary standard on vaccination and testing:

> I am disappointed in the court's decision, which is a major setback to the health and safety of workers across the country. OSHA stands by the Vaccination and Testing Emergency Temporary Standard as the best way to protect the nation's workforce from a deadly virus that is infecting more than 750,000 Americans each day and has taken the lives of nearly a million Americans.
>
> OSHA promulgated the ETS under clear authority established by Congress to protect workers facing grave danger in the workplace, and COVID is without doubt such a danger. The emergency temporary standard is based on science and data that show the effectiveness of vaccines against the spread of coronavirus and the grave danger faced by unvaccinated workers. The commonsense standards established in the ETS remain critical, especially during the current surge, where unvaccinated people are 15–20 times more likely to die from COVID-19 than vaccinated people. OSHA will be evaluating all options to ensure workers are protected from this deadly virus.
>
> We urge all employers to require workers to get vaccinated or tested weekly to most effectively fight this deadly virus in the workplace. Employers are responsible for the safety of their workers on the job, and OSHA has comprehensive COVID-19 guidance to help them uphold their obligation.

Regardless of the ultimate outcome of these proceedings, OSHA will do everything in its existing authority to hold businesses accountable for protecting workers, including under the Covid-19 National Emphasis Program and General Duty Clause.

2.12.2 Voluntary Protection Program

The evolution of OSHA started in early 1980s in the chemical manufacturing industry. There was a push to begin to allow companies to be more involved in their safety oversight. This evolved into what is now called the Voluntary Protection Program or VPP. The following is extracted from the OSHA VPP website (OSHA 2022b):

> The Voluntary Protection Programs (VPP) promote effective worksite-based safety and health. In the VPP, management, labor, and OSHA establish cooperative relationships at workplaces that have implemented a comprehensive safety and health management system. Approval into VPP is OSHA's official recognition of the outstanding efforts of employers and employees who have achieved exemplary occupational safety and health.
>
> What Is the Authority for VPP?
>
> The legislative underpinning for VPP is Section (2)(b)(1) of the Occupational Safety and Health Act of 1970, which declares the Congress's intent "to assure so far as possible every working man and woman in the Nation safe and healthful working conditions and to preserve our human resources – (1) by encouraging employers and employees in their efforts to reduce the number of occupational safety and health hazards at their places of employment, and to stimulate employers and employees to institute new and to perfect existing programs for providing safe and healthful working conditions.
>
> How Does VPP Work?
>
> In practice, VPP sets performance-based criteria for a managed safety and health system, invites sites to apply, and then assesses applicants against these criteria. OSHA's verification includes an application review and a rigorous onsite evaluation by a team of OSHA safety and health experts.
>
> OSHA approves qualified sites to one of three programs:
>
> Star: Recognition for employers and employees who demonstrate exemplary achievement in the prevention and control of occupational safety and health hazards the development, implementation and continuous improvement of their safety and health management system.

Merit: Recognition for employers and employees who have developed and implemented good safety and health management systems but who must take additional steps to reach Star quality

Demonstration: Recognition for employers and employees who operate effective safety and health management systems that differ from current VPP requirements. This program enables OSHA to test the efficacy of different approaches.

When Did VPP Begin?

- 1979 – California began experimental program
- 1982 – OSHA formally announced the VPP and approved the first site.
- 1998 – Federal worksites became eligible for VPP.

How Has VPP Improved Worker Safety & Health?

Statistical evidence for VPP's success is impressive. The average VPP worksite has a Days Away Restricted or Transferred (DART) case rate of 52% below the average for its industry(1). These sites typically do not start out with such low rates. Reductions in injuries and illnesses begin when the site commits to the VPP approach to safety and health management and the challenging VPP application process.

How Does VPP Benefit Employers?

Fewer injuries and illnesses mean greater profits as workers' compensation premiums and other costs plummet. Entire industries benefit as VPP sites evolve into models of excellence and influence practices industry-wide.

How Does VPP Benefit OSHA?

OSHA gains a corps of ambassadors enthusiastically spreading the message of safety and health system management. These partners also provide OSHA with valuable input and augment its limited resources.

Another benefit to OSHA is a safety and health advocacy group that came into existence as a result of the VPP, the Voluntary Protection Program Participants' Association (VPPPA). The VPPPA is a nonprofit organization founded in 1985. As part of its efforts to share the benefits of cooperative programs, the VPPPA works closely with OSHA and State Plan States in the development and implementation of cooperative programs. The VPPPA also provides expertise to these groups in the form of comments and stakeholder feedback on agency rulemaking and policies. Additionally, the Association provides comments and testimony to members of Congress regarding legislative bills on health and safety issues.

The VPP program works and allows companies to utilize the safety skills of their employees and managers to craft a safety program that fits their organization. The OSHA VPP website also presents several success stories. The following is one:

Duro-Last, Inc.

Duro-Last, Inc., the world's largest manufacturer of custom-fabricated single-ply roofing systems, is Oregon OSHA's newest Voluntary Protection Program (VPP) Star Site. Star Site status represents the highest achievement for companies participating in the VPP, and the company achieved Star Site status on Nov. 17, 2016. OSHA's VPP recognizes employers and workers in the private industry and federal agencies who have implemented effective safety and health management systems and maintain injury and illness rates below national Bureau of Labor Statistics averages for their respective industries.

In addition to single-ply roofing systems, the company produces custom-fabricated parapets, stacks, and curb flashings. The Duro-Last vinyl roof membrane is a proprietary thermoplastic formulation, consisting of polyvinyl chloride resins, plasticizers, stabilizers, biocides, flame retardants, and ultraviolet absorbents. The company manufactures roofing systems for approximately 4,000 authorized roofing contractors in North America.

Duro-Last's Grants Pass facility employs 44 people and began working toward Oregon OSHA's Safety and Health Achievement Recognition Program (SHARP) in 1997. At that time, the company, which employs 760 people corporate-wide, had a Days Away, Restricted or Transferred (DART) rate of about 7.2 in an industry that was averaging in the high 3s and low 4s.

SHARP recognizes small business employers who have used OSHA's On-site Consultation Program services and operate an exemplary injury and illness prevention program. Acceptance into SHARP from OSHA is an achievement of status that singles a company out among its business peers as a model for worksite safety and health. OSHA's On-site Consultation Program offers free and confidential safety and occupational health advice to small and medium-sized businesses in all states across the country and several territories, with priority given to high-hazard worksites. On-site Consultation services are separate from enforcement and do not result in penalties or citations. Consultants from state agencies or universities work with employers to identify workplace hazards, provide advice on compliance with OSHA standards, and assist in establishing injury and illness prevention programs.

Over the next four years, management had the safety committee increasingly play a key role in guiding safety and health program decisions, and employees became more involved in those decisions. Steadily, the company improved its safety and health management systems. Meanwhile, its DART rate decreased. When the company qualified for SHARP status in 2001, it had a DART rate of 1.7. The industry was averaging 3.7. The company graduated from SHARP in 2006, with a DART rate of 1.3. By contrast, the industry average was 4.1.

During its SHARP journey, the company noted to Oregon OSHA consultants that the process not only improved the health and safety of its workers, but also bolstered the company's production and profitability. The SHARP process positioned the company for further success as part of the VPP program.

Indeed, obtaining VPP recognition requires an extensive application process. That process helped Duro-Last further solidify its corporate commitment to worker safety and health. The VPP certification is designed for exemplary worksites with comprehensive, successful safety and health management systems.

"Duro-Last Roofing has always been very conscious of our employees' safety and always puts safety first," said Tim Hart, vice president of western operations.

Hart has long recognized that a strong alliance with Oregon OSHA would benefit the company and its employees, customers, and vendors. Likewise, Oregon OSHA understands the importance of building relationships with employers to advance workplace safety and health programs. The relationship between Duro-Last and Oregon OSHA has lasted more than 29 years. Oregon OSHA consultants have toured the Grants Pass facility to check on workplace safety, health, and ergonomics and to help create effective safety programs. When Duro-Last expanded and purchased the 75,000-square-foot facility in the North Valley Industrial Park area of Grants Pass, the association with Oregon OSHA continued.

"OSHA's consultative services have proven to be invaluable over the years," said Hart. "This is especially true since we are now a SHARP graduate and in our recent recognition as a VPP Star Site."

A longtime member and supporter of the nonprofit Oregon SHARP Alliance, Duro-Last's commitment to continuous improvement through partnerships and tours of other VPP sites helped improve safety, such as installation of strobe lights (front and back) on forklifts, shadow-boarding, updated emergency action plans, and improvements to the near-miss portal. From Sept. 7, 2007, to Jan. 6, 2016, Duro-Last achieved 500,000 hours of working accident-free. Moreover, the company maintained a DART rate of 0.0 from 2013 through 2015.

"We are proud of the work we have done with OSHA that has allowed us to learn the best practices of other SHARP and VPP companies and helped us understand the importance for continuous improvement in safety programs," said Hart. "We thank Oregon OSHA for recognizing the dedication our Grants Pass employees have to safety. Without their safety mindset, this recognition would not have been possible."

Duro-Last is headquartered in Saginaw, Michigan, with additional manufacturing facilities in Grants Pass; Jackson, Mississippi; Sigourney, Iowa; and Carrollton, Texas. On June 20, 2017, the company opened its sixth manufacturing facility, located in Ludlow, Massachusetts. The company's manufacturing facility in Carrollton, Texas, will be the next facility to apply for VPP, with the goal of getting Star Site status for all Duro-Last facilities across the country.

Companies that embrace programs like VPP demonstrate that they have an engaged workforce in safety and have an excellent safety culture. The VPP program requires companies to look inward and find those areas of their organizations that need improvements and then seek to improve those areas. Will a VPP Star site ever have a serious accident? Hopefully, no, but VPP companies will actively fix those circumstances surrounding the accident.

2.13 Human Performance Improvement (HPI)

In Chapter 1 Shane Bush and Ron Farris talked about human performance improvement (HPI) and the importance of this concept to worker safety. This brief introduction provides just the tip of iceberg on the topic. According to the International Atomic Energy Agency (IAEA) (IAEA 2005):

> HPI is the systematic process of discovering and analyzing important human performance gaps, planning for future improvements in human performance, designing and developing cost-effective and ethically justifiable interventions to close performance gaps, implementing the interventions, and evaluating the financial and non-financial results.
>
> Human performance improvement (HPI) or Human performance technology (HPT) is the systematic process of discovering and redressing human performance gaps. Generic models describing HPI have been developed which are independent of domain or organization type and embed HPI into the broader context of Change Management. In the nuclear domain, adequate use of HPI techniques is one of the key prerequisites for ensuring safety and efficiency of nuclear facilities. Considering in particular Nuclear Power Plants (NPPs), in the past decade significant improvements have been made in both the operational and safety performance of NPPs based upon human performance improvement.
>
> Human performance improvement (HPI), also referred to as Human performance technology (HPT), is the systematic process of discovering

and analyzing important human performance gaps, planning for future improvements in human performance, designing and developing cost-effective and ethically justifiable interventions to close performance gaps, implementing the interventions, and evaluating the financial and non-financial results.

In the nuclear domain, adequate use of HPI techniques is one of the key prerequisites for ensuring safety and efficiency of nuclear facilities. Event-free performance in the nuclear industry requires an integrated view of human performance; how well managers, supervisors and facility staff function, the alignment of organizational processes and values in achieving facility operational and safety goals, and the behavior of individuals.

2.14 Chapter Summary

We traveled a great amount of safety and health history in this chapter. We started in a time with very little consideration to safety and health of the individual to stellar safety organizations, with VPP Star status. There are still so many safety and health issues in the United States and around the world that need attention. The case studies in the next seven chapters demonstrate how events can happen and the potential catastrophic consequences. For example, in Chapter 4 there is case study about the ammonium nitrate fertilizer explosion in Lebanon. How many of these types of explosions need to occur before there is inherit knowledge that storing huge amounts of this chemical is dangerous? I hope these case studies provide insights into how to prevent future events.

References

A&A (2021). The advantages of imported foods. https://www.aacb.com/advantages-of-imported-foods/ (accessed December 2021).

Basaraba, S. (2020). A guide to longevity throughout history. https://www.verywellhealth.com/longevity-throughout-history-2224054 (accessed December 2021).

Anders Bergström, Laurent Frantz, Ryan Schmidt, Erik Ersmark, Ophelie Lebrasseur, Linus Girdland-Flink, et al. (2020). Origins and genetic legacy of prehistoric dogs, Science. https://www.science.org/doi/10.1126/science.aba9572 (accessed December 2021).

Biggers, A. (2017). Polonium-210: why is Po-210 so dangerous?, Medical News Today. https://www.medicalnewstoday.com/articles/58088#effects (accessed December 2021).

Black, R. (2021). Ancient DNA reveals the oldest domesticated dog in the Americas, Smithsonian Magazine, February.

Blood, M. and Ronayne, K. (2021). US News, Battle Over Masks, Vaccines Roils California Recall Election. https://www.usnews.com/news/politics/articles/2021-08-13/battle-over-masks-vaccines-roils-california-recall-election (accessed December 2021).

Bureau of Labor Statistics (BLS) (2021). https://www.bls.gov (accessed December 2021).

Callaway, E. (2020). Ancient dog DNA reveals 11,000 years of canine evolution. *Nature* 587 (7832): 20.

Carr, S. (2014). Radium girls. https://www.mun.ca/biology/scarr/Radium_Watch-Dial_Painters.html (accessed December 2021).

Case Western University (2021). The civil war: sex and soldiers. https://artsci.case.edu/dittrick/online-exhibits/history-of-birth-control/contraception-in-america-1800-1900/the-civil-war-sex-and-soldiers/ (accessed December 2021).

Center for Disease Control (CDC) (2022). Understanding mRNA COVID-19 vaccines. https://www.cdc.gov/coronavirus/2019-ncov/vaccines/different-vaccines/mRNA.html?s_cid=11344:mrna%20vaccine:sem.ga:p:RG:GM:gen:PTN:FY21 (accessed January 2022).

Center for Disease Control and Prevention (CDC). (2021). Radiation emergencies. https://www.cdc.gov/nceh/radiation/emergencies/measurement.htm (accessed December 2021).

Clayton, E., et al. (2009). A re-introduction to "anomalies of criticality," Report CHPRC-00301-FP.

Cleveland Clinic (2021). Does the grapefruit diet work? Examining the eating plan's fat-burning claims, https://health.clevelandclinic.org/grapefruit-diet/ (accessed December 2021).

Consumer product Safety Commission (CPSC) (2018). BSN SPORTS recalls rubber critter toys due to violation of federal lead paint ban (recall alert). https://www.cpsc.gov/Recalls/2018/BSN-SPORTS-Recalls-Rubber-Critter-Toys-Due-to-Violation-of-Federal-Lead-Paint-Ban-Recall-Alert (accessed December 2021).

Consumer product Safety Commission (CPSC) 2021. Report of deaths and injuries involving off-highway vehicles with more than two wheels. https://cpsc-d8-media-prod.s3.amazonaws.com/s3fs-public/2021-Report-of-Deaths-and-Injuries-Invoving-Off-Highway-Vehicles-with-more-than-Two-Wheels.pdf?VersionId=nm0R8oYxTu.mRD7KNRvNp95ogHw2SYOD (accessed December 2021).

Deron, B., ATI (2015). 10 Awful jobs you'll be glad no longer exist. https://allthatsinteresting.com/awful-jobs/9 (accessed December 2021).

Eli Whitney Museum and Workshop (2021). The factory https://www.eliwhitney.org/7/museum/about-eli-whitney/factory (accessed December 2021).

Environmental Protection Agency (EPA) (2021). Radioactivity in antiques. https://www.epa.gov/radtown/radioactivity-antiques (accessed December 2021).

Food and Drug Administration (FDA) (2021a). Part I: The 1906 food and drugs act and its enforcement. https://www.fda.gov/about-fda/changes-science-law-and-regulatory-authorities/part-i-1906-food-and-drugs-act-and-its-enforcement (accessed December 2021).

Food and Drug Administration (FDA) (2021b). Grapefruit juice and some drugs don't mix. https://www.fda.gov/consumers/consumer-updates/grapefruit-juice-and-some-drugs-dont-mix (accessed December 2021).

Grandjean, E. (1988). *Fitting the Task to the Man: A Textbook of Occupational Ergonomics*. Taylor and Francis.

Healthy Children (2021). The history of antibiotics. https://www.healthychildren.org/English/health-issues/conditions/treatments/Pages/The-History-of-Antibiotics.aspx (accessed December 2021).

History Matters (2021). "The natural tie between master and apprentice has been rent asunder": an old apprentice laments changes in the workplace, 1826. http://historymatters.gmu.edu/d/6622/ (accessed December 2021).

Science History Institute (2021). Gerhard Domagk. https://www.sciencehistory.org/historical-profile/gerhard-domagk (accessed December 2021).

Idaho Department of Education (2021). Next Step Idaho https://nextsteps.idaho.gov/resources/apprenticeships (accessed December 2021).

International Atomic Energy Agency (IAEA) 2005. Human Performance Improvement in Organizations: Potential Applications in the Nuclear Industry, IAEA-TECDOC-1479, IAEA, Vienna.

Longfellow, R. (2017), United States Department of Transportation, Highway History. Stagecoach accidents. https://www.fhwa.dot.gov/infrastructure/back0309.cfm (accessed January 2022).

Mayo Clinic (2021). Gangrene. https://www.mayoclinic.org/diseases-conditions/gangrene/symptoms-causes/syc-20352567 (accessed December 2021).

Minnesota Department of Agriculture (2017). China: Top Market for U.S. Ag Products. https://www.mda.state.mn.us/sites/default/files/inline-files/profilechina.pdf (accessed December 2021).

Museum of Radiation and Radioactivity (2021). Shoe-fitting fluoroscope (ca. 1930-1940). https://www.orau.org/health-physics-museum/collection/shoe-fitting-fluoroscope/index.html (accessed December 2021).

Nader, R. (1965). *Unsafe at Any Speed: The Designed-In Dangers of the American Automobile*. New York, NY: Grossman Publishers.

National Environmental Health Association (2015). The hidden danger of vintage toys. https://www.neha.org/node/1310 (accessed December 2021).

National Highway Traffic Safety Administration (NHTSA) (2021). Takata airbag spotlight. https://www.nhtsa.gov/equipment/takata-recall-spotlight (accessed January 2022).

National Library of Medicine (2021). Genetically engineered foods. https://
    medlineplus.gov/ency/article/002432.htm (accessed December 2021).

New York Times (1940). 16,000 MET DEATH AT WORK IN 1939; 106,000 Suffered
    Permanent Impairment and 1,407,000 Temporary Disability HIGHEST TOLL ON
    FARMS Building Industry Second in Number of Fatalities, Wholesale-Retail Trade
    Third, June 1940.

Newman, S. (2021). National Public Radio (NPR), Russia fatally poisoned a
    prominent defector in London, a court concludes. https://www.npr.org/2021/09/
    21/1039224996/russia-alexander-litvinenko-european-court-human-rights-putin
    (accessed December 2021).

Northeast Waste Management Officials' Association (2021). Mercurochrome. http://
    www.newmoa.org/prevention/mercury/projects/legacy/personalcare.cfm
    (accessed December 2021).

Nuclear Regulatory Commission (NRC) (2020). Uses of radiation. https://www.nrc
    .gov/about-nrc/radiation/around-us/uses-radiation.html (accessed December
    2021).

Occupational Safety and Health Administration (OSHA) (2022a). COVID-19
    vaccination and testing ETS. https://www.osha.gov/coronavirus/ets2 (accessed
    January 2022).

Occupational Safety and Health Administration (OSHA) (2022b). US Department of
    Labor, Occupational Safety and Health Administration. OSHA.gov accessed
    January 2022.

Olmsted, L. (2016). Exclusive book excerpt: honey is world's third most faked food,
    forbes. https://www.forbes.com/sites/larryolmsted/2016/07/15/exclusive-book-
    excerpt-honey-is-worlds-third-most-faked-food/?sh=19dbac084f09 (accessed
    December 2021).

Oregon (2021). https://www.history.com/topics/westward-expansion/oregon-trail
    (accessed December 2021).

Oregon Health & Science University (OHSU) (2016). Mercury, marriage, and
    magic bullets. https://www.ohsu.edu/historical-collections-archives/mercury-
    marriage-and-magic-bullets#:~:text=Mercury%20was%20in%20use%20by,
    including%20pioneering%20attempts%20in%20rhinoplasty (accessed December
    2021).

Perrin, C. (2017). *The Apprenticeship Model: A Journey toward Mastery*. Classical
    Academic Press.

Phillips, T. (2008). Genetically modified organisms (GMOs): transgenic crops and
    recombinant DNA technology. *Nature Education* 1 (1): 213.

Roos, D., History (2020). When new seat belt laws drew fire as a violation of personal
    freedom. https://www.history.com/news/seat-belt-laws-resistance (accessed
    January 2022).

Roser, M. and Ritchie, H. (2022). Hunger and undernourishment. https://
    ourworldindata.org/hunger-and-undernourishment (accessed January 2022).

Rozsa, L., et al. (2020). Washington Post, The battle over masks in a pandemic: an all-American story. https://www.washingtonpost.com/health/the-battle-over-masks-in-a-pandemic-an-all-american-story/2020/06/19/3ad25564-b245-11ea-8f56-63f38c990077_story.html (accessed January 2022).

Sinclair, U. (1906). The Jungle, Doubleday, Page, and Company.

Softkey Multimedia (1995). Oregon trail game.

Sorokanvich, R. (2017). Yes, road and track, The Chevrolet Corvair really was a handful to drive. https://www.roadandtrack.com/car-culture/classic-cars/a13733502/yes-the-chevrolet-corvair-really-was-a-handful-to-drive/ (accessed January 2022).

State of Delaware Office of Highway Safety (2007). Delaware Law Enforcement Shares Top 10 excuses heard for not wearing a seat belt.

Taylor, F. (1929). *The Principles of Scientific Management*. Harper and Brothers.

The Center for Auto Safety (2021). Ford Pinto fuel tank. https://www.autosafety.org/ford-pinto-fuel-tank/ (accessed December 2021).

The Royal Society (2021). What is genetic modification (GM) of crops and how is it done?. https://royalsociety.org/topics-policy/projects/gm-plants/what-is-gm-and-how-is-it-done/#:~:text=GM%20is%20a%20technology%20that,will%20inherit%20the%20new%20DNA (accessed January 2022).

Ungoed-Thomas, J., Leake, J., and Wired (2021). The honey detectives are closing in on China's shady syrup swindlers. https://www.wired.co.uk/article/honey-fraud-detection (accessed January 2022).

United Nations (2021). Department of Economic and Social Affairs, Population Dynamics. https://population.un.org/wpp/ (accessed December 2021).

United States Department of Agriculture (2009). Imports From China and Food Safety Issues, Economic Information Bulletin Number 52.

United States Department of Agriculture (USDA), Federal Safety and Inspection Service (FSIS) (2022). https://www.fsis.usda.gov/recalls-alerts/sas-foods-enterprises-inc.-recalls-frozen-fully-cooked-beef-and-chicken-empanada (accessed January 2022).

Veggie Cage (2022). The tomato. https://www.tomato-cages.com/tomato-history.html#:~:text=The%20Tomato%20History%20has%20origins,sail%20to%20discover%20new%20lands (accessed January 2022).

Wetherell, S.(2019). A brief history of wheat. https://sustainablefoodtrust.org/articles/a-brief-history-of-wheat/ (accessed January 2022).

Wilson-Slack, K. (2018). The Masonic Philosophical Society, Apprentice, Journeyman, and Master: The Medieval Guild. https://blog.philosophicalsociety.org/2018/01/10/apprentice-journeyman-and-master-the-medieval-guild/ (accessed December 2021).

World Nuclear Association (2021). The cosmic origins of uranium. https://world-nuclear.org/information-library/nuclear-fuel-cycle/uranium-resources/the-cosmic-origins-of-uranium.aspx (accessed December 2021).

Wounds International (2011). Iodine made easy. https://www.woundsinternational.com/uploads/resources/03983a959705d34a896ec163f7fc34d9.pdf (accessed December 2021).

3

Chemical Manufacturing

3.0 Introduction

Chemical manufacturing is conducted in every state. In almost every state there have been catastrophic accidents as well. A complete book could be written just on accidents in chemical manufacturing facilities. This section focuses on the three accidents that have had significant impact on the industry.

This section will present the following chemical manufacturing case studies:

- DuPont La Porte, TX, Methyl Mercaptan Release – November 15, 2014
- BP Texas City Refinery Explosion – March 23, 2005
- T2 Laboratories, Inc. Explosion, MCMT Explosion – December 19, 2007

A description of the Process Safety Management (PSM) process will be provided before the case studies to better understand the safety requirements for the process industry.

3.1 Process Safety Management

3.1.1 Introduction

Humans have used chemicals for thousands of years. In fact, food is made up of chemicals that our body needs. One of the first chemicals that were most likely sought by humans was salt (sodium chloride) (Butler 2018). It is essential for life and the ancient Egyptians found it was useful for the preservation of food. From ancient times till the development of refrigeration, salt was the most common way of preserving food. The US Food and Drug Administration (FDA) very recently released new guidance on the amount of salt that people should consume. It is significantly lower than the guidelines from previously (FDA 2021). However, salt is an essential nutrient our body needs to function.

Impact of Societal Norms on Safety, Health, and the Environment: Case Studies in Society and Safety Culture, First Edition. Lee T. Ostrom.

Alcoholic beverages were some of the first chemicals humans sought to produce. Of course, fruit and vegetable matter will naturally be fermented by environmental yeast cells and produce alcohol. The first leavened breads were raised by natural yeasts and bacteria. A recent BBC article focused on a 13,500-year-old brewery site found near Haifa, Israel (BBC 2018). The brewing site was found near a burial site for semi-nomadic hunter-gatherers. Though, it wasn't fine quality beer or, of course, not a distilled spirit, it appeared to be a real drinkable form of beer, in a gruel-like form, according to the article. The beer appeared to be purposely produced and not just a by-product of making bread (Botha 2013).

Soap was one of the first synthesized chemicals produced. The Babylonians began making soap around 4,800 years ago (Soap History 2021). The soap the Babylonians made was by boiling fats with wood ashes. Wood ash contains variable amounts of Potassium Hydroxide. Potassium Hydroxide is an alkaline and animal fats or fatty acids are acidic in nature. In the right proportion the two react to form soap and Glycerin. Figure 3.1 shows the chemical reaction.

The ancient people probably discovered this process by accident. However, they did formalize it and the same basic chemistry is used today in many commercial and homemade soaps.

Refined sugar was first developed about 2500 years ago in India. Of course, documented honey consumption by humans has been for well over 9000 years. Probably more like 10s of thousands of years. Bears and other wild animals have consumed honey since bears have existed. Why wouldn't humans? The first sugars

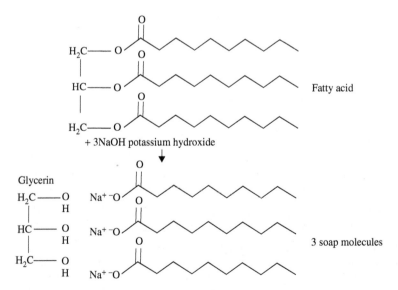

Figure 3.1 Chemistry of soap making. Source: Drawing by Ostrom.

were most likely produced from sugar cane, as they are now. However, sugar beets are an important source of sugar. Sugar was produced in the Mediterranean region starting around the thirteenth century (The Conversation 2015).

One interesting chemical combination that was developed in the seventh century was Greek Fire. The exact chemical composition has been lost to history. It was an incendiary weapon used primarily by the Byzantines for approximately seven centuries. It is thought to have been a combination of a light petroleum component and/or naphtha and possibly quicklime, sulfur, and potassium nitrate. Historians feel this chemical was responsible for Constantinople not falling into enemy hands for those approximately 700 years (Cartwright 2017).

In the history of chemical development gun powder was developed sometime in the ninth century by Chinese alchemists. Gun powder consists of three components: Charcoal for fuel, Saltpeter (potassium nitrate) as an oxidizer, and sulfur to lower the reaction temperature. All explosives and modern smokeless gun powders contain the same basic components but are different chemicals (Chemistry 2015). For instance, ammonium nitrate explosives use the ammonium nitrate as the oxidizer and fuel oil or some other fuel as the reducing agent. Several major accidents have occurred with the production of gun powder and with the storage and transportation of ammonium nitrate fertilizers. Some of these accidents will be discussed in Chapter 4.

By 1800 alchemists and more formal chemists began to synthesize all sorts of chemicals from raw materials. These include most all the organic chemicals we use today for solvents and fuels and hundreds of pharmaceuticals. Everyday chemical and pharmaceutical companies bring new products to market. In fact, a Smithsonian article (Daley 2017) reports that approximately 10 million new chemical compounds are synthesized each year. However, the testing of these new chemicals lags, due to lack of funding. The amount of production of chemical compounds is immense. The chemical plant I worked for the early 1980s produced 5 million lb of phosgene each year and we were a small producer. Gasoline production is tremendous on a weekly basis. Approximately 0.239 million barrels of gasoline was produced for the week ending October 8, 2021 (Trading Economics 2021).

The manufacturing of chemicals has numerous risks. Explosion potential, releases of toxic chemicals, and environmental damage are all possibilities. PSM was developed to help control the risks.

3.1.2 Process Safety Management

On the surface it appears that most of the aspects of the PSM regulation and standards mostly relate to physical aspects of a facility. However, it is humans that either ensure a facility adheres to the regulations and standards or fails to do so. The examples in this section demonstrate what can happen when PSM is not implemented or not followed.

Numerous American Petroleum Institute (API) standards are referenced by the Occupational Safety and Health Administration (OSHA) in this section. These standards can be purchased from API at www.api.org.

Unexpected releases of toxic, reactive, or flammable liquids and gases in processes involving highly hazardous chemicals (HHC) have been reported for many years, in various industries using chemicals with such properties. Regardless of the industry that uses these HHC, there is a potential for an accidental release any time they are not properly controlled, creating the possibility of disaster. This section will discuss numerous accidents and the safety culture of the organizations involved (OSHA 2021).

To help ensure safe and healthful workplaces, OSHA issued the PSM of HHC standard (29 CFR 1910.119), which contains requirements for the management of hazards associated with processes using HHC.

PSM is addressed in specific standards for the general and construction industries. OSHA's standard emphasizes the management of hazards associated with HHC and establishes a comprehensive management program that integrates technologies, procedures, and management practices.

OSHA developed several PSM publications, based on the industry. These are:

- Process Safety Management for Petroleum Refineries. OSHA Publication 3918, (2017).
- Process Safety Management for Explosives and Pyrotechnics Manufacturing. OSHA Publication 3912, (2017).
- Process Safety Management for Small Businesses. OSHA Publication 3908, (2017).
- Process Safety Management for Storage Facilities. OSHA Publication 3909, (2017).
- PSM Covered Chemical Facilities National Emphasis Program (NEP). OSHA Directive CPL 03-00-021, (January 17, 2017).
- Clearing Piping Systems with Natural Gas Letter. OSHA, (August 27, 2010).
- Process Safety Management - Small Business Advocacy Review Panel. OSHA is initiating a Small Business Advocacy Review Panel in order to get feedback on several potential revisions to OSHA's Process Safety Management Program (PSM) standard.
- Petroleum Refinery Process Safety Management National Emphasis Program. OSHA Directive CPL 03-00-010, (August 18, 2009).
- Refinery NEP First Year Inspection Results. OSHA.
- Process Safety Management Depends on You!. OSHA Publication 3315, (2009).

The following are excerpts from the OSHA Publication 3918 and will provide good insight into some of the accidents that have occurred and are discussed in this section of the book.

The OSHA Publication 3198 (3198) highlights areas of the Process Safety Management standard (PSM) where OSHA issued the most citations during the Petroleum Refinery Process Safety Management National Emphasis Program (NEP). These areas include:

- Process Safety Information (PSI)
- Process Hazards Analysis (PHA)
- Operating Procedures
- Mechanical Integrity (MI)
- Management of Change (MOC)

Since the PSM standard was promulgated by OSHA in 1992, no other industry sector has had as many fatal or catastrophic incidents related to the release of HHC as the petroleum refining industry (SIC 2911, NAICS 32411). In response to this large number of fatal or catastrophic incidents, OSHA initiated CPL 03-00-004, the Petroleum Refinery Process Safety Management National Emphasis Program (NEP), in June 2007. The purpose of the NEP was to verify refinery employers' compliance with PSM. After reviewing the citations issued for violations of the PSM standard under the NEP, OSHA discovered many common instances of non-compliance in the petroleum refinery industry.

OSHA recommends refineries review these common instances of non-compliance to ensure that they do not exist in their own PSM programs.

3.1.2.1 Process Safety Information

Employers are required to compile written PSI. The compilation of written PSI enables the employer and the employees involved in operating the process to identify and understand the hazards posed by those processes involving HHC. PSI must include information pertaining to the hazards of the HHC used or produced by the process, information pertaining to the technology of the process, and information pertaining to the equipment in the process. Complete and accurate compilation of PSI is critical to the effective implementation of all other aspects of the PSM.

PSI provisions of the standard also require that all equipment in PSM-covered processes comply with recognized and generally accepted good engineering practices (RAGAGEP). The PSM standard allows employers to select the RAGAGEP they apply in their covered processes. Examples of RAGAGEP include widely adopted codes, consensus documents, non-consensus documents, and internal standards. Furthermore, where the design codes, standards, or practices used in the design and construction of existing equipment are no longer in general use, the employer must determine and document that the equipment is designed, maintained, inspected, tested, and operating in a safe manner.

Regarding the PSI element, during inspections under the NEP OSHA issued many citations for violations of the PSM standard related to the (i) RAGAGEP,

(ii) piping and instrumentation diagrams (P&IDs), and (iii) relief system design basis.

During NEP inspections, OSHA found instances where employers' written PSI did not contain information about:

- Missing relief devices,
- Undersized safety relief valves,
- Incorrect relief valve set points,
- High back pressure on relief valves, and
- Relief devices in an inaccessible location (for further information, see the Section 3.1.2.2.3).

During NEP inspections, OSHA also found deficiencies in the positioning of intervening valves. According to the American Society of Mechanical Engineers (ASME) Boiler and Pressure Vessel Code, Section VIII (Code), an example RAGAGEP, "There shall be no intervening stop valves between the vessel and its pressure relief device or devices, or between the pressure relief device or devices and the point of discharge, except: (1) when these stop valves are so constructed or positively controlled that the closing of the maximum number of block valves possible at one time will not reduce the pressure relieving capacity provided by the unaffected pressure relief devices below the required relieving capacity; or (2) under conditions set forth in Appendix M." If intervening valves are closed in the event of an uncontrolled pressure increase, then the designed relief path will be blocked. As a result, pressure can rise instantly which can rupture pipes and vessels. The consequences can include facility damage, injury, and death. Appropriate, engineering, and administrative controls should be utilized and kept up to date in order to prevent such unsafe conditions.

3.1.2.1.1 Facility Siting RAGAGEP During NEP inspections, OSHA found many instances where petroleum refineries did not document facility siting RAGAGEP to control toxic and/or fire and explosion hazards in buildings and structures housing employees.

OSHA found several instances where refineries did not comply with RAGAGEP, including but not limited to API RP-752: Management of Hazards Associated with Location of Process Plant Permanent Buildings 3rd Edition, December 2009 and API RP-753: Management of Hazards Associated with Location of Process Plant Portable Buildings, 1st Edition, June 2007. Specifically, some petroleum refineries could not document that the locations of control rooms, portable buildings, and other areas frequented by employees were adequately protected from fire, explosion, or toxic release. In many cases separating these structures away from the covered process will be protective; however, in some cases a fixed structure may need more protection, such as positive pressure ventilation, structural reinforcement, or intervening blast walls. Employers can also restrict access of non-essential employees to prevent unnecessary exposure to the hazards of the covered process.

API RP-752, an example of an RAGAGEP for safe building siting, suggests that petroleum refineries should:

(a) Select a credible release scenario that refiners should not change unless the employer can reasonably demonstrate that the original scenario was unrealistic.
(b) Model the release and consequences.
(c) Locate personnel away from process areas consistent with safe and effective operations.
(d) Minimize the use of buildings intended for occupancy in close proximity to process areas.
(e) Restrict the occupancy of buildings in close proximity to process areas.
(f) Design, construct, install, modify, and maintain buildings intended for occupancy to protect occupants against explosion, fire, and toxic material releases; and
(g) Manage the use of buildings intended for occupancy as an integral part of the design, construction, maintenance, and operation of a facility.

3.1.2.1.2 *Piping and Instrumentation Diagrams*

The PSM standard requires employers to maintain P&IDs. P&IDs show the interconnection of process equipment, the instrumentation used to control the process, and they provide engineers, operators, and maintenance employees with information on how to maintain and modify the process. P&IDs must accurately demonstrate the physical sequence of equipment and systems, and how these systems are connected. Without accurate, complete, and up-to-date P&IDs, engineers and operators can be misinformed:

- During the PHA process,
- When creating or modifying operating, maintenance, and repair procedures,
- When generating work permits,
- During new equipment installation, and
- When troubleshooting or maintaining a process.

During NEP inspections, OSHA found that many petroleum refineries failed to maintain accurate, complete, and up-to-date P&IDs for the equipment in the process.

PSM is a performance-based standard and not all P&IDs contain the same information. Employers should be sure that employees understand what to expect on their P&IDs based on the refinery's PSM program. An example of an industry practice for the format and content of P&IDs is the Process Industry Practices, PIC 001, P&IDs Documentation Criteria.

During NEP inspections, OSHA cited several instances where petroleum refineries did not check to ensure that tags on their equipment matched what was written in the P&ID, or that all P&IDs at a facility shared the same notation system. Such errors and inconsistencies may lead to confusion or an incident when maintaining or repairing process equipment. An example of an industry practice notation system that employers can use is the Instrumentation, Systems, and Automation Society's (ISA) ANSI/ISA-S5.1 Instrumentation Symbols and Identification American National Standard.

When OSHA found P&ID deficiencies, such as those identified above, the problems were frequently systemic and encompassed many of these deficiencies. Failure to ensure the accuracy of facility P&IDs may also indicate problems with the employer's MOC program.

3.1.2.1.3   *Relief System Design and Design Basis* The PSI provisions of the standard require employers to keep accurate, complete, and up-to-date documentation of relief system design and design basis. It is important that a facility can demonstrate, in writing, why a relief system was designed in the selected manner. With this information employees remain informed of current relief design systems and any future changes to an associated process can be made appropriately. OSHA found that many refineries did not have this documentation. Potential sources of guidance for relief system design can be found in API 520:

- Sizing, Selection, and Installation of Pressure-Relieving Devices in Refineries and/or Chemical Center for Process Safety (CCPS).
- Guidelines for Pressure Relief and Effluent Handling Systems, 2nd Edition. Additionally, a potential source of guidance on relief system design basis is API 521: Pressure-relieving and Depressurizing Systems.

3.1.2.2   Process Hazards Analysis

A PHA is an organized and systematic effort to identify and analyze the significance of potential hazards associated with the processing and handling of HHC. The PHA must be appropriate to the complexity of the process and must identify, evaluate, and control the hazards involved in the process. The PHA must be in writing and must identify:

- Any previous incident which had a likely potential for catastrophic consequences in the workplace,
- Engineering and administrative controls applicable to the hazards,
- Consequences of failure of engineering and administrative controls,
- Facility siting,
- Human factors, and
- A qualitative evaluation of a range of the possible safety and health effects of failure of controls on employees in the workplace.

The PHA team may make recommendations for additional safeguards to adequately control identified hazards or to mitigate their effects, or these may be generated by post-PHA evaluations of the team's findings. Safeguards may include inherently safer or passive approaches to hazard control, new engineering controls (e.g. improved fire detection and suppression systems) or administrative controls (e.g. new operating procedures, inventory control measures, or separation of HHC into different storage areas).

During NEP inspections, OSHA frequently found PHA deficiencies in (i) recommendation resolution, (ii) facility siting, and (iii) human factors analysis.

3.1.2.2.1 *Recommendation Resolution*

When a PHA team finds a hazard that has not been properly addressed, and makes a recommendation to resolve the hazard, 29 CFR 1910.119(e)(5) states that "[t]he employer shall

establish a system to promptly address the team's findings and recommendations; assure that the recommendations are resolved in a timely manner and that the resolution is documented…" (emphasis added). Examples of PHA recommendations that were not resolved include:

- Rerate and protect coils from overpressure,
- Evaluate safety relief valves to ensure discharge to safe locations,
- Remove a 2-in. hazardous materials line that is no longer in use,
- Update P&IDs,
- Evaluate the need for a check valve downstream of a block valve; and
- Evaluate the potential effects of an increase in throughput in a crude unit, including potential effects on the relief system.

Employers must "establish a system" to ensure that PHA team recommendations are promptly resolved. Failure to establish such a system was a leading cause of PHA citations, during NEP inspections. In many instances, the unresolved PHA recommendations were over five years old.

OSHA urges petroleum refineries to promptly review their PHA findings, resolve any outstanding recommendations, and review their systems for tracking PHA recommendation resolutions. If an employer discovers an outstanding PHA recommendation, OSHA strongly recommends that the employer review other PHAs (older PHAs in the same process and other PHAs from other, similar covered processes) to identify any other outstanding PHA recommendations. Resolution of outstanding recommendations includes either the implementation of the recommendation or documentation that determination that no actual hazard exists (i.e. corrective action is not necessary). During NEP inspections, OSHA discovered systematic problems with multiple instances of failure to resolve findings and recommendations.

CCPS recommends that the "hazard identification and risk analysis team" (PHA team) provide a report with the rationale for any recommendations so that management can determine if they are appropriate. 20 CCPS further recommends that employers "formally resolve" the recommendations made by the PHA team, either by implementing the recommended resolution, implementing an alternative hazard reduction resolution, or documenting the rationale for rejecting the recommendation. 21 This information will be used by future PHA teams and OSHA to assess current process hazards and whether the safeguards in place adequately protect worker safety and health.

3.1.2.2.2 Facility Siting Facility siting hazards were a common basis for PSM citations during NEP inspections. In some cases, OSHA found instances where a facility siting analysis was completely omitted by the PHA team. In other instances, OSHA found that the PHA did not adequately evaluate whether temporary structures were properly sited. However, the most common facility siting citations involved permanent structures.

Many PHA teams did not address proper spacing of equipment or possible vehicle impacts to equipment or piping. Furthermore, OSHA found that some PHAs did not evaluate whether control rooms were protected by adequate separation or building construction from explosion, fire, toxic material, or high overpressure hazards.

PHA teams also failed to evaluate whether various locations (operator's break room, control room, parking lots, and abandoned administrative buildings) were safe from process releases.

3.1.2.2.3 Human Factors The PSM standard requires the employer's PHA team to evaluate human factors during its PHA. CCPS defines human factors as "a common term given to the widely-recognized discipline of addressing interactions in the work environment between people, a facility, and its management systems." The basic principle of assessing human factors is to determine whether the employer "fit the task and environment to the person rather than forcing the person to significantly adapt in order to perform their work."

Very few PHA citations resulted from the PHA team completely omitting the human factors review. Instead, the majority of citations resulted because the PHA overlooked specific human factors issues. During NEP inspections, OSHA found that some petroleum refinery PHAs had failed to address:

- Inadequate or unsafe accessibility to process controls during an emergency,
- A lack of clear emergency exit routes, or
- Inadequate or confusing labeling on equipment, procedures, and/or P&IDs.

Specific examples include:

- Inadequate alarm management, which required operators to perform multiple mental calculations and be aware of an unrealistic number of simultaneous alarms in a unit,
- Requiring operation of a bypass valve that required crawling across a pipe rack, and
- Locating emergency isolation valves where operators are in harm's way during an emergency.

During the four years of NEP inspections, OSHA also found that many refineries have over-relied on administrative controls to address human factors concerns. Over-reliance on administrative controls puts a burden on employees, invites confusion, and can increase the risk of failure-on-demand in an emergency.

Moreover, administrative controls may be inappropriate as the only protection against hazards with potentially severe consequences.

Risk assessment and reduction techniques (such as a Layers of Protection Analysis) that use the hierarchy of controls can be effective in identifying the level and extent of safeguards necessary and appropriate to protect workers.

3.1.2.3 Operating Procedures

PSM-covered petroleum refineries are required to develop and implement written operating procedures that provide clear instructions for safely conducting activities involved in each covered process consistent with the PSI. Operating procedures must provide clear instructions not only to specify the steps for normal operations, but also for upset conditions, temporary operations, safe work practices, and emergency shutdown.

Operating procedures must address the basic hazards that are or could be encountered in the process. During NEP inspections, many operating procedures citations resulted from a complete absence of written operating procedures.

However, even when operating procedures existed, OSHA found that they were not always accurate or implemented as written. Over the lifetime of a unit, operating activities may begin to deviate from the original written procedure. Sometimes deviations can produce the desired result, other times it can place workers in hazardous conditions. During NEP inspections, OSHA found many instances where operators deviated from the written operating procedures.

To prevent this, management and operators should meet to review the effectiveness of existing procedures and revise them as necessary. A strong employee participation plan can facilitate this interaction. Operators are often the first to realize when a procedure is unrealistic or unattainable, but they must be encouraged to approach management and technical staff with these issues instead of finding their own creative, and potentially dangerous, solutions.

Regardless, employers must periodically review operating procedures and certify them annually. CCPS also recommends that employers consider using event-based review periods – such as reviewing shutdown procedures prior to a planned turnaround, and using lessons learned from recent significant events – rather than waiting until the next review cycle.

In addition to initial startup, normal operations, and temporary operations, employers must develop and implement written operating procedures for emergency shutdown, emergency operations, normal shutdown and startup, and safe work practices. The findings from NEP inspections demonstrated a need to review (i) emergency shutdown procedures and (ii) safe work practices.

3.1.2.3.1 *Emergency Shutdown Procedures* Emergency shutdown procedures are an important component of workplace safety. There are several requirements that must be met.

First, emergency shutdown procedures must identify conditions that should be recognized by personnel in all affected unit(s) as emergency conditions. Examples of conditions that require emergency shutdown include (but are not limited to) failure of process equipment, loss of electrical power, loss of instrumentation, loss of containment, severe weather conditions, fires, and explosions.

Second, in the event of an emergency, qualified operators must be assigned shutdown responsibility to ensure the emergency shutdown is executed in a safe and timely manner. CCPS further recommends identifying another individual (or team) to manage all activities that are not emergency response related, such as accounting for personnel and responding to questions from the media.36

During NEP inspections, OSHA found many instances where employers did not identify conditions (failure of process equipment, loss of electrical power, loss of instrumentation, loss of containment, severe weather condition, fire, explosion, etc.) that required emergency shutdown and failed to designate appropriate responsible for emergency shutdown procedures. OSHA urges refineries to review their emergency shutdown procedures and their assignments of shutdown responsibility to minimize hazards in the workplace in the event of an emergency.

3.1.2.3.2 *Safe Work Practices* The Operating Procedures section of PSM also requires employers to develop and implement safe work practices that will control hazards during normal operations. During NEP inspections, OSHA found that many petroleum refineries were deficient in the following areas:

- Controlling entry of motorized equipment into ignition source-controlled areas,
- Controlling personnel access to process units,
- Line breaking and equipment opening practices,
- Hot work permitting,
- Lock-out and tag-out (LOTO) practices,
- Vehicle collision control, and
- Housekeeping.

During NEP inspections, OSHA issued several citations in which more than one of the above deficiencies was present. For instance, a facility allowed motorized equipment to enter operating units that contained flammable materials, and personnel entry was not controlled to refinery areas. In this instance, the refinery was deficient in both controlling access and restricting motorized equipment from ignition source-controlled areas when the entry was not performed using a safe work practice procedure.

Another example involved a facility where contract employees and vehicles were not controlled when entering and exiting process areas. Vehicles also entered electrically classified areas without a vehicle permit or hot work permit, presenting a risk of fire or explosion. In this instance, the refinery had failed to control access, restrict motorized equipment from ignition source-controlled areas, or ensure that a hot work permit had been issued.

OSHA strongly encourages petroleum refinery industry employers to review their safe work procedures. In many cases, OSHA regulates these non-routine activities through existing prescriptive standards, such as:

- 29 CFR 1910.146 Permit-required confined spaces,
- 29 CFR 1910.147 The control of hazardous energy (lockout/tagout),
- 29 CFR 1910.252 Welding, Cutting, Brazing, and
- 29 CFR 1910.307 Hazardous (classified) locations.

3.1.2.4 Mechanical Integrity

The MI element of the PSM Standard requires employers to create written procedures to maintain the ongoing integrity of process equipment, train for process maintenance activities, inspect and test process equipment, correct equipment deficiencies, and perform quality assurance. MI programs must address pressure vessels, storage tanks, piping systems (including piping components such as valves), pumps, relief and vent systems and devices, emergency shutdown systems, and controls (including monitoring devices and sensors, alarms, and interlocks).

During NEP inspections, OSHA found MI compliance issues, including (i) equipment deficiencies, (ii) inspection, testing, and maintenance procedures, (iii) resolving anomalous data, and (iv) ensuring site-specific inspection and testing.

3.1.2.4.1 *Equipment Deficiencies* Failure to correct equipment deficiencies that are outside acceptable limits39 is one of the leading causes of PSM non-compliance in the petroleum refinery sector. Non-compliance for equipment deficiencies broke down into four major groups:

- Lack of proper maintenance or repair,
- Inappropriate installation (such as inappropriate sizing),

- Missing protective system (such as not including relief devices), and
- Insufficient structural support.

Equipment most cited for deficiencies were relief devices, followed by piping circuits, pressure vessels, and alarm systems.

Other examples of equipment cited for violations of the PSM MI requirements that OSHA found during NEP inspections include:

- A broken gate valve caused a level gauge to not work properly, which rendered visual verification of liquid level for the vessel ineffective. This deficiency went uncorrected.
- The installation of an engineered clamp failed to correct a deficient piece of process piping, which was a 90-degree elbow that was outside acceptable limits. The employer continued to use the leaking 90-degree elbow as part of a piping circuit that conveyed waste hydrogen sulfide gas.
- Hydrogen sulfide monitors were not inspected and tested on a regular basis to correct deficiencies in alarms that were outside acceptable limits due to bad sensors, loose wiring, or monitors that needed to be replaced. Work orders were not managed by a tracking system to ensure that deficiencies were fixed in a timely manner. Some work orders marked "fix today" or "ASAP" were not fixed for a week or longer.
- Six relief systems in an alkylation unit were incorrectly sized and were not corrected in a timely manner when the deficiencies were reported. No MOC was performed to justify the decision to delay replacing the deficient systems.
- Grounding cables were removed from equipment, such as a heat exchanger and pump motors, but were not replaced.
- Excessive vibration was observed on motors with visible movement of structural steel decking and supports. Also, two 1" pipes and one 4" pipe containing flammable liquid were not adequately supported.

According to CCPS, "designing and maintaining equipment that is fit for its purpose and functions when needed is of paramount importance to process industries."40 OSHA urges petroleum refiners to reevaluate their own MI program, focusing on their procedures for correcting equipment deficiencies and ensuring that they do not have any lingering equipment deficiencies. CCPS also recommends that employers develop and implement an equipment deficiency management process.

3.1.2.4.2 *Inspection, Testing, and Maintenance Procedures* A compliant MI program will have written procedures for the on-going integrity of process equipment. During NEP inspections, OSHA found compliance issues in written inspection, testing, and repair procedures.

Citations issued for violations of the PSM standard's Inspection, Testing, and Maintenance provisions often referenced a complete lack of inspection and testing procedures or found the written procedures in use to be inadequate. The most cited types of equipment for non-compliant inspection or testing procedures were piping circuits, pressure vessels, relief devices, and monitoring alarms. Basic items such as the number and location of thickness measurements, welder certification requirements, and positive materials testing must follow the employer's MI procedures and be appropriately documented.

Examples of non-compliant inspection and testing procedures include:

- Not establishing and implementing adequate procedures to clearly set the specific number and locations of thickness measurement locations (or condition management locations) for each pressure vessel and sections of piping, and to address anomalous inspection data, such as increasing thickness measurement values for pressure vessels.
- Failure to list specific requirements for welder qualifications in the inspection and testing program when the procedure establishes that all systems that are inspected must contain the welders' qualifications.
- Not establishing and implementing adequate procedures to inspect, repair, and maintain pressure vessels, piping, and other process equipment, which led to an inadequate evaluation of a leaking tube in a crude heater unit and a multitude of inadequately recorded piping inspections.
- Not updating procedures to reflect a recently initiated practice of changing the piping inspection interval.
- Failure to inspect for corrosion under insulation, which resulted in pressure vessels, process equipment, and piping in the sulfur unit to be subjected to corrosive environments without assurance that exterior corrosion would be found during routine inspections.

One common non-compliant condition found during inspections was corrosion under insulation. Insulated vessels and piping circuits can harbor advanced corrosion sites under the insulation. Many critical factors may affect corrosion under insulation, such as equipment under cyclic temperature processes, poor design allowing moisture to collect, insulating materials holding moisture, and certain local environments or emissions that can accelerate corrosion. 44 Petroleum refineries must be prepared for this type of corrosion and adequately inspect under insulation.

The MI section of the PSM standard also requires that "inspection and testing procedures shall follow recognized and generally accepted good engineering practices," and "the frequency of inspections and tests of process equipment shall be consistent with applicable manufacturers' recommendations and good engineering practices, and more frequently if determined to be necessary by prior operating experience." The most cited inspection and testing RAGAGEP deficiencies involved piping and pressure vessels. During NEP inspections, OSHA found (i) insufficient thickness measurements and (ii) unacceptable/unestablished inspection frequencies for these types of equipment, based on the employer's selected RAGAGEP or lack thereof. Moreover, a significant portion of citations were issued due to inadequate or complete lack of inspection of covered process equipment.

Thickness Measurements RAGAGEP   Citations with respect to thickness measurements involved:

- Actual corrosion rates exceeding the expected rate and being uncorrected or accounted for.
- Failure to properly inspect multiple pressure vessels, specifically citing a lack of ultrasonic thickness testing.
- Failure to inspect a boiler at appropriate location points, as specified by the employer's RAGAGEP.
- Failure to ensure an appropriately certified individual performed pressure testing.
- Failure to calculate corrosion rates to determine appropriate thickness measurement intervals.

Employers are reminded that in the lifetime of a pipe, especially those used in corrosive service, thickness measurements must be taken at predetermined locations, called condition monitoring locations (CMLs), to establish the integrity of the pipe. For instance, API 570: Piping Inspection Code: In-service Inspection, Rating, Repair, and Alteration of Piping Systems, an example RAGAGEP, states that CMLs must be taken not just at various locations along the length of a pipe, but at susceptible areas of deterioration – such as injection points, mixing points, and dead legs. The final number of CMLs is determined by considering the potential worker health and safety impacts, expected corrosion rates, and system complexity of the piping system. 48

Inspection Frequency RAGAGEP   The most cited equipment for non-compliant inspection frequencies (of any type, not only thickness measurements) have been piping circuits followed by pressure vessels, relief devices, and monitoring alarms. As part of the inspection program, an appropriate inspection frequency must be established for equipment to determine whether pipe/vessel thickness is decreasing as expected. API 570 identifies three classes of piping services and recommends a thickness measurement inspection frequency based on the class. For example, Class 1 includes:

- Flammable,
- Pressurized services that may rapidly vaporize and explode upon release,

- Hydrogen sulfide,
- Anhydrous hydrogen chloride,
- Hydrofluoric acid
- Piping over water of public throughways, and
- Flammable services operating above their auto-ignition temperature.

Employers must follow their selected RAGAGEPs for appropriate equipment inspection frequency requirements, unless prior operating experience shows more frequent inspections are necessary. Recommended sources include the equipment manufacturer, an employer's corporate/company-specific procedures, and industry/consensus standards.

API publications are examples of piping/vessel inspection RAGAGEP. Specifically, API 510: Pressure Vessel Inspection Code, API 570: Piping Inspection Code: In-service Inspection, Rating, Repair, and Alteration of Piping Systems, and API 580: Risk-Based Inspection.

3.1.2.4.3 Resolve Anomalous Data During MI inspection and testing, if it is believed that a testing result is inaccurate or irregular, the uncharacteristic result must be retested, not disregarded. Upon retesting, should the result remain irregular, or anomalous, action must be taken to determine the cause of the anomalous condition and then the issue must be resolved. According to CCPS' Guidelines for MI Systems "[inspection, testing, and preventive maintenance] data should be reviewed for anomalies so that suspicious information can be verified or corrected."

Inspections conducted under the NEP found multiple instances where employers failed to address inconsistencies in testing measurements.

In some cases, thickness measurements were increasing, and the employer took no action to investigate the anomaly. CCPS recommends that employers establish a means to efficiently analyze data and highlight anomalies. Also, personnel should be empowered to identify abnormal component parts. Any anomalies must be taken seriously to prevent an incident. This same principle for regarding resolving anomalous inspection data is addressed in API 570 (2016), Section 6.5.4, Data Analysis.

3.1.2.4.4 Ensure Proper "Site-Specific" Inspections and Tests It is also important that all inspections and testing be tailored to the specific equipment and environment of the process equipment.

General industry standards and corporate-wide, or "boiler-plate," procedures will not be applicable in all situations and may be lacking necessary hazard information. As such, employers must tailor site-specific MI procedures that address their unique PSM covered process(es).53 CCPS notes that "simply assimilating information published in standards may not be appropriate... some failure modes are very service- or process specific."54 Employers must train each employee involved in maintaining the on-going integrity of process equipment in an overview of that process and its hazards and in the procedures applicable to the employee's job tasks to ensure that the employee can safely perform the job tasks.

During NEP inspections, OSHA found that all instances where equipment was not inspected or tested (or was inspected or tested inadequately) were the result of an inadequate site-specific inspection or test procedure. Examples include:

- Employer failed to inspect and test H2S monitors on a regularly scheduled basis and failed to inspect and calibrate flow alarms, temperature alarms, flame detectors, auxiliary burners, level alarms, shutdown alarms, and control valve alarms. The monitoring devices and alarms were inspected only when there was a malfunction.
- Employer failed to adequately inspect and test process equipment by failing to ensure that the pressure gauges for a butylene feed coalescer and recycle isobutane coalescer were calibrated on a periodic basis.
- Employer failed to adequately inspect and test process piping, such as a stream vapor, a stream draw, a diesel draw, a naphtha overhead line, a naphtha/crude exchanger inlet, and a stream reflux line.
- Employer failed to adequately inspect and test heat exchangers, resulting in nuts being loose or missing on anchor bolts. Employer also failed to inspect steel vessel supports and failed to inspect grounding cables for deficiencies.

Employers must ensure that their inspection and testing procedures are specific to their covered processes; otherwise, equipment may be incidentally omitted from the MI program. OSHA states in the non-mandatory appendix of PSM "The first step of an effective mechanical integrity program is to compile and categorize a list of process equipment and instrumentation for inclusion in the program."

CCPS also states that facility personnel need to establish boundaries and develop a list of equipment to include in their MI program.

3.1.2.5 Management of Change

The MOC section of the PSM standard requires the employer to implement written procedures to manage changes (except for "replacements in kind") to process chemicals, technology, equipment, procedures, and changes to facilities that affect a covered process. The MOC procedure requires descriptions of the technical basis for the change, impact on safety and health, modifications to operating procedures, necessary time for change, and appropriate authorizations. Any employee who will be impacted by the change must be informed and trained appropriately before the unit/process can restart.

During NEP inspections, OSHA found MOC non-compliance for changes in (i) equipment design, (ii) operating procedure, (iii) regular maintenance/repair, (iv) facilities, and (v) excessive time limits for temporary changes.

3.1.2.5.1 Changes in Equipment Design Listed are some examples of OSHA citations where an MOC was not utilized when there was a change in equipment design:

- Installing a control valve bypass,
- Installing a spill guard berm under a fracturing tank along with proper grounding and bonding,
- Changes to an alarm set point, and
- Changes to materials of construction.

If any piece of equipment is changed to equipment with different specifications from the design in the PSI (i.e. not a "replacement-in-kind"), an MOC must be utilized. Likewise, changes in design such as chemicals used or increases/decreases in operating parameters outside their range described in the PSI also require MOC. Such changes may result in new hazards or necessitate new or additional safeguards and/or procedures.

3.1.2.5.2 Changes in Operating Procedures OSHA found that some refineries did not utilize MOC when there was a change in operating procedures. Examples include:

- Changing procedures for the manual addition of methanol to a chloride injection tank; and
- Procedures for installing a new type of relief device (different from the original).

If operating procedures are changed, MOC is required to assess the potential hazards introduced by the change. Additionally, MOC ensures proper training on the new operating procedures for effected personnel prior to start-up of the process or affected part of the process.

3.1.2.5.3 Changes in Inspection and Test and Maintenance Procedures During NEP inspections, OSHA found MOC was not used when there was a change in maintenance procedures. Examples include:

- Changing inspection intervals for piping circuits,
- Changing the number of thickness measurement locations (or CMLs) on a pipe, and
- Changing maintenance procedures following a change (not replacement-in-kind) in process equipment.

Maintenance procedures will dictate preventive maintenance intervals and repair procedures. Like operating procedures, maintenance intervals or regular repair procedures need to be changed, MOC must be utilized.

3.1.2.5.4 *Changes in Facilities* If an existing structure is being modified (such as: upgrading ventilation, changing exit locations, or re-enforcing the structure), MOC must be initiated. Also, for newly installed facilities within or near a PSM covered process, MOC must be created. NEP inspection OSHA citation examples include:

■ Installing a light wood or metal shed structure near a Hydrocracker Unit; and
■ Changes to a control room located within a PSM covered process unit.

Employers must initiate MOC for these types of changes.

3.1.2.5.5 *Time Limitations on Temporary Changes* Finally, MOCs were not initiated when there were temporary changes, including when using temporary supports during installation of a new vessel or piping circuit, or using a shed or break area as a temporary control room during construction or repair of the main control room.

Temporary changes, which are usually initiated while a permanent change is being made, must be properly assessed through the MOC process for the permanent change. OSHA also found that some employers create MOC procedures that fail to define or fail to adhere to the time limit of the temporary change. CCPS notes that "if temporary changes are permitted, the MOC review procedures should address the allowable length of time that the change can exist, and the procedure should include a process to confirm the removal of temporary changes or restoration of the change to the original condition within the time period specified in the approved change request."

3.2 DuPont La Porte, TX, Methyl Mercaptan Release – November 15, 2014

We encounter methyl mercaptan every day in our lives. The chemical is found in our blood, brain, and feces, and is the partial odorant of bad breath and flatus, to use a polite term. It is found in numerous foods, including some nuts and various cheeses. It is a colorless gas and the chemical formula for methyl mercaptan is CH_3SH. It is an organosulfur compound and the simplest thiol. The chemical is used as on odorant in natural gas and propane because these gases are odorless. A gas leak might go undetected if methyl mercaptan was not added to the fuels.

Methyl mercaptan is produced by reacting methanol (methyl alcohol) with hydrogen sulfide.

$$CH_3OH + H_2S \rightarrow CH_3SH + H_2O$$

The LD50 of methyl mercaptan is 61 mg/kg is mice. In humans, occupational exposure to methyl mercaptan may induce headache, nausea, vomiting, eye irritation, chest tightness and wheezing, dizziness, diplopia, and a productive cough. The current OSHA PEL is 10 ppm (20 mg/m^3) ceiling. The IDLH level is 150 ppm (CDC 1994). NFPA ratings for methyl mercaptan:

NFPA Health Rating: 4 – Materials that, under emergency conditions, can be lethal.

NFPA Fire Rating: 4 – Materials that rapidly or completely vaporize at atmospheric pressure and normal ambient temperature or that are readily dispersed in air and burn readily.

NFPA Instability Rating: 1 – Materials that in themselves are normally stable but that can become unstable at elevated temperatures and pressures.

Methyl mercaptan is a useful chemical in the manufacture of other chemicals. For example, it is used as a precursor for making pesticides. The following accident description and analysis pertains to a methyl mercaptan leak in an E.I. DuPont de Nemours and Company (DuPont) pesticide manufacturing plant in La Porte, TX. The PPG chemical plant I worked at in my early career days was in La Porte as well.

3.2.1 Accident Description and Analysis

The following accident and analysis are excerpted from the Chemical Safety Board (CSB) report (CSB 2019).

On November 15, 2014, approximately 24,000 lb of highly toxic methyl mercaptan was released from an insecticide production unit (Lannate® Unit) at the DuPont chemical manufacturing facility in La Porte, TX. Lannate has the chemical formula $C_5H_{10}N_2O_2S$. It is an acute toxin, with an LD50 17 mg/kg in rats. A lethal dose in humans is 12–15 mg/kg (NIH 2021).

The release of methyl mercaptan killed three operators and a shift supervisor inside a manufacturing building. They died from a combination of asphyxia and acute exposure (by inhalation) to methyl mercaptan.

The CSB determined that the cause of the highly toxic methyl mercaptan release was the flawed engineering design and the lack of adequate safeguards. Contributing to the severity of the incident were numerous safety management system (SMS) deficiencies, including deficiencies in formal process safety culture assessments, auditing and corrective actions, troubleshooting operations, MOC, safe work practices, shift communications, building ventilation design, toxic gas detection, and emergency response. Weaknesses in the DuPont La Porte SMS resulted from a culture at the facility that did not effectively support strong process safety performance.

The highly toxic methyl mercaptan release resulted from a long chain of PSM system implementation failures stemming from ineffective implementation of the PSM system at the DuPont La Porte facility.

The CSB investigation viewed the chain of implementation failures as starting with the flawed engineering design of the US$20 million nitrogen oxides (NO_x) reduced scrubbed (NRS) incinerator, a capital project implemented in 2011. DuPont La Porte had long-standing issues with vent piping to this incinerator because the design did not address liquid accumulation in waste gas vent header vapor piping to the NRS, and DuPont La Porte did not fully resolve the liquid accumulation problem through hazard analyses or MOC reviews. Instead, to deal with these problems, daily instructions had been provided to operations personnel to drain the liquid from these pipes to the atmosphere inside the Lannate manufacturing building without specifically addressing the potential safety hazards this action could pose to the workers. DuPont La Porte's instructions did not specify additional breathing protection for this task. On the night of the incident, not realizing the hydrate blockage in the methyl mercaptan feed piping was cleared, workers went to drain the liquid from the waste gas vent piping. They did not know that high pressure in the waste gas vent piping was related to the fact that liquid methyl mercaptan was flowing through the methyl mercaptan feed piping and into the waste gas vent piping.

The chain further developed when the ineffective building ventilation system failed to be addressed after DuPont auditors identified it as a safety concern about five years before the incident. DuPont La Porte's management system did not resolve the PSM recommendation (i.e. did not take corrective action) to address the building ventilation system. The ventilation design for the manufacturing building was based on flammability characteristics and did not take into consideration toxic chemical exposure hazards, even though the building contained two highly toxic materials, chlorine, and methyl mercaptan. DuPont La Porte records indicated that the manufacturing building's dilution air ventilation design was based on providing sufficient ventilation to ensure that the concentration of flammable gases did not exceed 25% of the lower explosion limit (LEL). At the time of the incident, neither of the manufacturing building's two rooftop ventilation fans was working, despite an "urgent" work order written nearly a month earlier. Even had the fans worked, they probably would not have prevented a lethal atmosphere inside the building due to the large amount of toxic gas released.

DuPont La Porte's installation of a methyl mercaptan detection system inside the manufacturing building added another link to the chain. Neither the workers nor the public was protected by DuPont's toxic gas detection system on the night of the incident. The building where the workers died was not equipped with an adequate toxic gas detection system to alert personnel to the presence of dangerous chemicals. First, DuPont La Porte set the detector alarms well above safe exposure limits for workers. Second, DuPont La Porte relied on verbal communication of alarms that automatically displayed on a continuously manned control board. Finally, DuPont La Porte did not provide visual lights or audible alarms for the manufacturing building to warn workers of highly toxic gas concentrations inside it. When a release caused a detector to register a concentration above the alarm limit, the toxic gas detection system did not warn workers in the field about the potential leak and the need to evacuate. Among other factors, this detection system contributed to workers' growing accustomed to smelling the methyl mercaptan odor in the unit. Additionally, when the toxic gas detectors triggered alarms, DuPont La Porte personnel investigated potential methyl mercaptan leaks without using respiratory protection. Personnel normalized unsafe methyl mercaptan detection practices by using odor to detect the gas, further deteriorating the importance or effectiveness of utilizing instrumentation in response to alarms signaling potential toxic gas releases.

In the spring of 2014, the chain propagated when DuPont La Porte's interlock program did not require verification that interlocks that had been bypassed for turnaround maintenance were returned to service before the plant resumed operating. Because of this ineffective program, a bypassed interlock caused acetaldehyde oxime (AAO), a critical raw material, to become diluted, leading to a shutdown of the Lannate Unit days before the November 2014 incident. During the shutdown, water entered the methyl mercaptan feed piping and, due to the cold weather, formed a hydrate (an ice-like material) that plugged the piping and prevented workers from restarting the unit. Furthermore, DuPont La Porte did not establish adequate safeguards after a 2011 DuPont La Porte process hazard analysis identified hydrate formation in this piping, revealing yet another link in the chain.

When the hydrate formed, lacking safeguards to control the potential safety hazards associated with dissociating (breaking up) the hydrate (such as using heat tracing to prevent the hydrate from forming a solid inside the piping or developing a procedure to dissociate the hydrate safely), DuPont workers went into troubleshooting mode. Ineffective hazard management while troubleshooting the plugged methyl mercaptan feed piping formed yet another link in the chain and allowed liquid methyl mercaptan to flow into the waste gas vent header piping toward the NRS incinerator, a location where it was never intended to go. As discussed earlier, DuPont La Porte did not fully resolve liquid accumulation in the waste gas vent header by the NRS incinerator. Consequently, DuPont La Porte workers dealt with the common problem of liquid accumulation in the waste gas vent header on a routine basis by draining the liquid (line breaking) without an engineered solution or without ensuring the use of safety procedures or personal protective equipment (PPE).

However, when the liquid drain valves were opened on November 15, 2014, flammable, and highly toxic methyl mercaptan flowed onto the floor and filled the manufacturing building with toxic vapor.

Once the methyl mercaptan release began, an ineffective emergency response program at La Porte contributed to the extent and duration of the chemical release, placed other workers in harm's way, and did not effectively evaluate whether the chemical release posed a safety threat to the public.

This chain of events illustrates DuPont La Porte's ineffective implementation of its PSM system. The individual components of this PSM system exhibited cross-cutting weaknesses, resulting in the deaths of four workers and leading to the ultimate decision to close the DuPont La Porte facility.

The following is a more detailed look at the methyl mercaptan release.

The November 15, 2014 incident occurred in DuPont La Porte's Lannate Unit, which produced insecticides. Part of the Lannate process occurred inside a closed manufacturing building (Figure 3.2).

Figure 3.2 Lannate® production building. Source: Chemical Safety Board (2019).

On Monday, November 10, 2014, an inadvertent chemical dilution caused operating difficulties that forced a shutdown of DuPont's Lannate Unit. In response, DuPont La Porte personnel adjusted the unit's control system to resume operations. On Wednesday, November 12, 2014, operators tried to restart the Lannate process. During the shutdown, the methyl mercaptan piping to the reaction section of the process had become plugged due to the formation of a clathrate hydrate (hydrate), halting restart. At that time, operators did not know the source of the plugging and began troubleshooting the process to try to clear the piping, unaware that the solid hydrate had formed.

Two days later, on Friday, November 14, 2014, the troubleshooting efforts to clear the plugging were still ongoing. That morning, the Lannate Unit Technical Team (Technical Team), composed of engineers and other employees experienced in the process, met with operations personnel to discuss troubleshooting options to clear the plugging. This meeting identified the likely scenario that water had entered the methyl mercaptan system, forming a solid hydrate.

DuPont's methyl mercaptan technical standard identifies the potential to form a hydrate at low temperatures. The technical standard states that methyl mercaptan "will form a hydrate with water, which is a solid below 40 °F (4.4 °C) per information provided by a [methyl mercaptan] supplier." Based on this understanding, the Technical Team asked that operators put hot water on the outside of the methyl mercaptan piping, under the insulation, to warm the piping and its contents to break up the plugging. The Technical Team also realized that when heated, methyl mercaptan would expand, requiring a safe place to vent to avoid over-pressuring the piping. To address this concern, operations personnel opened valves between the methyl mercaptan piping and a waste gas vent header, not recognizing that this alignment created a pathway for liquid methyl mercaptan to eventually release from the piping inside the manufacturing building when drain valves were opened.

In addition, when operations personnel opened other valves while troubleshooting, such as drain valves, methyl mercaptan was released to the atmosphere, causing a strong methyl mercaptan odor that could be smelled by site personnel. The odor threshold for methyl mercaptan is 0.002 ppm. During troubleshooting, methyl mercaptan was released both outside and inside the manufacturing building, triggering 32 methyl mercaptan gas alarms on the control panel throughout the 17 hours preceding the incident. Although methyl mercaptan is a toxic and flammable chemical, and at high enough concentrations lethal, some operations personnel did not respond in accordance with the nature of the emergency circumstances because they associated the alarms with the ongoing troubleshooting efforts. They did not perceive the methyl mercaptan alarms as signifying a serious hazard because they had normalized the methyl mercaptan odor within the Lannate Unit, as well as the detector alarms.

Early in the morning on Saturday, November 15, 2014, the hot water from the hoses warmed the hydrate plugging, causing it to dissociate, clearing the plugging. Due to the valve alignment, liquid methyl mercaptan flowed into the waste gas vent header, located inside the manufacturing building. The vent header, however, was not intended nor designed for liquid methyl mercaptan. At 2:51 a.m., alarms began to sound on the control system indicating high pressure in equipment inside the manufacturing building.

Operations personnel did not realize that liquid methyl mercaptan was causing the high pressure. Instead, they attributed the high pressure to a common, long-standing problem with process condensate, which DuPont La Porte personnel believed to be mostly water, accumulating in the vent header piping. The Lannate control room board operator (Board Operator) contacted the night shift supervisor (Shift Supervisor) and another operator (Operator 1) separately, asking them to help troubleshoot the high-pressure situation. Manually draining the vent header inside the manufacturing building, a long-standing practice, was the typical approach used to remove process condensate from the vent header, as it successfully reduced system pressure many times in the past. The Shift Supervisor and Operator 1 each went separately to the manufacturing building, likely in order to drain liquid from the vent header piping manually.

Sometime between 3:01 and 3:13 a.m., a worker (likely the Shift Supervisor) manually opened two sets of drain valves on the vent header piping, located on the third floor of the manufacturing building (Figure 3.3). But instead of the expected condensate composed mostly of water, the liquid methyl mercaptan that had filled the piping escaped out of the valves, vaporized, and killed the Shift Supervisor.

Figure 3.3 Waste gas vent header. Source: Chemical Safety Board (2019).

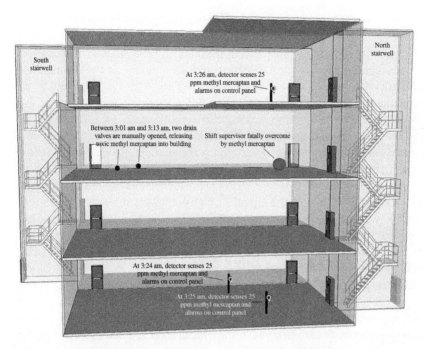

Figure 3.4   Location of shift supervisor overcome by methyl mercaptan. Source: Chemical Safety Board (2019).

Between 3:24 and 3:26 a.m., three methyl mercaptan detectors inside the manufacturing building sensed at least 25 parts per million (ppm) of methyl mercaptan and triggered alarms at the control panel (Figure 3.4). The Board Operator – still focused on reducing the high pressure – did not realize a major chemical release was occurring inside the manufacturing building.

At about 3:30 a.m., Operator 1 made an urgent call for help over the radio. Various personnel interpreted her communication in different ways, generally as either "We need help!" or "I need help on the fourth floor!" The Board Operator tried to get more information through radio communication, but neither the Shift Supervisor nor Operator 1 ever responded.

Two operators (Operator 2 and Operator 3) who were in the control room and heard the distress call ran to the manufacturing building to help. Another operator (Operator 4) saw Operator 2 and Operator 3 running to the manufacturing building, and he followed them. None of these operators knew of the major release of toxic methyl mercaptan inside of the manufacturing building; therefore, they did not wear any respiratory protection when they ran into the manufacturing building. The manufacturing building lacked automatic visual or audible alarms to alert fieldworkers or prevent them from entering a potentially toxic atmosphere. In addition, the building's ventilation fans were not working, a situation that, in

DuPont's operating procedures, required "restricted access" to the building. But DuPont La Porte's procedures did not define "restricted access" or require that operators wear respiratory protection in the building when access was restricted, even though a toxic chemical release could accumulate in the unventilated building.

Operator 2, Operator 3, and Operator 4 each entered the south stairway and took different a route inside the manufacturing building to try to find Operator 1 (Figure 3.5). Operator 2 went to the third floor, where he was fatally overcome by methyl mercaptan. Operator 4 went to the second floor. He walked about 10 ft and hit what he later described as a "wall" of methyl mercaptan, but he managed to retreat to the stairwell. Operator 3 went to the fourth floor but did not find anyone. He then announced on the Lannate Unit's public-address system that he did not see anyone on the fourth floor. The Board Operator responded that the Shift Supervisor and Operator 1 may be on the third floor. Operator 3 then began to feel light-headed. He made his way to the stairwell and lost consciousness while descending from the fourth floor.

The Board Operator then tried to reach the Shift Supervisor, Operator 1, and Operator 2 over the radio, but they did not respond. An operator in the control room, Operator 6 (the brother of Operator 2), then grabbed three 5-minute escape respirators. Other operators in the control room warned him not to enter the manufacturing building because they did not know what was going on or where the nonresponsive operators were. Operator 6, however, took the escape respirators and rushed to the manufacturing building.

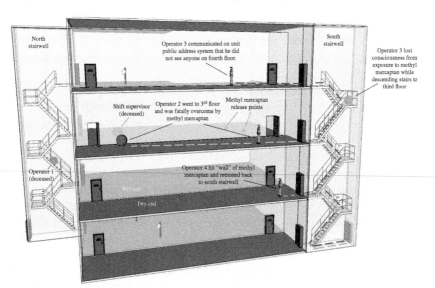

Figure 3.5 Locations of operators. Source: Chemical Safety Board (2019).

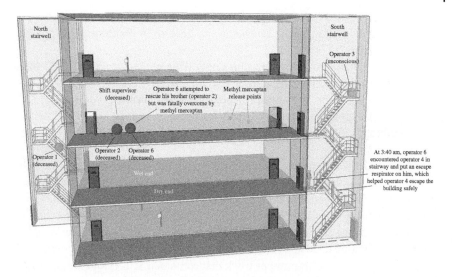

Figure 3.6 Locations of operator 6. Source: Chemical Safety Board (2019).

At about 3:40 a.m., Operator 6 encountered Operator 4 in the south stairway of the manufacturing building and put an escape respirator on him (Figure 3.6). The breathing air helped Operator 4 recover, and he exited the manufacturing building safely. A worker – likely Operator 6 – then manually activated the manufacturing building fume release alarm, an alarm intended to alert area workers of a toxic chemical release in the building. Operator 6 then went to the third floor of the manufacturing building, at some point after helping Operator 4.

At about 3:50 a.m., the Board Operator called for the plant emergency response team (ERT) to respond to the manufacturing building, stating over the intercom system, "We need rescue people, and there's people missing." This announcement started a chain of miscommunication. To the ERT, a request for "rescue" meant specifically high-angle or confined-space rescue. Therefore, the ERT responded to the request for help by gathering only technical rescue gear (e.g. harnesses and ropes), not knowing that there was a major toxic chemical release.

At 3:57 a.m., the Board Operator called the security guard at the main entrance to ask the guard to call 9-1-1. The Board Operator communicated only those workers were missing and requested rescue. The guard, in turn, gave limited information to the 9-1-1 operator:

> 9-1-1 Operator: Harris County 911. What is the location of your emergency?
> Caller: This is DuPont on Strang Road .... They had an emergency out here. They got four people missing.... I was just called by the Lannate® supervisor [Board Operator] that he has four people missing.

9-1-1 Operator: So, are they somewhere there, or you don't know where they're at? Caller: I don't know, ma'am. I am up in the front. I just got the phone call.

9-1-1 Operator: Okay, you don't now their names or anything? Caller: No ma'am, just they needed rescue.

9-1-1 Operator: They need rescue? Caller: Yes ma'am.

9-1-1 Operator: And you don't know their names, no descriptions, nothing? Caller: No ma'am. I just know that they're workers.

At 3:58 a.m., site emergency responders from the ERT arrived at the scene with only their technical rescue gear. They quickly realized that there was an ongoing chemical release and that they needed additional PPE. At 4:05 a.m., the Incident Commander, who had responded to the initial rescue request, called for the ERT to come to the scene with bunker gear and SCBAs – the PPE necessary to enter an area with a toxic and flammable chemical release. When ERT members attempted to start the mini-pumper truck that contained SCBAs and radios; however, it would not start and could not make it to the incident scene.

The Incident Commander also requested safety data sheets for chemicals processed inside the manufacturing building, including methyl mercaptan. The ERT, however, did not initially use air monitoring equipment to determine what chemical was leaking. According to DuPont La Porte's Lannate emergency response plan, area personnel control air monitoring equipment for their unit. Because the Shift Supervisor (Process Coordinator) who oversaw process operations was a victim of the incident and was not available to fulfill the assigned role of providing chemical hazard and unit-specific information about each of the processes to the ERT during an emergency, air monitoring was delayed during the early hours of the incident. Meanwhile, site personnel recognized the release by odor and believed that methyl mercaptan was the chemical leaking inside the building.

At around 4:10 a.m., the Incident Commander established a hot zone around the manufacturing building. The hot zone's boundaries, however, were not clearly communicated or marked by the ERT.

At 4:12 a.m., an emergency operations center coordinator made a second 9-1-1 call, but he had not yet been informed by ERT personnel what chemical was leaking, or whether the leak posed a public threat, and was therefore unable to relay this critical information to the 9-1-1 operator:

9-1-1 Operator: Harris County 911. What is the location of your emergency?

Caller: This is in La Porte, Texas.... We have an emergency at the DuPont Plant in La Porte.... We have a possible casualty five, is what my medics are telling me. We have some injuries. We need La Porte EMS for transportation....

9-1-1 Operator: Okay, sir. How did this happen? Is it chemical-related?

Caller: I am not certain. I just know … they're doing a rescue—they're doing a rescue right now. We have some injured people. I'm not sure if there's any chemicals involved or not. I am just relaying the message.

Caller: I have just got some more information. We have five people unaccounted for. We have had a chemical release … I am not sure what chemical, in one of our buildings. Five people are unaccounted for. Rescue team is trying to reach them at this point. That's all the information I can give you right now.

…

9-1-1 Operator: Okay sir. And can you tell me, is this any risk to the public? Is it going to be a possible escaping from your premises?

Caller: No ma'am, it is not. 9-1-1 Operator: No threat? Caller: No ma'am.

9-1-1 Operator: And we don't know what kind of chemical it is Caller: No ma'am. As soon as I find out, I will let you know. I've got my team trying to determine that right now."

At about 4:15 a.m., ERT members assigned an operator (Operator 7) to bring them SCBAs because the ERT mini-pumper truck holding SCBAs could not respond to the scene. DuPont stored additional SCBAs outside a building closer to the manufacturing building.

Assisting the ERT as requested, Operator 7 went alone to retrieve these SCBAs and unknowingly walked into the path of the methyl mercaptan being released from the manufacturing building (Figure 3.6). The hot zone was not clearly identified or communicated to plant personnel. As a result, when Operator 7 went to retrieve the SCBAs, she was unaware that she was entering a potentially hazardous area. When Operator 7 went outside, though unable to smell methyl mercaptan, she felt ill. Realizing that she might be in a dangerous position, she retreated to the control room.

At about 4:20 a.m., Operator 3 regained consciousness and managed to exit the manufacturing building.

ERT members saw him near the manufacturing building and escorted him to safety.

Beginning at about 4:25 a.m., the first ERT entry team entered the manufacturing building. Although the ERT used SCBAs to protect themselves from toxic exposure, they did not monitor the atmosphere during this building entry to assess explosive hazards. Portions of the building likely had an explosive atmosphere during the incident, and air monitoring could have alerted personnel to this potential hazard during this entry. The ERT's post-incident critique stated "metering and [air] monitoring equipment were maintained by the unit. Since the technical knowledge base was incapacitated, there was a delay in establishing [air] monitoring."

During the building entry, an ERT member found drain valves from the waste gas vent header on the third floor open, with a gas (later identified as methyl mercaptan) flowing from them. The ERT members also found the Shift Supervisor on the floor of the third story, about 30 ft north of the drain valves. They also found Operator 6 next to his brother (Operator 2), and both men were unresponsive. Operator 2 had a five-minute escape respirator bag over his head. This respirator was one of the three escape packs brought into the building by Operator 6. Operator 6 had a 30-minute SCBA air bottle in front of him and the mask on his face, but he had not connected the mask to the air bottle. Methyl mercaptan likely incapacitated Operator 6 while he was trying to don his SCBA. A postmortem examination determined that the Shift Supervisor, Operator 2, and Operator 6 were killed by asphyxia through exposure to methyl mercaptan.

During the release, the ERT used a plume dispersion model to predict the methyl mercaptan plume (cloud) size and concentration profile releasing from the manufacturing building. To predict the methyl mercaptan plume effectively, the modeling software required an estimate of the methyl mercaptan release rate. The emergency response personnel, however, did not have an available method to estimate the methyl mercaptan release rate. Lacking other data sources and based on the perceived low odor of methyl mercaptan outside the manufacturing building, they estimated a release rate of 10 pb/h. With this input, the model predicted that methyl mercaptan would not leave DuPont property at concentrations harmful to the public. The CSB determined, however, that the emergency response personnel greatly underestimated the release rate. It was not possible for the CSB to determine the exact off-site methyl mercaptan concentration immediately after the release; however, the CSB concluded that the total release of about 24,000 lb of methyl mercaptan created the potential for a dangerous concentration of methyl mercaptan to have been released off-site.

Beginning at 5:08 a.m., three external firefighter groups arrived on-site, and at 5:15 a.m. the ERT conducted its second entry into the manufacturing building. On the third floor, an ERT responder closed the open drain valves from which he had observed vapor (methyl mercaptan) escaping. At about 5:40 a.m., emergency responders established and barricaded the hot zone. Until this point, the hot zone had not been marked clearly. At 5:57 a.m., the Incident Commander requested Harris County personnel to perform air monitoring off-site.

At 6:02 a.m., approximately three hours after the release started, DuPont La Porte personnel turned off the methyl mercaptan storage tank pump, which had been operating since the last startup attempt, before the release began (Figure 3.7). Shutting off this pump significantly slowed the methyl mercaptan release rate (Figure 3.8).

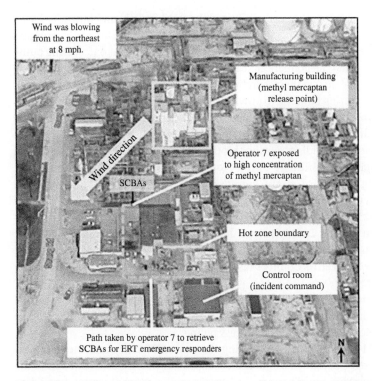

Figure 3.7 Layout of the plant and wind direction. Source: Chemical Safety Board (2019).

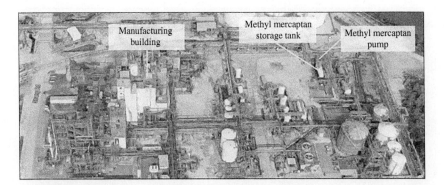

Figure 3.8 Location of the methyl mercaptan storage tank and pump, in relation to the manufacturing building. Source: Chemical Safety Board (2019).

From 6:10–9:30 a.m., the ERT conducted the third, fourth, and fifth ERT building entries to search for Operator 1, who was still missing. Due to the complex layout and emergency responders' lack of knowledge of the manufacturing building, the emergency responders had difficulty finding Operator 1. They had no maps to reference during their search, and DuPont La Porte did not have cameras in the manufacturing building to provide a view of the different floors of the building. To facilitate the search efforts, the ERT assigned area operators – including the Board Operator – to draw building maps to help emergency responders navigate the manufacturing building. During this time, ERT responders continued closing valves throughout the Lannate Unit to stop the release of methyl mercaptan.

At 6:14 a.m., a Harris County Sheriff's sergeant noticed an "intense smell" in Deer Park (west of La Porte), and at 6:30 a.m., he communicated, "the odor on [Highway] 225 [is] strong" (Figure 3.9).

At 8:07 a.m., Harris County's Hazardous Materials Team performed off-site air monitoring to measure the concentration of toxic methyl mercaptan outside of the DuPont fence line. Harris County's instrumentation did not detect methyl mercaptan in the air.

At 10:07 a.m. – seven hours after the release began – DuPont activated the methyl mercaptan tank emergency isolation valve to isolate the methyl mercaptan storage tank completely from the leaking process piping.

Between 11:15 and 11:55 a.m., during the sixth ERT entry into the manufacturing building, Operator 1 was found unresponsive in the north stairwell of the manufacturing building. The coroner later determined that Operator 1 had died from asphyxia and acute exposure to methyl mercaptan.

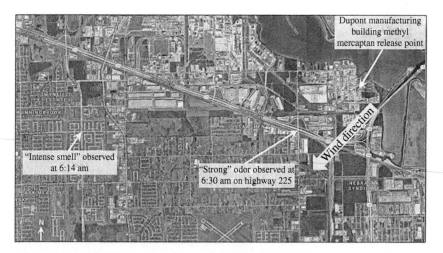

Figure 3.9 Location of plant in relation to the surrounding communities. Source: Chemical Safety Board (2019).

Highly toxic methyl mercaptan was released into the manufacturing building for about 40 minutes before operations personnel in the control room realized the extent of the emergency in the building, and for more than six additional hours before the release was fully controlled. The CSB identified factors that led to this delay in recognizing a major chemical release, including:

1. **The Board Operator was focused on process equipment high-pressure events that he believed were critical.** Operations personnel became aware of the high-pressure events when insecticide manufacturing equipment venting into the waste gas vent header began to exhibit high-pressure alarms. At this point, operations personnel shifted their attention from the problem of plugging in the methyl mercaptan feed system to the current problem of high pressure in the waste gas vent header piping to the NRS incinerator. DuPont La Porte operations staff did not correlate the high-pressure problem with the troubleshooting efforts to unplug the methyl mercaptan feed piping through hydrate dissociation. Rather, they attributed the high-pressure events to the routine problem of liquid accumulation in the waste gas vent header piping. The Board Operator directed personnel to help troubleshoot the high-pressure problem. The operating pressure of various pieces of process equipment inside the manufacturing building, however, continued to increase, in some equipment exceeding the high-scale detection limits of process instrumentation. The Board Operator was concerned about the potential for equipment overpressure and the release of toxic chemicals to the atmosphere outside the manufacturing building. Even though methyl mercaptan detectors were beginning to alarm, the Board Operator either was unaware of them or had normalized them, not realizing the alarms were alerting him that a significant release had started to occur inside the building. Unaware that a significant methyl mercaptan release had begun, the Board Operator remained focused on controlling the high pressure he was observing inside the equipment.

   Lannate Unit personnel accepted methyl mercaptan releases and the associated alarms as normal because of methyl mercaptan's low odor threshold, a history of frequent alarms, and a lack of hazard recognition that methyl mercaptan could be lethal.

   Methyl mercaptan is a highly toxic chemical, but its odor can be detected at very low concentrations that are not considered a health risk to people. Additionally, the Lannate Unit used methyl isocyanate (MIC) and chlorine, which are generally accepted as being more hazardous (toxic) than methyl mercaptan. MIC has an LD50 of 120 mg/kg and an IDLH level of 20 ppm. Chlorine has an LD50 of 850 mg/kg. As a result, even though Lannate Unit personnel were trained on methyl mercaptan toxicity hazards, personnel at the site became accustomed to smelling methyl mercaptan daily without experiencing negative health effects, leading workers to become less wary of methyl mercaptan leaks and the risks associated with them. According to one DuPont worker, there was a perception that methyl mercaptan (which workers called "MeSH") was not dangerous:

> And I smell MeSH all the time … but it never … I smelled MeSH, but I never felt I couldn't breathe. Like whether it's, there's—whether it's a small leak or anything, it just never—I didn't think MeSH was that dangerous. I thought it, you know, but it is.

During the troubleshooting efforts before the major release began, smaller releases of methyl mercaptan had caused the site to have a strong odor of methyl mercaptan. Therefore, when the major release began, site personnel did not realize there was a major release, in part because they were accustomed to smelling the chemical.

2. **Personnel were not wearing personal methyl mercaptan detectors.** DuPont La Porte had some personal methyl mercaptan detectors available, but they were never issued to operators to wear in the field. Such detectors could have warned Operator 2, Operator 3, and Operator 4 (personnel who initially responded to Operator 1's distress call) of the high methyl mercaptan concentration inside the manufacturing building.

3. **DuPont La Porte did not have audible or visual alarms on or in the manufacturing building to alert field personnel of a methyl mercaptan release automatically.** DuPont La Porte relied on verbal communication of alarms, which automatically displayed on a continuously manned control board.

The delay in recognizing this event as a major toxic chemical release that was becoming a deadly concentration of methyl mercaptan inside the building contributed to the response of Operator 2, Operator 3, and Operator 4 to Operator 1's distress call without respiratory protection. According to DuPont La Porte's Lannate emergency planning and response manual, if a sensor detected a methyl mercaptan leak and initiated an alarm, personnel were required to notify a shift supervisor, sound a plantwide fume release alarm, and wear appropriate PPE (including SCBA) during any attempts to stop the leak. The manufacturing building, however, including the building's overall construction, detectors, and alarms, was not designed such that personnel could be readily notified of a major chemical release in the manufacturing building. Without being made aware of the major chemical release inside the building, Operator 2, Operator 3, and Operator 4 entered without respiratory protection. As a result, Operator 2 was killed by methyl mercaptan exposure, and Operator 3 and Operator 4 were exposed to dangerous concentrations of methyl mercaptan.

On the night of the incident, the Board Operator's request for the ERT to come to the scene triggered the start of emergency response operations. From this initial request for ERT aid to the final response activities, the La Porte site's emergency response efforts were characterized by miscommunication, disorganization, and a lack of situational awareness. The following key emergency response weaknesses are discussed in this section:

- Based on DuPont La Porte's emergency response plan, emergency response personnel relied on the expertise of the shift supervisor, who was the designated process coordinator. The Shift Supervisor, however, was a victim of the incident, and no one else on the shift was designated to backfill the role of process coordinator. This gap led to weaknesses in several aspects of the emergency response operations, including the initial assessment of the problem and of the resources needed for an effective response.
- There was a delay in ERT deployment and readiness to respond to the release, due to both a misunderstanding of the type of emergency and the failure of an ERT mini-pumper truck that stored SCBAs to start.
  Searching for the missing workers put the ERT responders at high risk. ERT responders entered a hazardous and potentially explosive atmosphere without performing air monitoring. Additionally – unbeknownst to them – they entered an area of the manufacturing building that DuPont La Porte had previously identified as a collapse risk in the event of an explosion.
- DuPont La Porte's emergency planning and response manual lacked a building map or floor plans to aid emergency responders in navigating the manufacturing building. As a result, the ERT experienced difficulties navigating the manufacturing building while searching for a missing operator; they resorted to assigning unit operators, including the Board Operator, to draw maps of the building to help guide them. This assignment diverted the Board Operator's attention from other critical tasks, such as viewing process data on the control room computer screen to try to identify the source of the release.
- The methyl mercaptan release was not controlled until three hours after it began. Personnel did not evaluate process data to determine the cause of the release until hours into the response efforts. Analysis of process data from the methyl mercaptan storage tank earlier during the release may have revealed that the methyl mercaptan storage tank level was continually dropping and that isolating the tank could stop the release.
- Emergency responders established a hot zone that was not clearly marked or communicated, thereby allowing an operator to be potentially exposed to a dangerous concentration of methyl mercaptan when tasked to assist the ERT.
- The ERT underestimated the quantity of toxic methyl mercaptan released. Because DuPont La Porte lacked fence-line monitors and the ERT did not

conduct air monitoring to determine the concentration of methyl mercaptan leaving the site, the ERT did not have accurate values to use in its dispersion model. Instead, ERT members estimated a small quantity of methyl mercaptan based on verbal descriptions of the release's downwind odor. As a result, the incident command did not issue warnings to the community surrounding the La Porte facility to take protective actions from the possibility of toxic concentration of methyl mercaptan exiting the site during the release. Furthermore, because air monitoring had not occurred during the release, neither DuPont La Porte nor the CSB can pinpoint the exact methyl mercaptan concentration that left the site, or what the potential health risk was to the public, if any.

- DuPont has long been perceived as a safety leader and has asserted that its history of process safety extends over two centuries, from its gunpowder manufacturing through its present-day production of chemicals. Over the course of this time, DuPont's PSM system evolved with the emergence of federal regulations, such as the PSM standard and the RMP rule, as well as the decision to engage in Responsible Care.

DuPont developed several methods that it uses to manage and integrate its process safety systems, including its proprietary Safety Perception Surveys and PSM and Risk Model. Its corporate process PSM and implementation process involves four steps: (i) building a safety culture, (ii) management leadership and commitment, (iii) implementing a comprehensive PSM program, and (iv) operational discipline [54, p. 2]. According to DuPont, the effective implementation of these steps will result in a robust PSM system [55, p. 10]. Consequently, DuPont La Porte used both the Safety Perception Surveys and the Process Safety Management and Risk Model in implementing its PSM system.

Despite having robust written process PSM (i.e. MOC and PHA), over the course of five years, DuPont had major process safety incidents – including those at its Belle, West Virginia; Buffalo, New York; and La Porte, Texas sites, each of which resulted in fatalities. The CSB found that DuPont La Porte's ineffective implementation of its integrated PSM system contributed to the severity of the highly toxic material release in the manufacturing building at La Porte.

DuPont's corporate PSM standard lists four key steps for the implementation and sustainability of an effective PSM program:

1. Establishing an effective and sustainable unified safety culture,
2. Providing management leadership and commitment,
3. Implementing a comprehensive PSM program, including organizational learning and continuous improvement to integrate new risk management knowledge, learnings, and practices into existing site and corporate systems, and
4. Achieving operational excellence through operational discipline.

This section breaks down these steps into two parts: (i) Building a Safety Culture (step one); and (ii) Using the PSM and Risk Model (steps two through four).

DuPont's corporate PSM standard states that process safety culture a "determines the manner by which PSM is implemented and managed at each site as part of both individual and group values and behaviors to enable sound decision making and continuous improvement." Therefore, DuPont links the effectiveness of PSM implementation to the process safety culture of a facility.

Moreover, previous CSB investigations also support DuPont's corporate standard that links effective PSM systems with a strong process safety culture.

According to DuPont, establishing a safety culture is the first step in developing a PSM system because a system will be only as effective as its safety culture permits. Additionally, DuPont's corporate PSM standard recommends periodic site evaluations of process safety culture. DuPont's standard further recommends that sites use the results of process safety culture assessments to highlight strengths and develop potential improvement strategies.

In Essential Practices for Creating, Strengthening, and Sustaining Process Safety Culture, CCPS states,

> A strong, positive process safety culture enables the facility's [PSM system] to perform at its best. This gives the facility its best chance to prevent catastrophic fires, explosions, toxic releases, and major environmental damage
> [56, p. 2].

For more than a decade, DuPont developed and honed a safety culture program to reduce its OSHA total recordable injury rate by assessing and improving occupational (personal) safety, using its Bradley Curve and Safety Perception Survey tools. These tools, however, focus on only one aspect of safety culture – personal safety.

DuPont La Porte used the Safety Perception Survey without doing a process safety culture assessment as recommended in the corporate PSM standard. Because this survey did not formally assess process safety culture perceptions, DuPont La Porte never evaluated its process safety culture. Had its efforts included a focus on perceptions of process safety as well as personal safety in its Safety Perception Survey, or had it performed a separate process safety culture assessment with the intent of improving process safety culture as recognized in DuPont's corporate PSM standard, DuPont La Porte likely would have been more aware of potential process safety issues and better positioned to prevent or mitigate future process safety incidents. CCPS recognizes the importance of a strong process safety culture in preventing major accidents.

DuPont La Porte did not use any type of robust, formal process safety culture assessment. Process safety culture can affect whether a site has a sense of vulnerability in terms of PSM risks; complacency or overconfidence; and transparent, timely, and thorough responses to PSM concerns, action items, and issues, including leadership measures to prevent a "check-the-box" mentality (i.e. simply accomplishing a task as the objective rather than ensuring a high degree of focus on risk management and prevention). The Safety Perception Surveys conducted at the DuPont La Porte facility before the November 2014 incident were designed to lower OSHA total recordable injury rates. DuPont did not intend for these surveys to measure or address the perception of process safety performance.

As part of DuPont La Porte's 2012 Safety Perception Survey, comments were collected and organized by job position. Some of these comments, from different levels of the organization, raised safety culture concerns. For example, one hourly worker (operator) stated,

> Some fi[rs]t line [supervisors] turn the other way if a worker put[s] production over safety.

A professional employee expressed concerns about PSM initiatives at the site:

> Some days it seems like the focus is more about creating documentation and reviewing that documentation than on targeting those areas that will have the biggest impact on improved PSM performance. From my perspective, I feel like "real" PSM has suffered in some respects and that our processes may actually be less safe than they were before the initiatives.

Safety concerns were expressed by multiple supervisors in the survey comments:

> I hate to say it but not all employees put the same value on safety. That includes [wage roll] employees and first line supervisors. DuPont employee actions demonstrate to me that safety rules are suggestions and do not have to be followed.
>
> DuPont has always been a strong leader in safety principles for the industry that we work in. The [La Porte] plant has a strong leadership for safety yet sometimes the message does not always get across to the employe[es]. Some still feel that production is above safety yet all managers have stated and demonstrated that production is far down the list when it comes to safety.
>
> I believe our 'passion' for safety has slipped over the past 10 years. I believe this has impacted [our] ability to consistently achieve safety excellence. I interact with safety professionals from many companies in our area and, as a company, we are no longer the safety leader we were once thought of as.

Despite concerns about its safety culture expressed by multiple levels of personnel within the organization, DuPont La Porte did not use these comments to take corrective actions that would institute process safety culture change at the DuPont La Porte facility.

In 2014, DuPont La Porte hired an organizational consultant to increase production in the Lannate Unit. Safety concerns resembling the ones from the 2012 Safety Perception Survey were raised again. According to an interview of a front-line supervisor in July 2014, four months before the incident,

> …we just are not as good as we think we are…. not in pay, benefits, safety, etc.… We used to sell the safety program. We spend more time on driving incidents down, to keep the number down. We just have to make it look good. Don't need to report every detail. Long slow slide—may not even notice how far we have fallen.

This supervisor also noted in that interview that DuPont La Porte was at its most vulnerable point because younger engineers were most impressed by improved production, and "safety first" was not always the practice.

Had DuPont La Porte formally assessed its process safety culture and taken effective corrective actions, cultural weakness that contributed to significant and long-standing process safety program deficiencies, which the CSB determined were causal to the November 15, 2014, incident (listed below), may have been identified, communicated, and addressed. These were:

- **Poor hazard analysis practices.** Site personnel did not perform a hazard analysis on the troubleshooting techniques that they used leading up to the incident. Additionally, the DuPont La Porte site did not adequately analyze hazards associated with the manufacturing building's ventilation system in PHAs. A PHA on the ventilation system could have evaluated the ability of the fans to handle toxic releases and established more robust management systems to protect workers when a fan was broken.
- **Poor design of the manufacturing building.** DuPont housed the Lannate process inside a manufacturing building. The building design increased hazards for the workforce by confining toxic chemicals indoors. The building's ventilation design also did not consider chemical toxicity hazards inside the building.
- **Failure to maintain safety-critical equipment.** Two "PSM critical" ventilation fans were not operational at the time of the incident [1, pp. 33-35]. DuPont recognized that a ventilation fan breakdown could result in a high-consequence event. Both fans, however, exhibited poor reliability. One fan had been down since June 2014, five months prior to the incident, due to an electrical problem. On October 20, 2014, the other fan was making a noise significant enough that DuPont operators turned it off and wrote an "urgent" work order to have

it repaired. Even had the fans worked, they probably would not have prevented a lethal atmosphere inside the building due to the large amount of toxic gas released.

- **Failure to develop written procedures for operator actions.** DuPont La Porte did not establish a formal procedure for draining liquid from the waste gas vent header piping, aside from its line-breaking policy (Section 6.5). In addition, when the DuPont Technical Team met to discuss how to clear the plugging in the piping system, they did not develop a written procedure for the techniques used to clear plugging in equipment, even though the plugging formed where it had never been before.
- **Lack of hazard recognition.** The methyl mercaptan detectors installed in the manufacturing building were not intended for worker safety – the alarms were set to alarm at 25 parts per million (ppm), which was above OSHA's ceiling limit of 10 ppm. Operators also normalized smelling methyl mercaptan because of its low odor threshold and perceived lack of negative health impacts at those low concentrations, even though it is a toxic chemical. Although formally trained, numerous employees lacked a working knowledge of methyl mercaptan's hazards and did not recognize that the chemical could be both lethal and explosive. As a result, operators often confirmed methyl mercaptan leaks by purposely smelling for the chemical's odor without understanding the dangers of this practice. Furthermore, this method of leak confirmation is a poor, unsafe, and unreliable practice that could have been replaced by using portable monitors set at appropriate levels.

3.2.2 DuPont's Initiation of Process Safety Culture Assessments

DuPont La Porte's safety culture assessment program (the Safety Perception Survey) before the November 2014 incident focused on lowering its OSHA total recordable injury rate. Even as the industry concentrated on improving process safety following the 2005 BP Texas City incident, and DuPont's consultant arm developed process safety questions for the survey, DuPont La Porte lagged and did not incorporate DuPont's process safety questions into its Safety Perception Survey. Furthermore, DuPont La Porte did not formally assess aspects of its culture beyond the Safety Perception Survey.

In 2017, however, DuPont required each site to assess its process safety culture periodically and added process safety questions to the Safety Perception Surveys. This shift in incorporating process safety culture questions in its Safety Perception Surveys is important because, as explained because the accuracy of a site's culture assessment may be called into question if all aspects of culture are not considered as part of the assessment. Furthermore, DuPont practice has become more aligned with its corporate philosophy that a PSM program is only as effective as its safety culture. The CSB views the adoption of process safety culture questions in DuPont's Safety Perception Survey as a positive development.

The DuPont La Porte incident was the CSB's third investigation of a fatal DuPont incident in five years. On September 30, 2015, the CSB issued a <u>presentation</u>, a <u>safety video</u>, and an *Interim Recommendations* report [1], which detailed needed safety improvements at the DuPont La Porte facility and issued recommendations for implementation before resuming operations of the Lannate Unit [2]. Due to the amount of methyl mercaptan stored at the DuPont La Porte site, the Lannate process was covered by the PSM standard of the U.S. Occupational Safety and Health Administration (OSHA) and the Risk Management Plan (RMP) rule of the US Environmental Protection Agency (EPA). In the Interim Recommendations, the CSB identified significant PSM deficiencies at the La Porte facility, including delayed maintenance of safety-critical equipment, lack of written procedures, poor hazard analysis practices, lack of implementation of important inherently safer design concepts, and a lack of hazard recognition.

In response to the CSB findings and recommendations, DuPont committed to making changes at the La Porte facility. DuPont La Porte, however, was not able to address all these findings and recommendations, because in the spring of 2016 DuPont announced that it would not reopen the Lannate Unit. The company had determined that "significant changes in market conditions during the period of the shutdown [would] persist over the long term and that the cost required for the restart was not a long-term viable and cost-efficient option for the DuPont Crop Protection business." In 2017, DuPont dismantled and removed from the DuPont La Porte facility the buildings and equipment that had been associated with its insecticide and herbicide business units, including the Lannate Unit.

Despite the closure of the insecticide and herbicide units, the CSB determined that the DuPont La Porte incident offers important lessons for the chemical industry relating to the following areas:

- ***Emergency Response.*** The emergency response efforts at the DuPont La Porte facility during the toxic chemical release were disorganized and placed at risk operators, emergency responders, and potentially the public. Chemical plants need a robust emergency response program to mitigate emergencies and to protect the health of workers, emergency responders, and the public.
- ***DuPont's PSM System.*** DuPont created its own corporate PSM system that integrated its internal safety requirements with those of the American Chemistry Council's Responsible Care® program and those required by regulations under the EPA RMP rule and the OSHA PSM standard. Over the course of five years, despite implementing its corporate PSM system, DuPont experienced three major process safety incidents.

In addition to developing an integrated PSM system, DuPont established a program to implement it. According to DuPont, the implementation of the PSM system can be broken down into two parts: (i) Building a Safety Culture and (ii) Using the PSM and Risk Model. Although DuPont's corporate standard recommended that its sites assess their process safety culture, DuPont La Porte had not formally evaluated process safety culture at its facility before the November 2014 incident. DuPont La Porte used a proprietary Safety Perception Survey that focused on personal or occupational safety but did not evaluate or assess the process safety culture. Because the Safety Perception Survey did not reasonably evaluate all safety aspects of culture, it could not help identifying the significant process safety weaknesses at the DuPont La Porte facility, leaving the site vulnerable to potential process safety incidents. While measuring worker perceptions of personal safety is important, a safety culture assessment program should also provide an effective gauge of process safety. The second tool that DuPont used to implement its PSM system was the PSM and Risk Model, the company's visual representation of an effective PSM system, showing the implementation and interaction of Management Leadership and Commitment, Comprehensive PSM Program, and Operational Discipline.

- *DuPont La Porte's Process Safety Management Deficiencies.* The CSB identified significant process safety deficiencies at the DuPont La Porte site that contributed to the incident. DuPont's corporate PSM system did not identify, prevent, or mitigate these deficiencies. A company must effectively implement a PSM system and its corresponding programs to reap the accompanying process safety benefits.
- *DuPont La Porte's Employee Incentive Program.* The DuPont La Porte bonus structure may have disincentivized workers from reporting injuries, incidents, and "near misses." Ensuring that employees can report injuries or incidents in accordance with the Occupational Safety and Health (OSH)Act and OSHA regulations, without fear of discrimination, retaliation, or other adverse consequence is central to protecting worker safety and health, and aiding accident prevention.

3.2.3 Summary of Safety Culture Findings

There are so many safety culture issues that came together in this event to end up with a major release or methyl mercaptan and the five fatalities. The complete CSB report details more of the actions or lack of actions DuPont, the shift supervisor, and operators took that led to the release and the deaths. The major issues in this event were:

- the lack of proper environmental air sampling,
- the lack of understanding on the part of the operators concerning the toxic nature of methyl mercaptan,
- the improper storage of SCBAs,

- the shift supervisor entering the plant and turning valves without understanding the potential consequences,
- the lack of a proper PSM,
- operators responding to an emergency without proper PPE,
- lack of adequate communication between operators, plant staff, and outside emergency response,
- and, operators responding to emergencies without proper PPE over the years have resulted to numerous deaths.

In many emergency events fellow workers respond without taking adequate precautions and they wind up being victims also. This event is a perfect example. A mine fire that happened in Siberia killed 52 miners and rescuers on November 25, 2021 (NPR 2021). The rescuers died trying to save and/or recover injured or dead miners. Unfortunately, this recent incident is a good example of what happens when emergency personnel die in their duty to rescue others. A NIOSH study found that 60% of confined space deaths were rescuers (Koester 2018; NIOSH 2021).

Overall, NIOSH investigations of confined space incidents found:

- 85% of the time a SUPERVISOR was present.
- 29% of the dead were SUPERVISORS.
- 31% had WRITTEN Confined Space Entry PROCEDURES.
- 0% used the WRITTEN PROCEDURES.
- 15% had Confined Space TRAINING.
- 0% had a RESCUE PLAN.
- 60% of "WOULD-BE" RESCUERS died.
- 95% were AUTHORIZED by supervision.
- 0% of the spaces were TESTED prior to entry.
- 0% were VENTILATED.

People mostly want to help, but they need to be trained to do so safely.

3.3 BP Texas City Refinery Explosion – March 23, 2005

3.3.1 Introduction

The BP explosion at the Texas City refinery was thoroughly investigated by the OSHA and by the Chemical Safety Board (CSB 2007). The following presents the highlights of that investigation, and the safety culture attributes that contributed to the accident.

On March 23, 2005, at 1:20 p.m., the BP Texas City Refinery suffered one of the worst industrial disasters in recent US history. Explosions and fires killed 15 people and injured another 180, alarmed the community, and resulted in financial losses exceeding US$1.5 billion. The incident occurred during the startup of an isomerization (ISOM) unit when a raffinate splitter tower was overfilled; pressure relief devices opened, resulting in a flammable liquid geyser from a blowdown stack that was not equipped with a flare. The release of flammables led to an explosion and fire. All the fatalities occurred in or near office trailers located close to the blowdown drum. A shelter-in-place order was issued that required 43,000 people to remain indoors. Houses were damaged as far away as three-quarters of a mile from the refinery.

3.3.2 Texas City

Texas City is a city in Galveston County in the US state of Texas. Located on the southwest shoreline of Galveston Bay, Texas City, is a busy deep-water port on Texas' Gulf Coast. Its history starts in the late 1800s. The area was originally called Shoal Point. In the 1890s a trio of brothers saw the potential for the area to become a port and convinced a group of investors to put up money to buy 10,000 acres. Once purchased they renamed the area Texas City. The investors in the 1890s formed the Texas City Improvement Company (TCIC) and began to layout the future city. The city was officially incorporated in 1911. Many improvements to the infrastructure of the area occurred in those early years. These included dredging a shipping channel, bringing in the railroad, and having utilities brought into the future city. The city also became the home of the Second Division of the Army that was stationed there in 1913 to guard the coast during the Mexican Revolution. In addition, the Wright Brothers trained military pilots in Texas City and the city claims to birthplace of the Airforce. Texas City began to have refineries 1908 and soon had three.

Texas City has been home to many disasters. These include natural disasters in the form of hurricanes and man-made disaster in the form of infamous explosions. Disastrous hurricanes have struck Texas City numerous times. Researching this topic, found too many hurricanes to list in this chapter.

The most notable man-made disaster prior to the 2005 BP event, occurred on April 16, 1947 when the French Ship, Grandcamp, loaded with Ammonium Nitrate exploded (Scher 2007). It is considered the worst disasters in Texas. The French-owned vessel, carrying explosive ammonium nitrate produced during wartime for explosives and later recycled as fertilizer, caught fire early in the morning, and while attempts were being made to extinguish the fire, the ship exploded. The entire dock area was destroyed, along with the nearby Monsanto Chemical Company, other smaller companies, grain warehouses, and numerous oil and chemical storage tanks. Smaller explosions and fires were ignited by flying debris, not only along the industrial area, but throughout the city. Fragments of iron, parts of the ship's cargo, and dock equipment were hurled into businesses, houses, and public buildings. A fifteen-foot tidal wave caused by the force swept the dock area. The concussion of the explosion, felt as far away as Port Arthur, damaged or destroyed at least 1000 residences and buildings throughout Texas City. The ship SS High Flyer, in dock for repairs and also carrying ammonium nitrate, was ignited by the first explosion; it was towed 100 ft from the docks before it exploded about sixteen hours later, at 1:10 a.m. on April 17. The first explosion had killed twenty-six Texas City firemen and destroyed all of the city's fire-fighting equipment, including four trucks, leaving the city helpless in the wake of the second explosion. No central disaster organization had been established by the city, but most of the chemical and oil plants had disaster plans that were quickly activated. Although power and water were cut off, hundreds of local volunteers began fighting the fires and doing rescue work. Red Cross personnel and other volunteers from surrounding cities responded with assistance until almost 4000 workers were operating; temporary hospitals, morgues, and shelters were set up.

Probably the exact number of people killed will never be known, although the ship's anchor monument records 576 persons known dead, 398 of whom were identified, and 178 listed as missing. All records of personnel and payrolls of the Monsanto Company were destroyed, and many of the dock workers were itinerants and thus difficult to identify. Almost all persons in the dock area-firemen, ships' crews, and spectators-were killed, and most of the bodies were never recovered; sixty-three bodies were buried unidentified. The number of injured ranged in the thousands, and loss of property totaled about US$67 million. Litigation over the Texas City disaster was finally settled in 1962, when the US Supreme Court refused to review an appeals court ruling that the Republic of France, owner of the Grandcamp, could not be held liable for any claims resulting from the explosion. The disaster brought changes in chemical manufacturing and new regulations for the bagging, handling, and shipping of chemicals. More than 3000 lawsuits involving the United States government, since the chemicals had originated in US ordnance plants, were resolved by 1956, when a special act passed by Congress settled all claims for a total of US$16.5 million. Some temporary housing was built and donated to the city, and other housing, docks, warehouses, and chemical plants were rebuilt by 1950. Public commemoration of the event began in June of 1947, when the bodies of the unidentified dead were buried together in a memorial cemetery and park, and in 1991 a new section was added to the park.

3.3.3 Description of the BP Refinery

The enormous BP Amoco Oil Refinery in Texas City was built in 1934 by a company called Pan-American Refinery, which was at the time a subsidiary of Standard

Oil of Indiana (CSB 2007; Pop History Dig 2005). The success of the refinery and the jobs it provided caused the population of Texas City to skyrocket.

During the World War II, the port and the refinery became very important to the war effort, as they manufactured materials for the military. Pan-American Refinery eventually became Amoco, which ran the Texas City refinery for many years. However, the company failed to make many safety improvements to the facility. In 1991, they opted against replacing outdated blowout drums to save money. BP acquired Amoco for US$61 billion in 1999. Like Amoco, BP avoided renovations to cut costs. The Texas City refinery became part of BP with the 1998 merger with Amoco. It is a large, highly complex refinery processing 475,000 barrels of oil per day. The refinery employed approximately 2150 BP staff and contractor numbers that varied between 1000 and 3000 each day.

BP sold the refinery to Marathon Petroleum Corporation in 2012 for approximately US$2.4 billion. The sale of its Texas City, Texas refinery and a portion of its retail and logistics network in the Southeast United States to Marathon Petroleum Corporation for an estimated US$2.4 billion. This comprises approximately US$0.6 billion in cash, US$1.1 billion for the estimated value of hydrocarbon inventory and an earn-out arrangement payable over six years of US$0.7 billion (based on assumed future margins and refinery throughput).

BP stated that "The completion of this divestment is a major milestone in the refocusing of our U.S. fuels portfolio," said Iain Conn, chief executive of BP's global refining and marketing business. "Together with the sale of our Carson, California refinery, which we also expect to close this year, the divestment of Texas City allows us to focus BP's U.S. fuels investments on our three northern refineries, which are crude feedstock advantaged, and their associated marketing businesses."

Subsequently, Marathon sued BP stating that the plant was in disrepair. Marathon Petroleum Corp alleged BP Plc failed to deliver a Texas oil refinery and three products terminals in the condition promised under a US$2.4-billion sales agreement signed in 2012, according to a federal lawsuit (BP 2013).

Marathon took over the refinery in Texas City, Texas, and terminals when the transaction closed on Feb. 1, 2013, and began finding problems that breached the sale agreement, according to the lawsuit.

"After assuming operation of the refinery, Marathon Petroleum discovered that, in numerous respects, the refinery and the terminals were not in compliance with environmental laws," according to the lawsuit.

BP also failed to maintain PSI on 3756 pressure vessels at the refinery, Marathon alleged.

BP began a project in 2010 to compile the PSI on the pressure vessels, "but it abandoned this project in April 2012 having only completed documentation for 555 of the vessels," according to the lawsuit.

Marathon also said BP planned to carry out an overhaul of an aromatics recovery unit prior to the sale being complete but did not do so after signing the sale agreement, according to the lawsuit.

Marathon also wanted to terminate payments, as of July 31, 2016, it has been making to BP for servicing retail stations Marathon received as part of the sale, according to the suit (ref).

3.3.4 The Accident

On the morning of March 23, 2005, the raffinate splitter tower in the refinery's ISOM unit was restarted after a maintenance outage. The ISOM plant isomerization at the site was designed for the conversion of low octane hydrocarbons, through various chemical processes, into hydrocarbons with higher octane ratings that could then be blended into unleaded gasoline. One component of this ISOM site was a unit called the raffinate splitter. Figure 3.10 shows the diagram of the plant. Figures 3.11–3.13 show other critical plant components involved in the explosion. When operational, this 170-ft (50 m) tall tower was used to separate out lighter hydrocarbon components from the top of the tower (mainly pentane and hexane), which condensed and were then pumped to the light raffinate storage tank, while the heavier components were recovered lower down in the splitter, then pumped to a heavy raffinate storage tank. It had an operational capacity of 45,000 barrels (7200 m^3) per day. During the startup, operations personnel pumped flammable liquid hydrocarbons into the tower for over three hours without any liquid being removed. This was contrary to startup procedure instructions. Critical alarms and control instrumentation provided false indications that failed to alert the operators of the high level in the tower.

Consequently, unknown to the operations crew, the 170-ft (52-m) tall tower was overfilled, and liquid overflowed into the overhead pipe at the top of the tower.

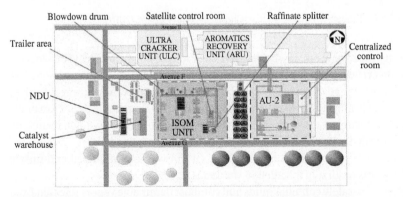

Figure 3.10 Diagram of the plant. Source: Chemical Safety Board (2007).

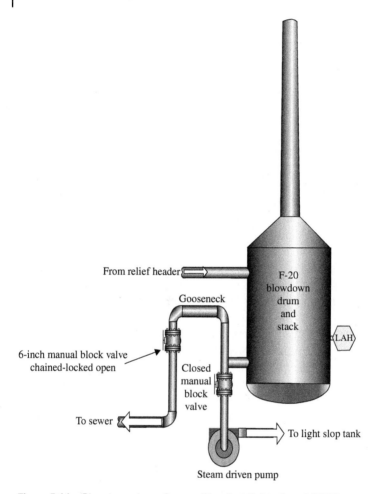

Figure 3.11 Blowdown drum. Source: Chemical Safety Board (2007).

The overhead pipe ran down the side of the tower to pressure relief valves located 148 ft (45 m) below. As the pipe filled with liquid, the pressure at the bottom rose rapidly from about 21 lb/in. (psi) to about 64 psi. The three pressure relief valves opened for six minutes, discharging a large quantity of flammable liquid to a blowdown drum with a vent stack open to the atmosphere. The blowdown drum and stack overfilled with flammable liquid, which led to a geyser-like release out the 113-foot (34 m) tall stack. This blowdown system was an antiquated and unsafe design; it was originally installed in the 1950s and had never been connected to a flare system to safely contain liquids and combust flammable vapors released from the process.

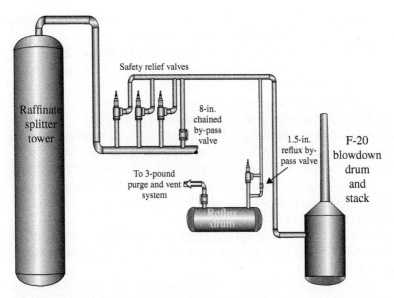

Figure 3.12 Diagram of relationship of reflux drum to blowdown drum. Source: Chemical Safety Board (2007).

1. The released volatile liquid evaporated as it fell to the ground and formed a flammable vapor cloud. The most likely source of ignition for the vapor cloud was backfire from an idling diesel pickup truck located about 25 ft (7.6 m) from the blowdown drum (Figure 3.11). Figure 3.12 shows the calculated overpressure wave caused by the ignition of the vapor. In contrary, 40 psi can be produced by 20 kg TNT exploding 6 meters away. In this situation 50% or more

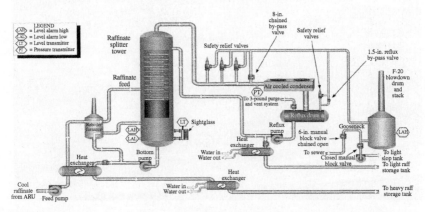

Figure 3.13 Drawing of raffinate process. Source: Chemical Safety Board (2007).

of casualties will sustain pulmonary damage with pressures of 80 psi or more, with overpressures of 200 psi being uniformly fatal in open-air blasts (blast).

The 15 employees killed in the explosion were contractors The ISOM startup procedure required that the level control valve on the raffinate splitter tower be used to send liquid from the tower to storage. However, this valve was closed by an operator and the tower was filled for over three hours without any liquid being removed. This led to flooding of the tower and high pressure, which activated relief valves that discharged flammable liquid to the blowdown system. Underlying factors involved in overfilling the tower included:

- The tower level indicator showed that the tower level was declining when it was overfilling. The redundant high-level alarm did not activate, and the tower was not equipped with any other level indications or automatic safety devices (Figure 3.14 and 3.15).
- The control board display did not provide adequate information on the imbalance of flows in and out of the tower to alert the operators to the dangerously high level.
- A lack of supervisory oversight and technically trained personnel during the startup, an especially hazardous period, was an omission contrary to BP safety guidelines. An extra board operator was not assigned to assist, despite

Figure 3.14 Remains of the pickup truck. Source: Chemical Safety Board (2007).

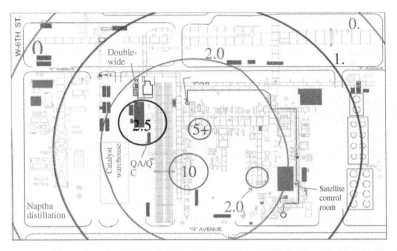

Figure 3.15 Calculated over pressure wave. Source: Chemical Safety Board (2007).

    a staffing assessment that recommended an additional board operator for all ISOM startups.

- Supervisors and operators poorly communicated critical information regarding the startup during the shift turnover; BP did not have a shift turnover communication requirement for its operations staff.
- ISOM operators were likely fatigued from working 12-hour shifts for 29 or more consecutive days.
- The operator training program was inadequate. The central training department staff had been reduced from 28 to 8, and simulators were unavailable for operators to practice handling abnormal situations, including infrequent and high hazard operations such as startups and unit upsets.
- Outdated and ineffective procedures did not address recurring operational problems during startup, leading operators to believe that procedures could be altered or did not have to be followed during the startup process.

2. The process unit was started despite previously reported malfunctions of the tower level indicator, level sight glass, and a pressure control valve.
3. The size of the blowdown drum was insufficient to contain the liquid sent to it by the pressure relief valves. The blowdown drum overfilled, and the stack vented flammable liquid to the atmosphere, which fell to the ground and formed a vapor cloud that ignited. A relief valve system safety study had not been completed.
4. Neither Amoco nor BP replaced blowdown drums and atmospheric stacks, even though a series of incidents warned that this equipment was unsafe. In 1992, OSHA cited a similar blowdown drum and stack as unsafe, but the

citation was withdrawn as part of a settlement agreement and therefore the drum was not connected to a flare as recommended. Amoco, and later BP, had safety standards requiring that blowdown stacks be replaced with equipment such as a flare when major modifications were made. In 1997, a major modification replaced the ISOM blowdown drum and stack with similar equipment, but Amoco did not connect it to a flare. In 2002, BP engineers proposed connecting the ISOM blowdown system to a flare, but a less expensive option was chosen.

5. Occupied trailers were sited too close to a process unit handling highly hazardous material. All fatalities occurred in or around the trailers.

6. In the years prior to the incident, eight serious releases of flammable material from the ISOM blowdown stack had occurred, and most ISOM startups experienced high liquid levels in the splitter tower. Neither Amoco nor BP investigated these events.

7. BP Texas City managers did not effectively implement their pre-startup safety review policy to ensure that nonessential personnel were removed from areas in and around process units during startups, an especially hazardous time in operations.

8. Cost-cutting, failure to invest and production pressures from BP Group executive managers impaired PSM at Texas City.

9. The BP Board of Directors did not provide effective oversight of BP's safety culture and major accident prevention programs. The Board did not have a member responsible for assessing and verifying the performance of BP's major accident hazard prevention programs.

10. Reliance on the low personal injury rate at Texas City as a safety indicator failed to provide a true picture of process safety performance and the health of the safety culture.

11. Deficiencies in BP's mechanical integrity program resulted in the "run to failure" of process equipment at Texas City.

12. A "check the box" mentality was prevalent at Texas City, where personnel completed paperwork and checked off on safety policy and procedural requirements even when those requirements had not been met.

13. BP Texas City lacked a reporting and learning culture. Personnel were not encouraged to report safety problems and some feared retaliation for doing so. The lessons from incidents and near-misses, therefore, were generally not captured or acted upon. Important relevant safety lessons from a British government investigation of incidents at BP's Grangemouth, Scotland, refinery were also not incorporated at Texas City.

14. Safety campaigns, goals, and rewards focused on improving personal safety metrics and worker behaviors rather than on process safety and management safety systems. While compliance with many safety policies and procedures

was deficient at all levels of the refinery, Texas City managers did not lead by example regarding safety.

15. Numerous surveys, studies, and audits identified deep-seated safety problems at Texas City, but the response of BP managers at all levels was typically "too little, too late."
16. BP Texas City did not effectively assess changes involving people, policies, or the organization that could impact process safety.

On August 17, 2005, the CSB issued an urgent safety recommendation to the BP Group Executive Board of Directors that it convene an independent panel of experts to examine BP's corporate SMS, safety culture, and oversight of the North American refineries. BP accepted the recommendation and commissioned the BP US Refineries Independent Safety Review Panel, chaired by former Secretary of State James Baker, III ("Baker Panel"). The scope of the Baker Panel's work did not include determining the root causes of the Texas City ISOM incident.

"The Report of the BP U.S. Refineries Independent Safety Review Panel" was issued January 16, 2007. The Baker Panel Report found that "significant process safety issues exist at all five U.S. refineries, not just Texas City," and that BP had not instilled "a common unifying process safety culture among its U.S. refineries." The report found "instances of a lack of operating discipline, toleration of serious deviations from safe operating practices, and [that an] apparent complacency toward serious process safety risk existed at each refinery." The Panel concluded that "material deficiencies in process safety performance exist at BP's five U.S. refineries."

The Baker Panel Report stated that BP's corporate SMS "does not effectively measure and monitor process safety performance" for its US refineries. The report also found that BP's over-reliance on personal injury rates impaired its perception of process safety risks, and that BP's Board of Directors "has not ensured, as a best practice, that BP's management has implemented an integrated, comprehensive, and effective PSM system for BP's five US refineries." The report's 10 recommendations to BP addressed providing effective process safety leadership, developing process safety knowledge and expertise, strengthening management accountability, developing leading and lagging process safety performance indicators, and monitoring by the Board of Directors the implementation of the Baker Panel's recommendations.

3.3.5 Trailer Siting Recommendations

On October 25, 2005, the CSB issued two urgent safety recommendations. The first called on the API to develop new guidelines to ensure that occupied trailers and similar temporary structures are placed safely away from hazardous areas of process plants; API agreed to develop new guidelines. A second recommendation to API and the National Petrochemical and Refiners Association (NPRA) called for both to issue a safety alert urging their members to take prompt action to ensure that trailers are safely located. API and NPRA published information on the two recommendations, referring to the CSB's call for industry to take prompt action to ensure the safe placement of occupied trailers away from hazardous areas of process plants.

3.3.6 Blowdown Drum and Stack Recommendations

On October 31, 2006, the CSB issued two recommendations regarding the use of blowdown drums and stacks that handle flammables. The CSB recommended that API revise "Recommended Practice 521, Guide for Pressure Relieving and Depressuring Systems," to identify the hazards of this equipment, to address the need to adequately size disposal drums, and to urge the use of inherently safer alternatives such as flare systems.

The CSB issued a recommendation to OSHA to conduct a national emphasis program for oil refineries focused on the hazards of blowdown drums and stacks that release flammables to the atmosphere and on inadequately sized disposal drums. The CSB further recommended that states that administer their own OSHA plan implement comparable emphasis programs within their jurisdictions.

3.3.7 Additional Recommendations from July 28, 2005, Incident

The CSB also made two recommendations as a result of its investigation of the July 28, 2005, incident in the Resid Hydrotreating Unit (RHU) of the BP Texas City refinery, one of two incidents after the March 23, 2005, incident. The RHU had a major fire that resulted in a shelter-in-place for 43,000 people and a reported US$30 million in plant property damage. In October 2006, the CSB released a Safety Bulletin on the findings of its investigation of the incident, available at www.csb.gov. working in and around temporary trailers that had been previously sited by BP as close as 121 ft (37 m) from the blowdown drum.

3.3.8 Summary of Safety Culture Issues

The CSB report (2007) noted numerous safety and safety culture issues with the BP refinery. Refineries run safely every day in the United States and around the world. Many times, there are many equipment maintenance issues and safety operating philosophy issues that go undetected until there is a major problem. Refineries contain so much explosive fuel that when the Swiss Cheese holes line up and major fire/explosion and fatalities can happen (Reason 1990, 2008). This case study provides a good example of how many things can go wrong before a major incident occurs. Maintenance issues and associated safety culture problems need to be fixed immediately so major incidents do not occur.

3.4 T2 Laboratories, Inc. Explosion – December 19, 2007

As a young safety professional working in a chemical plant, I witnessed one night a batch reactor having similar issue to what is being described in the following event. I do not remember the exact chemical, but it was in the chloroformate family. I was called into the plant in the event a release or fire/explosion were to happen. We sat in the control room and watched the pressure and temperature gauges rise and waited for the cooling jacket to kill the reaction. After several hours, the polymerization reaction slowed, and the pressure and temperature of the reactor reduced to a "safe" level. The batch of chemicals was ruined, but everything was safe. The description of the runaway chemical reaction that follows was primarily obtained from the CSB investigation report (CSB 2009) and news reports.

3.4.1 T2 Laboratories, Inc.

T2 Laboratories, Inc. (T2) was a small privately-owned corporation located in Jacksonville, Florida, that began operations in 1996. A chemical engineer (Scott Gallagher) and a chemist (Marion "Mike" Wyatt) founded T2 as a solvent blending business and co-owned it until the incident. From 1996 to 2001, T2 operated from a warehouse located in a mixed-used industrial and residential area of downtown Jacksonville. T2 blended and sold printing-industry solvents. It also manufactured Methylcyclopentadienyl manganese tricarbonyl (MCMT or MMT). This is an organomanganese compound used as an octane-increasing gasoline additive. The Ethyl Corporation originally developed MCMT in the late 1950s. T2 manufactured and sold MCMT under the trade name Ecotane. MCMT is a stable compound and is commonly used as an antiknock agent in gasoline. It is also used as an additive in unleaded gasoline (Sigma-Aldrich SDS 2021). The flash point for MCMT is:

204 F or 96 C.
The NFPA Rating: Health: 4; Flammability: 1; Instability: 0

The manufacturing process for the compound is complex and involves several steps utilizing volatile chemicals and sodium metal. Sodium metal is pyrophoric and reacts vigorously with water to form sodium hydroxide and hydrogen gas (Fisher Scientific 2014). The NFPA rating for sodium metal:

Health: 3; Flammability: 4; Instability: 3 and no water should be used to try to extinguish a sodium metal fire, it will cause a hydrogen explosion.

The other chemicals used in the process include an ether-based solvent, Methylcyclopentadiene dimer (Millipore-Sigma 2021), and an anhydrous manganous salt.

Methylcyclopentadiene dimer has a flash point of 78.8 F or 26 C. and has an NFPA Rating:

Health: 0; Flammability: 3; Instability: 0, indicating it is quite flammable.

There are several chemical intermediary steps in this chemical synthesis. Including the Methylcyclopentadiene dimer being cleaved, this generates hydrogen gas. The reaction can become runaway and adequate cooling needs to be applied to ensure the process remains stable. There was not enough cooling for the reactor in this event and a runaway reaction ensued.

In 2001, T2 leased a 5-acre site in a north Jacksonville industrial area and began constructing an MCMT process line. In January 2004, T2 began producing MCMT in a batch reactor. By December of 2007, MCMT production was the primary business operation. On the day of the incident, T2 employed 12 people and was producing its 175th MCMT batch (Batch 175).

3.4.2 Event Description

At 1:33 p.m. on December 19, 2007, an explosion and fire destroyed T2 Laboratories, Inc. (T2). The explosion, which was felt and heard 15 miles away in downtown Jacksonville, killed four T2 employees, including a co-owner, Robert Scott Gallagher, a 49-year-old father of five (Figure 3.16). The co-owner of Jacksonville's T2 Laboratories spent his life's last frantic minutes trying to stop a runaway chemical reaction he didn't fully understand (Florida Times-Union 2009).

The explosion killed or injured 32, including four T2 employees and 28 members of the public at surrounding businesses.

Debris from the explosion was found up to one mile away, and the blast damaged buildings within one quarter mile of the facility. The City of Jacksonville subsequently condemned buildings used by four of the businesses surrounding T2. Three of these businesses relocated operations while their buildings were repaired; the remaining business, a trucking company adjacent to T2, permanently closed due to lost business.

3.4.3 Events Leading Up to the Explosion

On the evening of December 18, 2007, a night shift process operator cleaned and dried the reactor in preparation for a new MCMT batch. At about 7:30 a.m. on

Figure 3.16 Aerial photograph of T2 taken December 20, 2007. Source: Chemical Safety Board (2009).

December 19, the day shift process operator began manufacturing Batch 175 from the control room adjacent to the process line. He likely followed the routine batch procedures, loading the reactor with the specified quantities of raw materials using an automated process control system. An outside operator hand-loaded the reactor with blocks of sodium metal, then sealed the reactor. At about 11:00 a.m., the process operator began heating the batch to melt the sodium and initiate the chemical reaction, while monitoring the temperature and pressure on the process control screen. Figure 3.17 shows a drawing of the reactor vessel.

Once the sodium melted, at 210°F (98.9°C), the process operator likely started the mixer (agitator). Mixing the raw materials increased the reaction rate, creating heat. Heat from the reaction and the heating system continued raising the temperature in the reactor. At a reaction temperature of about 300°F (148.9°C), the process operator likely turned off the heating system as specified in the procedure, but heat from the reaction continued increasing the mixture temperature.

At a temperature of about 360°F (182.2°C), the process operator likely started cooling, using the process control system, as specified in the procedure. However, the mixture temperature in the reactor continued to increase.

At 1:23 p.m., the process operator had an outside operator call the owners to report a cooling problem and request they return to the site. Upon their return, the owner/chemical engineer went to the control room to assist and the owner/chemist searched for the plant mechanic. Employees indicated that after visiting the control room, the owner/chemical engineer went to the reactor. He told an outside operator – who was coming to the control room to investigate multiple process alarms sounding – that he thought there would be a fire and

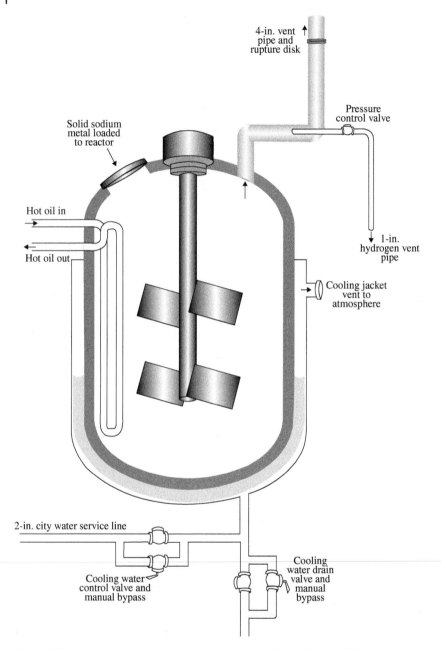

Figure 3.17 Drawing of the reactor. Source: Chemical Safety Board (2009).

motioned employees to get away. The owner/chemical engineer then returned to the control room.

At 1:23 p.m., that employee had a co-worker phone Gallagher and Wyatt, who had left the plant together that day, and asked them to come back to fix a cooling problem.

The partners arrived within minutes. Gallagher headed for the control room while Wyatt went to find the plant's mechanic.

As Wyatt searched, Gallagher checked the control room, then the reactor 50 ft away, then warned employees to get away as he returned to the control room and the employee inside, where alarms were sounding.

At 1:33 p.m., just 10 minutes after the call for help, the plant blew up.

By 1:33 p.m., the reactor's relief system could no longer control the rapidly increasing temperature and pressure of the runaway reaction. Eyewitnesses from nearby businesses reported seeing venting from the top of the reactor and hearing a loud jet engine-like sound immediately before the reactor violently ruptured, its contents exploding. The explosion killed the owner/chemical engineer and process operator who were in the control room (50 ft from the reactor) and two outside operators who were leaving the reactor area (Figure 3.18). Another outside operator and the plant mechanic were injured. The owner/chemist was sheltered from the force of the explosion by a shipping container but suffered a non-fatal heart attack during the incident.

Responding to the explosion and subsequent fire were the Jacksonville Fire and Rescue Department (JFRD); US Naval Air Station Mayport Fire Department;

Figure 3.18 Control room. Source: Chemical Safety Board (2009).

Jacksonville International Airport Fire Department; Jacksonville Sheriff's Office (JSO); City of Jacksonville Environmental Resource Management Division; City of Jacksonville Planning and Development Department; Florida State Fire Marshal; Florida Department of Environmental Protection (FDEP); US Bureau of Alcohol, Tobacco, Firearms, and Explosives (ATF); US Environmental Protection Agency (EPA); US Department of Homeland Security, Chemical Security Compliance Division (DHS-CSCD); US Department of Labor, Occupational Safety and Health Administration (OSHA); and American Red Cross (ARC).

The explosion killed four T2 employees and injured 32 workers at T2 and surrounding businesses. All the people at T2 during the incident – eight T2 employees and one truck driver making a delivery – were injured or killed. The four fatally injured employees died of blunt force trauma as a result of the explosion. Another T2 employee was critically injured and hospitalized for several months.

The CSB conducted a community survey of the surrounding businesses to characterize injuries and structure damage (Figure 3.17). At the nine businesses within 1900 ft of the reactor, the explosion injured 27 workers. Of those, 11 suffered lacerations and contusions, seven reported hearing loss, and five fell or were thrown by the force of the blast. The CSB survey team photographed and catalogued 32 damaged structures, some as far as 1700 ft from T2.

The explosion leveled the plant, propelling debris in all directions. Two large steel support columns from the reactor structure traveled about 1000 ft along Faye Road in both directions. A 2000-lb section of the 3-in.-thick reactor head (Figure 3.19) impacted railroad tracks adjacent to T2, pushing a rail out of place, before impacting and damaging a building about 400 ft from the reactor. The explosion threw piping from inside the reactor hundreds of feet onto the other businesses and wooded areas surrounding T2. The 4-in. diameter agitator shaft from the reactor was thrown about 350 ft across Faye Road in two large pieces that imbedded in a sidewalk and the ground (Figure 3.19).

3.4.4 Analysis of the Accident

The incident at T2 was one of the most energetic explosions investigated by the CSB. The CSB estimated the energy release of this explosion to be equivalent to 1400 lb of TNT.

Businesses near T2 sustained heavy damage. Structures sustained window damage from the overpressure up to 1700 ft away from T2. The explosion destroyed a trucking company office trailer located 250 ft from the reactor (Figure 3.20). If trucking company employees had been in the trailer at the time of the incident, it is likely that they would have been seriously injured or killed. Two warehouses located about 400–500 ft from the reactor both sustained heavy damage; nine of the 25 employees at these two businesses were injured.

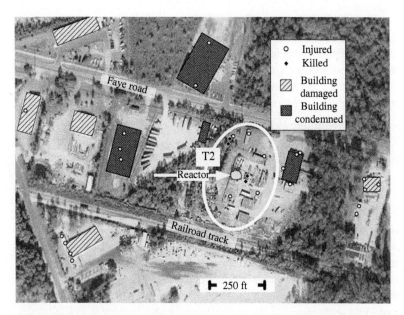

Figure 3.19 Injury and business locations. Source: Chemical Safety Board (2009).

Figure 3.20 Portion of the 3-in.-thick reactor. Source: Chemical Safety Board (2009).

The blast occurred due to a runaway chemical reaction that generated high temperature and pressure in the reactor. The CSB determined insufficient cooling to be the only credible cause for this incident, which is consistent with witness statements that the process operator reported a cooling problem shortly before the explosion. The T2 cooling water system lacked design redundancy, making it susceptible to single-point failures including:

- Water supply valve failing closed or partially closed.
- Water drain valve failing open or partially open.
- Failure of the pneumatic system used to open and close the water valves.
- Blockage or partial blockage in the water supply piping.
- Faulty temperature indication.
- Mineral scale buildup in the cooling system.

Interviews with employees indicated that T2 ran cooling system components to failure and did not perform preventive maintenance. On at least one prior occasion since 2006, the reactor cooling drain valve had failed during operations and required repair. Additional credible cooling system failures include formation of mineral scale inside the jacket that could interfere with system heat removal capacity or loose scale blocking the inlet/drainpipe and causing it to stick open. Although a control system malfunction or operator error might also contribute to insufficient reactor cooling, there is no evidence to indicate that either of these occurred.

Based on the reaction chemistry, raw material concentration abnormalities were not credible failure scenarios. Of the raw materials used, only an increase in the amount of sodium would have accelerated the reaction rate. Sodium was hand-loaded by operators in the form of one four-drum pallet per batch, making amount variation extremely unlikely. Varying local concentrations of chemicals within the reactor would reduce rather than accelerate reaction rate, since uniform distribution of the three metalation reaction raw materials results in the maximum reaction rate.

Heat was applied to the mixture using the hot oil system; had heating continued beyond 300°F (148.9°C), the CSB calculated that the cooling system would have easily overwhelmed it. The capacity of the cooling system was more than 10 times greater than the maximum capacity of the hot oil system.

CSB conducted tests in a closed (sealed) test cell, two exothermic reactions were observed using the T2 recipe. The first exothermic reaction occurred at about 350°F and was the desired reaction between the sodium and the MCPD. A second and more energetic exothermic reaction occurred when the temperature exceeded 390°F (198.9°C); this reaction was between the sodium and the diglyme solvent.

The pressure and temperature rise during the second exothermic reaction were about 32,000 psig per minute (2206 bar per minute) and 2340°F per minute (1300°°C per minute), respectively, and burst the test cells.

Using the data obtained from these tests, the CSB determined that it is unlikely that an overpressure relief device of any size set at 400 psig could have prevented failure of the reactor once the second exothermic reaction began. Failure could only be prevented by relieving at a lower pressure during the first exothermic reaction and allowing the MCPD and diglyme solvent to boil and vent, removing both heat and reactants. Had T2 set its 4-in. reactor rupture disk at 75 psig, rather than the 400 psig used, the runaway reaction likely would have been relieved during the first exothermic reaction, precluding the second exothermic reaction. This could have prevented the catastrophic reactor failure that occurred.

3.4.5 Process Development

The president of Advanced Fuel Development Technologies (AFD) asked T2 to consider manufacturing MCMT in 1998. Although both of the T2 owners had prior chemical industry experience, neither had previously worked with reactive chemical processes; they waited two years before agreeing to pursue the project. Upon T2's agreement, the AFD president, a PhD chemist with more than 20 years' experience, provided patent literature and research support to T2's owner/chemist. The owner/chemist then duplicated and tested the chemistry described in the patents and created a three-step process for making MCMT in the laboratory. Between 2000 and 2001, the owner/chemist ran about 110 test batches of MCMT in a 1-l reactor.

In 2001, T2 leased its Faye Road site, in an area zoned for heavy industry. With financial support from a number of investors including AFD, T2 designed and constructed a full-scale MCMT production plant. T2 hired consulting engineers to assist in the process design, control system engineering, and project management. Due to limited funding, T2 purchased and refurbished used equipment, including the 2450-gal high-pressure batch reactor used for the three-step MCMT reactions.

On January 9, 2004, T2 began manufacturing its first full-scale MCMT batch (Batch 1) in the new process line. Batch 1 produced an unanticipated exothermic reaction in the first step; T2 noted the anomaly, adjusted the batch recipe and production procedures to include reactor cooling in the first step, and began a new production batch. Between February and May 2004, T2 manufactured nine more MCMT batches, adjusting the recipe and procedures between batches. Yields varied from no saleable product to about 70% saleable product. Batch 5 resulted in an uncontrolled (runaway) exothermic reaction in the first step. In Batch 10, the temperature also increased beyond expectations – though not as severely as in Batch 5 – due to the exothermic reaction.

On May 24, 2004, following Batch 11, T2's owner/chemical engineer sent a memo to investors declaring successful plant startup and full-scale MCMT production. Although all of the investors had ended their involvement by late 2004, T2 continued producing and selling MCMT on an irregular basis, relying upon sales to pay for more raw materials. On July 28, 2005, producing Batch 42, T2 increased batch size by one-third. T2 manufactured batches from 2005 until the incident (Batch 175) at this larger size.

A process operator ran each step of the batch reaction from a control room adjacent to the process line. The T2 plant manager, a chemical engineer, began working as a process operator in 2004. In 2006, T2 hired two additional chemical engineers to run the process throughout multiple weekday shifts. Each batch required about 48 hours to manufacture. By December 2007, client demand had increased, and T2 produced three batches a week.

3.4.6 Manufacturing Process

T2 manufactured MCMT in three steps that occurred sequentially within a single process reactor (Figure 3.21). The National Annealing Box Company of Washington, Pennsylvania, originally constructed the reactor in 1962 for an internal pressure of 1200 psig. T2 purchased the reactor in 2001 and contracted a

Figure 3.21 Agitator Shaft. Source: Chemical Safety Board (2009).

firm specializing in pressure vessels to refurbish, modify, and test the reactor. The modifications included replacing and adding of piping nozzles and reducing the maximum allowable working pressure from 1200 to 600 psig.

A 4-in. vent pipe that made two 90° pipe bends before connecting to a 4-in. rupture disk provided overpressure protection for the reactor. Employees stated that the rupture disk was set at 400 psig. A pressure control valve installed in a 1-in. vent pipe, which branched off of the 4-in. vent pipe below the rupture disk, controlled reactor pressure.

The MCMT process required both heating and cooling. A heating system circulated hot oil through 3-in. piping installed around the inside of the reactor. A cooling jacket covered the lower three quarters of the reactor. A pipe from the city water system connected to the bottom of the jacket through a control valve and a common supply/drain connection. Water was injected into the jacket and allowed to boil; steam from the boiling water vented to atmosphere through an open pipe connected to the top of the jacket.

3.4.7 Summary Safety Culture Issues

The safety culture aspects in this event might not be as straight forward as other in other events. However, there are many. These include:

1. The two owners did not have prior experience with this type of chemical process, prior to their investment into MCMT chemical manufacturing process.
2. It was noted in the writeup by CSB that components of the cooling system were run to failure and there was no periodic maintenance of the system. Also, the cooling system might not have had the capacity to kill a runaway reaction.
3. There was no emergency plan for dealing with runaway chemical reactions. There were not signaling devices to let operators in the plant know that a major problem was happening.
4. The reactor system was not properly constructed to handle the pressures generated by the runaway chemical reaction. For instance, the rupture disk was designed to fail at 400 psig. This is close to the predicted failure pressure of the vessel – 600 psig. This leaves very little safety margin.
5. It appears that a risk assessment was not conducted of the equipment or the manufacturing process.
6. The location of the plant was too close to other businesses. A siting analysis should have been performed.

The chemical plant I worked at in La Porte, TX had an oil field company next door. We found out that their explosive magazine was on our plant boundary. We informed them that they should move it because we had glass pipe in the plant with highly toxic chemicals running through them.

3.5   Final Thoughts for This Chapter

Unfortunately, these three incidents are not unique. They involve poor safety culture and the lack of necessary equipment and periodic maintenance. Equipment maintenance in every industry needs to be done or disasters can occur. PSM principles must be implemented, and risk assessments need to be done and corrective actions taken to help ensure disasters do not happen. The cost of doing the right thing before a disaster is much less than cleaning up the consequences of one. Human life should never have a price tag associated with it.

References

BBC (2018). World's oldest brewery' found in cave in Israel, say researchers https://www.bbc.com/news/world-middle-east-45534133 (accessed November 2021).

Botha, J. (2013). Alcohol, beer, beer news, origin, research, wine, World Wine News. https://smartaboutwine.wordpress.com/2013/06/05/cheers-eight-ancient-drinks-uncorked-by-science-by-nbcnews-com/ (accessed November 2021).

BP (2013). BP completes sale of Texas City refinery. https://www.bp.com/en/global/corporate/news-and-insights/press-releases/bp-completes-sale-of-texas-city-refinery-and-related-assets-to-marathon-petroleum.html (accessed November 2021).

Butler, S. (2018). Off the spice rack: the story of salt. https://www.history.com/news/off-the-spice-rack-the-story-of-salt (accessed November 2021).

Cartwright, M. (2017). Greek fire world history encyclopedia. https://www.worldhistory.org/Greek_Fire/ (accessed November 2021).

Center for Disease Control (CDC) (1994). Methyl mercaptan. https://www.cdc.gov/niosh/idlh/74931.html/ (accessed November 2021).

Chemical Safety Board (CSB) (2007). Refinery explosion and fire, Report NO. 2005-04-I-TX.

Chemical Safety Board (CSB) (2019). Toxic chemical release at the DuPont La Porte Chemical Facility, Investigation Report, June, 2019.

Chemistry (2015). What is the oxidation mechanism of gunpowder?. https://chemistry.stackexchange.com/questions/35680/what-is-the-oxidation-mechanism-of-gunpowder (accessed November 2021).

Chemistry Safety board (CSB) and T2 Laboratories Inc. (2009). Reactive chemical explosion, Final Report.

Daley, J. (2017). Science is falling woefully behind in testing new chemicals. https://www.smithsonianmag.com/smart-news/science-falling-woefully-behind-testing-new-chemicals-180962027/ (accessed November 2021).

Fisher Scientific (2014). Sodium metal safety data sheet. https://www.fishersci.com/store/msds?partNumber=S25556A&productDescription=SODIUM+METAL+100G&vendorId=VN00115888&countryCode=US&language=en (accessed November 2021).

Florida-Times-Union (2009). Report: Owners of Jacksonville's T2 Lab never knew risks of deadly explosion. https://www.jacksonville.com/story/news/2009/09/15/report-owners-of-jacksonvilles-t2-lab-never-knew-risks-of-deadly-explosi/15973372007/ (accessed November 2021).

Koester, C., OSHAOLINE (2018). We must change the statistics of confined space injuries and fatalities. https://ohsonline.com/articles/2018/08/01/we-must-change-the-statistics-of-confined-space-injuries-and-fatalities.aspx (accessed December 2021).

Millipore-Sigma (2021). Methylcyclopentadiene dimer. https://www.sigmaaldrich.com/US/en/product/aldrich/129828 (accessed December 2021).

National Institute of Health (NIH) (2021). Lannate. https://pubchem.ncbi.nlm.nih.gov/compound/Lannate (accessed November 2021).

National Institute of Occupational Safety and Health (NIOSH) (2021). Confined spaces. https://www.cdc.gov/niosh/topics/confinedspace/default.html (accessed December 2021).

National Public Radio (NPR) (2021). Death and destruction follow a Siberian coal mine firen. https://www.npr.org/2021/11/25/1059263530/death-and-destruction-follow-a-siberian-coal-mine-fire (accessed December 2021).

Occupational Safety and Health Administration (OSHA) (2021). Process safety management. https://www.osha.gov/process-safety-management/sbrefa (accessed November 2021).

Reason, J. (1990). *Human Error*. Cambridge University Press.

Reason, J. (2008). *The Human Contribution: Unsafe Acts, Accidents and Heroic Recoveries*. CRC Press.

Scher, L. (2007). *The Texas City Disaster*. Bearport Publishing.

Soap History (2021). http://www.soaphistory.net/ (accessed November 2021).

The Conversation (2015). A history of sugar – the food nobody needs, but everyone craves. https://theconversation.com/a-history-of-sugar-the-food-nobody-needs-but-everyone-craves-49823#:~:text=The%20first%20chemically%20refined%20sugar,important%20centres%20for%20sugar%20production (accessed November 2021).

The Pop History Dig (2005). Texas City disaster. https://www.pophistorydig.com/topics/tag/texas-city-oil-refinery-history/ (accessed December 2021).

Trading Economics (2021). United States gasoline production. https://tradingeconomics.com/united-states/gasoline-production (accessed November 2021).

United States Food and Drug Administration (FDA) (2021). Guidance for industry: voluntary sodium reduction goals. https://www.fda.gov/regulatory-information/search-fda-guidance-documents/guidance-industry-voluntary-sodium-reduction-goals (accessed November 2021).

4

Chemical Storage Explosions

4.0 Introduction

The case studies that are provided here are all examples of energetic chemicals that under the right conditions can cause massive explosions. There are no secrets to how reactive these chemicals are. Ammonium nitrate (AN) (NH_4NO_3) is an oxidizer and is used as a fertilizer, but it is also a large component of explosives. The Grandcamp ship explosion in Texas City, TX, should be the tattooed on every safety professional, emergency manager and shipper's brain to remind them how potentially explosive AN is. The CSB (2013) report about the West Fertilizer Company explosion that is presented at the end of this chapter provided a great discussion on how AN decomposes. The following provides some of that information:

> "AN has a melting point between 311 and 337 °F (155 and 169 °C). It begins to rapidly decompose at a significant rate soon thereafter. When it is exposed to high heat and pressure, AN experience endothermic (heat-absorbing) and exothermic (heat-producing) reactions simultaneously, causing the compound to split into its constituent molecules and transforming it from solid state to molten, or liquefied, state. When AN decomposes or breaks down under thermal conditions, at least seven unique reactions can occur at varying temperatures, with different heat outputs and rates of reaction. Some reactions can produce toxic and detonable by-products. All of the reaction pathways begin with the AN splitting into gaseous ammonia (NH_3) and nitric acid (HNO_3), although that step is usually not explicit.
>
> In the following main exothermic reaction (Eq. (4.1)), which can occur in conditions up to 482 °F (250 °C), AN yields nitrous oxide and water:
>
> $$NH_4NO_3(s) \rightarrow N_2O(g) + 2H_2O(g) \tag{4.1}$$

(Continued)

Impact of Societal Norms on Safety, Health, and the Environment: Case Studies in Society and Safety Culture, First Edition. Lee T. Ostrom.
© 2023 John Wiley & Sons, Inc. Published 2023 by John Wiley & Sons, Inc.

> **(Continued)**
>
> Above 482 °F (250 °C), a reversible endothermic reaction (Eq. (4.2)) takes place at a significant rate, splitting the AN to form ammonia and nitric acid:
>
> $$NH_4NO_3(s) \leftrightarrow NH_3(g) + HNO_3(g) \qquad (4.2)$$
>
> This endothermic reaction is accompanied by a number of exothermic reactions between gaseous ammonia (NH_3) and nitric acid (HNO_3) that vary by degree, depending on reaction conditions. As previously described, AN is in a liquid or molten state, which is aerated with off-gases such as nitrogen oxides (NO, NO_2), water vapor, and nitrous oxide (N_2O). This bubbly liquid is much more sensitive to detonation than solid prills or unaerated liquid. Depending on the rate of these endothermic and exothermic reactions, detonation can occur. Conditions other than heat and pressure, such as pH levels and the presence of impurities, can also influence the rate of reaction.
>
> An example of fueling AN to produce a blasting agent is the addition of fuel oil at around 6% by weight to produce AN/fuel oil (ANFO). ANFO may be used for mining and other purposes. Moreover, the military uses a mixture of fuel-rich trinitrotoluene (TNT), AN, and sometimes aluminum to produce a more effective explosive than TNT alone."

On the other hand, ammonium perchlorate (AP) decomposes before it melts. Mild heating results in production of hydrogen chloride, nitrogen, oxygen, and water (Eq. (4.3)).

$$4NH_4ClO_4 \rightarrow 4HCl + 2N_2 + 5O_2 + 6H_2O \qquad (4.3)$$

The combustion of AP is quite complex and is widely studied. AP crystals decompose before melting, even though a thin liquid layer has been observed on crystal surfaces during high-pressure combustion processes. Strong heating may lead to explosions. Complete reactions leave no residue. Pure crystals cannot sustain a flame below the pressure of 2 MPa (NFPA 2022; Boggs 1970).

AP is a Class 4 oxidizer that can undergo an explosive reaction for particle sizes over 15 μm and is classified as an explosive for particle sizes less than 15 μm.

All chemicals with hazardous properties should be stored safely. There should be no exceptions.

4.1 Port of Lebanon – August 4, 2020

The explosion that occurred on August 4, 2020 is a prime example of what happens when too much of a highly energetic material is stored in one place, near a highly populated area. Most of the news articles are on the political aftermaths of the explosion, even more so than the humanitarian issues caused by the explosion. The explosion killed 135–218 people and injured approximately 5000–7000, depending on which news source provided the information (Relief Web 2022; BBC 2021). It also caused 300,000 people to lose their homes.

It theorized that the explosion was caused by a fire in an unsafe warehouse that was storing 2750 tons of AN. The material had been there for six years. Always in these events the question is where does the fuel come from? However, there is always something these strong oxidizers can use as fuel. It could be wood, metal, organic solvents, grain, or dust. Once the temperature of the fire reaches a certain point AN will disintegrate violently.

The BBC (2021) reports that the AN arrived at the port in 2013. It was there because a Moldovan-flagged cargo vessel made an unscheduled stop and was then banned from leaving because of a legal dispute over unpaid fees and ship defects. A court order in 2014 required the ship's cargo to be unloaded was unloaded to a port warehouse.

The heads of the port and customs authority said their warnings about the danger posed by the AN and calls for it to be removed were repeatedly ignored.

This is a common theme in some of these major events. People with knowledge of the hazards are ignored because the people in power have usually some monetary goals. The 2750 tons of AN have a value of approximately US$550 per ton. In today's world market the total value of the fertilizer would be approximately US$1.6 million.

Storing 2750 tons of AN so close to a large city or even by a small town shows the total lack of understanding as to how dangerous this chemical is.

It is always a catastrophe when anyone is killed or displaced by an incident like this. However, it always seems that those who can least afford it are hurt the worst. Figure 4.1 shows the devastation from the explosion.

4.1.1 PEPCON Explosion – May 4, 1988

The PEPCON explosion on May 4, 1988, demonstrates the power of chemicals of the class of oxidizers. The chemical that was being manufactured and stored at the PEPCON facility in Henderson, NV, was AP.

The NASA safety share flyer on the PEPCON event states that AP has been used for decades as a powerful oxidizer that is then mixed with finely ground aluminum and other combustible materials to create solid propellants for launch vehicles and military weapons. AP production involves several hazardous compounds, and many were present in bulk at the facility (NASA 2012).

AP is a colorless, odorless, inorganic, crystalline compound with the molecular formula ClH_4NO_4 and CAS number 7790-98-9. It imparts a bitter and salty

Figure 4.1 Devastation of the Port of Beirut. Source: Rashid khreiss/Unsplash.

taste to water and does not readily burn but will burn if contaminated by combustible materials. When powdered into particles smaller than 15 μm in diameter or if powdered into larger particles but thoroughly dried, AP is currently classified as a division 1.1 explosive. AP is considered acutely toxic and can be harmful if swallowed, cause serious eye irritation, skin irritation, and may cause respiratory tract irritation if its dust is inhaled (The Chemical Company 2022).

The NFPA 704 Codes are:

Health Hazards – 1 – Can cause significant irritation.
Fire Hazard – 0 – Will not burn under typical fire conditions.
Instability – 4 – Readily capable of detonation or explosive decomposition or explosive reaction at normal temperatures and pressures.
Special – Possesses oxidizing properties.

Until the PEPCON event, AP had been tested in small scale detonation scenarios and was classified as a class-4 oxidizer (NASA 2012).

Although classified as less hazardous than mixed fuel, AP greatly accelerates the explosive properties of combustible material.

The US Fire Administration/Technical Report Series reports that series of explosions on May 4, 1988, near the city of Henderson, Nevada, claimed two lives, injured approximately 327 people, including 15 firefighters, and caused damage estimated over 100 million dollars. The explosions affected a large portion of the metropolitan Las Vegas area and caused the activation of disaster plans by several

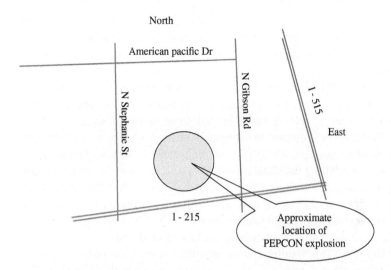

Figure 4.2 Approximate location of PEPCON plant. Source: Drawing by Ostrom, 2022.

agencies. Considering the magnitude of the explosions, the loss of only two lives, and the fact that only a few of the injuries were critical, can be described as very fortunate, especially, since it was in the town of Henderson. Figure 4.2 shows the approximate location of the plant in the town. Now, in that location there are apartments and many businesses.

The incident presented tremendous risk and unusual challenge to the fire departments involved, but they managed the incident with relatively minor casualties. The lives of most of the plant employees were saved by their decision to evacuate the plant, prior to the major explosions (Homeland Security 1988).

AP was the only product manufactured at the PEPCON facility. The process uses several Hazmat, including anhydrous ammonia, hydrochloric acid, nitric acid, and various chlorate compounds. These chemicals were shipped to the site primarily by rail and were present in bulk quantities. At the time of the explosion, PEPCON was one of two major producers of AP worldwide. The other producer, Kerr-McGee, was located less than 2 mi away – well within the blast effect area of the PEPCON explosion.

PEPCON operated in Clark County, NV, located approximately 10 mi southeast of downtown Las Vegas, NV, near the City of Henderson. Henderson was a growing suburb with a 1988 population of approximately 50,000 and had been a center of production for the commercial and defense industry since World War II. In 1988, all NASA launch activities had been indefinitely halted by the Challenger disaster on January 28, 1986, and subsequent investigative activities of the Rogers Commission. The stand down froze PEPCON's AP shipping yet had no effect on

PEPCON's AP contract orders with Space Shuttle Solid Rocket Booster (SRB) manufacturer Morton Thiokol. Over the next 15 months following the Challenger disaster, PEPCON had accumulated a stockpile of over 4000 tons of the oxidizer (NASA 2012).

On May 4, 1988, PEPCON employees were repairing the plant's wind-damaged steel and fiberglass drying structure when, at 11:30 a.m. Pacific Daylight Time (PDT), sparks from a welding torch ignited some of the fiberglass building material. The workers attempted to extinguish the fire with a hose line, but then charged a second hose line, which reduced the water pressure, and allowed the fire to grow. Exacerbating the situation was the profusion of AP residue on many of the drying structure's surfaces. The fire spread quickly to the 55-gal AP storage drums stacked next to the drying structure. Within 10–20 minutes (estimated by PEPCON employees) from the time of the spark, the first AP explosion occurred.

Welding/cutting activities in manufacturing facilities is defined as "hot work" by OSHA 29 CFR 1917.152. Welding related activities are associated with numerous fires and explosions each year. The massive explosions at the PEPCON manufacturing facility outside Henderson, Nevada on May 4, 1988, are an example of such an event. The welding might have caused the fire that led to the explosion, but there were many safety culture factors that laid the groundwork for the explosion.

No alarm or audible announcement system was installed at the facility, and no sprinkler system existed in the processing structures, except for the administration building. According to policy, personnel were advised to evacuate the premises if they observed a fire beyond beginning stages; however, there was no formal evacuation plan.

Of the 77 PEPCON employees, 75 escaped, running or driving away from the facility through the desert. Two were killed; one, confined to a wheelchair, was unable to escape in time, the other, who was also disabled, stayed behind to alert emergency dispatchers of the situation. Employees at the neighboring Kidd Marshmallow plant also evacuated and contacted authorities. Fire response teams dispatched to the area immediately following the calls, which began at 11:51 a.m. PDT. The closest Clark County units were a little over 5 mi away and saw smoke from their station. The Fire Chief of the City of Henderson, 1.5 mi away, saw smoke and ordered his units to dispatch to the PEPCON plant.

At 11:53 a.m. PDT, multiple 55-gal drums exploded into a giant fireball – the first of two major blasts, approximately 100 ft in diameter. The shockwave shattered vehicle windows of the fire response units, halting their approach. PEPCON employees escaping in vehicles met the responders and warned of potential, larger explosions, sending the responders in the same direction as the escaping workers. Four minutes later, at 11:57 a.m. PDT, the fire reached the storage area of the facility that held large, aluminum 5000-lb shipping containers loaded with AP, resulting in the largest explosion.

Figure 4.3   PEPCON explosion. Source: NASA, 2012.

Witnesses reported a shockwave visibly travelling across the ground toward them (Figure 4.3). Many were temporarily blinded by the immediate flash and others lacerated by flying glass. Henderson responders were effectively incapacitated. Of the 4000 tons of AP stored at the plant, it is estimated that approximately 1500 tons were consumed in the subsequent explosion. Figure 4.4 shows the aftermath of the explosion.

Little fuel remained after this second major explosion and flames diminished rapidly. At 12:59 p.m. PDT, the damaged facility gas line that continued to feed the remaining flames was closed off. Clark County responders staged 1.5 mi away and aided the injured Henderson responders. Several additional area fire departments responded with mutual aid, but no attempt was made to approach or fight the fires, which were beyond the departments' suppression capability.

The PEPCON facility and neighboring Kidd Marshmallow plant were completely obliterated. The shockwaves had damaged buildings in Henderson; shattering windows, cracking walls, and breaking doors. The affected area spanned a 10-mi radius from the PEPCON facility.

Smoke rose several thousand feet on the thermal column created by the fire and travelled downwind and eastward over residential and commercial zones of Henderson. The smoke was seen from 100 mi away. The two major blasts measured 3.0 and 3.5, respectively, on the Richter scale at observatories in California and Colorado. Investigators surveying the damage in the surrounding communities estimated the blast as similar to a 1-kt air blast nuclear detonation.

AP is a powerful oxidizer which is mixed with combustible materials to produce rocket fuels. The mixed fuels provide a high energy release at a very rapid rate of combustion.

Figure 4.4 Aftermath of the explosion. Source: NASA (2012).

The rate of combustion is controlled by the mixture, particle size, and moisture content. Finely ground aluminum frequently is used as the combustible component.

Without the combustible component, AP is classified as an oxidizer and considered less hazardous than a mixed fuel. Normal combustion will be greatly accelerated in its presence, and contamination of AP with an organic product will create explosive mixtures.

It is normally shipped in large aluminum "tote" containers, each holding several thousand pounds of the white granular material. Several hundred of the aluminum "totes" were stored in one area of the plant awaiting shipment, along with a smaller number of fiber drums. A quantity of nearly empty drums was also on hand to be refilled.

Another storage area at the plant contained several thousand 55-gal plastic drums of the product waiting final blending to customer specifications. The plastic drums were not used to ship the product outside the plant.

The product (AP) apparently had not been tested for mass (large quantity) detonation prior to this fire, and its classification was based on small scale tests. Although not previously considered to be explosive, this incident obviously gives testimony to the fact that AP can explode.

According to the US DOT Emergency Response Guidebook (the "yellow book" carried by many fire companies), the precautions for AP are somewhat different depending on its particle sizes. There are different guide numbers, Hazmat numbers, and instructions for particle size under 45 μm and over 45 μm. A firefighter approaching a vehicle or plant might well not know which instructions to follow without detailed knowledge gained in pre-fire planning. At the Henderson plant, particle sizes were 90 μm and above, according to Deputy Chief Pappageorge of the Clark County Fire Department.

The information in the guidebook is intended to apply to quantities that would be encountered in transportation. For both particle sizes, the guide clearly notes the possibility of fire or explosion. For the smaller particle size, it provides the same information as a Class A explosive. It advises to stop traffic and evacuate to one mile away and not to fight the fire.

For the larger particle size product, the guidebook suggests the use of master streams from a safe distance and warns of irritating or poisonous gases that may be produced from a fire.

For either particle size, the firefighters are warned of the explosion potential. Bulk quantities as large as those normally stored in a plant or fixed site are not covered by the manual, but the warnings are dire enough to preclude approaching a major fire involving the product. However, the guidebook may not make it clear enough that even large particle size AP can explode catastrophically.

In addition to the chemicals at the plant, a 16-in., high-pressure (300 psi) natural gas transmission line ran underneath the plant and also supplied the plant through a pressure reducing assembly.

The fire is reported to have originated in or around a drying process structure in the PEPCON plant between 1130 and 1140. The steel frame with fiberglass walls and roof structure had been damaged in a windstorm and employees were conducting repairs using a welding torch at the time. The fire spread rapidly in the fiberglass material, accelerated by AP residue in the area.

As employees attempted to fight the fire with hose lines, the flames spread to 55-gal plastic drums containing the product that was stored next to the building.

The employee efforts at extinguishment were unsuccessful, and they abandoned the effort when the first of a serious of explosions occurred in the 55-gal drums. The time between ignition and the first explosion has not been determined exactly; it was estimated at 10–20 minutes. When the control efforts were abandoned, most of the plant employees evacuated the area by running or driving away. Approximately 75 managed to evacuate, leaving only the two who were killed in subsequent larger explosions. One of these victims stayed behind to call the Clark County Fire Department and the other was confined to a wheelchair and was unable to leave the area. The first explosion also alerted employees of the nearby marshmallow factory, and they also evacuated the area.

The fire continued to spread in the stacks of filled 55-gal plastic drums and created an extremely intense fireball. The first of two major explosions then occurred in the drum storage area. The fire continued to spread and reached the storage area for the filled aluminum shipping containers. This resulted in an even larger, second major explosion, approximately four minutes later. Very little fuel remained after the second explosion and the flame diminished rapidly except for the flame plume created when the high-pressure natural gas line beneath the plant was ruptured in one of the explosions. The gas line was shut off at 1259 hours by the gas company, at a valve about a mile away, eliminating the fuel for this fire.

A huge column of smoke rose from the plant and was carried downwind to the east, over most of the residential and business areas of Henderson. The smoke rose on the thermal column to an altitude of several thousand feet and was spotted almost 100 mi away.

All told, seven explosions occurred involving various containers of AP, with the two largest occurring in the plastic drums and then the aluminum containers. These two explosions were measured at 3.0 and 3.5 on the Richter scale at an observatory in California! Over eight million pounds of the product were consumed in the fire and explosions. A crater estimated at 15 ft deep and over 200 ft long was left in the storage area.

The Clark County Fire Department had received numerous telephone calls reporting the fire after the first (small) explosion, including the call from the plant employee. These calls had begun at 1151 and a first alarm assignment was dispatched immediately. The closest Clark County units had a response of over five miles and could see the heavy smoke from their station.

At approximately the same time, the huge column was spotted by the fire chief of the city of Henderson who was leaving the main fire station, approximately 1.5 mi north of the PEPCON facility. The chief immediately ordered his units to be dispatched and headed toward the scene. As he approached within a mile of the plant, he could see a massive white and orange fireball, approximately 100 ft in diameter, and dozens of people running across the desert toward him. He advised his dispatcher to call for mutual aid assistance although he was still unaware of the nature of the fire.

As he approached the scene at 1154, the first of the two major explosions occurred. The shock wave shattered the windows of his car and showered the chief and his passenger with glass. The driver of a heavily damaged vehicle coming away from the plant advised the chief of the danger of further, even larger explosions. With this information, the chief turned around and headed back toward his station. The other Henderson companies en route to the scene stopped where they were on their own volition after the explosion (about 1 mi away).

Approximately four minutes after the first major explosion, the second large explosion occurred. Witnesses reported that this explosion created a visible shock wave coming toward them across the ground. Several videotape recordings of the explosions were made by people in the area, graphically demonstrating the movement of the shock wave.

The second major explosion virtually destroyed the chief's car. The chief and his passenger were cut by flying glass, but he was able to drive the damaged vehicle to a hospital to seek treatment. The windshields of the responding Henderson Fire Department apparatus were blown in, and the drivers and officers were injured by the shattered glass. The Henderson Fire Department was essentially totally incapacitated by the second major explosion. The injuries consisted of numerous cuts from flying glass, but did not require hospitalization.

The Clark County response was upgraded to a third alarm while units were still en route. Several area fire departments also responded on mutual aid. The Clark County units staged 1.5 mi from the scene and provided assistance to the injured Henderson firefighters. From this distance they attempted to size-up the situation. Both the PEPCON facility and the neighboring marshmallow plant had been destroyed in the explosions prior to their arrival. The magnitude of the fire in the PEPCON facility was beyond any fire suppression capability, and flames also were visible in the rubble of the marshmallow plant. The only hydrants were in the immediate area of the two involved plants, but there was no water supply due to the loss of electrical power to the pumps. Recognizing the danger and futility of operations, no attempt was made to approach or to fight the fire.

A command post was established more than 2 mi from the scene at a location that afforded a view of the involved area. Responding fire and medical units were staged as an assessment was made of the situation.

Both the PEPCON and marshmallow manufacturing facilities were virtually destroyed. Damage with a 1.5 mi radius was heavy, including destroyed cars, structural damage to buildings and downed power lines. Within three miles there was extensive window breakage and moderate structural damage. Many structures had damage to suspended ceilings and overhangs, windows and doors, exterior details, and cracked walls.

Damage extended for a radius of up to 10 mi. Buildings were damaged throughout Henderson including over US$100,000 damage to the main fire station and heavy structural damage to a warehouse next door. Hundreds of windows were shattered, doors were blown off their hinges, walls cracked, and scores of people were injured by flying glass and debris. At Las Vegas' McCarran International Airport seven miles away, windows were cracked and doors were pushed open. A Boeing 737 on final approach was buffeted by the shock wave.

The fire departments in the area were heavily committed to the actual incident scene and had little involvement with damage assessment or other activities away from the immediate area.

One of the major challenges faced by the Clark County Fire Department in this incident was the management of information. The department itself had an urgent need for information on what had happened, was happening, and could happen, to formulate a plan for operations and evacuation. This required consultation with fire department personnel, plant management, and experts from other agencies, under extremes of stress and uncertainty.

The Federal Emergency Management Agency (FEMA) led the development of the National Incident Management System (NIMS) to help overcome the problems with communication and information management at all levels of government, during emergencies (FEMA 2022).

While the process of planning and evaluation was taking place, there were immediate and constant pressures from the local news media for details and for information to broadcast to the public concerning the dangers and actions that should be taken. The time required to gather and analyze information resulted in some incorrect information being broadcast and caused widespread public confusion. At the same time the national news media were calling for more details. The Clark County Fire Department's public information officer responded and established an official source of media information within an hour after the explosion.

Many residents were in near panic from rumors of several different scenarios and dangers. Radio and television stations quickly devoted their air time to the situation, but lacked a source of accurate information during the first hour. Conflicting information was broadcast and, as a result, people in the area reported confusion about whether to stay indoors to avoid the smoke, evacuate, go to shelters, or take some other action. The confusion extended to schools in the area, with some keeping children inside and others sending students home.

Telephone lines were overloaded with people checking on each other's welfare, seeking advice from different sources, or trying to report conditions to emergency response agencies and the news media. The 9-1-1 telephone system was rapidly overloaded with concerned callers, many seeking instructions, and the cellular telephone system was overloaded.

This emphasizes the need to establish working lines of communication with the news media. But even with a good relationship, the ability to provide accurate information and to depend upon the news media to convey instructions to the public becomes very uncertain in an incident of this magnitude. In this case several inaccurate reports, including "confirmed reports of 9–14 dead" were broadcast.

There was also a concern for the safety of news gathering personnel who approached closer to the involved area than fire department personnel would venture, including helicopters circling within the danger zone.

4.1.2 Lessons Learned

1. Land development decisions must consider risks of disasters.
   The potential destructive power from an incident of this type needs to be evaluated in land use decisions. The encroachment of residential and commercial development into the area around the PEPCON plant contributed significantly to the injuries and damage. The magnitude of the incident was much greater than had been contemplated by urban planners or in pre-incident planning.

2. Need for triage outside hospitals: Large numbers of even minor injuries can overwhelm a medical facility once inside.
   Damage and injuries spread over a large area present unusual challenge to emergency services, which are accustomed to incidents occurring in a well-defined area. Large numbers of injured presented themselves to hospitals. Triage centers need to be set up outside hospitals to prevent overloads within the hospitals when the EMS cannot "capture" most victims at the site of the incident. Fortunately, in this incident there were not large numbers of seriously injured waiting for assistance and relying on public agencies for treatment and transportation. This potential needs to be considered more than it has been by local communities.

3. Disaster mutual aid plans should be established, practiced, and kept up to date.
   The value of established mutual aid and interagency coordination procedures was demonstrated once again. Many of these relationships came because of Las Vegas' experience from major fires at the MGM Grand and Hilton Hotels in 1980 and 1981. Communities should review their disaster coordination plans and make sure they are up to date.

4. Hazmat incidents require size-up from a safe distance.
   The need for a "stand back and assess the situation" strategy for some Hazmat incidents was well demonstrated. Had fire units continued at full speed to the scene they probably would have been destroyed.

5. Public information needs to be accurate and timely in a disaster.
   In spite of a good relationship and established procedures, dealing effectively with the media was a major problem in the early stages of the incident. Misinformation by the media and rumors among the public created near panic. Since the aftermath was not dangerous, it did not matter much here whether people stayed indoors or not. But in an environmentally serious incident, clear information should be given to the public as soon as possible on what to do, even if that must be changed as conditions change. The departments did a good job in providing information to the media as it became available, but the media did not wait for good information.

6. Evacuation plans and implementation must consider human nature and the media.

The roads in the area were virtually gridlocked by spectators going toward the scene and residents fleeing. Media misinformation, the failure to broadcast adequate, specific requests to stay away from the scene, and the inability to control the roads early enough, exacerbated the traffic situation. The need for traffic control around high hazards should be considered in disaster plans.

7. The problem of assessing the immediate risk of Hazmat releases and products of combustion on the surrounding area needs additional research and development.

   Determining the risk of explosion and the risk of toxic fumes to the public and to their firefighters can be extremely difficult. Some aspects of risk assessment are too specialized to be covered in general Hazmat training courses. Special expertise needs to be called into play when unusual or exotic Hazmat is known to be present in a plant or other location. The "worst case" situation needs to be anticipated.

8. Industrial safety plans need to be established, kept up-to-date, and understood by all employees.

   The employees in the chemical plant averted a life loss catastrophe by fleeing immediately after the first ("small") explosion and fireball. There was no alarm system and no evacuation plan, according to the employees. Plants such as this should have explicit emergency plans, and all employees should be trained.

9. Safety of the disabled needs to be considered in high-hazard occupancies.

   One of the two fatalities in this fire was a wheelchair-bound employee who obviously could not just run across the desert or jump in his car, as most others did. Society now encourages and assists the disabled to visit and work in a much wider range of occupancies, which exposes them to new risks. The disabled need to be given realistic appraisal of their risks in a potential emergency and their alternatives for escape or refuge, even though the probability of the event occurring is low. Volunteers might be assigned to disabled individuals to help them to escape by car or truck, or by being wheeled or carried to a safe distance. This may suffice for most emergencies, but in the face of a catastrophic explosion the value of such assignments may be moot.

10. The section on AP in the DOT Emergency Response Guidebook needs to be revised.

    The danger of explosion from large (over 45 µm) particle size AP when exposed to flames seems to be much greater than indicated in the manual. The small particle size might be viewed as a Class A explosive and the large particle size version as almost that dangerous.

4.1.3 Safety Culture Issues

Welding/cutting activities in manufacturing facilities is defined as "hot work" by OSHA 29 CFR 1917.152. Welding related activities are associated with numerous fires and explosions each year. The massive explosions at the PEPCON manufacturing facility outside Henderson, Nevada, on May 4, 1988, are an example of such an event. The welding might have caused the fire that led to the explosion, but there were many safety culture factors that laid the groundwork for the explosion.

An article by Sahagun (1989) in the Los Angeles Times presents several different scenarios for what caused the fire and explosion. These range from a natural gas leak to the welding operation to a dryer used to dry the AP. The facts are that there was too much storage of the chemical, the facility was built from combustible materials, and there was no fire suppression system to slow an incipient fire. The article also said that the Nevada Division of Occupational Safety and Health (NDOSH) and the Clark County Fire Department failed to ensure the plant followed safety and fire codes and regulations.

When I look at a map of where the plant was in Henderson, I am amazed there were not more fatalities and injuries. At that time the plant was in a relatively empty part of town, but still relatively close to populated areas. Plants with such potential for fires and explosions need to be located away from populated areas. However, developers are constantly looking for new land to build on and these new developments creep up on these hazardous plants.

4.2 PCA DeRidder Paper Mill Gas System Explosion, DeRidder, Louisiana – February 8, 2017

There are 135 paper mills in operation in the United States (IBIS World 2021).

Bajpai states that "Pulp and paper mills are highly complex and integrate many different process areas including wood preparation, pulping, chemical recovery, bleaching, and papermaking to convert wood to the final product. Processing options and the type of wood processed are often determined by the final product. The pulp for papermaking may be produced from virgin fiber by chemical or mechanical means or may be produced by the repulping of paper for recycling. Wood is the main original raw material. Paper for recycling accounts for about 50% of the fibers used – but in a few cases straw, hemp, grass, cotton, and other cellulose-bearing material can be used. Paper production is basically a two-step process in which a fibrous raw material is first converted into pulp, and then the pulp is converted into paper. The harvested wood is first processed so that the fibers are separated from the unusable fraction of the wood, the lignin. Pulp making can be done mechanically or chemically. The pulp is then bleached and further processed, depending on the type and grade of paper that is to be produced. In the paper factory, the pulp is dried and pressed to produce paper sheets. Post-use, an increasing fraction of paper and paper products is recycled. Non recycled paper is either landfilled or incinerated" (Bajpai 2015).

From personal experience, there is a great amount of foul-smelling odor that comes from some paper and pulp mills, depending on the manufacturing process being used. Wisconsin Department of Health Services says that one type of odor comes from a special technique – called kraft pulping – which uses heat and chemicals to pulp wood chips for making paper. This reaction produces gaseous sulfur compounds called "total reduced sulfur" or TRS gases. The odor associated with TRS gases is typically described as "rotten cabbage" or "rotten eggs." The US Environmental Protection Agency (EPA) finds these odors are a "nuisance" but not a health concern, at levels normally found in the environment. However, there are times when the normal levels are exceeded (EPA 2022).

Non-condensable gases, usually referred to as TRS or Dilute Vent Gases, are by-products of the kraft pulp process. These gases, mostly sulfurous, are extremely malodorous and flammable (Wisconsin Department of Health Services 2021). Because venting non-condensable gases directly into the atmosphere is prohibited for environmental reasons, pulp mills usually collect and incinerate them in a waste or recovery boiler, a lime kiln, or an incinerator. In the past, continuous sampling and gas monitoring was a problem for three key reasons:

The environment that needs to be sampled is often saturated with water vapor. Combined with corrosive aspect of the TRS gases, most sensors will stay operational for only a short period of time.

Most sensors can only operate at ambient, as opposed to process temperatures, so they are positioned far away from incinerators, using a snaking pipe arrangement and pumps to cool samples. This convoluted network can lead to many samples' delivery and maintenance problems, particularly when TRS is saturated with water vapor.

Many flammable gas monitors are accurate for only a small group of gases. The varied gases present in pulp mill applications often exceed the accuracy capabilities of many sensors. It's crucial to find a monitor that is accurate for all the gases that are present.

The WestRock MSDS for TRS states that "Total Reduced Sulfur Gases (TRS) are formed as off-gases in the Kraft pulping and recovery process. The gaseous mixture consists mainly of hydrogen sulfide (H_2S), methyl mercaptan (MM), dimethyl sulfide (DMS) and dimethyl disulfide (DMDS) gases. TRS gas emissions are generated in the process of Kraft pulping or acidification of liquors and vary in their concentrations and relative percentages greatly, depending on where in the process the emissions are generated. Gases are generated in multiple stages in the process and can be in higher concentrations in areas such as the digester and stripper off gas streams and in lower concentrations in areas such as tank air spaces and smelt dissolving tanks. There are many other processes, vessels, and process streams which generate TRS gas emissions in the mills (e.g. Black Liquor). TRS gases can exist in liquid phase depending on pressure and temperature (WestRock 2015)."

The NFPA ratings for TRS are:

NFPA Rating (Scale 0–4): Health = 4 Fire = 4 Reactivity = 0

Turpentine is another by product of paper production. Turpentine has the chemical formula C10H16. However, various other related chemicals called terpenes can be distilled off with turpentine. The NFPA ratings for Turpentine are:

Health – 1; Flammability – 3; and Reactivity – 0

The flash point for Turpentine is 95 °F or 35 °C.

4.2.1 PCA DeRidder Mill

The DeRidder mill was built in 1969. In October 2013, PCA acquired Boise Inc., the former owner and operator of the DeRidder mill. This acquisition furthered PCA's capability to manufacture both packaging and paper products (CSB 2018).

The PCA DeRidder mill consists of a pulp mill area, where pulp is made from softwood chips and recycled corrugated cardboard, and a paper mill area, where containerboard paper is produced from the pulp (Figure 4.5) (CSB 2018).

The PCA website says that the DeRidder mill has two containerboard machines. The No. 1 machine produces linerboard grades 26# through 52#, including many high-performance and specialty grades. The No. 3 machine can produce both corrugating medium and lightweight linerboard.

PCA has the capacity to annually produce about 4.8 million tons of kraft linerboard (the outer layers of a corrugated container) and semi-chemical corrugating medium (the fluted center layer) for PCA's own converting facilities, as well as for numerous domestic and international customers (PCA 2022).

The major components of the plant are shown in Figures 4.6 and 4.7.

4.2.2 The Explosion

The sequence of events that led to the explosion are listed in the following text, condensed from the CSB report:

> The mill was undergoing its annual planned maintenance outage at the time of the accident. The foul condensate tank likely contained water, a layer of flammable liquid turpentine on top of the water, and an explosive vapor space containing air and flammable turpentine vapor.

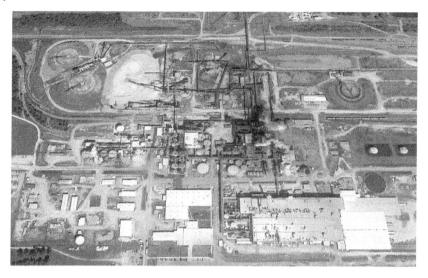

Figure 4.5 PCA DeRidder Plant. Source: Chemical Safety Board (2018).

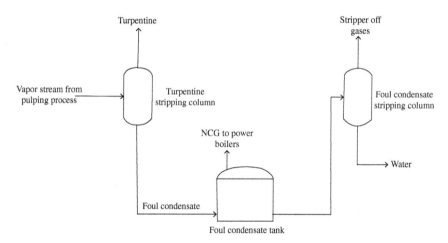

Figure 4.6 Diagram of components of the PCA DeRidder Mill. Source: Chemical Safety Board (2018).

On Wednesday, February 8, 2017, at approximately 11:05 a.m., a foul condensate tank, part of a non-condensable gas system, exploded. The explosion killed three people and injured seven others. All 10 people were working at the mill as contractors. The explosion also heavily damaged the surrounding process. The foul condensate tank traveled approximately 375 ft and over a six-story building before landing on process equipment.

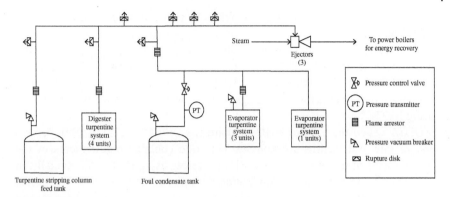

Figure 4.7 Diagram 2 of components of PCA DeRidder Mill. Source: Chemical Safety Board (2018).

The CSB report stated that "Although some oxygen in the vapor space of the foul condensate tank is normal, by design the flammable vapor inside the tank should be kept above its explosive limit, also known as its upper explosive limit (UEL)." Though an MSDS for Turpentine vapor shows there is no UEL (Fisher Scientific 2021). An MSD by W.M. Barr also shows no UEL for Turpentine vapors (W.M. Barr 2021).

According to the CSB report the oxygen present should be insufficient to support combustion. Air likely entered the foul condensate tank through a vacuum relief device on the tank's roof. During typical operation, automatic controls continually cycled the liquid level inside the tank, creating routine periods of low pressure. These low-pressure conditions were relieved by the vacuum relief device, which pulled air into the tank. During the annual outage, air also likely entered the tank as its contents cooled. This cooling created another low-pressure condition within the tank. These sources of air ingress allowed air to mix with turpentine vapor and thereby form an explosive mixture in the tank's vapor space.

On the day of the incident, contractors supporting the annual outage work made repairs by welding, (a type of hot work) on water piping above and de-coupled (disconnected) from the foul condensate tank. This hot work appears to be the probable source of ignition, although other possible ignition sources could not be definitively excluded. In Chapters 3 and 5 contain more information about the importance of hot work procedures.

Federal regulation requires many types of chemical facilities that process highly hazardous substances to have a process safety management (PSM) system in place to protect the workforce and the public from catastrophic incidents (OSHA 2021). OSHA describes the need for a PSM program as: "Unexpected releases of toxic, reactive, or flammable liquids and gases in processes involving highly hazardous chemicals (HHCs) have been reported for many years, in various industries using chemicals with such properties." Regardless of the industry that uses these HHCs, there is a potential for an accidental release any time they are not properly controlled, creating the possibility of disaster. To help ensure safe and healthful workplaces, OSHA has issued the PSM of HHCs standard (29 CFR 1910.119), which contains requirements for the management of hazards associated with processes using HHCs.

OSHA issued the PSM of HHC standard (29 CFR 1910.119) in 1991, to help ensure safe and healthy workplaces. This rule contains requirements of the management of hazards associated with processes using HHCs. Additionally, in January 2017, OSHA issued a new National Emphasis Program to further protect workers' health and safety in certain industries that pose high risks to people and the environment.

These regulations were put in because of several incidents:

- Bhopal, India, gas leak in 1984
- Phillips' explosions and fire in 1989 in Pasadena, TX
- Cincinnati, OH chemical plant explosion in 1990
- Sterlington, LA explosion, and fire in 1991

PSM is addressed in specific standards for the general and construction industries. OSHA's standard emphasizes the management of hazards associated with HHCs and establishes a comprehensive management program that integrates technologies, procedures, and management practices.

The PSM standard applied to only two processes at the DeRidder mill. These protective rules neither apply to the foul condensate tank nor to most of the non-condensable gas system (process equipment that included the foul condensate tank).

During the outage, there was no planned or actual work directly on the foul condensate tank. However, work was planned on water piping located above the foul condensate tank. Months earlier, that piping shifted off its supports and cracked at the intersection of an 8-in. line and a 3-in. line that was connected to the foul condensate tank (Figure 4.8), resulting in a significant leak that was visible in a pre-incident video taken by a mill employee who planned the repair work (Figure 4.8).

Although not required by Federal regulation, good-practice guidance recommends developing and implementing a robust safety management system to manage the hazards related to generating, collecting, and treating of non-condensable gases. PCA did not voluntarily apply its PSM system to the non-condensable gas system. Such an approach could have helped PCA to identify, evaluate, and control the non-condensable gas system process hazards that led to the explosion.

The CSB determined that air entered the foul condensate tank and mixed with the tank's flammable turpentine vapor. An explosive atmosphere within the tank was created when outside air was able to enter the tank.

Figure 4.8 Repaired pipe.
Source: Chemical Safety Board
(2018).

Ongoing hot work repairs above the tank likely ignited this vapor and caused the explosion.

The CSB documented its causal determination throughout this report based on a review of the evolution of mill activities related to its non-condensable gas system, which includes the foul condensate tank; a history of similar explosions in various process industries; and an analysis of all available physical, documentary, and testimonial evidence.

The CSB determined that the conditions enabling the explosion to occur could have been mitigated by a broader application of the company's PSM systems and by the use of other established industry good safety practices. Specifically, PCA did not take the following actions:

- Evaluate the majority of the non-condensable gas system, including the foul condensate tank, for certain hazards. The DeRidder mill never conducted a process hazard analysis to identify, evaluate, and control process hazards for the non-condensable gas system.
- Expand the boundaries of its PSM program beyond the units covered by safety regulations.
- Effectively apply the hierarchy of controls to the selection and implementation of safeguards that the company used to prevent a potential non-condensable gas explosion.

- Evaluate inherently safer design options that could have eliminated the possibility of air entering the non-condensable gas system, including the foul condensate tank.
- Establish which mill operations group held ownership of, and responsibility for, the foul condensate tank.
- Apply important aspects of industry safety guidance and standards, including the Technical Association of the Pulp and Paper Industry Technical Information Paper 0416-09, National Fire Protection Association Standard 69, National Fire Protection Association Standard 68, and International Society of Automation Standard 84. All these standards seek to prevent or control explosions, such as non-condensable gas explosions.

As a result of its investigation, the CSB is issuing safety guidance to the pulp and paper industry and a safety recommendation to PCA. The CSB is also reiterating a previous recommendation issued to the Occupational Safety and Health Administration, as a result of the CSB's 2002 investigation of Motiva Enterprises, to cover atmospheric storage tanks under the PSM standard.

4.2.3 Safety Culture Summary

The NFPA reports that (Ahrens 2021):

- US fire departments responded to an average of 4580 structure fires involving hot work per year in 2014–2018. These fires caused an average of 22 civilian deaths, 171 civilian injuries, and US$484 million in direct property damage per year.
- Of the fires involving hot work, 43% occurred in or on homes, including one- or two-family homes and apartments or other multifamily homes, while 57% occurred in or on non-home properties.
- While welding and cutting torches were the leading types of equipment involved in hot work fires, this varied by occupancy, as follows:
  o Welding torches were involved in 40% of the non-home hot work fires but only 31% of the home fires associated with hot work.
  o Cutting torches were involved in one-quarter of the non-home fires (25%), twice the percentage of home fires (13%).
  o Heat-treating equipment was involved in 19% of the non-home hot work fires compared with 8% of the hot work fires in homes.
  o Soldering equipment was involved in more than five times the number of hot work home fires (33%) compared to only 6 percent of such fires in non-home properties.
- The peak areas for home fires involving hot work were wall assemblies or concealed spaces (16%) and bathrooms or lavatories (13%).

- Exterior roof surfaces (12%) and processing or manufacturing areas (11%) were the leading areas of origin for non-home fires.
- Leading items first ignited included the following:
  - Structural members or framing in one-quarter of the home fires (24%) but only 8% of the non-home fires.
  - Insulation within structural areas in 20% of the home hot work fires but only 9% of the non-home fires.
  - Flammable or combustible liquids or gases, filters, or piping in 17% of the non-home fires but only 5% of the home fires.
  - Exterior roof coverings or finishes in 10% of the non-home fires and 6% of the home fires.
- From 2001 to 2018, five firefighters were fatally injured in four unintentional fires started by torches.

4.3 West Fertilizer Explosion – April 17, 2013

The West fertilizer explosion is another example of an AN storage fire and explosion incident (CSB 2016). The CSB report contains a very detailed analysis of the explosion and fire. The abbreviation the CSB used in this report is Fertilizer Grade AN (FGAN). The reason for this will become clear later in this section. This incident occurred in the small town of West, TX, about 80 mi south of Dallas, TX. The name of the company was the West Fertilizer Company (WFC). The town of West had a population of 2800. The WFC stored a tremendous amount of agricultural chemicals because of the demand by local farmers. Table 4.1 contains a list of the chemicals being stored on the day of the event.

Table 4.1 Chemicals stored at the West Fertilizer Company on April 17, 2013.

Fertilizer name	Amount (tons)
FGAN (fertilizer building)	40–60
FGAN (railcar)	100
Anhydrous ammonia	17
Potash	45
Diammonium phosphate	70
Diammonium phosphate and potash	25
Ammonium sulfate	60–70
Zinc sulfate	17.5

Source: Data from Chemical Safety Board (2016).

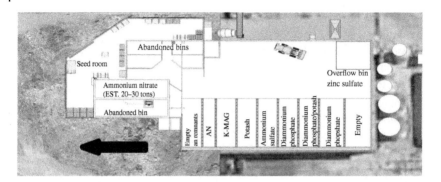

Figure 4.9 Layout of the West Fertilizer Company Building. Source: Chemical Safety Board (2016).

As has been discussed many times in this book, developments had been creeping up on the property boundary of the WFC. The CSB report writes that a park was created less than 150 ft, an apartment complex, the nearest aggregation of homes were about 370 ft, West Intermediate School (WIS) was a little more than 200 ft, and West High School (WHS) was about 500 ft from the plant property line.

Figure 4.9 shows the basic layout of the WFC.

4.3.1 The Fire and Explosion

The following event description is excerpted from the CSB (2016) report. On April 17, 2013, at approximately 7:29 p.m., citizens reported signs of smoke and fire at the WFC facility to the local 911 dispatch center. Within 20 minutes, a massive explosion occurred, killing 15 people, and sending a blast wave through the town that damaged or destroyed many buildings and homes. The fire was witnessed from several vantage points by different individuals associated with the West Police Department, Dallas Fire-Rescue Department, and volunteer fire departments (VFDs) from West, Abbott, Bruceville-Eddy, Mertens, and Navarro Mills.

One of the first responders to the incident was a West Police Department officer who was on routine patrol that evening. The officer reported that he smelled smoke as he was driving through the city park but was not able to identify the exact location of the smoke until he encountered a concerned citizen who advised him that smoke was venting from the highest portion of the WFC building. The officer advised the dispatch center of the smoke and requested that the West Volunteer Fire Department (WVFD) be dispatched to the WFC facility. Once the officer arrived on scene, he witnessed flames that were visible through the wall, extending upward from the lower level to the upper level of the northeast corner of the two-story fertilizer storage building. Then he called dispatchers again and asked them to inform the WVFD that the smoke had escalated to a structure fire.

The WVFD contacted the officer via radio and requested that he establish traffic control to prevent citizens from driving over the fire hoses once the fire engines arrived and laid down fire hoses. The officer agreed but notified the WVFD that he needed to evacuate the city park first. As the officer proceeded to the city park, the responding West firefighters drove past him, heading toward the facility. Once the officer reached the city park, he used his public address system to order an evacuation of the park. After the park was evacuated, he left the area to establish traffic control on the north end of the fertilizer facility. There was no traffic control at the south end toward WHS, so the officer asked a nearby resident to assist by using his truck to block that intersection. At this time, the officer contacted the police chief and another officer who had called to determine whether he needed assistance. The officer asked the police chief to establish traffic control by the WIS and requested that the other officer relieve the resident who was helping near the high school.

Numerous citizens began parking their cars at WHS to watch the fire. The WVFD truck left the WFC facility and headed toward the police officer. The manager of the WFC arrived on scene to assist the WVFD. Via radio traffic, the officer learned that the entire fertilizer storage building was engulfed in flames, and shortly thereafter he saw and felt the explosion. The officer was briefly disoriented and then unsuccessfully attempted via radio and cell phone to notify the dispatch center of the explosion. An injured member of the public and an injured firefighter approached the officer, who assisted them. The possibility of further explosions or toxic releases was a concern because of the anhydrous ammonia pressure vessels on the south side of the WFC property. Based on this information, the officer decided to evacuate homes within a 1-mi radius. Because the officer's patrol car would not start, he proceeded on to alert people to evacuate. By the time the officer made his way to Reagan Street, he had become aware that other emergency responders had initiated the evacuation of the northern portion of the city.

Emergency dispatchers paged the WVFD, and firefighters responded to the scene with two fire engines, two initial attack apparatus or brush trucks and a water tender truck at various times. Dispatchers also paged mutual aid personnel from neighboring counties, including Abbott, which responded. Many of the firefighters also responded by using their personally owned vehicles (POVs). According to eyewitness accounts, the fire intensified very quickly and was described as a rolling fire that moved from the northeast end of the fertilizer building in the seed storage area north of the office toward the southern end of the building.

Five firefighters arrived on scene in two fire engines at different times. The first fire engine arrived on scene and staged east of the burning structure while one of the brush trucks staged to the north of the first fire engine. Four other firefighters directed water using two 1.5-in. hoses from the first fire engine's internal tank onto the fire through the northeast doorway of the bagged fertilizer room, where fire was present. Once the second fire engine arrived on scene, the two firefighters from that fire engine began laying 1000 ft of 4-in. hose line from the fire hydrant near the high school 1600 ft away toward the fertilizer facility. After laying all the hose lines from the second fire engine, they discovered that the hose was approximately 700 ft short of the length needed to effectively fight the fire. After assessing the situation, one firefighter arranged to take the first fire engine, which had a better pump with greater pressure capabilities and additional hose that would allow him to continue to reverse-lay the lines. However, rather than resuming where the first fire engine ran out of hose, the firefighter went back to the fire hydrant near the high school to connect the first engine to the hydrant without laying the additional length of hose needed to supplement the hose that had already been laid from the second engine. He saw flames 40–50 ft high coming out of the cupola atop the fertilizer storage building and out of the door on the northeast corner of the building. Before the firefighter could make his way back to the end of the hose run, the explosion occurred. Before the explosion, the WVFD assistant chief arrived at the WFC facility, spoke with the police officer on scene, and advised him to begin evacuating nearby homes. He also made a radio request to the dispatch center, asking for a ladder truck to set up at the West Terrace Apartments in case a fire started there, but a ladder truck was not available. The WVFD chief and assistant fire chief were assessing the situation just before the explosion and were considering a total evacuation, even though neither believed that the FGAN would explode.

Based on that CSB conducted after the incident, the WVFD came to understand that it did not have enough water to effectively fight the fire. Accordingly, the WVFD was considering the appropriate course of action, possibly standing down, letting the structure burn, and focusing on evacuation.

On the evening of the incident, a group of volunteer firefighters from neighboring city fire departments including Bruceville-Eddy, Mertens, and Navarro Mills, who were taking an Emergency Medical Technician (EMT)–Basic class at the West Emergency Medical Services (EMS) building, responded to the fire. The West EMS facility is located a few blocks west of the WFC facility. When these volunteer firefighters heard the sirens activated in the city, they immediately made their way to the site. In addition, an ambulance responded with two EMTs and a volunteer firefighter. According to interviews that CSB conducted with emergency responders, radio and cell phone capabilities at the scene were limited after the explosion. Following the explosion, officials established two different staging areas. The first staging area, at the high school football field about 0.25 mi from the blast site, was used as a triage area for injured residents. Injured personnel and residents were relocated from the football field to the second staging area, at the community center about 1 mi away. After the explosion at

Figure 4.10 Photos of the West Fertilizer Company Explosion. Source: Chemical Safety Board (2016).

approximately 8:15 p.m., additional volunteer firefighters from the neighboring cities of Abbott, Bruceville-Eddy, Mertens, and Navarro Mills responded to the WFC facility. Figure 4.10 shows still photos of the WFC explosion as it unfolded.

4.3.2 Injuries and Fatalities

The explosion at the WFC facility fatally injured 12 emergency responders and 3 members of the public. All the fatalities except one resulted from fractures, blunt force trauma, or blast force injuries sustained at the time of the explosion. Two fatally injured members of the public lived at a nearby apartment complex while the third resided at the nursing home and died from injuries brought on by the trauma of the explosion shortly after the incident. The incident resulted in more than 260 injured victims, including emergency responders and members of the public.

4.3.3 Safety Culture Summary

Three incidents in this chapter are all related in that there was a general perception that AN and AP could not explode. How can this thinking still be taking place when before any of these events AN was being used as an explosive. AN or FGAN, along with fuel oil and other chemicals was used in the Oklahoma City bombing by Timothy McVeigh (FBI 2022). The pervasive thought that "this could never happen here or to me" is so dangerous. Bad accidents and events happen every day because of this attitude. Chemists don't always understand how a chemical will react under various conditions (Department of Education 2014). However, AN and AP have well known reactions. They provide the oxygen for explosions, and they can decompose violently on their own. They should not be stored in large quantities, in facilities built from combustible materials. Also, proper fire protection/suppression equipment should be in place to combat incipient fires (NFPA 2022).

Finally, first responders should be informed of all chemicals in a facility and the hazardous nature of those chemicals. Many years ago, one of my emergency management classes went to every business in our local community and collected the names of and quantities of hazardous material located on their property. The students put this into binders and supplied the binders to the fire department. This effort was quite an eye opener as far as the types of materials in our town.

The morning I am finishing this chapter there is a fire burning at a Winston-Salem, North Carolina fertilizer plant and the town has evacuated 6000 people in the event the plant explodes (Boyette 2022). AN is dangerous stuff.

References

Ahrens, M. (June 2021). NFPA, Structure Fires Caused by Hot Work. https://www.nfpa.org/-/media/Files/News-and-Research/Fire-statistics-and-reports/US-Fire-Problem/Fire-causes/osHotWork.ashx.

Bajpai, P. (2015). *Management of Pulp and Paper Mill Waste*. Springer International.

Barr, W.M. and Company (2021). Klean Strip Pure Gum Spirits Turpentine, MSDS. https://www2.pcad.edu/Facilities/health_safety/SDS/Painting/Solvents/W.M.%20Barr/klean%20strip%20pure%20gum%20turpentine.pdf (accessed December 2021).

BBC (2021). Beirut port explosion investigation suspended for second time. https://www.bbc.com/news/world-middle-east-58705687 (accessed January 2022).

Boggs, T.L. (1970). Deflagration rate, surface structure and subsurface profile of self-deflagrating single crystals of ammonium perchlorate. *AIAA Journal* 8 (5): 867–873. https://doi.org/10.2514/3.5780. Bibcode:1970AIAAJ...8..867B.

Boyette, C. (2022). CNN, 6,000 urged to evacuate as North Carolina fertilizer plant fire threatens an ammonium nitrate explosion. https://www.cnn.com/2022/02/01/us/winston-salem-fertilizer-plant-fire/index.html (accessed February 2022).

Chemical Safety Board (CSB) (2016). West Fertilizer Explosion and Fire, Investigation Report.

Chemical Safety Board (CSB) (2018). Non-Condensable Gas System Explosion at PCA DeRidder Paper Mill.

Department of Education, Government of Queensland, Australia (2014). Safety alert, unpredictable science experiments. https://education.qld.gov.au/initiativesstrategies/Documents/unpredictable-science-experiments.pdf.

Environmental Protection Agency (2022). Paper mill odor. EPA.gov (accessed January 2022).

FBI (2022). History, Oklahoma City Bombing. https://www.fbi.gov/history/famous-cases/oklahoma-city-bombing (accessed January 2022).

Federal Emergency Management Agency (FEMA) (2022). National emergency management system, NIMS. https://www.fema.gov/emergency-managers/nims (accessed January 2022).

Fisher Scientific (2021). Turpentine Oil MSDS. https://fscimage.fishersci.com/msds/24580.htm (accessed December 2021).

Homeland Security (May 1988). U.S. Fire Administration/Technical Report Series, Fire and Explosions at Rocket Fuel Plant, Henderson, Nevada, USFA-TR-021.

IBIS World (2021). Paper mills in the US – number of businesses 2002–2027. https://www.ibisworld.com/industry-statistics/number-of-businesses/paper-mills-united-states/ (accessed January 2022).

NASA (2012). The PEPCON ammonium perchlorate plant explosion, November 2012 Volume 6 Issue 9, 2012.

NFPA (2022). Standard for the Installation of Sprinkler Systems.

OSHA (2021). Process Safety Management. https://www.osha.gov/process-safety-management (accessed December 2021).

PCA (2022). DeRidder containerboard mill. https://www.packagingcorp.com/deridder-containerboard-mill (accessed January 2022).

Relief Web (2022). Lebanon: Beirut port explosions – Aug 2020. https://reliefweb.int/disaster/ot-2020-000177-lbn (accessed January 2022).

Sahagun, L. (1989). Los Angeles times, hazards at nevada plant blamed for explosion. https://www.latimes.com/archives/la-xpm-1989-04-28-mn-2000-story.html (accessed January 2022).

The Chemical Company (2022). Ammonium perchlorate. https://thechemco.com/chemical/ammonium-perchlorate/ (accessed January 2022).

Westrock (2015). Total reduced sulfur gases MSDS. https://www.westrock.com/-/media/pdf/safety-data-sheets/wr0013trsgases.pdf (accessed January 2022).

Wisconsin Department of Health Services (2021). Pulp and paper industry odors. https://www.dhs.wisconsin.gov/air/pulpodors.htm (accessed January 2022).

5

Dust Explosions and Entertainment Venue Case Studies

5.0 Introduction

According to a study by the Chemical Safety Board, Dust explosions are a serious problem in American industry. Over the last 28 years there have been approximately 3500 combustible dust explosions, 281 of these have been major incidents resulting in the deaths of 119 workers and another 718 workers sustained injuries.

There were 13 reported agricultural dust explosions in the United States in 2005 resulting in 2 fatalities and 11 injuries.

The Occupational Safety and Health Administration (OSHA) has developed a set of physical properties to describe combustible dusts (OSHA 2009). These are:

- MIE, the minimum ignition energy, which predicts the ease and likelihood of ignition of a dispersed dust cloud.
- MEC, the minimum explosible concentration, which measures the minimum amount of dust dispersed in air required to spread an explosion. (The MEC is analogous to the lower flammable limit (LFL) or lower explosive limit (LEL) for gases and vapors in air).
- K_{st}, the dust deflagration index, measures the relative explosion severity compared to other dusts. The larger the value for K_{st}, the more severe the explosion (see Table 5.1). K_{st} provides the best "single number" estimate of the anticipated behavior of a dust deflagration

The concept of a major explosion due to dust was hard for most people to imagine. However, dust explosions happen many times a year. Several sources discussed one of the first documented dust explosion occurred in Turin, Italy in 1785 (Beyond Discovery 2015; Hughes 2021; ATEX 2021). This dust explosion was investigated extensively. The wheat flour explosion in Mr. Giacomelli's Bakery in Turin was a comparatively minor one. The considerations related to the low moisture content of the flour due to dry weather are still highly relevant. The same applies to the

Impact of Societal Norms on Safety, Health, and the Environment: Case Studies in Society and Safety Culture,
First Edition. Lee T. Ostrom.

Table 5.1 Examples of K_{st} values for different types of dusts.

Dust explosion class[a]	$K_{st\ (bar.m/s)}$[b]	Characteristic[b]	Typical material
ST 0	0	No explosion	Silica
ST 1	>0 and $\leq$200	Weak explosions	Powdered milk, charcoal, sulfur, sugar, and zinc
ST 2	>200 and $\leq$300	Strong explosion	Cellulose, wood flour, and poly methyl acrylate
ST 3	>300	Very strong explosion	Anthraquinone, aluminum, and magnesium

The actual class is sample specific and will depend on varying characteristics of the material such as particle size or moisture.
a) NFPA 68, Standard on Explosion Prevention by Deflagration Venting.
b) OSHA CPL 03-00-008 – Combustible Dust National Emphasis Program.

observation of a primary explosion causing a secondary explosion by entrainment of dust deposits.

When the Academy of Science of Turin heard about Mr. Morozzo's investigations, they asked him to prepare a written account of his findings. The following are observations made by Mr. Morozzo's that I have edited to make them more readable.

At 6 p.m. on December 14, 1785, in the house of Mr. Giacomelli, a baker in this city, an explosion occurred in his shop. The shop overlooked the street. The noise of the explosion was as very loud and was compared to a large firecracker, and it was heard at a considerable distance away. The explosion, had a very bright flame, which lasted only a few seconds, was seen in the shop. It was immediately observed, and the fire proceeded from the flour warehouse that was situated over the back shop. At the time a boy was employed to stir flour by the light of a lamp. The boy had his face and arms scorched by the explosion. The boy's hair was burned, and it was more than a couple weeks before his burns were healed. He was not the only victim of this event. Another boy, who happened to be upon a scaffold, in a little room on the other side of the warehouse saw the flame and made his passage that way. He thought the house was on fire. He jumped down from the scaffold and broke his leg.

The flour warehouse that was located above the back shop was 6 ft high by 6 ft wide, and about 8 ft long. It was divided into two parts, by a wall and had an arched ceiling extends over both. The pavement of one part was raised about two feet higher than that of the other. In the middle of the wall was an opening through which the flour was conveyed from the upper chamber into the lower one.

The boy who was employed in the lower chamber was collecting flour to supply the bolter below. He dug out the sides of the opening to make the flour fall from the upper chamber, so it would fall into the area where he was. As he was digging a great quantity of flour fell, followed by a thick cloud. The flour immediately caught fire from the lamp hanging on the wall and caused the explosion.

The flame pushed itself in two directions. It propagated into the upper chamber of the warehouse and into a very small room above it. The greatest fire took place in the smaller chamber. The flame front went in the direction of a small staircase that leads into the back shop, where another explosion occurred. This rattled the building and threw down frames of the windows that looked into the street. The baker saw the fire before he felt the shock of the explosion.

The baker said he had never had flour so dry as in that year (1785), during which the weather had been remarkably dry. There having been no rain in Piedmont for the space of five or six months. He attributed the accident that had happened in his warehouse to the extraordinary dryness of the corn.

The fire was not entirely new to the baker. He said when he was a boy, he witnessed a similar conflagration. It took place in a flour warehouse, where they were pouring flour through a long wooden trough, into a bolter, while there was a lamp with a flame on one side of the room. However, in this case, the fire was not followed by an explosion.

Mr. Morozzo connected the fact that the unusually dry flour contributed to the fire and explosion.

This section discusses the safety culture issues associated with dust explosion case studies. Please read the references for more detailed information about the explosions.

5.1 Dust Explosion Information and Case Studies

In this section dust explosions involving metal dust, sugar dust, and flour will be presented and the safety culture issues associated with these explosions.

Dust explosions can occur if the right conditions exist for any of the compounds listed in Table 5.2.

Table 5.3 is a partial listing of dust explosions that have occurred in the last 20 years.

Table 5.2 List of compounds with dust explosion potential.

Agricultural products	Parsley (dehydrated)	Lead stearate
Egg white	Peach	Methyl-cellulose
Milk, powdered milk, nonfat, dry	Peanut meal and skins	Paraformaldehyde
	Peat	Sodium ascorbate
Soy flour starch, corn Starch, rice starch, wheat	Potato, Potato flour Potato starch	Sodium stearate
		Sulfur
	Raw yucca seed dust Rice dust	**Metal dusts**
Sugar		Aluminum
Sugar, milk sugar, beet	Rice flour, rice starch, rye flour, semolina	Bronze
Tapioca, whey, wood flour		Iron carbonyl
Agricultural dusts	Soybean dust, spice dust	Magnesium
Alfalfa, apple, beet root	Spice powder, sugar (10x), sunflower	Zinc
Carrageen, carrot		**Plastic dusts (poly)**
Cocoa bean dust, cocoa powder, coconut shell dust	Sunflower seed dust, tea Tobacco blend, tomato, walnut dust, wheat flour	Acrylamide (poly)
		Acrylonitrile (poly)
Coffee dust		Ethylene
Corn meal, cornstarch	Wheat grain dust, wheat starch, xanthan gum	Epoxy resin
Cotton	Carbonaceous dusts, charcoal, activated	Melamine resin
Cottonseed, garlic powder		Melamine, molded
Gluten	Charcoal, wood	(Phenol-cellulose)
Grass dust	Coal, bituminous coke, petroleum, lampblack	Melamine, molded
Green coffee		(wood flour and mineral filled
Hops (malted)	Lignite	phenol-formaldehyde)
Lemon peel dust, lemon pulp, linseed	Peat, 22%H20	(Poly)methyl acrylate (poly)
	Soot, pine, Cellulose, Cellulose pulp, Cork	
Locust bean gum, malt		Methyl acrylate, emulsion polymer
Oat flour	Corn	
Oat grain dust	Chemical Dusts	Phenolic resin (poly)
Olive pellets	Adipic acid	Propylene
Onion powder	Anthraquinone	Terpene-phenol resin, urea-formaldehyde/ cellulose, molded
	Ascorbic acid	
	Calcium acetate	(Poly)vinyl acetate/ ethylene copolymer
	Calcium stearate	
	Carboxy-methylcellulose	(Poly)vinyl alcohol (poly)
	Dextrin	Vinyl butyral (poly)
	Lactose	Vinyl chloride/ethylene/ vinyl acetylene suspension Copolymer
		(Poly)vinyl chloride/vinyl acetylene emulsion copolymer

Table 5.3 Listing of dust explosions.

Company and type of event	Event description
Didion Milling Company Explosion and Fire	On May 31, 2017, combustible dust explosions at the Didion Milling facility in Cambria, Wisconsin, killed five of the 19 employees working on the night of the incident. The other 14 were injured. The investigation is ongoing.
Imperial Sugar Company dust explosion and fire	On February 7, 2008, a huge explosion and fire occurred at the Imperial Sugar refinery northwest of Savannah, Georgia, causing 14 deaths and injuring 38 others, including 14 with serious and life-threatening burns. The explosion was fueled by massive accumulations of combustible sugar dust
US Ink Fire throughout the packaging building	On October 9, 2012, at approximately 1:15 p.m. Eastern Standard Time (EST), a flash fire caused burn injuries to seven workers, including three who sustained third-degree burns, at the US Ink/Sun Chemical Corporation ink manufacturing facility in East Rutherford, New Jersey. Before October 2012, the facility used a wet scrubber system 7 to collect particulate materials during the dry material charging stages of the ink mixing process. However, the scrubbing system deteriorated over the years. The new dust collection system was installed to improve the management of particulate material and produce an overall improvement in the operating conditions of the black ink production process. A flammable mixture consisting of hydrocarbons and combustible dusts accumulated in the ductwork during the start-up of US Ink's dust collection system. The mixture spontaneously ignited leading to a series of events that caused a flash fire.
Incident Description: Hoeganaes Corporation Fatal Flash Fires	Three combustible dust incidents over a six-month period occurred at the Hoeganaes facility in Gallatin, TN, resulting in fatal injuries to five workers. At the third incident on May 27, 2011, the trench involved contained many pipes including nitrogen and hydrogen supply and vent pipes for band furnaces. In addition to housing the pipes, the trench also acted as a drain for cooling water used in the band furnaces. At the time of the incident, this water came out of the furnaces how and drained directly onto the pipes and into the trench. Hoeganaes did not regularly inspect the pipes in the trench. The design and maintenance of this trench should have addressed the issue of slow corrosion over time caused by the hot water runoff and solids accumulation. Hoeganaes did not have a procedure to inspect piping within the trench to ensure that corrosion had not compromised the piping systems which would allow an uncontrolled release of hydrogen.

(continued)

Table 5.3 (Continued)

Company and type of event	Event description
AL Solutions Fatal Dust Explosion	An explosion ripped through the New Cumberland A.L. Solutions titanium plant in West Virginia on December 9, 2010, fatally injuring three workers. The workers were processing titanium powder, which is highly flammable, at the time of the explosion.
Combustible Dust Hazard Investigation	In 2003, the CSB launched investigations of three major industrial explosions involving combustible powders. These explosions – in North Carolina, Kentucky, and Indiana – cost 14 lives and caused numerous injuries and substantial property losses. The Board responded by launching a nationwide study to determine the scope of the problem and recommend new safety measures for facilities that handle combustible powders. The CSB issued its final report at a public meeting in Washington, DC, on November 9, 2006, calling for a new OSHA regulatory standard designed to prevent combustible dust fires and explosions.
West Pharmaceutical Services Dust Explosion and Fire	On January 29, 2003, an explosion and fire destroyed the West Pharmaceutical Services plant in Kinston, North Carolina, causing six deaths, dozens of injuries, and hundreds of job losses. The facility produced rubber stoppers and other products for medical use. The fuel for the explosion was a fine plastic powder, which accumulated above a suspended ceiling over a manufacturing area at the plant and ignited.
CTA Acoustics Dust Explosion and Fire	On February 20, 2003, an explosion and fire damaged the CTA Acoustics manufacturing plant in Corbin, Kentucky, fatally injuring seven workers. The facility produced fiberglass insulation for the automotive industry. CSB investigators have found that the explosion was fueled by resin dust accumulated in a production area, likely ignited by flames from a malfunctioning oven. The resin involved was a phenolic binder used in producing fiberglass mats.
Hayes Lemmerz Dust Explosions and Fire	On the evening of October 29, 2003, a series of explosions severely burned two workers, injured a third, and caused property damage to the Hayes Lemmerz manufacturing plant in Huntington, Indiana. One of the severely burned men subsequently died. The Hayes Lemmerz plant manufactures cast aluminum automotive wheels, and the explosions were fueled by accumulated aluminum dust, a flammable byproduct of the wheel production process.

Source: CSB (2018)/Public domain.

The case studies in this section are:

- AL Solutions, December 9, 2010
- Imperial Sugar, February 7, 2008
- Crowd Surge, Bar and Night Club Fires
- New Taipei Water Park Fire, June 22, 2015, Dust Explosion

5.2 AL Solutions December 9, 2010

The following description of the accident is excerpted from the CSB Report (2014). Please note that the full report contains a much more detailed analysis of this fatal accident.

AL Solutions processes titanium and zirconium scrap metal into pressed compacts that aluminum producers use as alloy additives. AL Solutions obtains scrap from titanium and zirconium manufacturers, and the end user adds the pressed compacts to furnaces or molten metal to increase the strength of aluminum alloys.

In 2006, AL Solutions purchased Jamegy, Inc., a metal producer based in New Cumberland, West Virginia. Jamegy founded the New Cumberland facility and operated it before AL Solutions.

At the time of the incident, AL Solutions owned and operated two processing facilities. The primary office and production facility was in New Cumberland, West Virginia.

AL Solutions also has a facility for milling in Washington, Missouri. In 2010, AL Solutions employed 23 workers at the New Cumberland facility and two at the Washington facility. After the 2010 incident, AL Solutions constructed a new manufacturing facility in western Pennsylvania.

5.2.1 Facility Description

The New Cumberland facility (shown in Figure 5.1) lies on the east bank of the Ohio River in the Northern Panhandle of West Virginia, approximately 40 mi west of Pittsburgh, Pennsylvania. The New Cumberland site contains a main production.

The New Cumberland production facility operated 24 hours a day, 7 days a week, and contained processing equipment for metal milling, blending, pressing, and water treatment. Separated from the warehouse and office area by an access road, the production building was the site of the December 2010 explosion.

A variety of suppliers shipped scrap titanium and zirconium to AL Solutions in 55-gal drums (Figure 5.2). The metal typically arrived packed in water, but it also was packaged with salt or an inerting agent, such as argon gas, to reduce the risk of explosion during transit.

AL Solutions took the raw material from the drums and milled it for several hours in a batch (or lot) while submerging it under water. During milling, a blade reduced the metal particle size and removed the oxide surface layer of the raw

Figure 5.1 Overhead view of the AL Solutions site. Source: U.S. Chemical Safety Board (CSB) (2014).

Figure 5.2 Drum of scrap. Source: U.S. Chemical Safety Board (CSB) (2014).

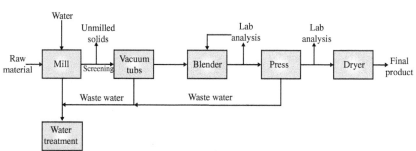

Figure 5.3 Flow of milling process. Source: CSB (2014)/Public domain.

material. When milling was completed, operators decanted the water and some fine metal particulates (known as fines) and then sent the water to treatment tanks (Figure 5.3).

The milled metal was transferred from the milling tank and screened to remove any large solids. These solids were either returned to the mill for further size reduction or discarded.

Figure 5.4 Photo of compacted titanium/zirconium. Source: U.S. Chemical Safety Board (CSB) (2014).

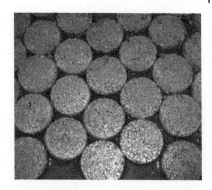

Operators placed the milled and screened metal in tubs and then vacuumed excess water into water treatment tanks. Next, operators blended the metal to ensure uniform composition in the lot.

Once the material met the laboratory specifications, it proceeded to the hydraulic presses. The blended material was pressed into 3-in.-diameter compacts (Figure 5.4). At the time of the incident, one press was used to press zirconium compacts, and two presses were used for titanium compacts. An oven dried the compacts to remove any remaining water. The laboratory analyzed the density and percent moisture of two compacts from each dryer lot before the compacts were wrapped in aluminum foil and sent to the customer.

On a typical shift, four operators worked in the production building: two press operators, one blender operator, and the shift supervisor. The shift supervisor oversaw the mill and water treatment. Another operator worked in the warehouse to dry and wrap the compacts. During the day, the plant manager, and an engineer responsible for operations were onsite to assist with any operational issues. During the day shift, two laboratory workers analyzed the blended samples and created recipes for the blenders.

Titanium (symbol Ti) is a widely used metal with unique flammability characteristics. Fine titanium particulates/powders are easily ignited in air and can ignite spontaneously at elevated oxygen concentrations or high pressures. The autoignition temperature for titanium powder is approximately 420 °F (250 °C) (US Research Nanomaterials, Inc. 2021). Titanium metal has an autoignition temperature of approximately 2128 °F (1200 °C). According to the Hazardous Materials Identification System (HMIS), titanium fines and powders have a flammability hazard rating of 3 or 4 (on a scale of 0 to 4, with 4 the highest hazard) (HMIS 2021). The AL Solutions titanium swarf materials safety data sheet (MSDS) states that titanium particles dispersed in a cloud can explode. The MSDS recommends having procedures in place to keep the powder away from static discharges, sparking equipment, and other ignition sources. Class D sodium chloride extinguishers are intended for use on combustible metal fires involving titanium and zirconium.

The MSDS also states that large quantities of water can be used to quench small titanium fires, but water should not be used on large fires because water can react with the burning metal at high temperatures to produce explosive hydrogen gas.

Titanium powder has low toxicity, but excessive inhalation can cause respiratory irritation or acute respiratory distress. The primary route of exposure is inhalation and ingestion, especially during processes that produce metal particulates (US Research Nanomaterials, Inc. 2021).

5.2.2 Zirconium

The metal zirconium (symbol Zr) also carries a significant flammability hazard. Zirconium is much more hazardous than titanium. The hazard ratings for zirconium are (Sigma-Aldrich 2014):

HMIS
  Health hazard: 1
  Chronic Health Hazard
  Flammability: 3
  Physical hazards: 3
NFPA rating
  Health hazard: 1
  Fire: 3
  Reactivity Hazard: 3

According to the zirconium MSDS, dust clouds with very small concentrations of zirconium (less than $100\,g/m^3$) are explosible. The MSDS recommends keeping the powder wet to avoid explosion hazards. If zirconium metal particulates ignite, the MSDS advises letting the material burn out and not fighting the fire; the MSDS also notes that fires in wet metal zirconium fines can result in an explosion. The MSDS recommends keeping zirconium fines either extremely dry (less than 5% water) or extremely wet (more than 25% water). Spontaneous explosions of moist, finely divided zirconium scrap have occurred during handling. Zirconium powder can cause respiratory and digestive irritation if inhaled or ingested. Zirconium can also self-heat.

5.2.3 Description of the Incident

Around noon on the day of the incident, the day shift operators returned to work from lunch. Two operators were running the three presses making titanium and zirconium compacts, and another operator was at the blender, mixing a batch of zirconium. The shift supervisor was changing the mill blade in the adjacent milling room.

Three electrical contractors were also onsite, running conduit in a hydraulic room adjacent to the blending and press room. These contractors were performing preparatory work for a maintenance outage planned for the next day.

At about 1:20 p.m., immediately before the explosion, an electrical contractor located about six feet outside a partially open door heard a loud noise that he characterized to the CSB investigators as a "metallic failure ... like something popped ... or fell." He then heard a "woof ... just how you'd light your gas grill" and "a big boom." The shift supervisor in the mill room heard a loud bang and seconds later noticed an orange glow or flame coming from the blending and press room. At about the same time, a second electrical contractor working in the hydraulic room heard an explosion in the neighboring blending and press room and then saw a fireball moving rapidly into the hydraulic room through the blending and press room door. The fireball burned his head, neck, arms, and hand as he exited the production building. The third electrical contractor was in the restroom, where he heard "an angry noise," felt a strong wind enter through a door, and then saw orange sparking flame at the ceiling.

The shift supervisor told CSB investigators that he noticed the "air was sparking" after the explosion. He had previously experienced this phenomenon at AL Solutions and knew that it signified airborne metal was burning. After the explosion, the supervisor ran outside and around the production building. At about this time, employees in the main office building, who heard the explosion, called 911 to request emergency assistance. The plant manager was walking from the warehouse to the production building at the time of the explosion and witnessed the event. Some employees and contractors reported hearing a second explosion minutes after the initial explosion, which might have been caused by a propane tank rupture from a forklift inside the building.

The explosion and fire severely burned the zirconium press operator. The supervisor, an electrical contractor, and the plant manager provided aid to the press operator until emergency personnel arrived.

A nearby volunteer firefighter heard the explosion and immediately proceeded to the scene. The New Cumberland Volunteer Fire Department (VFD) arrived minutes later. At approximately 2:30 p.m., an airlift transported the injured contractor to a hospital. By the time the VFD arrived, the building water deluge system had activated. The firefighters attempted to access the building through the office, but the fire was too intense, and they could not enter. Upon entering the original explosion area, firefighters discovered two deceased operators at the inside locations indicated in Figure 5.5.

The two operators in the blending and press room died at the scene, and the zirconium press operator died three days following the incident from severe burn injuries. The explosion and subsequent fire caused minor blast damage to doors, walls, and interior windows as well as more substantial thermal damage throughout the production area (illustrated in Figure 5.6). The explosion caused thermal damage to the wall and overhead ceiling area adjacent to the blender (shown in Figure 5.6). Equipment damage included a lift truck, the blender, and the press

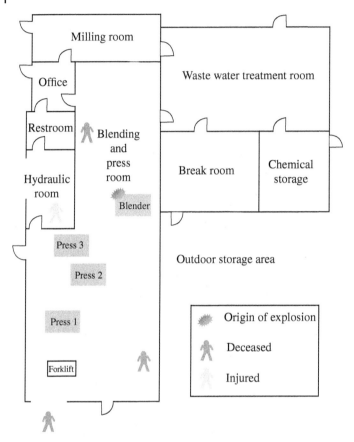

Figure 5.5 Locations of operators. Source: CSB (2014)/Public domain.

Figure 5.6 Damage inside production area. Source: U.S. Chemical Safety Board (CSB) (2014).

feed conveyor. The explosion propelled papers, desks, and lockers from the office into the parking lot outside of the production building.

The CSB investigators observed and documented the production building after the incident and concluded that the fire damage and deformations caused by the explosion overpressure were consistent with a metal dust explosion. Most solid organic materials (and many metals and some nonmetallic inorganic materials) will burn or explode if finely divided and dispersed in sufficient concentrations. Even seemingly small quantities of accumulated dust can cause catastrophic damage (CSB 2006).

5.2.4 The Origin of the Explosion

Metal particulates or dusts can produce a flash fire or explosion, using metal as the fuel and oxygen in the air as the oxidizer. As metals become finely divided in milling, blending, or crushing operations, freshly exposed surfaces on the fines or dust particles can become highly reactive. Finely divided metals such as titanium and zirconium can become pyrophoric and spontaneously combust in air (Pape and Schmidt 2009).

The AL Solutions explosion likely initiated when particulates ignited in the blender that was processing zirconium. Sparks or heat produced by metal-to-metal contact between the blender blades and the blender sidewall ignited the zirconium. Mechanical impacts, such as the blender blades against the sidewall, can produce potential ignition sources where metal-to-metal contact occurs. In rotating machinery, repeated impacts can result in hot spots with temperatures high enough to result in ignition (Eckhoff 2003).

The pre-explosion sound of metal failure or popping reported by AL Solutions employees likely originated in the blender, which showed evidence of the most substantial metal deformation after the incident. Both end walls were deformed inward and had visual indications of the blade scraping or scoring on the wall (Figure 5.7). Residual burned zirconium was detected in the blender beneath

Figure 5.7 Scraping on wall of blender. Source: U.S. Chemical Safety Board (CSB) (2014).

Figure 5.8 Blender wall. Source: U.S. Chemical Safety Board (CSB) (2014).

Figure 5.9 Ceiling above blender. Source: U.S. Chemical Safety Board (CSB) (2014).

the shaft. Severe burn damage was also seen in the press on the other side of the production room (shown in Figure 5.8) and there were indications of two adjacent cracks in one of the blender sidewalls. Investigators examined the production building wall and ceiling adjacent to the blender and found visual evidence of a burning dust cloud projected upward toward the ceiling (displayed in Figure 5.9).

Investigators examined the production building wall and ceiling adjacent to the blender and found visual evidence of a burning dust cloud projected upward toward the ceiling (displayed in Figure 5.9).

AL Solutions employees noted mechanical problems with the blender in the days before the explosion. Blender paddles were striking the sidewall of the blender and scoring the housing. To address these problems, maintenance personnel adjusted the blender blades to increase clearance from the blender wall, but this action did not permanently address the issue with the metal-to-metal contact. When the problem returned, maintenance personnel decided to replace parts of the blender to resolve the issue. While the operators were disassembling the blender, they discovered a large groove cut in the shaft and an 8–12-in. crack in the sidewall of the blender, most likely caused by the paddle blades scraping the sidewall. The maintenance worker welded this crack, and operators continued to use the blender.

On the morning of the incident, an operator requested that maintenance replace a worn paddle in the blender. A maintenance employee retrieved a paddle from an old blender and attached it to the blender in the production building about two hours before the incident. All the maintenance performed on the blender only temporarily addressed the problems caused by the metal-to-metal contact, such as the scoring on the sidewall and worn paddles. AL Solutions did not repair or replace the blender to avoid exposing combustible metal dusts to sparks or heat produced by the mechanical impact from the paddles.

Based on production operations at the time of the incident, the blender likely contained a substantial quantity of zirconium. Indications of burned residue and char on the ceiling above the blender (shown in Figures 5.8 and 5.9) and the presence of burning deposits on the wall behind the blender suggest that the burning zirconium particulates lofted from the blender after the initial explosion, engulfing the room in fire. After the incident, the blender lid, intended to be closed during production, was found left in the open position. The scraping of the blender blades against the sidewalls of the blender likely ignited the zirconium powder early in the incident sequence. The cloud of combustible zirconium resulted in a deflagration. Hot gases expanded, producing the "woof" or "wind" observed by two witnesses. The burning zirconium dust cloud also ignited open drums and tubs of titanium and zirconium at several locations in the production building, propagating the fire.

The CSB commissioned combustible dust testing of materials from the AL Solutions facility to determine whether the metal powder contributed to the fire and explosion. CSB investigators collected titanium and zirconium particulate samples for testing, both from drums of raw (received) materials and from in-process materials stored in drums. The testing determined that zirconium and titanium samples in use at AL Solutions were combustible and could produce a fire or metal dust deflagration.

Combustible dust is dust that will burn, and Class II combustible dusts possess a higher explosion severity ratio. OSHA requires dust-ignition-proof electrical equipment where Class II combustible dusts are present. The AL Solutions production building equipment had ignition-proof enclosures on motors and electrical wiring to prevent dusts in the production atmosphere from igniting because of electrical arcs and heat produced by moving parts or mechanical impacts. The metal-to-metal contact occurring in the blender posed a similar hazard, and it was open to the production atmosphere, where combustible dusts were present.

5.2.5 AL Solutions Dust Management Practices

CSB investigators learned through interviews that AL Solutions employees were generally aware that the metal dust was combustible and that a spark on the metal could cause a fire. In 2007, AL Solutions hired a laboratory to conduct burning rate testing on samples of zirconium and titanium scrap, sludge, and fines. The testing indicated that zirconium sludge and titanium scrap would propagate combustion and have a burning rate high enough to be considered "ignitable."

The AL Solutions safety manual listed several requirements for safe storage and handling of titanium and zirconium. However, the CSB found that management did not enforce these requirements. AL Solutions and the previous owner designed and installed production equipment with enclosures intended to limit the accumulation of combustible metals on external surfaces. The blender and the press conveyor had metal lids, and the plan was to leave the lids closed whenever possible to isolate the equipment and prevent dispersion of dust. Management did not enforce this practice at the facility, and the blender, conveyor belt lid, and storage drums were regularly left open during operation. In addition, to limit the quantity of flammable metal in the blending area, only metal currently in production was to be allowed in the production area. All other material was to be kept in a separate warehouse or in outside storage. However, AL Solutions employees commonly left barrels of titanium and zirconium in the production building, even if they were not in use (illustrated in Figure 5.10).

AL Solutions also did not enforce industrial hygiene practices related to handling metal dusts, although this lack of action was not causal to the incident. The AL Solutions titanium and zirconium MSDSs recommended using a high-efficiency particulate respirator when handling the materials, but management did not enforce the requirement.

AL Solutions management had procedures in place to remove dust and to control fire and explosion hazards at the facility. Operators used spark-resistant tools and wore 100% cotton clothes, which, although not flame resistant, were intended to reduce the risk of sparks from metal-to-metal contact or static electricity.

Figure 5.10 Drums of material. Source: U.S. Chemical Safety Board (CSB) (2014).

No open flames were allowed in the production building, and all smoking was restricted to an area outside the break room. The electrical equipment in the production building was designed to reduce the chance of an explosion. Motors in the production room were designed to not provide an ignition source for any metal present.

AL Solutions regularly washed down the equipment with water as its primary means to eliminate dust from process equipment and areas. This approach was intended to keep the metal powder damp and to remove any accumulated powder on process surfaces. All process equipment, walls, and floors were intended to be washed at regular intervals with the hoses available in the production building. AL Solutions expected operators to clean their areas and equipment at the end of every shift, whenever any metal accumulated, or when any equipment was changing from handling titanium to processing zirconium.

Operators washed metal powder to the floor and then into troughs that ran along the floor of the production building, and they cleaned these troughs weekly. AL Solutions did not use a dust collection system to remove zirconium and titanium dusts generated during processing, as recommended by industry standards for the safe handling of combustible metal dusts and powders. AL Solutions instead relied on the use of water as the only method to eliminate dust fires and explosions by keeping dust moist and avoiding dust accumulation on equipment and flat surfaces.

In the event of a metal fire, the AL Solutions safety manual instructed operators to evacuate the building, alert management, and call the fire department. The AL Solutions Emergency Plan recommended that employees avoid fighting metal dust fires with fire extinguishers, but noted that operators could attempt to fight nonmetal incipient fires with a fire extinguisher. The AL Solutions facility was not equipped with Class D fire extinguishers appropriate for fighting metal fires, so the company instructed its personnel to let metal fires burn and wait for the fire department.

5.2.6 Water Deluge System

The main production building had a water deluge system that was set to activate upon detection of high temperatures or high concentrations of hydrogen gas. Sensors located in the blender and press feed belt conveyor system would activate the deluge of water if readings reached a designated alarm point (illustrated in Figure 5.11). Operators also could manually activate the deluge system by turning a valve located in the blending and press room.

5.2.7 Safety Audits

The CSB found that facility safety and insurance audits did not adequately identify and address metal dust hazards at the New Cumberland facility. Jamegy,

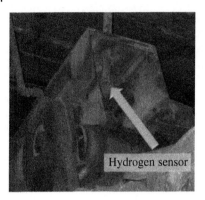

Figure 5.11 Location of hydrogen sensor. Source: U.S. Chemical Safety Board (CSB) (2014).

Hydrogen sensor

the previous owner of the facility, engaged a consultant to conduct a risk assessment of the Jamegy titanium process in 1998 and 2006. The purpose of the risk assessments was to identify events that could cause fatalities or major equipment damage at the site. Fires and explosions due to titanium or zirconium dust ranked among the major events considered during these assessments. In both 1998 and 2006, the consultant recommended that Jamegy acquire an automatic fire control system for the blender and press equipment to augment the water deluge system on the ceiling of the production building. Both risk assessments also recommended audits to address the accumulation of metal powder in the press and blender area. Although the risk assessments considered the risk that water pouring onto burning metal could generate an explosion, the consultant made no recommendations to address the presence of the automatic water deluge fire suppression system in the production area.

The AL Solutions property risk insurer conducted insurance audits at the AL Solutions facility in 2008 and 2009. These insurance audits recognized the hazards posed by titanium and zirconium dust. However, the resulting audit reports show that the insurance companies considered the AL Solutions dust control methods of washing down the metal powder to be acceptable. The 2008 audit commended the facility on its "wet process producing no dust." The 2009 audit declared that incidents are effectively controlled by "good housekeeping," "established raw material storage practices," and "water added during manufacturing process to control dust." The fire protection system is also mentioned as good process control, even though it is not advisable to use water to fight a titanium or zirconium fire.

The audits reference a 2006 incident when metal dust ignited inside a milling tank at the New Cumberland facility. The audits address the changes made to the milling tank after this incident, but they do not mention what (if any) changes were implemented to reduce the risk of metal dust ignition.

Neither of the insurance audits made any recommendations to AL Solutions that it should change its process design or dust management systems. The insurance audits also did not reference industry consensus standards for dust, such as National Fire Protection Association (NFPA) 484, *Standards for Combustible Metals*, the standard for processing and handling titanium and zirconium metal. A comprehensive process hazard analysis was not recommended or performed despite the hazardous nature of the process and despite previous incidents.

5.2.8   Hydrogen Explosion

CSB investigators determined that an ignition of zirconium dust likely caused the fatal explosion; however, the presence of flammable hydrogen acting in addition to the metal dust cannot be ruled out. Both titanium and zirconium dusts are water reactive, burn vigorously compared to other metals, and can produce hydrogen gas in the presence of heat.

5.2.9   Previous Fires And Explosions

Before the 2010 incident, the New Cumberland facility had experienced two fatal explosions involving the ignition of metal dust. From 1993 until the December 2010 incident, the New Cumberland VFD responded to at least seven fires at AL Solutions. The AL Solutions milling facility in Missouri also had a fire that required a response from the local fire department.

The CSB learned that several other fires occurred at the New Cumberland facility that did not result in a fire department response. In fact, almost all employees (except for the newest employees) reported to CSB investigators that they had witnessed one or more fires in the production building. Despite the frequent incidence of fires during operation, AL Solutions continued to use a housekeeping approach as its principle means to minimize dust accumulations rather than adopting more robust engineering controls. The NFPA 484 Standard recommends that affirmative steps of dust control, such as dust collection equipment, should be implemented above housekeeping.

NFPA 484 requires the use of flame-resistant clothing to reduce the severity of injuries from flash fires. AL Solutions provided its operators with 100% cotton work shirts and pants that were not flame resistant. NFPA 484-2009 stated that personnel handling dry titanium powder must wear non-sparking shoes and non-combustible or flame-retardant clothing without pockets, cuffs, laps, or pleats in which powder can accumulate. A similar requirement was listed for zirconium. The titanium and zirconium chapters also required the installation and use of dust collection systems to control metal dust accumulations near process equipment and dust-producing operations. The AL Solutions production building was not equipped with a dust collection system.

AL Solutions acknowledged the hazards of titanium and zirconium powder in its product MSDSs and company safety plan. Several previous incidents also had demonstrated the hazards of these materials. However, management did not enforce company policies to reduce the risk of metal dust explosions and fires, and its dust management practices did not align with industry consensus standards, such as NFPA 484. The CSB determined that the operators, supervisors, and engineers at AL Solutions were not familiar with NFPA 484 and that the process design, construction, and fire prevention practices did not consider the provisions of the NFPA standard.

Though engineering controls such as dust collection systems should be used to prevent accumulations, housekeeping operations may be a secondary means of control. In the absence of a dust collection system, the AL Solutions facility relied primarily on water sprays and wash-downs to control and reduce accumulations of fugitive titanium and zirconium powders. As noted previously, water sprays are not recommended for zirconium and titanium fires and are particularly hazardous when in contact with molten or burning titanium because of a reaction that liberates explosive hydrogen gas.

Chapter 15 of NFPA 484, *Fire Prevention, Fire Protection and Emergency Response*, specifies the following important safety requirements for water cleaning operations:

Water Cleaning Requirements – The use of water for cleaning shall not be permitted in manufacturing areas unless the following requirements are met:

1. Competent technical personnel have determined that the use of water will be the safest method of cleaning in the shortest exposure time.
2. Operating management has full knowledge of, and has granted approval of its use.
3. Ventilation, either natural or forced, is available to maintain the hydrogen concentration safely below the LFL.
4. Complete drainage of all water and powder to a remote area is available.

NFPA informed the CSB that these requirements are intended to apply to water cleaning activities in all combustible metal operations, including those involving titanium and zirconium.

As a result of the AL Solutions investigation, the CSB makes the following findings:

1. The explosion in the production building was caused by combustible titanium and zirconium dusts that were processed at the facility.
2. The explosion likely originated in a blender containing milled zirconium particulates and ignited by frictional heating or spark ignition of the zirconium arising from defective blender equipment.
3. The hydrogen gas produced by the reaction of molten titanium or zirconium metal and water, possibly from wash-down operations or the water deluge system, may have also contributed to the explosion.

4. Testing conducted after the incident determined that zirconium and titanium samples collected from the AL Solutions facility were combustible and were capable of causing an explosion when lofted near heat or an ignition source.

5. AL Solutions did not mitigate the hazards of metal dust explosions through engineering controls, such as a dust collection system. Specifically, AL Solutions did not adhere to the practices recommended in NFPA 484 for controlling combustible metal dust hazards.

6. The West Virginia Area Office of OSHA did not conduct a Combustible Dust NEP inspection at AL Solutions before the 2010 incident, despite the company's history of metal dust incidents. The Combustible Dust NEP inspections are based on a randomized selection of facilities regardless of previous incidents, unless initiated by a complaint or referral.

7. Combustible dust incidents continue to occur throughout susceptible industries, but the next steps of the OSHA rulemaking process for promulgating a general industry combustible dust standard have been delayed.

5.2.10 Summary of Safety Culture Findings

This case study demonstrates that the company knew how flammable and explosive titanium and zirconium dusts were. The company had several incidents in the past and the plant where this incident occurred also had numerous fires. There was no secret as to the hazards the company was dealing with. Yet, the company management did not take all the necessary steps to mitigate future fires and explosions. The Throughout this book it is made clear that the science and engineering to prevent serious accidents and fatalities exists and is readily available in regulations, design standards, and guidelines. It is the implementation and enforcement of these that is lacking. Engineering facilities to be safe can be expensive, but the consequences of not doing this is much worse in terms of human life. Enforcement of safety regulations and standards is difficult because employers might not want to burden employees with wearing PPE or to perform difficult activities. However, again, the alternative is much worse. I have and still do have to enforce various safety protocols in my workplace. Safety professionals and managers must develop thick skin to manage employees who push back. That is part of a good safety culture.

5.3 Imperial Sugar Company, February 7, 2008

5.3.1 Sugar

There are numerous chemical types of sugar. The ones most familiar from a consumer point of view are classified into two broad categories:

Monosaccharides or simple sugars. These include:

- Glucose
- Fructose
- Galactose

Fructose, galactose, and glucose are all simple sugars, monosaccharides, with the general formula $C_6H_{12}O_6$. They have five hydroxyl groups (—OH) and a carbonyl group (C=O) and are cyclic when dissolved in water (Perez-Castineira 2020).

Disaccharides are double sugars, connected with a glycosidic bond. These include:

- Sucrose (Fructose and Glucose)
- Lactose (Galactose and Glucose)
- Maltose (two molecules of Glucose)

Sugar molecules combine in longer chains to form carbohydrates.

Sugar is very energetic. One gram of sugar contains 16 kJ of energy or 4 food calories. In comparison, fat contains 37 kJ of energy or 9 food calories. A calorie is enough energy to raise on gram of water one degree Celsius. A food Calorie (big C) is enough energy to raise one kilogram of water one degree Celsius.

Sugar burns easily. A simple example is how well a marshmallow burns when one is trying to toast it over a fire. So, instead of a marshmallow that weighs less than ounce, consider a sugar refinery with thousands of tons of sugar. Sugar dust is classified as an ST 1, as described in the first part of this chapter. However, it can create more powerful explosions depending on the particle size and the amount of moisture.

The following describes the explosion at the Imperial Sugar Factory on February 7, 2008.

5.3.2 Accident Description

The following case study is primarily excerpted from the CSB report (CSB 2009). The complete report contains much more information. Many of my students in safety classes I taught examined this accident from a risk perspective. It is a very tragic, but interesting event.

5.3.3 Synopsis of Events

On February 7, 2008, at about 7:15 p.m., a series of sugar dust explosions at the Imperial Sugar manufacturing facility in Port Wentworth, Georgia, resulted in 14 worker fatalities. Eight workers died at the scene and six others eventually succumbed to their injuries at the Joseph M. Still Burn Center in Augusta, Georgia. Thirty-six workers were treated for serious burns and injuries – some caused permanent, life altering conditions. The explosions and subsequent fires destroyed the sugar packing buildings, palletizer room, and silos, and severely damaged the bulk train car loading area and parts of the sugar refining process areas.

The Imperial Sugar manufacturing facility housed a refinery that converts raw cane sugar into granulated sugar. A system of screw and belt conveyors, and bucket elevators transported granulated sugar from the refinery to three 105-ft-tall sugar storage silos. It was then transported through conveyors and bucket elevators to specialty sugar processing areas and granulated sugar packaging machines. Sugar products were packaged in four-story packing buildings that surrounded the silos or loaded into railcars and tanker trucks in the bulk sugar loading area.

The CSB determined that the first dust explosion initiated in the enclosed steel belt conveyor located below the sugar silos. The recently installed steel cover panels on the belt conveyor allowed explosive concentrations of sugar dust to accumulate inside the enclosure. An unknown source ignited the sugar dust, causing a violent explosion. The explosion lofted sugar dust that had accumulated on the floors and elevated horizontal surfaces, propagating more dust explosions through the buildings. Secondary dust explosions occurred throughout the packing buildings, parts of the refinery, and the bulk sugar loading buildings. The pressure waves from the explosions heaved thick concrete floors and collapsed brick walls, blocking stairwell and other exit routes. The resulting fires destroyed the packing buildings, silos, palletizer building, and heavily damaged parts of the refinery and bulk sugar loading area.

The CSB investigation identified the following incident causes:

1. Sugar and cornstarch conveying equipment was not designed or maintained to minimize the release of sugar and sugar dust into the work area.
2. Inadequate housekeeping practices resulted in significant accumulations of combustible granulated and powdered sugar and combustible sugar dust on the floors and elevated surfaces throughout the packing buildings.
3. Airborne combustible sugar dust accumulated above the MEC inside the newly enclosed steel belt assembly under silos 1 and 2.
4. An overheated bearing in the steel belt conveyor most likely ignited a primary dust explosion.
5. The primary dust explosion inside the enclosed steel conveyor belt under silos 1 and 2 triggered massive secondary dust explosions and fires throughout the packing buildings.
6. The 14 fatalities were most likely the result of the secondary explosions and fires.
7. Imperial Sugar emergency evacuation plans were inadequate. Emergency notifications inside the refinery and packaging buildings were announced only to personnel using 2-way radios and cell phones. Many workers had to rely on face-to-face verbal alerts in the event of an emergency. Also, the company did not conduct emergency evacuation drills.

After the Imperial Sugar incident, the OSHA announced it intends to initiate rulemaking on a combustible dust standard. This was among several recommendations that the CSBs 2006 *Combustible Dust Study* made to address combustible dust workplace hazards in general industry.

5.3.4 Detailed Accident Scenario

At about 7:15 p.m. on February 7, 2008, a sugar dust explosion occurred in the enclosed steel conveyor belt under the granulated sugar storage silos at the Imperial Sugar Company sugar manufacturing facility in Port Wentworth, Georgia. Seconds later, massive secondary dust explosions propagated throughout the entire granulated and powdered sugar packing buildings, bulk sugar loading buildings, and parts of the raw sugar refinery. Three-inch thick concrete floors heaved and buckled from the explosive force of the secondary dust explosions as they moved through the four-story building on the south and east sides of the silos. The wooden plank roof on the palletizer building was shattered and blown into the bulk sugar railcar loading area. Security cameras located at businesses to the north, south, and west of the facility captured the sudden, violent fireball eruptions out of the penthouse on top of the silos, the west bucket elevator structure, and surrounding buildings.

When Garden City and Port Wentworth fire department personnel arrived minutes later, they were confronted with dense smoke, intense heat, ruptured fire water mains, and large amounts of debris strewn around the fully involved burning buildings. Workers at the facility had already started search and rescue efforts and injured workers were being triaged at the main gate guardhouse.

Eight workers died at the scene, including four who were trapped by falling debris and collapsing floors. Two of these fatally injured workers had reportedly reentered the building to attempt to rescue their co-workers but failed to safely escape. Nineteen of the 36 workers transported to Savannah Memorial Hospital who were severely burned were transported to the Joseph M. Still Burn Center in Augusta, Georgia, where six eventually succumbed to injuries, bringing the total fatalities to 14 workers – the last burn victim died at the burn center six months after the incident.

Thirty-six injured workers ultimately survived including some with permanent, life altering conditions. Approximately 85 other workers at the facility at the time of the incident were uninjured.

The major fires in the buildings were extinguished the next day, but small fires continued burning for many days. The granulated sugar fires in the 105-ft tall silos continued to smolder for more than seven days before being extinguished by a commercial industrial firefighting company. The packing buildings, granulated sugar silos, and palletizer room were destroyed. The bulk sugar loading area and parts of the refinery were severely damaged by the explosion and fires (Figure 5.12).

Figure 5.12 West bucket elevator tower; silos 3, 2, and 1; and south packing building destroyed by the February 7, 2008, sugar dust explosions and fires. Source: U.S. Chemical Safety Board (CSB) (2009).

5.3.5 The Chemical Safety Board Investigation

The CSB investigation team arrived at the Imperial Sugar facility the morning of February 8, 2008. The Bureau of Alcohol, Tobacco, Firearms, and Explosives (ATF), the Georgia State Fire Marshal's office, and the responding fire departments were conducting victim recovery operations, fire suppression inside the silos and other hot spots inside the buildings, and the cause and origin investigation.

The CSB team met with the Port Wentworth fire chief, the ATF lead investigator, the State Fire Marshal, OSHA investigators, and Imperial Sugar management personnel to explain the CSBs purpose and authority for investigating independently of the other agencies and organizations. The CSB and OSHA investigation supervisors coordinated witness interviews, evidence collection, and other investigation activities with the ATF lead investigator and the incident commander.

CSB investigation occurred over four months and examined and photographed the interior of the heavily damaged buildings and equipment and interviewed eyewitnesses who were working at the refinery the night of the incident, other company and contractor employees, and injured workers. The team examined engineering documents, worker training records, and equipment operation and maintenance records; witnessed the building demolition work; and examined equipment recovered from the wreckage before releasing the equipment to the company. Finally, the CSB commissioned testing to determine the combustibility characteristics of the granulated and powdered sugar, raw sugar, and cornstarch samples collected at the site.

Imperial Sugar Company, headquartered in Sugar Land, Texas, was incorporated in 1924. The company purchased the Port Wentworth facility from Savannah Foods and Industries, Inc. in December 1997. At the time of the incident, Imperial Sugar operated the Port Wentworth facility and a sugar manufacturing and packaging facility in Gramercy, Louisiana, and a warehousing operation in Ludlow, KY. The sugar manufacturing facilities received raw sugar and refined it into granulated sugar. Some granulated sugar was used to make powdered sugar, specialty sugars, and liquid sugar products. They packaged sugar products in capacities ranging from bulk tank and hopper railcar, to 100-lb bags, to small boxes and bags. Customers included industrial bakeries, and large chain and small grocers. In 2007, the company produced more than 1.3 million tons of sugar, making it one of the largest sugar refiners in the United States. More than 350 employees and contractors worked at the Port Wentworth facility, where annual sugar production exceeded 700 thousand tons.

Imperial Sugar Company is a public company, listed on the NASDAQ. The Imperial Sugar Company Board of Directors comprises eight members plus the board chairman. Board member backgrounds include investment banking, international diversified manufacturing, food and chemical manufacturing industries, and wholesale food distribution. The president and chief executive officer (CEO) is also a board member.

The corporate safety director of Imperial Sugar was responsible for overseeing workplace safety and facility security programs. The Port Wentworth management team had a safety and environmental manager who reported directly to the plant manager. Safety and product quality personnel at the refineries reported to the corporate technical services manager in Port Wentworth. The Port Wentworth safety coordinator reported to the safety and environmental manager.

Savannah Foods and Industries, Inc. began constructing the Port Wentworth facility in the early 1900s; granulated sugar production started in 1917. Over the years the facility added refining and packaging capacity, raw sugar, and product warehouses, and upgraded the steam/electric powerplant (Figure 5.13).

Refined sugar was stored in three silos, and then transferred to the bulk sugar truck and train loading area, to the packing buildings, and to the powdered sugar production equipment, which was in the south packing building. Packaged products were palletized and transferred to a warehouse for distribution to customers.

Dozens of screw conveyors, bucket elevators, and horizontal conveyor belts transported granulated sugar throughout the packing buildings (Figure 5.14). Although bucket elevators were enclosed and screw conveyors were covered, they were not adequately sealed to prevent releasing sugar dust and sugar into the work areas. Because the large open work areas were not equipped with airborne dust removal equipment, sugar dust accumulated on overhead conduit, piping, ceiling beams, lights, and equipment. The closed granulated sugar screw

Figure 5.13 Imperial Sugar facility before the explosion. Granulated sugar storage silos and packing buildings are circled. Raw sugar warehouses in lower right. Source: U.S. Chemical Safety Board (CSB) (2009).

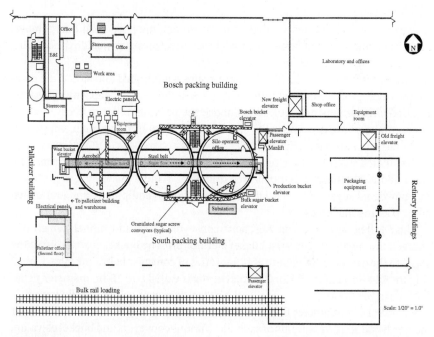

Figure 5.14 Packing buildings first floor plan. Source: CSB (2009)/Public domain.

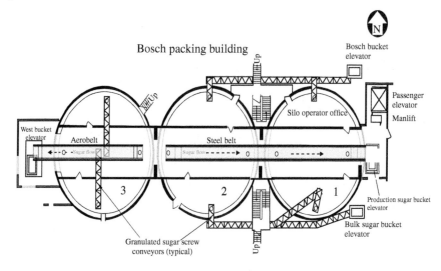

South packing building

Figure 5.15 Silo tunnel and conveyor plan. Source: CSB (2009)/Public domain.

conveyors located throughout the facility were not equipped with dust removal equipment and were not designed to safely vent overpressure outside the building if combustible dust inside the enclosure ignited.

Three 40-ft diameter, 105-ft-tall concrete silos conditioned and stored the granulated sugar produced in the refinery. Each 5-million-pound capacity silo sat on a raised circular concrete foundation above the packing building floor. Two belt conveyors were located under the silo floors in a 130-ft-long tunnel, which was 7½ ft tall and 12 ft wide (Figure 5.15).

Granulated sugar from the refinery entered silo 3 and then was transferred into silo 1 and 2. The sugar was transferred from silos 1 and 2 to the bulk sugar building, powdered sugar mills, specialty sugar production equipment, and granulated sugar packing machines.

Sugar exited silo 3 into an Aerobelt® conveyor in the silo tunnel, which discharged the sugar into the west bucket elevator pit. The bucket elevator lifted the sugar to the penthouse and onto another series of conveyor belts, into silos 1 and 2 (Figure 5.14 and Figure 5.15). Granulated sugar exited two 18-in. diameter penetrations in the floors of silo 1 and 2 onto a steel conveyor belt in the silo tunnel.

Separate 18-in. diameter floor openings fed sugar to the screw conveyors under the north and south quadrants of each silo. Various conveyors and bucket elevators distributed the sugar to the powdered sugar mills, the packaging equipment, and the bulk sugar building.

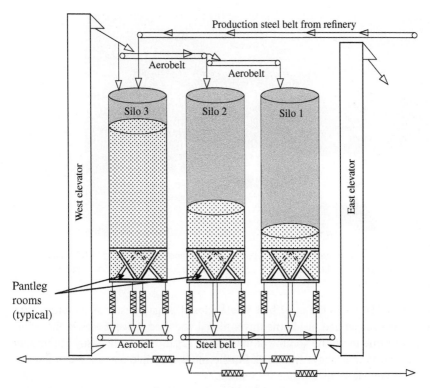

Figure 5.16 Granulated sugar supply and discharge through the silos. Source: CSB (2009)/Public domain.

A 32-in. wide, 80-ft-long steel conveyor belt (Figure 5.16) in the silo tunnel transported granulated sugar from silos 1 and 2 into the packaging production bucket elevator pit at the east side of silo 1. Similar steel belt conveyors were located in the penthouse above the silos.

For many years, granulated sugar on these conveyors was exposed to possible contamination from debris that could fall onto the sugar. In 2007, Imperial Sugar installed a stainless-steel frame with top and side panels to fully enclose each belt assembly to protect the granulated sugar from falling debris and reduce the possibility of intentional contamination. The top and side panels were removable to provide access for cleaning and conveyor maintenance. Although sugar dust was generated as the sugar flowed onto the belt and when flow blockages caused sugar to spill off the belt, these new enclosures were not equipped with a dust removal system and were not equipped with explosion vents.

During more than 80 years of operation, sugar dust released inside the large open volume of the silo tunnel most likely never accumulated to concentrations above the MEC. The MEC for sugar is 0.015 kg/m^3 (Laurent 2003)–0.035 kg/m^3

Figure 5.17 Granulated sugar steel conveyor belts above the silos, c. 1990. Heavy dust accumulation on conduit and spilled sugar on the floor. Source: U.S. Chemical Safety Board (CSB) (2009).

(Mills 2004). Once the steel belt was enclosed, sugar dust was contained and remained suspended inside the unventilated enclosed space. Because the enclosure was one-tenth the volume of the tunnel, sugar dust could easily accumulate to concentrations above the MEC. Furthermore, multiple potential dust ignition sources were identified inside the enclosure.

The four-story Bosch packing building located on the north side of the silos was a steel frame, corrugated steel-sided structure with poured concrete floors. Large doorways at the south-east corner on each floor gave forklift and personnel access to the south packing building. A steel-framed stairway enclosed in a concrete masonry block wall was located between silos 1 and 2 to provide access to each floor and the roof (see Figure 5.17).

Granulated sugar was transported by screw conveyors and a bucket elevator to the screens inside the "hummer room" located on the roof of the building. Screw conveyors transported the screened sugar to feed hoppers on the third floor above each packaging machine. The second floor contained granulated sugar packaging machines and paper and plastic packaging supplies. Conveyors transported the packaged sugar west to the palletizer room. Packaging supplies, including paper and plastic rolls, and pre-formed paper bags and boxes were stored on the first, third, and fourth floors.

5.3.6 South Packing Building

The south packing building was a four-story, steel-frame structure with 3-in.-thick poured concrete floors and brick exterior walls. The three silos formed the north wall. A steel-frame, brick wall enclosed stairway was located between silos 1 and 2 to provide access to the floors above. The east end of the packing building opened into the refinery areas. Heavy, semi-transparent plastic strip curtains were typically installed on the large passageways to contain airborne dust.

The second and third floors housed granulated, powdered, and soft (brown) sugar packaging equipment. Screw conveyors and bucket elevators transported the granulated sugar to hoppers above the packaging equipment throughout the packing building. Belt and roller conveyors transported the packaged sugar products to the palletizer room, some passing through large openings in the floors and west wall.

Container packaging supplies including paper, plastic, and cardboard were stored near the packaging equipment. Packaging machines filled paper and plastic bags and cardboard containers. The machines were equipped with Plexiglas® panels to protect workers from moving machine parts, hot surfaces, and other hazards. The enclosures also contained some of the sugar dust that was generated during container filling. Vacuum ductwork attached to the packaging machines removed some sugar dust from the packaging machines. Dust collectors located on the packing building roof removed sugar dust from air transferred from the packaging machines.

The powdered sugar production equipment was located on the fourth floor. Cornstarch was transported from an outdoor storage silo into hoppers above the grinding mills through a pressurized air transfer line. A small amount of cornstarch was added to the granulated sugar; then the mixture was fed into the grinding mills. The mills pulverized the granulated sugar/cornstarch mixture to the specified grain size, and then discharged the powdered sugar into hoppers above the packaging machines. The grinding mills were connected to the dry dust collectors system to remove airborne sugar and cornstarch dust.

5.3.7 Sugar Spillage and Dust Control

The CSBs investigation determined that sugar conveying and processing equipment were not adequately sealed; so significant quantities of sugar spilled onto the floors. Injury incident reports, internal correspondence, and other records identified significant accumulations of spilled sugar. Less than two months before the incident, an internal inspection performed by company supervisors and quality assurance personnel indicated that many tons of spilled sugar had to be routinely removed from the floors and returned to the refinery for reprocessing. Packaging operators and other employees also reported significant sugar dust escaping the packaging equipment into the work areas.

The cornstarch transport system, grinding mills, and powdered sugar packaging machines generated significant sugar dust in the work area. Cornstarch is used to help keep sugar from clumping. Workers reported that airborne sugar dust and spilled sugar in the powdered sugar processing and packaging work areas were a constant problem and that significant accumulations were typically seen on equipment and the floor. One worker told the CSB that he used a squeegee to clear a path on the floor through spilled powdered sugar to get to equipment he operated during his shift. Dust collection systems were attached to the packaging machines to remove sugar dust generated during container filling. Sugar screening equipment and the powdered sugar mills were also equipped with dust removal systems.

The dust transport ducts attached to the powdered sugar and cornstarch systems were connected to dry dust collectors. Dust transport ducts attached to granulated sugar equipment were connected to wet dust collectors. However, as reported in a January 2008 review of the dust handling system, the dust collection equipment was in disrepair, and some equipment was significantly undersized or incorrectly installed. Some dust duct pipes were found to be partially, and in some locations, filled with sugar dust.

5.3.8 Force of the Explosion

Shortly before 7:15 p.m. on February 7, 2008, the new Imperial Sugar Company CEO was touring the facility with three employees. Walking through the refinery toward the south packing building they were startled by what sounded and felt like a heavy roll of packing material dropped from a forklift somewhere in the packing building. Three to five seconds later a loud explosion knocked them backward, and debris was thrown through the large packing building doorway. The security guard at the gatehouse, and other workers in nearby buildings, also heard a loud explosive report.

Outside, massive flames and debris erupted above the packing buildings and silos.

Three-inch thick concrete floors in the south packing building heaved and buckled from the explosive force of the sugar dust-fueled explosions as they progressed through the packing buildings and into adjacent buildings. The wooden roof on the palletizer room was shattered and blown into the bulk sugar rail loading area. Workers in the packing buildings had little or no warning as walls, equipment, and furniture were thrown about. Superheated air burned exposed flesh. Workers attempting to escape struggled to find their way out of the smoke-filled, darkened work areas. Debris littered the passageways. Some exits were blocked by collapsed brick walls and other debris. The fire sprinkler system failed because the explosions ruptured the water pipes.

Intense fireballs advanced through the entire north and south packing and palletizer buildings as sugar dust, shaken loose from overhead surfaces, ignited. Heaving and buckling floors opened large crevasses. The piles of granulated and powdered sugar that had accumulated around equipment rained down and intensified the fires burning below. Fireballs advanced through enclosed screw conveyors and ignited fires in the refinery and bulk sugar building hundreds of feet from the packing buildings where the incident had begun. Violent fire-balls erupted from the facility for more than 15 minutes as spilled sugar and accumulated sugar dust continued fueling the fires.

5.3.9 Pre-explosion Sugar Dust Incident History

The Port Wentworth facility operated for more than 80 years without experiencing a devastating dust explosion, even though photographs dating from the 1970s through 2007 and internal correspondence dating to the early 1960s confirmed that significant sugar dust and spilled sugar in the packing buildings and silo penthouse were long-standing problems. In the year preceding the February 2008 explosion, quality audits and worker injury reports documented spilled sugar at "knee deep" levels in some areas.

Workers told the CSB investigators that small fires sometimes occurred in the buildings but were quickly extinguished without escalating to a major incident. Less than two weeks before the February incident, a small explosion in a dry dust collector on the roof of the packing building damaged the dust collector, but was safely vented through the deflagration vent panels. The dust collector had not yet been returned to service at the time of the incident.

Operators reported that the buckets inside bucket elevators throughout the facility sometimes broke loose and fell to the bottom of the elevator. However, they could recall only one incident, some 10 years earlier, where a falling bucket was thought to be the ignition source of a fire involving a bucket elevator in the raw sugar warehouse. That event did not escalate to a major facility fire or explosion.

5.3.10 Steel Belt Conveyor Modifications

The CSB investigators learned through worker interviews that conveying granulated sugar on the steel belt conveyor in the tunnel under silos 1 and 2 generated some sugar dust. When sugar lumps lodged in a silo outlet pipe and blocked the movement of the sugar on the belt, as operators reported sometimes occurred, sugar spilling off the belt would release airborne dust. Before the company enclosed the steel belt conveyor, the dust was released into the large volume of the tunnel. Airflow through the tunnel would also keep the airborne dust concentration low; thus, airborne sugar dust likely never accumulated to an ignitable concentration.

In 2007, Imperial Sugar enclosed the granulated sugar belt conveyors in the penthouse and the tunnel under the silos to address sugar contamination concerns. However, the company did not evaluate the hazards associated with generating and accumulating combustible dust inside the new enclosure.

They did not install a dust removal system to ensure that sugar dust did not reach the MEC inside the enclosure. Furthermore, the enclosures were not equipped with deflagration vents to direct overpressure safely out of the building if airborne sugar dust ignited.

5.3.11 Primary Event Location

For more than 80 years, the facility operated the open steel belt conveyor in the silo tunnel without incident. Less than a year after the enclosure was completed, however, the silos and packing buildings were destroyed by devastating explosions and fires. The CSB examined the blast patterns and damage inside the tunnel, at the east and west tunnel openings, and rooms inside the base of the silos directly above the silo tunnel (pantleg rooms, see Figure 5.18) and concluded that the primary dust explosion most likely occurred in the silo tunnel, approximately midway along the length of the steel belt.

Examination of the equipment in the silo tunnel found that every panel used to enclose the 80-ft-long steel belt conveyor had been violently blown off the support frame (Figure 5.19). Panels east of the approximate midpoint of the conveyor (directly under the silo 1–silo 2 abutment) were deflected eastward, and panels west of the midpoint were deflected westward. Pressure damage inside the pantleg rooms above the tunnel was greatest inside the silo 2 pantleg room and least inside silo 3 pantleg rooms. Major equipment damage was observed at the east entrance of the silo tunnel with all equipment deflections eastward, away from the tunnel. The equipment under silo 3 and outside the west end of the tunnel had damage patterns clearly indicating that a pressure wave traveled west out of the tunnel.

Pressure waves blew the wood walls and doors off the west and east tunnel entrances. Mangled steel belt cover panels were blown out of the east end of the tunnel (Figure 5.20). The pressure wave also traveled up between silos 1 and 2 and

Figure 5.18 Silo tunnel steel conveyor belt. Source: U.S. Chemical Safety Board (CSB) (2009).

Figure 5.19 Steel belt covers (arrows) crumpled from an initial dust explosion inside the steel belt enclosure (steel frame structure, right photo). Source: U.S. Chemical Safety Board (CSB) (2009).

Figure 5.20 Stainless steel cover panels (arrows) blown off the steel belt enclosure and out of the tunnel into the packing building. Source: U.S. Chemical Safety Board (CSB) (2009).

blew the brick walls that enclosed the south stairwell into the work areas on each floor of the south packing building (Figure 5.21).

5.3.12 Primary Event Combustible Dust Source

The CSB learned that, during the 3–4 days preceding the incident, workers were clearing sugar lumps lodged in the silo 1 sugar discharge holes above the steel belt. Through access ports in the pantleg room (Figure 5.22), they used steel rods to break and dislodge sugar lumps found above and inside the sugar discharge holes. Workers also used access ports located on the discharge chutes to break lumps that lodged on the steel belt, blocking sugar flow from the upstream discharge chutes (Figure 5.23).

During the silo 1 "rodding" activities, sugar continued to flow from silo 2 onto the steel belt upstream of the silo 1 discharge chutes. Based on the statements from

Figure 5.21 South stairwell brick walls blown into the packing building. Source: U.S. Chemical Safety Board (CSB) (2009).

Figure 5.22 Access port inside the pantleg room and steel rod used to break sugar lumps in the sugar outlet on silo floor. Source: U.S. Chemical Safety Board (CSB) (2009).

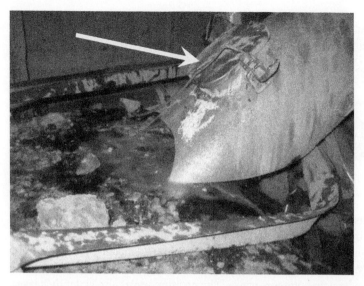

Figure 5.23 Limited clearance between sugar discharge chute and the steel belt would sometimes be blocked by large sugar lumps; lump-clearing access port (arrow) is on top of the chute. Source: U.S. Chemical Safety Board (CSB) (2009).

the workers who were clearing the sugar lumps from the silo outlets, it is highly likely that sugar lumps lodged between the moving steel belt and one or both silo 1 discharge chutes. Sugar lumps small enough to enter the discharge chute, but too large to pass between the end of the chute and the steel belt would create a "dam" that would cause sugar traveling from silo 2 to spill off the steel belt upstream of the silo 1 discharge chute.

Spilled sugar would then accumulate inside the covered conveyor, likely covering parts of the steel belt. Sugar dust released into the air from the sugar spilling off the conveyor, and sugar dust generated as the steel belt traveled through the sugar pile, below, could accumulate in the unventilated enclosure.

The CSB concluded that as granulated sugar spilled off the moving steel conveyor at the blocked outlet, sugar dust accumulated above the MEC inside the enclosed conveyor, and then ignited. The explosion triggered a series of secondary explosions that rapidly progressed through the packing buildings and palletizer room and into the bulk sugar station.

5.3.13 Secondary Dust Explosions

The primary dust explosion in the silo tunnel sent overpressure waves into the three silo pantleg rooms, and then out into the first floor of the Bosch building. Explosion pressure waves traveled between silos 1 and 2 and exited into the south stairwell, blowing the brick walls out into the packing building. Sugar dust

Figure 5.24 Three-inch thick concrete floor slabs lifted off the steel supports by the explosion pressure waves.

accumulations on elevated surfaces and spilled sugar on the floors around the packaging equipment contributed to the explosion energy.

The fireballs continued to be fueled by sugar dust dislodged from overhead equipment. Powdered and granulated sugar added fuel as sugar was thrown into the air by the advancing pressure waves. The pressure waves also violently heaved concrete floors upward throughout the south packing building (Figure 5.24). Spilled sugar that had accumulated on the floors around conveyors and packaging equipment rained into the rooms below adding more fuel to the advancing fireballs.

5.3.14 Ignition Sources

Combustible dust can be ignited by electrical sparks or static discharge, hot surfaces, open flames, or friction-induced hot surfaces or sparks. Controlling possible ignition sources wherever combustible dust is present helps minimize the chance of a fire or explosion.

5.3.15 Open Flames and Hot Surfaces

Imperial Sugar had policies and procedures to control open flames. Workers were permitted to smoke only in designated areas, which were away from the production or packaging equipment. The hot work permit procedure required the workers to control combustible materials and post a fire watch whenever they performed welding or other hot work activities.

The CSB obtained photographs taken in the fall of 2006 that showed large electric motors covered with sugar dust (Figure 5.25), a condition that could cause the motors to overheat and possibly ignite the spilled sugar or airborne sugar dust. These poor housekeeping practices likely contributed to, or directly resulted in, some of the small sugar fires that sometimes occurred.

Workers told the CSB investigators that occasionally, small incipient stage fires occurred in the packing buildings when sugar or packaging materials was ignited

Figure 5.25 Motor cooling fins and fan guard covered with sugar dust; large piles of sugar cover the floor. Source: Vorderbrueggen (2011)/John Wiley & Sons.

by a hot electrical device or a hot bearing. However, none of these earlier incidents ever resulted in a large fire or involved airborne dust. The CSB ruled out open flame as a possible ignition source for the February 2008 incident.

5.3.16 Ignition Sources Inside the Steel Belt Enclosure

The CSB concluded that the initiating event was most likely the ignition of an explosive concentration of sugar dust within the steel belt enclosure under silos 1 and 2. Multiple possible ignition sources were identified during examination of the damaged steel belt and discussions with operations and maintenance personnel. However, the extent of the destruction from the explosions and long-burning fires prevented identifying a likely ignition source for the primary explosion.

Because the four limit switches on the steel belt conveyor were located inside the newly installed enclosure, they were exposed to airborne sugar dust. Examination of two of the four switches that survived the fires showed that they were rated explosion proof. Therefore, the CSB concluded the switches most likely were not the ignition source of the primary dust explosion.

5.3.16.1 Hot Surface Ignition

Experiments have been conducted on various combustible dusts to measure the minimum ignition temperatures (MIT) of a dust cloud. Sugar MITs range from 360 to 420°C (680–788°F), depending on the test apparatus used. Similar experiments involving combustible dusts have been conducted to examine the influence of the time a combustible dust cloud remains in the test apparatus (i.e. residence time, from 0.1 to 0.4 seconds). Those experiments found that the MIT decreases as the residence time of the dust in the test furnace increases (Eckhoff 2003).

Airborne combustible dust was likely always present when sugar was being transported inside the enclosed steel belt conveyor, especially when blockages caused the granulated sugar to spill off the steel belt.

Operators told the CSB investigators that bearings on the steel belt roller supports sometimes malfunctioned and "got very hot." A hot bearing inside the enclosed steel belt conveyor could ignite airborne sugar dust, especially if the dust remained in contact with the hot surface for many seconds, as it likely did in the unventilated enclosure. Also, if sugar in contact with a hot surface begins to smolder, combustion gases that are released will mix with the airborne sugar dust and decrease the ignition temperature below the ignition temperature of pure sugar dust (Eckhoff 2003).

5.3.16.2 Friction Sparks

Although not ruled out, the least likely ignition source could have been from metal-to-metal sliding between the steel belt and a jammed support wheel or the steel enclosure frame. Testing has demonstrated that sparks from carbon steel sliding contact are unlikely to ignite combustible dust unless the contact speed exceeds 30.1 ft/s (9.2 m/s) (Dahn and Reyes 1986). The steel conveyor belt operating speed was estimated to be less than 5 ft/s (1.5 m/s).

5.3.16.3 Worker Training

Imperial Sugar conducted routine subject-specific worker training for employees and contractors. The company's "Specific Safety Rules" policy stated that new workers receive comprehensive safety rule training on the first day of employment. The policy also required workers to attend annual refresher safety training. A day-long, annual training session was conducted once each month. Workers typically attended the training during the month of their birthday, a scheduling practice that led to the safety training being called "Birthday Safety Training." Interviews showed that most, but not all, personnel received their safety training in a timely fashion in accordance with the "Specific Safety Rules" policy requirements.

"Birthday Safety Training" subjects in 2006 and 2007 included personal protective equipment, lockout/tagout procedures, confined space entry, fall protection, manlift operation, forklift operation, fire extinguisher operation, compressed air nozzle use, safe and at-risk behaviors, employee self (safety) audits, MSDSs, emergency communication, workers compensation, and maritime security. Interviews revealed minimal retention of the topics covered: while some employees said that the hazard of dust accumulation was included at some time in the training sessions, a review of more than 10 000 pages of "Birthday Safety Training" and "Plant Rules and Guidelines" tests revealed no such coverage since 2005. Furthermore, the training materials did not contain information about combustible dust, and no other documents were provided to verify that training on the hazard of dust accumulation occurred prior to 2005. Although "Birthday Safety Training" extensively covered site safety issues, the hazard of dust accumulation was not among them and was not included in the site's "Specific Safety Rules."

Imperial Sugar quality assurance and safety personnel were made aware of the OSHA combustible dust national emphasis program in early December 2007. A January 30, 2008, "Safety Focus of the Month" email from the safety manager and the *Written Program– Housekeeping & Material Storage Program* that was attached made no mention of combustible dust. Failure to understand the hazard and control sugar dust accumulations in the work areas by properly maintaining equipment and performing routine housekeeping led to the massive secondary explosions and fires that claimed the lives of the 14 workers and injured dozens.

Had the company trained the workers on the hazards of combustible sugar dust and adopted an effective dust control program, the secondary dust explosions that devastated the facility would likely not have occurred.

5.3.17 Evacuation, Fire Alarms, and Fire Suppression

The written emergency response procedure directed workers to use an intercom system to report an emergency. However, workers told the CSB investigators that the intercom system was not used inside the refinery or packing buildings. Rather, they had to rely on radios and cell phones carried by some personnel to announce or be alerted to an emergency. Having no audible or visual alarm devices in the work areas, some workers had no means of prompt notification in the event of an emergency.

The company had posted emergency evacuation routes but did not provide work location specific evacuation training to all employees and contractors and did not conduct evacuation drills.

Emergency evacuation lights and illuminated exit signs were provided throughout the facility. However, many workers reported that the explosions and fires caused the normal lighting and many emergency lights to fail. The emergency lights that worked did not provide adequate illumination – some workers had extreme difficulty finding their way out of the darkened buildings. Large, complex mechanical equipment and long package conveyors 3 to 4 ft above the floors impeded rapid and safe emergency egress in the darkened work areas.

A fire suppression sprinkler system was intended to protect each floor of the packing buildings and the palletizer room. The city water main and an onsite firewater tank and emergency diesel-driven pump provided water to the heat-activated sprinkler heads. However, typical of incidents involving explosive energy releases, the fire suppression piping system was heavily damaged in the initial explosions and rendered ineffective. Lack of water flow from the fire hydrants around the perimeter of the buildings caused by the ruptured sprinkler systems severely hampered firefighting efforts.

Portable fire extinguishers were located throughout the work areas. Wheeled carts containing as many as 16 portable fire extinguishers were located in sugar packaging areas, such as the south packing building third floor. Workers reported using fire extinguishers to extinguish small fires that sometimes occurred. However, the fire extinguishers were useless against the rapidly progressing fires that quickly engulfed the packing building.

5.3.18 Electrical Systems Design

The NFPA publishes NFPA 499, *Recommended Practice for the Classification of Combustible Dusts and Hazardous (Classified) Locations for Electrical Installations in Chemical Process Areas,* which is used to classify hazardous work locations, including locations where combustible dust is or might be present. Electrical devices installed in these "classified" locations should comply with applicable NFPA design standards to minimize ignition sources such as NFPA 654 – *Standard for the Prevention of Fire and Dust Explosions from the Manufacturing, Processing, and Handling of Combustible Particulate Solids.* The National Electric Code (NEC) and International Fire Code (IFC) incorporate these and other NFPA standards and guidelines. Local, state, and federal government agencies sometimes incorporate the NEC or IFC into building regulations or worker safety standards.

OSHA standard 1910 Subpart S contains the electrical equipment safety requirements to protect employees in their workplaces. The standard addresses work practices and electrical device maintenance requirements for hazardous locations including electrical devices and wiring installed in locations where combustible dusts or fibers may be present.

Imperial Sugar had no written policy or procedure to require classifying hazardous areas or require electrical devices rated for such locations at the Port Wentworth facility. Some locations at the facility contained significant accumulations of combustible sugar or cornstarch dust. The CSB investigators noted that some electrical devices were rated for use in hazardous work locations, but non-rated devices were also installed in the same location, frequently near the rated devices. The CSB investigators also observed many non-hazard classified electrical devices installed in dusty work areas and equipment enclosures. Furthermore, some electrical devices in dusty areas were poorly maintained, such as having missing covers or open doors on many breaker panels and other electrical enclosures.

The powdered sugar grinding mills motor control room located on the fourth floor of the south packing building was the only area that contained hazardous dust control features to protect the non-rated electrical switchgear. The room was equipped with a positive pressure air conditioning system and sealed doors to prevent sugar dust from entering the room and accumulating on or near the electrical control equipment.

The south packing building sustained extensive blast damage (report cover photo). The ceilings in areas where sugar dust was present were not protected with a suspended ceiling to minimize dust accumulation on rafters, conduit, conveyors, piping, and other equipment. Elevated areas were infrequently cleaned so sugar dust accumulated to dangerous levels. The secondary explosions were most likely fueled by significant accumulations of sugar dust on elevated surfaces. When the floors heaved from the first series of explosive overpressures, large piles of sugar on the floors would have been released into the areas below and ignited. Overpressure waves advancing horizontally through the packing buildings would also loft sugar piles into flaming fireballs.

5.3.19 Sugar Dust Handling Equipment

Transporting granulated and powdered sugar in mechanical conveying devices, and filling bags and boxes in the packaging machines generated sugar dust. To control dust, a duct system connected to wet dust collectors was designed to remove sugar dust from granulated sugar process equipment including some conveyors, bucket elevators, and packaging machines. Dust removal ducts were also located in the silos' penthouse and the tunnel under the silos. The dust collectors used water spray to remove the sugar dust from the air. The water from the collectors was returned to the refinery to recover the dissolved sugar. The roof mounted collectors exhausted the cleaned air outside the building.

The powdered sugar and cornstarch process equipment dust collection system used dry dust collectors. Filter bags inside six dust collectors located on the roof of the south packing building captured the sugar and cornstarch dust. Traveling "blow rings" used compressed air to dislodge the accumulated sugar from the filter bags. The dislodged sugar fell to the bottom of the filter chamber. It then passed through a rotary valve to containers below the dust collectors.

Explosion vent panels protected the dry dust collectors from excessive overpressure if sugar dust ignited inside a collector. In fact, 10 days before the February incident, a malfunction inside a powdered sugar atomizer resulted in a sugar dust fire and explosion inside one of the dry dust collectors. The explosion panels operated as intended and the dust fire safely vented to the atmosphere.

Although much of the granulated and powdered sugar processing equipment was connected to the dust collection system, it was inadequately maintained and did not effectively remove dust from the equipment. A report submitted to Imperial Sugar management by an independent contractor less than a week before the incident identified the following major problems with the two dust collection systems:

- Air flow in ducts significantly below the minimum dust conveying velocity
- Undersized fans – static pressure too low to transport sugar dust
- Fans operating well below the required performance curve
- Incorrectly installed duct piping
- Duct piping plugged with sugar

In that the dust collection systems were ineffective, the only solution for minimizing sugar dust accumulation to prevent a dust explosion would be timely and effective housekeeping practices to remove the sugar dust from the work areas.

5.3.20 Housekeeping and Dust Control

Workers told CSB investigators that sugar spillage and dust generation were constant problems in the packing buildings. They reported that sugar leaked from worn seals, lose or missing covers, and other breaches in the aging screw conveyors, bucket elevators, hoppers, and other bulk sugar transport devices.

Leaks in the pressurized air ducts used to transport cornstarch, and the occasional powdered sugar packaging machine malfunction, released dust into the work areas. In addition, sugar-filled paper or plastic containers sometimes tore open and spilled their contents as they traveled along the conveyors, generating more sugar dust. However, because the large work areas were typically not equipped with dust removal systems, the dust released from these sources would float in the air and settle on overhead piping, equipment, lights, ceiling support beams, and any other horizontal surface.

Packaging machines required frequent cleaning to operate properly. Workers routinely used water or steam, and sometimes industrial vacuum cleaners to remove sugar and sugar dust from equipment and work areas. But they also routinely used compressed air to remove accumulated granulated sugar or sugar dust from packaging equipment. Compressed air cleaning only lofted dust into the air where it would later accumulate on horizontal surfaces in the area. More frequent cleaning was needed to remove dust in the work areas, especially on elevated surfaces, before it accumulated to hazardous levels, but was not routinely performed.

Because Imperial Sugar manufactured and packaged food-grade sugar products, general area cleanliness (removal of trash and other debris) was necessary to prevent product contamination and control rodents. Spilled sugar also presented a slip or trip hazard, especially when wet, as workers reported was often the case throughout the facility.

Written housekeeping policies included planned daily, weekly, and monthly packaging area cleaning schedules, but workers reported that these policies were not effectively implemented. Pre-incident photographs of equipment and packaging areas, worker injury reports in 2006 and 2007, and a December 2007 quality assurance survey provided evidence that the housekeeping practices were inadequate – deep piles of spilled granulated and powdered sugar accumulated around and on equipment, and sugar dust accumulated on floors, equipment, and other elevated horizontal surfaces (Figure 5.26). Workers interviewed by the CSB investigators reported cleaning activities were seldom performed on hard-to-reach elevated surfaces and some powdered sugar packaging areas frequently had dense sugar dust in the air.

Figure 5.26 Deep piles of sugar accumulated on floors and equipment. Note shield installed above screw conveyor drive motor to deflect sugar dust. Source: U.S. Chemical Safety Board (CSB) (2009).

5.3.21 Imperial Sugar Management and Workers

Correspondence dating to as early as 1961 indicates that management and refinery personnel were aware of the explosive nature of sugar dust and the importance of minimizing dust accumulation. An August 1961 internal memo describes a dust explosion in the powdered sugar room. A September 1967 internal memo from a refinery engineer to the executive vice president and treasurer reported that:

> This dust problem has become so serious and dangerous in modern refineries… at present, we have so much to correct that is knowingly wrong, there is no need for outside help. We make a lot of dust in the plant and have had a very inefficient dust collecting system[;] consequently, it has been hopeless to try to keep the dry end of our plant clean. We have heavy accumulations of dust in several areas… we hope to improve the house keeping around the silos…
>
> By removing these heavy accumulations of dust on beams, sills, and walls, the fuel for a continuous explosion will be eliminated. This is the reason an explosion will travel from one area to another, wrecking large sections of a plant.

One year later, a September 1968 dust explosion occurred inside the mill room. Refinery workers successfully extinguished the fire, but not before significant fire and smoke damage occurred.

The CSB reviewed Imperial Sugar-published MSDSs for granulated and powdered sugar products, both of which warn:

> Explosion: NFPA Class 2 Group G Airborne sugar dust accumulation ignition temperature is 370°C. At airborne concentrations of 0.045 gm/L or higher, sugar dust accumulations are explosive.

Another MSDS on file at Imperial Sugar issued by a cornstarch supplier identified the explosion hazard associated with cornstarch dust and warned: "Avoid procedures which could cause a dust cloud to be formed."

Over the years, combustible dust incidents at sugar refineries worldwide, and small combustible dust incidents at the Port Wentworth facility and other Imperial Sugar facilities were known to company management and employees. However, management action to control dust generation and accumulation was ineffective; thus, combustible dust remained a major workplace hazard.

In 1998, an employee was severely burned by a sugar dust explosion in the powdered sugar mill room at the Imperial Sugar refinery in Sugar Land, Texas. Yet, more than eight years after that incident the corporate safety manager wrote in a memo to senior management: "Based on conversations with the quality team at each location, we did not have a formal policy for sanitation/housekeeping at any of our sites," Still, the draft sanitation/housekeeping policy attached to the memo did not discuss sugar dust.

Management did not take adequate action to correct the long-standing hazardous combustible dust conditions in their facilities.

Workers told the CSB investigators that occasionally, small fires occurred in the packing buildings when sugar or packaging material was ignited by a hot electrical device or overheated bearing. As recently as January 2008, one worker observed flames "3 ft high" before he and other workers in the area extinguished the fire with portable fire extinguishers. As noted in Section 3.6.3, the explosion panels on a dust collector were blown open when the dust inside the collector was ignited. None of these events resulted in a widespread fire or secondary dust explosion. Fires and even dust explosions occurred at the Imperial Sugar facilities and other sugar refineries without ever propagating into a secondary dust explosion or large fire. The unsafe work practices continued, and the combustible dust hazards were not abated.

The CSB concluded that the small events and near-misses caused company management, and the managers and workers at both the Port Wentworth, Georgia, and Gramercy, Louisiana, facilities to lose sight of the ongoing and significant hazards posed by accumulated sugar dust in the packing buildings. Imperial Sugar management and staff accepted a riskier condition and failed to correct the ongoing hazardous conditions, despite the well-known and broadly published hazards associated with combustible sugar dust accumulation in the workplace. The process wherein successful operations continue despite existing rejectable conditions or unsafe behaviors results in relaxing the minimum performance standard without basis is called "normalization of deviance." That is, an organization relaxes the criteria rather than identifying and correcting the underlying causes, so the deviations become the new normal operating condition.

The concept of normalization of deviance is applicable to any operating system. The Imperial Sugar incident involved known hazards and acceptance of a lesser performance standard based on operating history.

Imperial Sugar management was aware of the hazards associated with combustible sugar dust, but in the absence of any major catastrophic incident during many years of facility operation, the hazardous conditions went uncorrected. Management did not enforce adequate equipment design and maintenance practices to control sugar dust and spilled sugar. Furthermore, management did not ensure housekeeping activities were adequate to prevent sugar dust and spilled sugar from accumulating to unsafe levels in the workplace. Then, on February 7, 2008, conditions necessary to propagate a dust explosion aligned to cause the catastrophic explosions and fires that swept through the Port Wentworth facility.

5.3.22 Chemical Safety Board Key Findings

1. Imperial Sugar and the granulated sugar refining and packaging industry have been aware of sugar dust explosion hazards as far back as 1925.
2. Port Wentworth facility management personnel were aware of sugar dust explosion hazards and emphasized the importance of properly designed dust handling equipment and good housekeeping practices to minimize dust accumulation as long ago as 1958, but did not take action to minimize and control sugar dust hazards.
3. Over the years, the facility experienced granulated sugar and powdered sugar fires caused by overheated bearings or electrical devices in the packing building. However, none of these incidents resulted in a devastating sugar dust explosion or major fire before the February 2008 incident.
4. Company management and the managers and workers at both the Port Wentworth, Georgia, and Gramercy, Louisiana, refineries did not recognize the significant hazard posed by sugar dust, despite the continuing history of near misses.
5. The enclosure installed on the steel conveyor belt under silos 1 and 2 created a confined, unventilated space where sugar dust could easily accumulate above the MEC.
6. The enclosed steel conveyor belt was not equipped with explosion vents to safely vent a combustible dust explosion outside the building.
7. Company management and supervisory personnel had reviewed and distributed the OSHA *Combustible Dust National Emphasis Program* shortly after it was issued in October 2007 but did not promptly act to remove all significant accumulations of sugar and sugar dust throughout the packing buildings and in the silo penthouse.

8. The secondary dust explosions, rapid spreading of the fires throughout the facility, and resulting fatalities would likely not have occurred if Imperial Sugar had enforced routine housekeeping policies and procedures to remove sugar dust from overhead and elevated work surfaces and remove the large accumulations of spilled sugar throughout the packing buildings.

9. The Port Wentworth facility risk assessment performed by Zurich Services Corporation in May 2007 and the report submitted to Imperial Sugar management did not adequately address combustible dust hazards.

5.3.23 Summary of Safety Culture Findings

The safety culture failures that contributed to the tragic deaths are readily evident in the accident description. The primary safety culture failures were:

1. Lack of fully comprehending the amount of energy that the accumulated sugar had was a primary contributor to the event. As discussed at the beginning of this case study, sugar contains a great amount of energy, and, from all descriptions in the CSB report, there was tons of sugar everywhere in the facility. Management needed to ensure the workers were well trained/educated on the hazards.

2. Lack of ensuring granulated sugar and sugar dust did not accumulate on and in equipment and on the floors, walls, and support structures of the facility. The CSB report calls this housekeeping, which it is, but in this case, it goes beyond simple housekeeping. Management needed to ensure combustible sugar was not allowed to accumulate at all. It was a known hazard and was ignored until disaster struck.

3. Not properly assessing the risk of enclosing the conveyor system. A comprehensive failure mode and effect analysis would have helped to elucidate the risks with those plant modifications (Ostrom and Wilhelmsen 2019).

4. Not ensuring electrical components were up to code.

5. Emergency evacuation plans were not tested. Therefore, safe egress routes were not provided to the workers.

6. Not ensuring workers had a questioning attitude. Management needed to instill into workers that they could ask questions about safety related topics. The one worker described in the CSB report that noted he had to use a squeegee to clear sugar to walk to his machine. Also, the note that workers were seen with sugar up to their knees in sugar. Workers should not have to work in these conditions, and they should be able to question whether this is a correct way to work.

Primarily, it falls on management to ensure fire protection, safety, and health standards, regulations, and guidelines are followed. However, a workforce that is trained and educated on the hazards they are working with will also help prevent catastrophes. Also, empowering workers to be able to stop an operation once they have deemed it to be hazardous is also tantamount to ensuring safety.

5.4 Entertainment Venue Case Studies

This section primarily discusses the New Taipei Waterpark fire. However, the following introduction discusses several other tragic events at concerts and nightclubs.

5.4.1 Introduction

What can be said about tragic fires, stampede deaths, terrorist attacks, and explosions that happen during what should be family fun events? They are not only tragic, but these events also happen to unsuspecting families, couples, and single people seeking to enjoy an amusement park, a concert, or some other event. I can name a dozen or more events that happened because the organizers failed to perform any sort of risk assessment or to consider the possible negative outcomes that a lack of safety consideration could occur.

5.4.2 Crowd Surge Events

On December 3, 1979, 11 concert goers died of asphyxiation when a crowd rushed the gate when they thought the concert by The Who was about to begin. The venue was Riverfront Coliseum in Cincinnati, OH (King 2021). Courtney King writes about a survivor of the 1979 tragic concert and how she couldn't believe a similar event occurred 42 years later. This was the tragic Astroworld Travis Scott concert in November 2021. As of the writing of this book, 10 people died because of the crowd surge at the Travis Scott concert (AP 2021).

These are not the only crowd surge that events that have happened in recent years. Nine people died at the Roskilde Festival in 2000 when Pearl Jam was playing (Fricke 2000); At the Hillsborough soccer stadium in England, a human crush in 1989 led to nearly 100 deaths (BBC 2022); In 2015, a collision of two crowds at the hajj pilgrimage in Saudi Arabia caused more than 2400 (Almukhtar and Watkins 2016). These events were clearly preventable with proper crowd control techniques. Right after The Who concert in 1979 there was a movement to stop festival seating because of the issues with lack of crowd control. However, events occur every day with this type of uncontrolled crowds.

5.4.3 Fires at Bars and Nightclubs

Fires at bars and concert venues are far worse than crowd surge events. The Station Fire in Rhode Island, the Lame Horse nightclub fire in Perm, Russia, and the Colectiv nightclub fire in Bucharest, Romania are examples of the tragedies that happened.

The Station nightclub fire on February 20, 2003, in Rhode Island was caused by pyrotechnics and resulted in the death of 100 people and injured another 230. This disastrous fire was caused by pyrotechnics that caught the decorative willow twig ceiling on fire. The fire spread quickly once the ceiling ignited. The fire spread to the decorations on the walls. There were 300-plus patrons in the nightclub when the fire started. Ninety-four people died at the scene and most of the survivors incurred injuries. Sixty or more of the injured died in hospitals over the next several days, but those exact numbers were not available.

The Master of Ceremonies (MC) told patrons to evacuate soon after the fire started, but the fire spread so fast and the smoke that was created was so thick patrons were soon overcome. The emergency response was also lacking. Some of the victims laid on the cold tarmac or on snow for up to an hour without help (Heintz 2009; Levy 2009).

The Colectiv fire also was caused by pyrotechnics. Like the Station Fire, the pyrotechnics caught the polyurethane acoustic foam on fire. Sixty-four patrons died as the result of this fire.

These three nightclub fires were all caused by pyrotechnics. World communications aren't like they were before the telegraph and even before the Iron Curtain fell. News about disastrous events is readily available and the internet provides news coverage worldwide almost immediately. So, how can three disastrous fires, with such similar circumstances happen? The reason is that there was a lack of comprehension that pyrotechnics in a building could cause a disastrous fire. Pyrotechnics burn at 1200 °F (650 °C). Structural wood ignites at under 700 °F (370 °C). Plastic ignites at approximately 500 °F (260 °C) or less. There is a perception that pyrotechnics/fireworks are safe. Also, club owners and entertainers want to make a big impression.

These fires were also not unique. These are similar nightclub fire since the year 2000 (NFPA 2021):

- 2001 Canecão Mineiro nightclub fire in Belo Horizonte, Brazil;
- 2004 República Cromañón nightclub fire in Buenos Aires, Argentina;
- 2008 Wuwang Club fire in Shenzhen, China;
- 2009 Santika Club fire in Watthana, Bangkok, Thailand;
- 2013 Kiss nightclub fire in Santa Maria, Brazil.

5.4.4 The New Taipei Water Park Fire – June 2015

This event is unique in that it involves a non-industrial setting. However, the deaths and suffering it caused is just as tragic. The accounts of this event are mostly from news articles from around the world. In the previous two dust explosion events in this chapter the dusts that ignited were sugar and titanium/zirconium powders that were contained in the industrial processes where they caught fire and exploded. In this event, the colored cornstarch powder was deliberately lofted into the air above a crowd of up to 1000 people to help create a festive atmosphere.

The HMIS and NFPA ratings for cornstarch are (Natural Sourcing 2021):

Health Hazard – 0
Flammability – 1
Physical Hazard – 0

Cornstarch can and does form combustible and explosive mixtures in air, under the right conditions. Like sugar, it is a carbohydrate, and as such, is very energetic. The MEC for cornstarch is $0.04\,kg/m^3$ in air (Rhodes 1990).

Cornstarch is a very common ingredient in food and anyone who bakes has cornstarch in their homes.

The event, called the "Color Play Asia" occurred on June 27, 2015. The venue was the Formosa Fun Coast water park (Dust Safety Science 2021).

This event is unique in that it involves a non-industrial setting. However, the deaths and suffering it caused is just as tragic. The accounts of this event are mostly from news articles from around the world. In the previous two events in this chapter the dusts that ignited were sugar and titanium/zirconium powders that were not contained in the industrial processes where they caught fire and exploded. In this event, the colored cornstarch powder was deliberately lofted into the air above a crowd of up to 1000 people to help create a festive atmosphere.

The accounts vary, but between 1000 and 4000 people, mostly in their teens and twenties, attended an Asian Color Festival at the Formosa Fun Coast water park in New Taipei, Taiwan on June 27, 2015 (Dust Safety Science 2021). The school year had just closed and many of the attendees were students. This event was a well-advertised dance rave that included music and the possibility of a light and color show.

This event was to mimic the India's Festival of Colors. Cornstarch had been dyed with yellow, green, and other colors. The water park staff released many pounds of the powder into the crowd and then kept it suspended using blowers and compressed gas. Attendees closer to the stage were up to their ankles in the material. The suspended powder became a dense cloud above the attendees. The powder was then ignited by the hot stage lighting (BBC 2016). The ignited powder became fireballs that tour through the crowd, sending attendees running for their lives in terror. Most of the attendees were wearing swimming suits because the weather had been so hot, and this was a water park.

The cornstarch that had accumulated on the ground ignited as well and badly burned the bodies of those standing in it. Many of the attendees had burns over 70–90% of their bodies. The fire had been so hot and the burns so severe attendees' skin sloughed off of many of them.

Victims were transported to 37 different hospitals. Over 500 people were burned by the fire and 15 died from their injuries.

Lu Chung-chi, who headed Color Play and Juipo International Marketing and that rented the event site, admitted that he had bought three tons of the powder, which consisted of cornstarch and food coloring (Dust Safety Science 2021). The amount of potential energy in 3 tons of corn starch is approximately 178 million kJ of energy.

There were, of course, numerous lawsuits associated with the fire and explosion and the owner of Color Play Asia was jailed for four years and 10 months.

5.5 Safety Culture Summary

All the events described in this section were caused because the organizers did not thoroughly consider the potential negative outcomes. The crowd surge events were caused because of the lack of proper crowd control methods. These types of events have happened many times and when 50,000 plus people are all brought together, without structure disasters can happen. Fights can break out, the crowd can surge if the group think determines there is something they are missing or need to get closer to see, and a person's medical condition can become worse in a crowd when temperatures rise and there is a lack of breathing space.

Fires can happen anytime under the right conditions. The organizers did not consider that a fuel in the form or wood, plastic foams, or cornstarch could be easily ignited by pyrotechnics or hot stage lighting. Fire is a common hazard in every home, business, school, and even in churches. Fuels, in one form or another, exist everywhere. Oxidizer, in the form of oxygen, is everywhere. What we really need to control are ignition sources coming into contact with fuel, in the presence of oxygen.

References

Almukhtar, S. and Watkins, D. (2016). I'M dying, *The New York Times*. https://www.nytimes.com/interactive/2016/09/06/world/middleeast/2015-hajj-stampede.html (accessed December 2021).

AP. (2021). Travis Scott says he was unaware of deaths until after show. https://apnews.com/article/travis-scott-entertainment-business-lifestyle-arts-and-entertainment-8e0af18ed50e89865ce7271ed1303179 (accessed December 2021).

ATEX (2021). Appareils destinés à être utilisés en ATmosphères EXplosives. https://www.explosionhazards.co.uk/list-of-historic-explosions/ (accessed December 2021).

BBC (2016). Taiwan water park fire: Party organiser jailed for negligence. https://www.bbc.com/news/world-asia-36136712 (accessed December 2021).

BBC (2022). Hillsborough: Timeline of the 1989 stadium disaster. https://www.bbc.com/news/uk-england-merseyside-47697569 (accessed December 2021).

Beyond Discovery. 2015. 542 historical perspective wheat flour explosion in Turin 1785. https://www.beyonddiscovery.org/dust-clouds/542-historical-perspective-wheat-flour-explosion-in-turin-1785.html.

Chemical Safety and Hazard Investigation Board (CSB) (2006). Combustible Dust Hazard Study, Washington, DC.

Chemical Safety Board (CSB) (2014). AL Solutions, Inc., New Cumberland, WV, Metal Dust Explosion and Fire, Case Study.

Chemical Safety Board (CSB), Imperial Sugar Company (2009). Combustible Dust Explosion, Report No. 2008-05-I-GA.

Chemical Safety Board (CSB) (2018). Combustible Dust Timeline, https://www.csb.gov/combustible-dust-timeline/, 2018.

Dahn, J.C., Reyes, B.N. (1986). Determination of metal sparking characteristics and the effects on explosive dust clouds, Industrial Dust Explosions, ASTM Special Technical Publication 958. Philadelphia, PA.

Dust Safety Science (2021). New Taipei water park explosion. https://dustsafetyscience.com/new-taipei-water-park-explosion/ (accessed December 2021).

Eckhoff, R. (2003). *Dust Explosions in the Process Industries*, 3e, 64. Burlington, MA: Elsevier Science.

Fricke, D., Rolling Stone (2000). Nine dead at Pearl Jam concert. https://www.rollingstone.com/music/music-news/nine-dead-at-pearl-jam-concert-235167/ (accessed December 2021).

Hazardous Material Information System (HMIS) (2021). https://www.chemsafetypro.com/Topics/USA/Hazardous_Materials_Identification_System_HMIS.html (accessed November 2021).

Heintz, J. (December 4, 2009). Perm, Russia Nightclub Explosion Kills More Than 100: Reports. The Huffington Post. (accessed December 2021).

Hughes (2021). https://info.hughesenv.com/history-of-combustible-dust-explosions (accessed December 2021).

King, C. (2021). Survivor of 'The Who' concert at Riverfront: 'I can't believe it happened again'.

Laurent, A. (2003). Securite des procedes chimiques. *Tec et Doc* 237.

Levy, C., The New York Times (2009). Fire sweeps through club in Russia, Killing 100. https://www.nytimes.com/2009/12/05/world/europe/05russia.html (accessed December, 2021).

Mills, D. (2004). *Pneumatic Conveying Design Guide*, 577. Butterworth Heinemann.

Natural Sourcing, Safety Data Sheet (2021). Cornstarch. https://www.praannaturals
.com/downloads/msds/SDS_Corn_Starch_OTHCORNSTRUS701.pdf (accessed
December 2021).

NFPA (2021). The 10 deadliest nightclub fires in World History https://www.nfpa
.org/Public-Education/Staying-safe/Safety-in-living-and-entertainment-spaces/
Nightclubs-assembly-occupancies/Deadliest-public-assembly-and-nightclub-fires
(accessed December 2021).

OSHA (2009). Hazard Communication Guidance for Combustible Dusts, U.S.
Department of Labor Occupational Safety and Health Administration OSHA
3371-08.

Ostrom, L.T. and Wilhelmsen, C. (2019). *Risk Assessment Tools and Techniques and
Their Application*, 2e. Wiley.

Pape, R. and Schmidt, F. (2009). Fires and explosions: combustibility analysis of
metals. *Advanced Materials and Processes* 167 (11/12): 41.

Perez-Castineira, J. (2020). *Chemistry and Biochemistry of Food*. De Gruyter.

Rhodes, M.J. (1990). *Principles of Powder Technology*, 307. Wiley.

Sigma-Aldrich (2014). Zirconium MSDS. https://www.nwmissouri.edu/
naturalsciences/sds/z/Zirconium.pdf (accessed December, 2021).

US Research Nanomaterials, Inc. (2021). Titanium MSDS. https://n.b5z.net/i/u/
10091461/f/MSDS-MicroPowder/US5006.pdf (accessed November 2021).

Vorderbrueggen, J. (2011). Imperial sugar refinery combustible dust explosion
investigation. *Process Safety* 30 (1): 66–81.

6

University Laboratory Accident Case Studies

6.0 Introduction

Horrific accidents happen at universities. Of course, there are the typical industrial type accidents. However, many involve students working on funded projects, working in teaching laboratories, and some occur for students who are part of a club. The following presents some case studies of accidents that have occurred, and the safety culture attributes that contributed to the accidents.

6.1 My Experience at Aalto University

During the winter months of 2018, I was fortunate to do a Fulbright Specialist at Aalto University outside Helsinki, Finland. This activity was to conduct a safety review of the physics laboratories at the University. This was being performed because the University was concerned about some safety practices in the laboratories. The Physics Department and Aalto University conduct cutting edge research and the student populations includes students from all over the world. For instance, Figure 6.1 shows an experimental setup in which positrons are being directed toward targets for various experimental reasons. This section does not impugn the quality of the research or the quality of the faculty or students at Aalto University. It just points out laboratory safety issues that are common to universities and laboratories around the world.

The Aalto University physics laboratories conducted experiments utilizing large amounts of cryogenic materials, radioactive materials, and high energy non-ionizing radiation. The cryogenic materials included liquid nitrogen and liquid helium. Figure 6.2 shows a large Dewar of liquid helium.

Liquid helium boils are −452 °F (−270 °C). Whereas liquid nitrogen boils at −320 °F (−196 °C) and liquid oxygen boils at −297 °F (−183 °C). Obviously, they are all very cold materials when in their liquid state. Being odorless, colorless,

Impact of Societal Norms on Safety, Health, and the Environment: Case Studies in Society and Safety Culture, First Edition. Lee T. Ostrom.

Figure 6.1 Experimental setup for positron experiment. Source: Ostrom (2018).

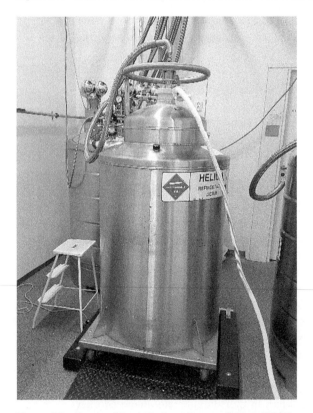

Figure 6.2 Dewar of liquid helium. Source: Ostrom (2018).

tasteless, and nonirritating, helium has no warning properties. Humans possess no senses that can detect the presence of helium. Although helium is nontoxic and inert, it can act as a simple asphyxiant by displacing the oxygen in air to levels below that required to support life. Inhalation of helium in excessive amounts can cause dizziness, nausea, vomiting, loss of consciousness, and death. Death may result from errors in judgment, confusion, or loss of consciousness that prevents self-rescue. At low oxygen concentrations, unconsciousness and death may occur in seconds and without warning. Personnel, including rescue workers, should not enter areas where the oxygen concentration is below 19.5%, unless provided with a self-contained breathing apparatus or air-line respirator. Liquid helium causes severe frost bite/cryogenic burn injuries almost instantaneously. Injuries to the skin and eyes are common with exposure to liquid helium. Personal protective equipment should always be worn when working around cryogenic materials. The same is true of liquid nitrogen.

If liquid helium or liquid nitrogen spill in a laboratory space, they can cause oxygen to be condensed out of the environmental air because these materials boil at a much lower temperature that oxygen. Liquid oxygen is a much more dangerous material, than either liquid helium or liquid nitrogen. The hazards associated with liquid oxygen are exposure to cold temperatures that can cause severe burns; over-pressurization due to expansion of small amounts of liquid into large volumes of gas in inadequately vented equipment; oxygen enrichment of the surrounding atmosphere; and the possibility of a combustion reaction if the oxygen is permitted to contact a non-compatible material. The low temperature of liquid oxygen and the vapors it releases not only pose a serious burn hazard to human tissue but can also cause many materials of construction to lose their strength and become brittle enough to shatter. The large expansion ratio of liquid-to-gas can rapidly build pressure in systems where liquid can be trapped. This necessitates that these areas be identified and protected with pressure relief. This expansion ratio also allows atmospheres of oxygen-enriched air to form in the area surrounding a release.

Fire chemistry starts to change when the concentration of oxygen increases to as little as 23%. Materials easily ignited in air not only become more susceptible to ignition, but also burn with added violence in the presence of oxygen. Oxygen fires are very intense. Ambulances have burned to the ground because a small amount of oil or other organic material was on the tubing or valves of an oxygen tank and a very intense fire started and then the oxygen actually started the tubing, oxygen bottle, or other metal oxidizing/burning. This type of fire progresses extraordinarily fast. Figure 6.3 shows the aftermath of such a fire. This fire resulted in the death of an Emergency Medical Technician (EMT) (Center for Disease Control [CDC] 1999). The materials that can burn include clothing and hair, which have air spaces that readily trap the oxygen. Oxygen levels of 23% can be reached very quickly and all personnel must be aware of the hazard. Any clothing that has been

Figure 6.3 Ambulance fire. Source: Center for Disease Control (CDC) – PHIL (1999).

splashed or soaked with liquid oxygen or exposed to high oxygen concentrations should be removed immediately and aired for at least an hour. Personnel should stay in a well-ventilated area and avoid any source of ignition until their clothing is completely free of any excess oxygen. Clothing saturated with oxygen is readily ignitable and will burn vigorously.

Many fires have occurred in hospital surgery suites when flammable intubation tubes or other materials are exposed to high levels of oxygen. Eunju Lee et al. (2012) discuss an endotracheal tube fire during a tracheostomy procedure.

Just as I was writing this section an oxygen fueled fire in the southeastern Iraqi city of Nasiriya killed over 90 people (CNN 2021).

The physics labs at Aalto University used a large amount of cryogenic liquids in their experiments and not all the time did researchers follow the standard safety procedures for these materials. I observed several seasoned researchers decant cryogenic materials without eye protection and without any gloves, let alone cryogenic protecting gloves.

Chapter 10 of this book presents detailed information about the safety culture survey I conducted during my visit. However, the following presents additional information about the safety culture at the school.

Figure 6.4 is a graph that shows the hazards in the laboratories that researchers were most fearful about. Chemical waste, cryogenic materials, flammable liquids, compressed gas, and toxic metals were the top five. This indicates that the researchers understand the primary hazards in their laboratories. Figure 6.5 shows the hazards that had caused injuries to researchers. These two graphs are not in contradiction to each other. There is some overlap and the top five hazards

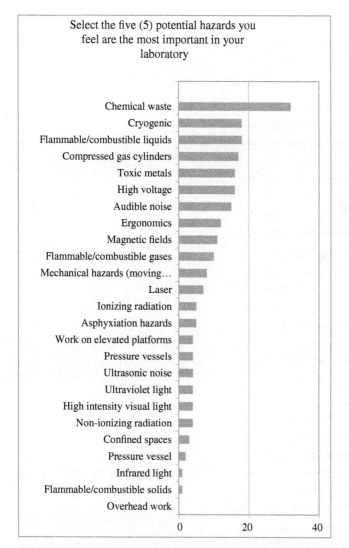

Figure 6.4 Hazards identified in Aalto University Laboratories. Source: Ostrom (2018).

in Figure 6.5 can cause more egregious injuries and, quite honestly, it is good there were not more injuries due to the top five hazards.

The Physics department also had a tremendous amount of chemicals. Figure 6.6 shows just a small amount of chemicals. An Aalto University employee was in the process of conducting an inventory of all the chemicals. Many incompatible

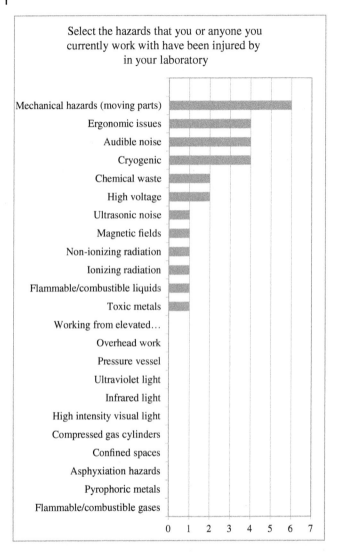

Figure 6.5 Hazards in Aalto University Laboratories that caused injuries. Source: Ostrom (2018).

chemicals were being stored close to each other when I first visited there. The employee was also in the process of separating the incompatible chemicals. One point that should be made about this situation or other chemical storage situation is to only buy the amount of chemical that is needed for the experiment and then at the end of the experiment the chemical(s) should gone. The reason

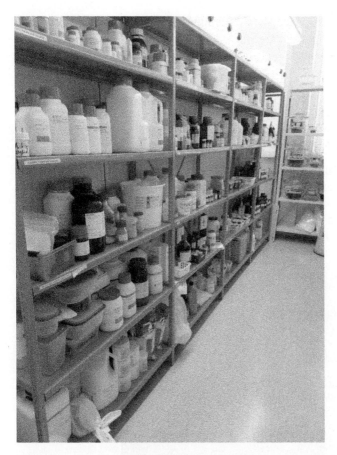

Figure 6.6 Chemical storage. Source: Ostrom (2018).

researchers buy large amounts of chemicals is because it is cheaper to buy in bulk and the researchers don't want to run out in the middle of an experiment.

However, once a chemical is open and not used it now becomes chemical waste. Most researchers do not want to use someone else's chemicals because they are not sure if the chemicals have been contaminated. The lesson here is to only buy what you need.

Single chemicals are less expensive to dispose of than chemical mixtures. Mixtures of chemicals are mixed waste and can be tremendously expensive to dispose of, especially if the mixed waste contains heavy metals.

A recent incident at my home university involved a professor wanted to order a small of uranium. The original cost was going to be approximately US$ 1000.

The disposal cost was quoted as $30,000! Consequently, the professor changed his mind and did not order the material.

Cleanliness of lab benches are always a problem. Figure 6.7 shows a lab bench mess a researcher left.

There are many questions associated with this mess:

1. Are the mixer and scale still usable?
2. What is the chemical? Is this now toxic waste?
3. Who did this and who is responsible for cleaning it up?

This is not uncommon. In a tour of a couple labs this week at my home university I talked with researchers about spills on the lab benches. In some cases, lab benches become discolored from spills and cannot be cleaned. However, it is not an active chemical and not a hazard. In other cases, like the one shown in Figure 6.8, there could be very real problems. I worked in a chemical plant in my early career

Figure 6.7   Lab bench mess. Source: Ostrom (2018).

Figure 6.8 Contaminated eyewash station. Source: Ostrom (2018).

and this experience taught me to keep my hands in my pockets in chemical labs, well, biological labs also.

Another major issue in laboratories is that safety shower/eyewash stations are not always available and/or maintained. Figure 6.9 shows an eyewash station that was not maintained. The water coming out of it will be contaminated with rust and could cause additional injury to a person's eyes. The safety shower in Figure 6.10 is associated with the eyewash station and it is blocked with the chair. In this case it is not a major problem, but in the case of someone with a strong chemical on their skin, it could mean the difference between a minor and a major injury. Figures 6.10–6.12 show other examples of safety shower/eyewash issues.

The aforementioned discussion about the physics laboratories at Aalto University presented many safety issues. All these issues are associated with the safety culture of the faculty and students. The case studies that will be presented in the remaining parts of this chapter all are due to safety culture issues, some like the discussion earlier.

Figure 6.9 Chair blocking eyewash station. Source: Ostrom (2018).

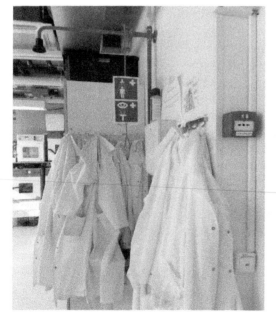

Figure 6.10 Lab coats blocking safety shower. Source: Ostrom (2018).

Figure 6.11 Disabled safety shower still present. Source: Ostrom (2018).

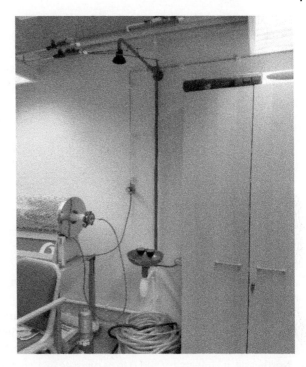

Figure 6.12 Electrical cord draped across eyewash. Source: Ostrom (2018).

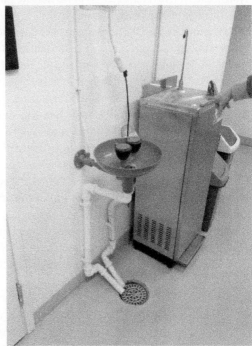

Once again, Aalto University is not unique. In fact, they developed this Fulbright Specialist because the University knew there was a problem in the Physics department.

Table 6.1 lists the laboratory accidents that have occurred in the United States from July 2001 to August 2018.

Summary table.

Events	Deaths	Injuries
300+	6	541

The following presents some case studies of accidents that have occurred, and the safety culture issues that contributed to the accidents.

6.2 Texas Tech University October 2008

The following is a description of the accident on January 7, 2010, and the description and analyses are excerpted from the US Chemical Safety Board (CSB 2011a,b).

In October 2008, Texas Tech (TTU) entered into a subcontract agreement with Northeastern University (NEU) to participate in a program titled "Awareness and Localization of Explosive-Related Threats" (ALERT), which was (and continues to be) funded by the US Department of Homeland Security (DHS). TTU's research focus was the detection of energetic materials that could represent a security threat. It included synthesizing and characterizing new potentially energetic materials. The incident on January 7 involved a fifth-year graduate student and a first-year graduate student he was mentoring. They began synthesizing a nickel hydrazine perchlorate (NHP) derivative. The amounts of NHP to be synthesized were on the order of 50–300 mg. Typical analytical techniques used in the laboratory to characterize new compounds' energetic properties included differential scanning calorimetry (DSC), drop hammer tests, and thermal gravimetric analysis (TGA). The students synthesizing the NHP decided they would need to make several batches of the compound to fully characterize the compound due to the amounts of compound needed to run each analytical test.

Additionally, they had concerns about reproducibility between batches. They wanted to synthesize a single batch of NHP that would provide enough compound to complete all the necessary characterization. They decided to scale

Table 6.1 List of laboratory accidents.

Incident date	Organization	City	State	Fatalities	Injuries
July 23, 2001	University of California, Irvine	Irvine	California	0	3
October 10, 2001	University of Utah	Salt Lake City	Utah	0	0
October 10, 2001	Genoa-Kingston High School	Genoa	Illinois	0	7
October 16, 2001	US Army Research Laboratory	Adelphi	Maryland	0	1
January 08, 2002	Los Alamos National Laboratory	Los Alamos	New Mexico	0	0
March 11, 2002	New Berlin West High School	New Berlin	Wisconsin	0	1
December 31, 2002	University of Washington Medical Center	Seattle	Washington	0	0
January 30, 2004	Federal Way High School	Federal Way	Washington	0	2
June 18, 2005	Huntington Beach High School	Huntington Beach	California	0	2
January 23, 2006	Western Reserve High School	Hudson	Ohio	0	4
January 24, 2006	Lansing Community College	Lansing	Michigan	0	0
January 26, 2006	Middle Township High School	Cape May County	New Jersey	0	5
January 31, 2006	Cornell University	Ithaca	New York	0	1
February 20, 2006	West Chester University	West Chester	Pennsylvania	0	2
February 24, 2006	University of Denver	Denver	Colorado	0	0
March 02, 2006	University of Idaho	Moscow	Idaho	0	1
March 07, 2006	Saratoga Springs High School	Saratoga Springs	New York	0	8
March 12, 2006	Monahans High School	Monahans	Texas	0	3
April 03, 2006	University of Maryland Biotechnology Institute	Baltimore	Maryland	0	1

(Continued)

Table 6.1 (Continued)

Incident date	Organization	City	State	Fatalities	Injuries
April 11, 2006	Southwest Minnesota State University	Marshall	Minnesota	0	1
April 20, 2006	Scripps Research Institute	Jupiter	Florida	0	0
April 21, 2006	Northwestern University	Evanston	Illinois	0	1
May 05, 2006	Massachusetts Institute of Technology	Cambridge	Massachusetts	0	0
May 09, 2006	Prosper High School	Dallas	Texas	0	1
May 16, 2006	Georgia Institute of Technology (Georgia Tech)	Atlanta	Georgia	0	1
June 08, 2006	Lafayette High School	James City	Virginia	0	1
June 16, 2006	Binghamton University	Vestal	New York	0	0
June 28, 2006	Purdue University	West Lafayette	Indiana	0	0
July 03, 2006	Cardiovascular Genetics	Salt Lake City	Utah	0	1
July 09, 2006	Washington Middle School	Aurora	Illinois	0	0
July 20, 2006	University of Colorado	Boulder	Colorado	0	3
July 20, 2006	University of Minnesota, Twin Cities	St. Paul	Minnesota	0	0
July 27, 2006	University of Colorado	Boulder	Colorado	0	1
August 01, 2006	University of Washington	Seattle	Washington	0	1
August 16, 2006	Virginia Tech	Blacksburg	Virginia	0	3
August 24, 2006	Rice University	Houston	Texas	0	0
August 27, 2006	Indiana University	Bloomington	Indiana	0	0
September 06, 2006	Georgia State University	Atlanta	Georgia	0	6
October 03, 2006	Hasbrouck Heights High School	Hasbrouck Heights	New Jersey	0	0
October 10, 2006	Oakland High School	Oakland	Oregon	0	0

Date	Institution	City	State	
October 11, 2006	Morrisville Pharmacy	Morrisville	Pennsylvania	0	1
October 31, 2006	Louisiana State University	Baton Rouge	Louisiana	0	2
November 01, 2006	Eastern Guilford High School	Greensboro	North Carolina	0	0
November 03, 2006	University of Kentucky	Lexington	Kentucky	0	7
November 12, 2006	Yale University	New Haven	Connecticut	0	1
December 03, 2006	Clinical Pathology Labs	Springville	Utah	0	1
December 19, 2006	Oak Ridge National Laboratory	Oak Ridge	Tennessee	0	0
December 21, 2006	University of Arkansas	Fayetteville	Arkansas	0	0
January 17, 2007	Pennsylvania State University	State College	Pennsylvania	0	0
January 18, 2007	South Fremont High School	Saint Anthony	Idaho	0	0
January 01, 2007	H.B. du Pont Middle School	Hockessin	Delaware	0	1
February 07, 2007	Mead Middle School	Mead	Washington	0	2
February 09, 2007	Dobyns-Bennett High School	Kingsport	Tennessee	0	0
February 20, 2007	James Madison University	Harrisonburg	Virginia	0	0
February 26, 2007	Duquesne University	Pittsburgh	Pennsylvania	0	0
March 20, 2007	Boston College	Boston	Massachusetts	0	1
March 27, 2007	North Carolina State University	Raleigh	North Carolina	0	1
March 30, 2007	Frontier Scientific	Logan	Utah	0	1
April 10, 2007	West Virginia University	Morgantown	West Virginia	0	3
April 30, 2007	Volusia County Health Department	Daytona Beach	Florida	0	1
May 01, 2007	University of Wisconsin, Milwaukee	Milwaukee	Wisconsin	0	0
May 07, 2007	CEMEX	Brooksville	Florida	0	1
May 22, 2007	Aspen Medical Clinic	Maplewood	Minnesota	0	1
May 24, 2007	University of Pittsburgh	Pittsburgh	Pennsylvania	0	1

(Continued)

Table 6.1 (Continued)

Incident date	Organization	City	State	Fatalities	Injuries
May 28, 2007	University of South Florida	Tampa	Florida	0	3
May 30, 2007	Texas A&M University	College Station	Texas	0	0
June 05, 2007	Rensselaer Polytechnic Institute	Troy	New York	0	3
June 18, 2007	Delaware State University	Dover	Delaware	0	2
July 18, 2007	Tennessee Technological University	Cookville	Tennessee	0	1
July 27, 2007	University of Arizona	Tucson	Arizona	0	0
August 03, 2007	State University of New York at Buffalo	Getzville	New York	0	0
August 06, 2007	Washington University in St. Louis	St. Louis	Missouri	0	0
August 10, 2007	Georgia Institute of Technology (Georgia Tech)	Atlanta	Georgia	0	1
August 15, 2007	Rich Products Corporation	Morristown	Tennessee	0	2
September 11, 2007	Ball State University	Muncie	Indiana	0	0
September 11, 2007	University of Florida	Gainesville	Florida	0	0
September 11, 2007	University of Vermont	Burlington	Vermont	0	15
September 16, 2007	ASTEC Charter Middle School	Oklahoma City	Oklahoma	0	2
September 17, 2007	University of Michigan	Ann Arbor	Michigan	0	0
September 19, 2007	Purdue University	West Lafayette	Indiana	0	0
October 01, 2007	Boston University	Boston	Massachusetts	0	0
October 08, 2007	Oakwood College	Huntsville	Alabama	0	3
October 14, 2007	University of California, Santa Barbara	Santa Barbara	California	0	0
October 19, 2007	Marshall University	Huntington	West Virginia	0	0
October 25, 2007	Yale University	New Haven	Connecticut	0	1
October 29, 2007	Drake University	Des Moines	Iowa	0	0

Incident Date	Organization	City	State	Fatalities	Injuries
October 30, 2007	Northwestern University	Evanston	Illinois	0	0
November 11, 2007	State of Illinois Police Crime Laboratory	Springfield	Illinois	0	0
November 16, 2007	Granite Hills High School	Apple Valley	California	0	1
November 27, 2007	University of North Carolina	Chapel Hill	North Carolina	0	2
December 06, 2007	Brigham Young University	Provo	Utah	0	0
December 19, 2007	T2 Laboratories, Inc.	Jacksonville	Florida	4	14
January 15, 2008	Craig Middle School	Craig	Colorado	0	8
January 16, 2008	Somers High School	Somers	New York	0	8
January 22, 2008	Fairhaven High School	Fairhaven	Massachusetts	0	0
February 01, 2008	Dickinson High School	Dickerson	North Dakota	0	0
February 05, 2008	University of Washington	Seattle	Washington	0	0
February 13, 2008	Southern Illinois University	Carbondale	Illinois	0	0
February 13, 2008	University of South Dakota	Vermillion	South Dakota	0	0
February 19, 2008	Johns Hopkins University	Baltimore	Maryland	0	0
February 22, 2008	University of Wisconsin, Madison	Madison	Wisconsin	0	0
March 05, 2008	Johns Hopkins University	Baltimore	Maryland	0	2
March 30, 2008	University of Texas	Austin	Texas	0	0
April 04, 2008	Western Middle School	Greenwich	Connecticut	0	0
April 07, 2008	University of Washington	Seattle	Washington	0	0
April 10, 2008	Florida A&M University	Tallahassee	Florida	0	1
May 13, 2008	Vintage High School	Vallejo	California	0	0
May 19, 2008	Huntington Park College-Ready Academy High School	Huntington Park	California	0	2

(Continued)

Table 6.1 (Continued)

Incident date	Organization	City	State	Fatalities	Injuries
May 21, 2008	Hudson High School	Hudson	Ohio	0	1
May 21, 2008	Mountain View High School	Bend	Oregon	0	0
May 21, 2008	University of South Florida	Tampa	Florida	0	6
May 27, 2008	University of Arizona	Tucson	Arizona	0	0
June 11, 2008	Massachusetts Institute of Technology	Cambridge	Massachusetts	0	1
June 12, 2008	University of Florida	Gainesville	Florida	0	0
June 23, 2008	University of California, Davis	Sacramento	California	0	2
June 24, 2008	Freemont High School	Freemont	California	0	0
July 16, 2008	University of Colorado	Boulder	Colorado	0	0
July 22, 2008	Colorado State University	Fort Collins	Colorado	0	0
August 08, 2008	Auburn University	Auburn	Alabama	0	0
August 11, 2008	Canisius College	Buffalo	New York	0	0
September 08, 2008	Lewisville High School	Lewisville	Texas	0	0
September 08, 2008	Michigan State University	East Lansing	Michigan	0	0
September 08, 2008	University of Alabama	Tuscaloosa	Alabama	0	1
September 09, 2008	Central Coast Pathology	San Luis Obispo	California	0	3
September 09, 2008	Clark Atlanta University	Atlanta	Georgia	0	0
September 12, 2008	Trinity College	Hartford	Connecticut	0	1
September 15, 2008	California State University, Chico	Chico	California	0	0
September 16, 2008	University of Southern Maine	Gorham	Maine	0	4
October 28, 2008	Penobscot Bay Medical Center	Rockport	Maine	0	2
Incident Date	Organization	City	State	Fatalities	Injuries

Date	Institution	City	State		
December 29, 2008	University of California, Los Angeles	Los Angeles	California	1	0
February 28, 2009	Boise State University	Boise	Idaho	0	1
March 09, 2009	Florida Medical Clinic	Zephyrhills	Florida	0	0
July 27, 2009	Wasatch Labs	Ogden	Utah	0	3
August 29, 2009	Eurand America, Inc.	Vandalia	Florida	0	2
September 08, 2009	Indiana University-Purdue University Indianapolis	Fort Wayne	Indiana	0	1
January 07, 2010	Texas Tech University*	Lubbock	Texas	0	1
May 11, 2010	Texas A&M University	College Station	Texas	0	2
June 02, 2010	Southern Illinois University	Carbondale	Illinois	0	1
June 28, 2010	University of Missouri	Columbia	Missouri	0	4
December 03, 2010	Northwestern University	Evanston	Illinois	0	1
January 17, 2011	Spectrum Microwave	Marlborough	Massachusetts	0	20
February 08, 2011	SynQuest Laboratories	Alachua	Florida	0	1
February 17, 2011	Oregon Health and Science University	Portland	Oregon	0	4
March 08, 2011	Southfield Lathrup High School	Lathrup Village	Michigan	0	3
March 10, 2011	Louisiana State University	Baton Rouge	Louisiana	0	1
March 16, 2011	Choice Dental Laboratory	St. Joseph	Michigan	0	1
April 26, 2011	Agilent Technologies	Santa Rosa	California	0	1
April 30, 2011	Aberdeen Proving Ground Laboratory	Aberdeen	Maryland	1	0
April 30, 2011	Front Range Community College	Longmont	Colorado	0	1
May 02, 2011	IMANNA Laboratory, Inc.	Rockledge	Florida	0	1
May 09, 2011	University of California, Berkeley	Berkeley	California	0	1
May 12, 2011	Clarkson University	Potsdam	New York	0	1

(Continued)

Table 6.1 (Continued)

Incident date	Organization	City	State	Fatalities	Injuries
May 18, 2011	Louisiana State University	Baton Rouge	Louisiana	0	0
June 20, 2011	Purdue University	West Lafayette	Indiana	0	6
June 25, 2011	Boston College	Chestnut Hill	Massachusetts	0	1
July 12, 2011	University of West Florida	Pensacola	Florida	0	2
July 20, 2011	New Life Worship Center	Smithfield	Rhode Island	0	4
August 02, 2011	Bradley University	Peoria	Illinois	0	0
August 17, 2011	University of Pittsburgh	Pittsburgh	Pennsylvania	0	1
September 02, 2011	Membrane Technology and Research, Inc.	Menlo Park	California	1	1
September 12, 2011	Geomet Technologies, LLC	Gaithersburg	Maryland	0	1
September 19, 2011	Harold L. Richards High School	Oak Lawn	Illinois	0	1
September 21, 2011	West Charlotte High School	Charlotte	North Carolina	0	1
September 26, 2011	University of Maryland	College Park	Maryland	0	2
September 27, 2011	University of Connecticut Health Center	Framingham	Connecticut	0	2
October 11, 2011	University of Florida	Gainesville	Florida	0	1
October 14, 2011	Texas Tech University	Lubbock	Texas	0	0
October 17, 2011	University of Florida	Gainesville	Florida	0	1
October 20, 2011	DeKALB Molded Plastics	Butler	Indiana	0	0
October 24, 2011	University of Arizona	Tucson	Arizona	0	0
October 24, 2011	University of California, Los Angeles	Los Angeles	California	0	0
October 25, 2011	E.E. Smith High School	Fayetteville	North Carolina	0	6
October 26, 2011	Kerr Middle School	Del City	Oklahoma	0	1
October 27, 2011	Texas Tech University	Lubbock	Texas	0	0

November 30, 2011	Bocchi Laboratories	Santa Clarita	California	0	2
December 01, 2011	Maple Grove High School	Maple Grove	Minnesota	0	4
December 21, 2011	University of Oregon	Eugene	Oregon	0	1
January 09, 2012	Scripps Research Institute	La Jolla	California	0	3
January 11, 2012	Carnegie Mellon University	Pittsburgh	Pennsylvania	0	0
January 11, 2012	University of Florida	Gainesville	Florida	0	1
January 13, 2012	David Douglas High School	Portland	Oregon	0	11
January 30, 2012	University of Cincinnati	Cincinnati	Ohio	0	1
January 30, 2012	University of Wisconsin, Madison	Madison	Wisconsin	0	1
February 06, 2012	South Carolina State University	Orangeburg	South Carolina	0	6
March 15, 2012	University of Florida	Gainesville	Florida	0	1
March 19, 2012	Reedsburg Area High School	Reedsburg	Wisconsin	0	3
March 20, 2012	University of Colorado	Boulder	Colorado	0	1
April 11, 2012	General Motors Technical Center	Warren	Michigan	0	1
April 12, 2012	Mililani High School	Mililani	Hawaii	0	1
April 22, 2012	Soule Road Middle School	Liverpool	New York	0	4
May 03, 2012	Midwest High School	Natrona	Wyoming	0	1
May 14, 2012	Colorado State University	Fort Collins	Colorado	0	1
June 29, 2012	Ventura Foods	St. Joseph	Missouri	0	1
July 02, 2012	BAE Systems	Radford	Virginia	0	1
July 10, 2012	Monsanto	Ankeny	Iowa	0	1
July 11, 2012	University of Minnesota, Duluth	Duluth	Minnesota	0	0
July 19, 2012	Organix, Inc.	Woburn	Massachusetts	0	1

(Continued)

Table 6.1 (Continued)

Incident date	Organization	City	State	Fatalities	Injuries
July 30, 2012	US Geological Survey, Oregon Water Science Center	Portland	Oregon	0	1
October 20, 2012	University of Akron	Akron	Ohio	0	5
October 24, 2012	Derry Area Schools	Derry	Pennsylvania	0	1
November 12, 2012	Eastern High School	Voorhees	New Jersey	0	7
November 19, 2012	Monolyte Laboratories, Inc.	Slaughter	Louisiana	0	0
February 01, 2013	Clovis North High School	Clovis	California	0	1
February 01, 2013	Washington State University	Pullman	Washington	0	1
February 09, 2013	Air Liquide America Specialty Gases, LLC	La Porte	Texas	1	1
February 12, 2013	Villanova University	Villanova	Pennsylvania	0	10
February, 26 2013	Massachusetts General Hospital	Charlestown	Massachusetts	0	1
April 10, 2013	Colorado College	Colorado Springs	Colorado	0	13
May 13, 2013	Energetic Materials Research and Testing Center	Socorro	New Mexico	0	3
June 04, 2013	University of Delaware	Newark	Delaware	0	1
June 17, 2013	St. Scholastica Academy	Covington	Louisiana	0	3
September 09, 2013	Roach Middle School	Frisco	Texas	0	3
October 03, 2013	Chapel Hill High School	Chapel Hill	Georgia	0	1
October 09, 2013	Dow Chemical Company	North Andover	Massachusetts	1	0
November 01, 2013	Syracuse University	Syracuse	New York	0	1
November 12, 2013	La Joya Community High School	Avondale	Arizona	0	4
November 12, 2013	University of Illinois	Urbana	Illinois	0	5

Date	Institution	City	State		
November 25, 2013	Lincoln Park High School	Chicago	Illinois	0	1
January 02, 2014	Beacon High School	New York	New York	0	2
January 08, 2014	Amgen, Inc.	South San Francisco	California	0	2
January 23, 2014	Auburn University	Auburn	Alabama	0	1
February 19, 2014	Northside College Prep High School	Chicago	Illinois	0	5
April 09, 2014	Tindley Accelerated School	Indianapolis	Indiana	0	1
June 09, 2014	Boise State University	Boise	Idaho	0	1
June 17, 2014	University of Minnesota, Twin Cities	Minneapolis	Minnesota	0	1
August 20, 2014	Bentley Laboratories	Edison	New Jersey	0	2
September 03, 2014	Terry Lee Wells Nevada Discovery Museum*	Reno	Nevada	0	13
September 15, 2014	STRIVE Preparatory School*	Denver	Colorado	0	4
October 18, 2014	University of Rochester	Rochester	New York	0	3
October 20, 2014	A Community of Faith Church*	Raymond	Illinois	0	4
October 31, 2014	UIC College Prep	Chicago	Illinois	0	2
July 30, 2012	US Geological Survey, Oregon Water Science Center	Portland	Oregon	0	1
October 20, 2012	University of Akron	Akron	Ohio	0	5
October 24, 2012	Derry Area Schools	Derry	Pennsylvania	0	1
November 12, 2012	Eastern High School	Voorhees	New Jersey	0	7
November 19, 2012	Monolyte Laboratories, Inc.	Slaughter	Louisiana	0	0
February 01, 2013	Clovis North High School	Clovis	California	0	1
February 01, 2013	Washington State University	Pullman	Washington	0	1

(Continued)

Table 6.1 (Continued)

Incident date	Organization	City	State	Fatalities	Injuries
February 09, 2013	Air Liquide America Specialty Gases, LLC	La Porte	Texas	1	1
February 12, 2013	Villanova University	Villanova	Pennsylvania	0	10
February 26, 2013	Massachusetts General Hospital	Charlestown	Massachusetts	0	1
April 10, 2013	Colorado College	Colorado Springs	Colorado	0	13
May 13, 2013	Energetic Materials Research and Testing Center	Socorro	New Mexico	0	3
June 04, 2013	University of Delaware	Newark	Delaware	0	1
June 17, 2013	St. Scholastica Academy	Covington	Louisiana	0	3
September 09, 2013	Roach Middle School	Frisco	Texas	0	3
October 03, 2013	Chapel Hill High School	Chapel Hill	Georgia	0	1
October 09, 2013	Dow Chemical Company	North Andover	Massachusetts	1	0
November 01, 2013	Syracuse University	Syracuse	New York	0	1
November 12, 2013	La Joya Community High School	Avondale	Arizona	0	4
November 12, 2013	University of Illinois	Urbana	Illinois	0	5
November 25, 2013	Lincoln Park High School	Chicago	Illinois	0	1
January 02, 2014	Beacon High School	New York	New York	0	2
January 08, 2014	Amgen, Inc.	South San Francisco	California	0	2
January 23, 2014	Auburn University	Auburn	Alabama	0	1
February 19, 2014	Northside College Prep High School	Chicago	Illinois	0	5
April 09, 2014	Tindley Accelerated School	Indianapolis	Indiana	0	1

Incident Date	Organization	City	State	Fatalities	Injuries
June 09, 2014	Boise State University	Boise	Idaho	0	1
June 17, 2014	University of Minnesota, Twin Cities	Minneapolis	Minnesota	0	1
August 20, 2014	Bentley Laboratories	Edison	New Jersey	0	2
September 03, 2014	Terry Lee Wells Nevada Discovery Museum*	Reno	Nevada	0	13
September 15, 2014	STRIVE Preparatory School*	Denver	Colorado	0	4
October 18, 2014	University of Rochester	Rochester	New York	0	3
October 20, 2014	A Community of Faith Church*	Raymond	Illinois	0	4
October 31, 2014	UIC College Prep	Chicago	Illinois	0	2
January 06, 2015	Newfield High School	Selden	New York	0	1
January 26, 2015	Kingsway Regional High School	Woolwich	New Jersey	0	17
February 02, 2015	Texas Tech University	Lubbock	Texas	0	4
February 24, 2015	Michigan State University	East Lansing	Michigan	0	1
February 28, 2015	Sandstone Elementary School	Billings	Montana	0	4
May 15, 2015	Apex High School	Apex	North Carolina	0	0
May 22, 2015	Lincoln High School	Tallahassee	Florida	0	2
June 20, 2015	United Taconite Laboratory	Forbes	Minnesota	0	0
August 25, 2015	Southern Illinois University	Carbondale	Illinois	0	1
September 15, 2015	University of Akron	Akron	Ohio	0	1
October 30, 2015	W.T. Woodson High School	Fairfax	Virginia	0	6
February 05, 2016	Slater High School	Slater	Missouri	0	1
February 29, 2016	Glendale Community College	Glendale	California	0	2
February 29, 2016	University of Rochester	Rochester	New York	0	1
March 09, 2016	Texas A&M University	College Station	Texas	0	1

(Continued)

Table 6.1 (Continued)

Incident date	Organization	City	State	Fatalities	Injuries
March 10, 2016	Texas Tech University	Lubbock	Texas	0	1
March 16, 2016	University of Hawaii	Honolulu	Hawaii	0	1
April 27, 2016	Hidden Lake High School	Westminster	Colorado	0	1
May 24, 2017	Perth Amboy High School	Perth Amboy	New Jersey	0	1
July 07, 2017	University of Maryland	College Park	Maryland	0	1
February 26, 2018	University of Utah	Salt Lake City	Utah	0	1
February 27, 2018	Frontage Laboratories Inc.	Exton	Pennsylvania	1	0
March 27, 2018	University of Utah	Salt Lake City	Utah	0	2
May 09, 2018	Merrol Hyde Magnet School	Hendersonville	Tennessee	0	10
May 15, 2018	Yellow School	Houston	Texas	0	12
May 29, 2018	University of Nebraska	Lincoln	Nebraska	0	2
June 30, 2018	Boston University	Boston	Massachusetts	0	1
July 17, 2018	Norris Labs	Bozeman	Montana	0	0
July 31, 2018	Dietary Pro Labs	Wausau	Wisconsin	0	1
August 07, 2018	Hanford Nuclear Reservation	Richland	Washington	0	2

Source: CSB (2018)/Public domain.

up the synthesis of NHP to make approximately grams of the compound. The PIs of the research were not consulted on the decision to scale up by the students. No written policies or procedures existed at the laboratory, departmental, or university levels that would have required the students to consult with the PIs before making decisions about changes to the experimental protocols. The two students had discovered that smaller amounts of the compound would not ignite or explode on impact when wet with water or hexane based on their experience. They assumed the hazards of larger quantities of NHP would be controlled in a similar manner to smaller batches. The senior student observed clumps in the product from the larger batch and believed the sample's uniform particle size was important. As a result, he transferred about half of the synthesized NHP into a mortar, added hexane, and then used a pestle to break up the clumps gently. No hazard evaluation was conducted on the improvised process to analyze the effectiveness of using either water or hexane to mitigate the potential explosive hazards associated with the quantity of NHP synthesized. The graduate student working on the clumps was wearing goggles but removed them and walked away from the mortar after he finished breaking the clumps. The decision to wear goggles was a personal choice that was based on how dangerous an activity was perceived to be by the students. The senior student working with NHP returned to the mortar but did not replace his goggles while he stirred the NHP "one more time." At this point, the compound detonated. The student lost three fingers, his hands and face were burned, and one of his eyes was injured.

6.2.1 Specifically, the CSB Found

- The physical hazards of the energetic materials research work were not effectively assessed and controlled at TTU.
- TTU's laboratory safety management program was modeled after OSHA's Occupational Exposure to Hazardous Chemicals in Laboratories Standard (29 CFR 1910.1450); yet, the Standard was created not to address physical hazards of chemicals, but rather health hazards as a result of chemical exposures.
- Comprehensive hazard evaluation guidance for research laboratories does not exist.
- Previous TTU laboratory incidents with preventative lessons were not always documented, tracked, and formally communicated at the university.
- The research-granting agency, DHS, prescribed no safety provisions specific to the research work being conducted at TTU at the time of the incident, missing an opportunity for safety influence.
- Safety accountability and oversight by the principal investigators, the department, and university administration at TTU was insufficient.

6.3 University of California Los Angeles – December 29, 2008

On December 29, 2008, 23 year-old research assistant Sheri Sangji was working on an organic chemistry experiment in a University of California Los Angeles laboratory (Benderly 2012; Kemsley 2018; CSB 2011a,2011b). She had graduated from Pomona College in May of that year and was in the process of applying to Law Schools across the country. The experiment she was conducting used *t*-butyl lithium in the process. *t*-Butyl lithium or *tert*-Butyl lithium is a chemical compound with the formula $(CH_3)3CLi$ (Sigma-Aldrich 2013). As an organolithium compound, it has applications in organic synthesis since it is a strong base, capable of deprotonating many carbon molecules, including benzene. The chemical is pyrophoric, meaning it will spontaneously ignite in air under room temperature conditions. The chemical is prepared commercially in flammable solvents. There are several safe ways to transfer this chemical. Most involve using an inert environment to ensure the chemical does not encounter air during the transfer. The safety data sheet (SDS) states that water should not be used to extinguish a fire involving *t*-butyl lithium (Sigma-Aldrich 2013).

Ms. Sanglii drew up the chemical into a 60 ml syringe and, at some point in the process, the syringe came apart or the chemical generated pressure the plunger blew out. The highly dangerous chemical spilled on her nitrile gloves and possibly her clothing and ignited or ignited instantaneously before contacting her gloves and clothing. The flash fire ignited her synthetic sweater, and she was engulfed in flame. A fellow researcher tried to extinguish the fire using a lab coat and water from the sink, although water is contraindicated for a *t*-butyl lithium fire.

Ms. Sanglii was not wearing a lab coat or apron, but she was possibly wearing safety goggles.

A Lab Manager article (2011) has eight rules for handling *t*-butyl lithium:

- All staff must always wear lab coats and safety glasses when working with hazardous chemicals in the laboratory.
- Never work alone when handling highly hazardous chemicals, especially organic lithium reagents. Always let others in the laboratory know when you are working with these solutions.
- Never work with hazardous chemicals unless there is an eyewash and safety shower nearby. Be sure to know where they are and the procedures to follow in the event of an emergency.
- Organic lithium compounds cannot be exposed to the atmosphere since they will react spontaneously with moisture in the air. Although purchased in bottles that self-seal, they will degrade after use and should be disposed of as hazardous waste within one month of opening.

- Only purchase the amount of the substance that you plan to use for each experiment.
- Review the safe procedures for handling highly reactive reagents.
- Work inside the fume hood with the horizontal sash positioned in front of you to protect you from any splash that may occur. If your fume hood does not have a horizontal sash, use a splash guard positioned in front of the bottle when drawing the liquid into the syringe.
- Go through your inventories and dispose of any opened containers of these reagents that you are not planning on using in the near future.

These precautions are also contained in the Sigma-Aldrich Material Safety Data Sheet (2013).

A similar accident occurred in February 2018, in a University of Utah Chemistry Department laboratory (Utah 2018). In this case two lab personnel incurred burns due to exposure to an air-reactive chemical. In this incident, the researcher conducting the experiment and their spotter, who had a fire extinguisher, each received burns. However, unlike the University of California Los Angeles (UCLA) accident, the researcher was wearing a flame resistant lab coat. Figure 6.13 shows

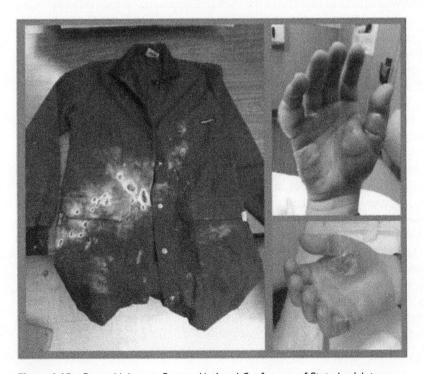

Figure 6.13 Burned lab coat. Source: National Conference of State Legislatures.

the lab coat and burns resulting from the accident. The lab worker's gloves were not appropriate for this chemical and he did incur burns to his hands.

6.4   University of Utah – July 2017

6.4.1   Utah, Report to the Utah Legislature Number 2019-06

A student was working on an experiment with sodium hydroxide. Sodium hydroxide is a base and is also called lye or caustic soda. Strong solutions of the chemical can dissolve animal flesh and is used for dissolving animal carcasses that have been infected with huff and mouth disease. However, more mild solutions are used for disinfecting food manufacturing equipment, in the manufacture of various consumer products like soaps and detergents. In fact, people making soap from fat add a solution of lye created from wood ash. Lye or sodium hydroxide is formed when wood ash that is mostly potassium carbonate, is mixed with water. The mixed solution is extremely alkaline. Lye reacts with fats to create soap (Simon 2016). If the chemical meets your skin, it begins to react with the oils and basically turns the fats in your skin into soap. That is the slick feeling one has on their skin after being exposed to sodium hydroxide/lye.

Lye is used in research because it is a strong base or alkaline, and as discussed earlier, it dangerous to handle. As the student carried a beaker across the lab at the University of Utah, some of it splashed and landed in his eye.

The Salt Lake Tribune article (Tanner 2019) says that the student, covered his face with one hand, he ran around the room looking for an emergency eyewash. There wasn't one. He dodged down hallway after hallway in the science building. Thirty seconds later, he was able to flush his face with water. But his cornea had already been severely burned.

Two months before that July 2017 accident, the lab was inspected and found to have nine major deficiencies, including the missing eyewash, no chemical hygiene plan, no safety instructions, improper chemical labels, and no spill kit. A year after, those issues were still not resolved.

A subsequent accident in August 2018 occurred and a student received burns on his legs and feet. A student was working with a 70% concentration of nitric acid when the 2.5-l bottle broke and spilled onto the student's leg and feet. The student was rushed to a nearby emergency shower and then taken to the University of Utah Hospital's emergency room for burns to his leg and feet.

Those injuries are documented in a new state audit report, released in 2019 (Utah 2019).

Table 6.2 Safety concerns on the University of Utah Campus.

Rank	Concern	Respondents	Percent
Top three hazards			
1	Chemical hazards	72	56%
2	Walking surfaces (Ice, uneven pavement, potholes)	56	43%
3	Indoor air quality (Mold, dust, odors, temperature)	53	41%
3	Vehicle/bicycle/pedestrian safety	53	41%
Other lab-related hazards			
5	Needle sticks and medical sharps Injuries	28	22%
10	Infectious research (Biological, infectious, or medical material)	22	17%
16	Infectious disease	13	10%
16	Radioactive materials	13	10%
21	Animal research – bites and scratches	8	6%

The findings from the report include:

1. Chemical Hazards are the leading safety concerns of lab personnel on the University of Utah campus. See Table 6.2.

1. Propensity for repeated major deficiencies
2. Prevalence of three major chemical hazard deficiencies
3. Failure to provide requested hepatitis B vaccinations
4. Not requiring health risk questionnaires for animal handlers

The Utah audit report states that:

> Our review of the University of Utah's lab safety system found repeat deficiencies in many laboratories between 2016 and 2018 that affected essential areas of lab safety. Similar repeat deficiencies in other institutions of higher education led to dismemberment and death. The primary reason for safety deficiencies in many University of Utah labs was inadequate oversight

from multiple levels of personnel, including university administration and the Department of Occupational and Environmental Health and Safety (OEHS). We found that the university's lab safety system suffers from inadequate oversight. Poor coordination from ineffective communication was occurring between university administration and OEHS. Therefore, each group lacked valuable feedback and guidance from the other, resulting in repeat deficiencies and unresolved safety concerns that affected critical safety issues.

University policy 3–300 sets ultimate responsibility for lab safety with the president. As the president sets top lab safety priorities, improved coordination between OEHS and university administration must occur in order to address those critical safety issues. The university president commissioned a Lab Safety Culture Task Force after a peer review found concerns in 2017. However, no recommendations have been implemented yet. We recommend that the university president direct administrators to prioritize and enforce the goal of eliminating repeat safety deficiencies from lab safety audits and inspections.

In May 2017, two months before the first incident, the research group was audited by OEHS. The audit identified the following nine major deficiencies:

- No chemical hygiene plan
- No updated chemical hygiene training records
- No safety data sheets (SDSs)
- No updated chemical inventory
- Respirator use without appropriate procedures
- Inappropriate container labeling
- Improper compressed gas storage
- Inappropriate chemical storage
- No spill kit

Auditors said the U. has known about serious deficiencies for years. It hired a consultant in 2017 to review its research practices. But it never put into place the recommendations to reduce risks, such as requiring staff to wear protective lab coats and restricting the volume of caustic chemicals that students can use. An audit found that the University knew about the serious deficiencies in lab safety for years. It hired a consultant who provided recommendations, such as wearing lab coats to reduce risks. However, these recommendations were not implemented and similar events continued to happen.

And if the state's flagship research institution continues to ignore the issues and delay fixes, the audit said, it's at risk for much worse injuries, maybe deaths.

"Any given one of them could be very serious or unfortunate," said Brian Dean, audit manager for the Office of the Utah Legislative Auditor General. "Ultimately, this system is broken. The department that is required to [oversee] this isn't tracking the problems."

The audit comes after a trio of high-profile tragedies at other research universities in the country; it was requested by state lawmakers who feared similar events could happen at the U. In 2008, a researcher at UCLA died after she spilled a chemical on her torso and the highly reactive liquid caught fire. In 2010, a graduate student at Texas Tech University lost three fingers from a chemical burn. In 2016, a lab assistant at the University of Hawaii had her arm amputated after an explosion.

In all three of those cases, like at the University of Utah, the schools knew about lab shortcomings beforehand and had failed to address them, the audit states.

Dean said the U. still has time to course-correct.

In a written response to the audit, U. President Ruth Watkins agreed with its findings. Ultimately, the report concludes that, as president since early 2018, she is responsible for ensuring the school's labs are safe. She said the U. has started adding and replacing fume hoods and emergency washes in its labs.

"The findings of this audit are of such importance," she noted, "that the university administration has already begun implementing changes to most effectively address the challenges and opportunities that were identified."

Watkins also said administrators will work to communicate better with the U.'s Department of Occupational and Environmental Health and Safety, which reviews the labs on campus each year. The audit dinged both for not coordinating. Staff members at Health and Safety weren't reporting the deficiencies to top leaders at the school and the school's leaders weren't asking for more information in annual presentations. Because of that breakdown, no necessary changes were being addressed.

As part of the fixes, the computer system for tracking research deficiencies will be updated so that it's easier to see which labs have issues and which ones aren't improving. Right now, there is no central platform for that data.

The school will report back to the Legislature in October. It currently points to its new Meldrum Innovation Lab in the College of Engineering as a model for safety. Students must complete a safety course to work in there.

The audit looked specifically at lab inspections between 2016 and 2018. When Health and Safety identified a problem one year, it persisted the next year in the same lab 49% of the time, auditors found. Half of all labs had one major deficiency. And the safety reviewers rarely stressed any urgency to fix them and never followed up.

Andrew Weyrich, the recently named vice president for research at the U., spoke to members of the legislative audit subcommittee Tuesday about how the school is restructuring the chain of command so those gaps won't happen.

> "The university in the past has looked and cataloged some of the deficiencies," he acknowledged. "But we do need to change the culture. We have a lot of labs that are great, but we have work to do."

The biggest problems highlighted in the audit were researchers not completing annual safety training and not updating their chemical spill protocols. Other issues included those working with blood-borne pathogens not having the hepatitis B vaccine and those working with animals not filling out health questionnaires.

Many people working in the labs don't wear protective equipment.

In 2017, just 12% reported wearing lab coats and 16% said they put on goggles. Those are both dips from the previous year. The audit calls that "clearly unacceptable."

In February 2018, a U. researcher was showing students an experiment at the front of a lab. The chemical she was using was highly reactive with air and caught on fire after she accidentally spilled it. She was wearing a lab coat that protected her torso but didn't have on the proper gloves and her hands were covered with burns and blisters. She was using the same chemical that killed the UCLA researcher in 2008.

The auditors implored the school to take precautions now before a more serious incident or a lawsuit. Students and staff working in labs often use chemicals that are hazardous, cancer-causing, flammable, and corrosive.

6.5 University of Hawaii – March 16, 2016

Severe accidents can happen in a flash, in this case, an explosion. In fact, this tragic event started with a flash. A static spark, to be specific and it resulted with a post-doc researcher losing her arm.

We all have experienced static shocks. The following explanation is rather simple, but there are more detailed explanations on the internet. Static electricity can build up anytime there are two objects that have been in contact and then separated. One of the materials must be more of an insulator or has high resistance to electrical current. When two solid objects come in contact, one object gives up electrons and become more positively charged and the object that gains electrons becomes more negatively charged. Therefore, an electrical imbalance occurs (Library of Congress 2021). These charges will build up on the surface of an object until the electrons find a path to be released or discharged. The path they create becomes a circuit. When we walk across a carpeted floor when the air is dry and then touch a doorknob, a screw on a light switch, or some other object that is grounded we can see a blue spark. This spark can be painful at times. The voltage that can be developed from shuffling one's feet on the carpet can be upwards of 20,000–25,000 V (Binns 2006). However, the amperage is low, less than 1 mA. The current, or amps, that are required to kill a person are between 0.1 and 0.2 A (PTJ 1966). There are a multitude of factors that come into play; amperage (A), whether a person's skin is wet, and the duration of the current. However, 15,000 V is enough to ignite a gas burner on a stove (Deziel 2021). A piezoelectric igniter on a barbeque generates about 800 V from a 1.5-V battery and a quartz crystal. Pushing the button causes a spark at 800 V to jump off the ignition rod and onto a ground. The spark passes through the gas and ignites it (Kate 2021).

In the tragic event that is described next, a static spark ignited an explosive gas mixture of hydrogen, carbon, dioxide, and oxygen. There are many issues in this event. One being the lack of adequate grounding. I recently reviewed an experimental set up at the Center for Advanced Energy Studies (CAES) for one of our students and my main concern was that the apparatus was not grounded. OSHA has several good writeups on grounding on their webpages. The following presents a summary of OSHA requirements on grounding.

6.5.1 Grounding (OSHA 2021)

The term "ground" refers to a conductive body, usually the earth. "Grounding" a tool or electrical system means intentionally creating a low-resistance path to the earth. When properly done, current from a short or from lightning follows this path, thus preventing the buildup of voltages that would otherwise result in electrical shock, injury, and even death.

There are two kinds of grounds; both are required by the OSHA construction standard:

> System or Service Ground: In this type of ground, a wire called "the neutral conductor" is grounded at the transformer, and again at the service entrance to the building. This is primarily designed to protect machines, tools, and insulation against damage.

> Equipment Ground: This is intended to offer enhanced protection to the workers themselves. If a malfunction causes the metal frame of a tool to become energized, the equipment ground provides another path for the current to flow through the tool to the ground.

There is one disadvantage to grounding: a break in the grounding system may occur without the user's knowledge. Using a ground-fault circuit interrupter (GFCI) is one way of overcoming grounding deficiencies.

6.5.1.1   Summary of Grounding Requirements

Ground all electrical systems. [For exceptions, see 29 CFR 1926.404(f)(1)(v)].

The path to ground from circuits, equipment, and enclosures must be permanent and continuous.

Ground all supports and enclosures for conductors. [For exceptions, see 29 CFR 1926.404(f)(7)(i)].

Ground all metal enclosures for service equipment.

Ground all exposed, non-current-carrying metal parts of fixed equipment. [For exceptions, see 29 CFR 1926.404(f)(7)(iii)].

Ground exposed, non-current-carrying metal parts of tools and equipment connected by cord and plug. [For exceptions, see 29 CFR 1926.404(f)(7)(iv)].

Ground the metal parts of the following non-electrical equipment:

Frames and tracks of electrically operated cranes.

Frames of non-electrically driven elevator cars to which electric conductors are attached.

Hand-operated metal shifting ropes or cables of electric elevators.

Metal partitions, grill work, and similar metal enclosures around equipment of over 1 kV between conductors.

6.5.1.2   Methods of Grounding Equipment

Ground all fixed equipment with an equipment grounding conductor that is in the same raceway, cable, or cord, or that runs with or encloses the circuit conductors (except for DC circuits only).

Conductors used for grounding fixed or moveable equipment, including bonding conductors for assuring electrical continuity, must be able to safely carry any fault current that may be imposed on them.

Electrodes must be free from nonconductive coatings, such as paint or enamel, and if practicable, must be embedded below permanent moisture level.

Single electrodes which have a resistance to ground greater than 25 Ω must be augmented by one additional electrode installed no closer than 6 ft to the first electrode.

For grounding of high voltage systems and circuits (1000 V and over), refer to 29 CFR 1926.404(f)(11).

A note that needs to be made is that bonding is not the same as grounding. A bond ensures two bodies are at the same electrical potential, but does not discharge a static charge, for instance, to ground. A proper ground to ensure a continuously discharge of a system to ground is achieved by attaching a wire to the device and then to a water pipe or an 8-ft-long copper clad steel rod embedded in the ground (NFPA 2020).

Another electrical term that is associated with the explosion at the University of Hawaii is "intrinsically safe electrical appliances." These devices are designed to not provide an ignition source. Intrinsically safe design is a protection technique applied to electrical and wiring products used in a hazardous location. In simple terms, intrinsically safe devices are limited in the amount of electrical and thermal output to prevent atmospheric ignition (NFPA 2020).

6.5.1.3 Event Description

The description of this event comes from news articles (Kemsley 2016a,b). Also, I was not able to secure permission to use photos from the event for publication. However, photos are available in these two news articles. A post-doctoral student, Thea Ekins-Coward, was running an experiment combining hydrogen, carbon dioxide, and oxygen gases from high-pressure cylinders into a tank at lower pressure. The gas mixture was an energy source or food for bacteria. These bacteria were being grown as an experiment to examine the possibilities to produce biofuels and bioplastics. The Post-Doc was working for the Hawaii Natural Energy Institute under the principle researcher Jian Yu. Kemsley's article said that the Post-Doc started working in the lab last fall, she purchased a 49-l steel gas tank, a different gauge not rated as intrinsically safe, a pressure-relief valve, and fittings, and she put them together. In the experiment Ekins-Coward would add the gases to the portable tank that would then be connected to a bioreactor. The mixture was 70% hydrogen, 25% oxygen, and 5% carbon dioxide for her experiments, the report says. The flammable range for hydrogen in air is 4–74% and 4–94% in oxygen rich environments (Rhodes 2011), therefore, the necessity to keep the two elements from mixing. All this mixture needs is an ignition source.

A similar set-up with a 3.8-l tank resulted in a "small internal explosion" the week before. The fire department that investigated the accident said that a small explosion occurred when Ekins-Coward pressed the off button on the pressure gauge. She also occasionally experienced static shocks when she touched the ungrounded tank. According to the article, her supervisor told her not to worry about static shocks.

The 49-l tank exploded when Ekins-Coward pressed the off button on the gauge on the day of the incident. The explosion severed her arm just below the elbow.

The University of Hawaii was originally fined US$ 115,500 for 15 workplace safety violations after a laboratory explosion in March on the university's Manoa campus. Postdoctoral researcher Thea Ekins-Coward, who worked for the Hawaii Natural Energy Institute, lost one of her arms in the explosion (Kemsley 2016b). The penalties in the citation were reduced from 15 to 9, and the total fine was reduced from US$ 115,500 to US$ 69,300, or about 40% (Trager 2016).

The safety violations cited by the Hawaii Occupational Safety & Health Division (HIOSH) include failing to do the following: reduce employee exposure to potential explosion and fire hazards, ensure safety practices were followed, perform periodic inspections to identify hazards, ensure employees wore appropriate personal protective equipment, make use of standard operating procedures, and require suitable exits from the laboratory.

Originally, HIOSH labeled all 15 violations as "serious" and assessed the maximum state penalty of US$ 7700 to each. The university were to fix the violations by October 21, 2016.

The original workplace safety violations identified by HIOSH were (Kemsley 2016b):

1. The employer failed to provide a safe workplace by reducing employee exposure to potential explosion and fire hazards.
2. The employer did not ensure that its safety practices were followed by employees and underscored through training, positive reinforcement, and a clearly defined and communicated disciplinary system.
3. The employer did not ensure periodic in-house inspections were being performed in Hawaii Natural Energy Institute laboratories to determine new or previously missed hazards.
4. Laboratory personnel working under the principal investigator did not use the required personal protective equipment at all times.
5. Two exit routes were not available in the laboratory to permit prompt evacuation of employees and building occupants.
6. The exit door did not swing out in the direction of exit travel.
7. The employer's emergency action plan(s) did not list the evacuation meeting point nor a way to account for the evacuees.
8. The employer did not review the emergency action plan when employees were initially assigned.
9. A fire prevention plan did not include specific provisions to address potential ignition sources in the presence of hydrogen and other flammable gases.
10. Activities performed in the laboratory by researchers with the potential exposure to explosion and fire hazards were not assessed for appropriate personal protective equipment.
11. Activities performed in the laboratory by researchers with the potential exposure to explosion and fire hazards were not assessed for appropriate glove protection to guard against static discharge and flame-retardant laboratory coats to guard against fire.
12. Where hazardous chemicals were used in the workplace, the employer did not carry out the provisions of a written Chemical Hygiene Plan, which were capable of protecting employees from health hazards associated with hazardous chemicals in that laboratory.
13. The employer's Chemical Hygiene Plan did not include the standard operating procedures relevant to safety and health considerations to be followed when laboratory work involved the use of hazardous chemicals.

14. The employer's Chemical Hygiene Plan did not include criteria to determine and implement controls relevant to the gas mixing operation (engineering controls, personal protective equipment, administrative).
15. The employer failed to review and evaluate the effectiveness of the Chemical Hygiene Plan at least annually and update it as necessary.

I was not able to find which violations were removed.

6.5.1.4 Summary of Safety Culture Issues

I work at a university and have seen the cavalier attitude to safety, fire protection, and chemical hygiene so many times among some faculty and students. I have invoked "stop work" orders multiple times in the last few years. Even though the amounts of chemicals are small in university experiments, as compared with industrial production facilities, the dangerous properties of chemicals do not go away. I did not receive any safety training when I was in undergraduate chemistry classes. In fact, not even in graduate level classes. I remember in my first organic chemical lab we were introduced to organic solvents like benzene, toluene, and various alcohols by smelling them. I guess this was safety training of sorts, so you could tell which chemical was which. At that time, we still mouth pipetted some chemicals. Well, times have changed. Every student who will be attending a chemistry class or any class that will be using chemicals must be trained in the hazardous nature of chemicals. PPE must be worn at all times when using any chemical, even small quantities. Lab coats must be worn and close-toed shoes. Major professors and supervisors must listen to their students and seek answers to their safety questions. At the University of Idaho, we have very competent safety and health professionals. I know at all the universities in the case studies earlier there are very competent safety and health professionals as well. In writing this chapter I looked through UCLA's safety manual and it is very thorough, and the processes provided in it can be reasonably applied (UCLA 2010). From my experience, many of the safety issues, arise from faculty, students, and other researchers not wanting to take the time to implement the safety procedures and wear the PPE. This can happen in industrial facilities as well. Instilling a positive safety attitude in students right as they start their academic careers is essential for the future researcher to be safety minded at their educational institution and at their future places of work.

The University of Idaho at Idaho Falls has a collaboration with the Idaho National Laboratory, Idaho State University and Boise State University in the CAES facility. Adherence to safety and health requirements is a must in the CAES facility. In fact, every project to perform in CAES is reviewed several times before the student of researcher is allowed to proceed. Sometimes this appears to be burdensome to researchers, but it works. Figure 6.14 shows the flow of review of projects.

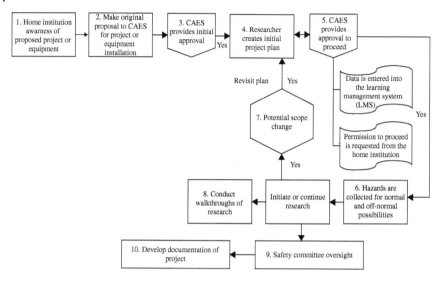

Figure 6.14   CAES project approval flow (Ostrom 2021).

References

Benderly, B. (2012). Are University Labs criminally dangerous? Scientific American. https://www.scientificamerican.com/article/are-university-labs-criminally (accessed December 2021).

Binns, C. (2006). Live science. https://www.livescience.com/4077-shocking-truth-static-electricity.html (accessed December 2021).

Center for Disease Control (CDC) (1999). Emergency medical technician receives serious burns from an oxygen regulator flash fire – South Carolina; Death in the Line of Duty…A summary of a NIOSH fire fighter fatality investigation, Ambulance Fire. https://www.cdc.gov/niosh/fire/reports/face9824.html (December 2021).

Chemical Safety Board (CSB) (2011a). Texas Tech University Chemistry Lab explosion. https://www.csb.gov/texas-tech-university-chemistry-lab-explosion (accessed November 2021).

Chemical Safety Board (CSB) (2011b). CSB releases new video on laboratory safety at academic institutions. https://www.csb.gov/csb-releases-new-video-on-laboratory-safety-at-academic-institutions (accessed December 2021).

CSB (2018). Laboratory incidents January 2001–July 2018. https://www.csb.gov/assets/1/6/csb_laboratory_incident_data.pdf (accessed December 2021).

CNN (2021). More than 90 people killed in fire at Iraqi hospital treating Covid-19. https://www.cnn.com/2021/07/12/middleeast/nasiriya-iraq-hospital-fire/index .html (accessed November 2021, patients).

Deziel, C. (2021). What causes the electric igniter on a stove to spark?, Home Guides. https://homeguides.sfgate.com/causes-electric-igniter-stove-spark-89082.html (accessed December 2021).

Kate. (2021) How many volts does a grill ignitor have?. https://antonscafebar.com/ %D1%81ook-outdoors/how-many-volts-does-a-grill-ignitor-have.html (accessed December 2021).

Kemsley, J. (2016a). Spark from pressure gauge caused University of Hawaii explosion, fire department says, Chemical and Engineering News. https://cen.acs .org/articles/94/web/2016/04/Spark-pressure-gauge-caused-University.html (accessed December 2021).

Kemsley, J. (2016b). University of Hawaii fined $115,500 for lab explosion, Chemical and Engineering News. https://cen.acs.org/articles/94/web/2016/09/University-Hawaii-fined-115500-lab.html (accessed December 2021).

Kemsley, J. (2018). 10 years after Sheri Sangji's death, are academic labs any safer? Chemical and Engineering News. https://cen.acs.org/safety/lab-safety/10-years-Sheri-Sangjis-death/97/i1 (accessed December 2021).

Lab Manager (2011). 8 Rules for the safe handling of *t*-butyl lithium. https://www .labmanager.com/lab-health-and-safety/8-rules-for-the-safe-handling-of-t-butyl-lithium-18346 (accessed December 2021).

Lee, E., Lee, S., Kim, J., and Son, Y. (2012). Endotracheal tube fire during tracheostomy. *Korean Journal of Anesthesiology* 62 (6): 586–587.

Library of Congress (LOC) (2021). How does static electricity work? https://www.loc .gov/everyday-mysteries/item/how-does-static-electricity-work (accessed December 2021).

NFPA 70 (2020). National electric code.

Occupational Safety and Health Administration (OSHA) (2021). Grounding. https:// www.osha.gov/electrical/hazards/grounding (accessed December 2021).

PTJ (1966). The fatal current. *Physical Therapy* 46 (9): 968–969. https://doi.org/10 .1093/ptj/46.9.968.

Rhodes, R. (2011). Explosive lessons in hydrogen safety. *Ask Magazine*, NASA.

Romboy, D., Deseret News (2019) 'Broken system' puts University of Utah lab workers at risk, audit says. https://www.deseret.com/2019/5/14/20673300/broken-system-puts-university-of-utah-lab-workers-at-risk-audit-says (accessed December 2019).

Sigma-Aldrich (2013). *tert*-Butyllithium solution, MSDS. https://westliberty.edu/ health-and-safety/files/2013/03/tert-Butyllithium-solution.pdf (accessed December 2021).

Simon, J. (2016). *Step by Step Soap Making: Material – Techniques – Recipes Paperback*, 1e. CreateSpace Independent Publishing Platform.

Tanner, C. (2019). A student burned his eye in a University of Utah lab. The U. knew about dangers beforehand, an audit finds, but didn't take action. Salt Lake Tribune. https://www.sltrib.com/news/education/2019/05/14/student-burned-his-eye (accessed December 2021).

Trager, R. (2016). University of Hawaii fines for lab explosion slashed by 40%, Chemistry World. https://www.chemistryworld.com/news/university-of-hawaii-fines-for-lab-explosion-slashed-by-40/1017564.article (accessed December 2021).

University of California (2010). Safety manual. chrome-extension:// efaidnbmnnnibpcajpcglclefindmkaj/viewer.html?pdfurl=https%3A%2F%http:// 2Fwww.chemistry.ucla.edu%2Fsites%2Fdefault%2Ffiles%2Fsafety%2Fdoc% 2FLaboratory_Safety_Manual_2013.pdf&clen=13248359&chunk=true (accessed December 2021).

Utah Public Radio (2018). 2 burned in a chemical explosion at University of Utah Lab. https://www.upr.org/utah-news/2018-03-28/2-burned-in-a-chemical-explosion-at-university-of-utah-lab (accessed December 2021).

7

Aviation Case Studies

7.0 Introduction

There are three basic types of aviation activities:

- Military
- Commercial
- General aviation

This section does not focus on military aviation. Nonmilitary flight operations are usually categorized as commercial or general aviation. The Commercial Air Transport (CAT) is defined as an aircraft operation involving transport of passengers, cargo, or mail for remuneration or hire (FAR 2021). It includes scheduled and non-scheduled air transport operations. Aerial Work (AW) is defined as an aircraft operation in which an aircraft is used for specialized services such as agriculture, construction, photography, surveying, observation and patrol, search and rescue, advertisement There are three categories that full under general aviation. They are: General Aviation (GA), AW and CAT. These are defined by the International Civil Aviation Organization (ICAO) (ICAO 2009 and 2010). Operations in AW operations are separated from general aviation by ICAO by this definition. AW is when an aircraft is used for specialized services such as agriculture, construction, photography, surveying, observation and patrol, search and rescue, and aerial advertisement. For statistical purposes ICAO includes AW within general aviation and has proposed officially extending the definition of general aviation to include AW, to reflect common usage. The proposed ICAO classification includes instructional flying as part of general aviation (non-aerial-work).

Impact of Societal Norms on Safety, Health, and the Environment: Case Studies in Society and Safety Culture,
First Edition. Lee T. Ostrom.
© 2023 John Wiley & Sons, Inc. Published 2023 by John Wiley & Sons, Inc.

The International Council of Aircraft Owner and Pilot Associations (IAOPA) refers to the category as GA/AW to avoid ambiguity. Their definition of GA includes (IAOPA 2021)

- Corporate Aviation: Company own-use flight operations (IAOPA 2021)
- Fractional Ownership Operations: Aircraft operated by a specialized company on behalf of two or more co-owners
- Business Aviation (or Travel): Self-flown for business purposes
- Personal/Private Travel: Travel for personal reasons/personal transport
- Air Tourism: Self-flown incoming/outgoing tourism
- Recreational Flying: Powered/powerless leisure flying activities
- Air Sports: Aerobatics, air races, competitions, rallies, etc.
- General aviation thus includes both commercial and non-commercial activities.

IAOPA's definition of AW includes but is not limited to (IAOPA 2021)

- Agricultural Flights, including crop dusting
- Banner Towing
- Fire Fighting
- Medical Evacuations
- Pilot Training
- Search and Rescue
- Sight Seeing Flights
- Skydiver Hoisting
- Transplant Organ Transports

CAT includes:

- Scheduled Air Services
- Non-Scheduled Air Transport
- Air Cargo Services
- Air Taxi Operations

Aviation accidents are relatively common. Though, most involve general aviation, rather than commercial aviation. Table 7.1 lists all the aviation accidents that occurred in the United States in January 2021. There are several accidents listed in this table that have fatalities, but do not provide findings. This is because those accidents are still being investigated.

Aviation accidents are very common and kill many people each year. This section discusses commercial and general aviation accidents and the safety culture of the personnel and organizations.

This section will cover the following case studies:

- General Aviation
  o My experiences as a pilot
  o Liberty Helicopter, March 2018

Table 7.1 Aviation accidents that occurred in the United States in January 2021.

Event date	City	State	N#	Highest injury level	Fatal injury count	Serious injury count	Minor injury count	Probable cause	Findings
2021-01-01	Marana	Arizona	N74823	Minor	0	0	1	The pilot's failure to maintain helicopter control, which resulted in a hard landing	Personnel issues – task performance – use of equip/info – aircraft control – pilot, aircraft – aircraft oper/perf/ capability – performance/control parameters – yaw control – not attained/maintained
2021-01-02	Santa Monica	California	N52NL	None	0	0	0	The pilot's failure to maintain directional control during the landing roll, which resulted in a ground-loop	Aircraft – aircraft oper/perf/ capability – performance/control parameters – directional control – not attained/ maintained, personnel issues – task performance – use of equip/info – aircraft control – pilot
2021-01-02	New Hudson	Michigan	N8347P	Fatal	3	0	0		
2021-01-02	Moab	Utah	N833RT	None	0	0	0		
2021-01-03	Portsmouth	New Hampshire	N489RS	None	0	0	0		

(continued)

Table 7.1 (Continued)

Event date	City	State	N#	Highest injury level	Fatal injury count	Serious injury count	Minor injury count	Probable cause	Findings
2021-01-04	O'Brien	Florida	N616PM	None	0	0	0	A partially engaged right wheel brake, which restricted the airplane's acceleration and the pilot's ability to pitch to takeoff attitude, and resulted in an aborted takeoff and subsequent runway excursion	Aircraft – aircraft systems – landing gear system – brake – malfunction, aircraft – aircraft oper/perf/ capability – performance/control parameters – airspeed – attain /maintain not possible, aircraft – aircraft oper/perf/ capability – performance/control parameters – pitch control – attain/maintain not possible
2021-01-04	Cash	Arkansas	N325GC	Fatal	2	0	0		
2021-01-06	Palm Springs	California	N737NB	Serious	0	1	0		
2021-01-08	Palouse	Washington	N4344B	Minor	0	0	1	The student pilot's improper landing flare and loss of directional control and the flight instructor's delayed remedial action resulted in a runway excursion and nose over.	Aircraft – aircraft oper/perf/ capability – performance/control parameters – landing flare – Incorrect use/operation, environmental issues – physical environment – Terrain – wet/ muddy terrain – contributed to outcome, personnel issues – action/decision – info processing/decision – decision making/judgment – student/ instructed pilot,

Date	Location	State	Registration	Injury				Narrative	Findings
2021-01-09	Cushing	Oklahoma	N4540C	None	0	0	0	The pilot's decision to takeoff from unsuitable terrain which resulted in an aerodynamic stall and subsequent impact with terrain.	personnel issues – task performance – use of equip/info – aircraft control – student/instructed pilot, personnel issues – psychological – attention/monitoring – monitoring other person – instructor/check pilot, personnel issues – action/decision – action – delayed action – instructor/check pilot. Personnel issues – action/decision – Info processing/decision – decision making/judgment – pilot, environmental issues – physical environment – runway/land/takeoff/taxi surface – soft surface – decision related to condition, personnel issues – task performance – use of equip/info – aircraft control – pilot, aircraft – aircraft oper/perf/capability – performance/control parameters – airspeed – not attained/maintained
2021-01-09	Warm springs	Oregon	N3RB	Fatal	1	0	0		
2021-01-09	Winthrop	Washington	N8612F	None	0	0	0	The pilot's failure to adequately monitor the environment which resulted in a collision with a snowbank and subsequent loss of control	Personnel issues – psychological – attention/monitoring – monitoring environment – pilot, personnel issues – task performance – use of equip/info – aircraft control – pilot, environmental issues – physical environment – terrain – sloped/uneven terrain – effect on operation

(continued)

Table 7.1 (Continued)

Event date	City	State	N#	Highest injury level	Fatal injury count	Serious injury count	Minor injury count	Probable cause	Findings
2021-01-09	Albany	Texas	N322SH	None	0	0	0		
2021-01-10	Old Bethpage	New York	N421DP	Serious	0	1	0		
2021-01-11	Vineland	New Jersey	UNREG	Fatal	1	0	0		
2021-01-12	Sparta	Michigan	N35718	None	0	0	0		
2021-01-12	Lafayette	Indiana	N552DR	None	0	0	0	The pilot's failure to maintain control of the helicopter shortly after liftoff into a hover, which resulted in a dynamic rollover when the left landing skid contacted the ground	Aircraft – aircraft oper/perf/ capability – performance/control parameters – Lateral/bank control – Not attained/ maintained, aircraft – aircraft oper/perf/capability – Performance/control parameters – Pitch control – not attained/maintained, personnel issues – task performance – use of equip/ info – aircraft control – pilot
2021-01-12	Poplar Grove	Illinois	N436T	None	0	0	0	The pilot's distraction and his failure to maintain a proper glidepath on the approach, which resulted in him landing short of the runway and impacting a snowbank	Personnel issues – psychological – attention/monitoring – monitoring other aircraft – pilot, aircraft – aircraft oper/perf/ capability – performance/control parameters – descent/approach/ glide path – not attained/ maintained, environmental issues – physical environment – terrain – snowy/icy terrain – contributed to outcome

Date	City	State	Registration	Injury				Narrative	Category
2021-01-13	Myrtle Beach	South Carolina	N862UP	Serious	0	1	0	An encounter with severe turbulence, which resulted in serious injury to an unseated passenger	Environmental issues – conditions/weather/phenomena – turbulence – clear air turbulence – effect on operation
2021-01-13	Spanish Fork	Utah	N7480F	None	0	0	0		
2021-01-13	Pell City	Alabama	N771JP	None	0	0	0		
2021-01-13	Columbia	South Carolina	N266DC	Fatal	1	0	0		
2021-01-14	Kelso	Washington	N5930R	None	0	0	0	The pilot's failure to maintain directional control during the landing roll, which resulted in a runway excursion and subsequent nose over	Personnel issues – task performance – use of equip/info – aircraft control – pilot, aircraft oper/perf/capability – performance/control parameters – directional control – not attained/maintained
2021-01-15	Hollywood	Florida	N404N	None	0	0	0		

(continued)

Table 7.1 (Continued)

Event date	City	State	N#	Highest injury level	Fatal injury count	Serious injury count	Minor injury count	Probable cause	Findings
2021-01-16	Elko	Nevada	N2828M	Minor	0	0	1	The pilot's unstable approach to land, which resulted in a loss of directional control and subsequent nose over. Contributing to the accident was the pilot's decision to continue an approach in sun glare conditions	Environmental issues – conditions/ weather/phenomena – light condition – glare – effect on personnel, aircraft – aircraft oper/perf/capability – performance/control parameters – descent/approach/glide path – not attained/maintained, personnel issues – task performance – use of equip/ info – aircraft control – pilot, aircraft – aircraft oper/perf/ capability – performance/control parameters – directional control – not attained/ maintained, personnel issues – action/decision – info processing/decision – decision making/judgment – Pilot
2021-01-17	Richmond Hill	Georgia	N919DC	None	0	0	0		
2021-01-17	St. Petersburg	Florida	N82746	None	0	0	0	A total loss of engine power as a result of the loose fuel line B-nut connected to the engine driven fuel pump	Aircraft – aircraft systems – fuel system – fuel distribution – malfunction

Date	City	State	Registration	Severity				Narrative	Findings
2021-01-19	Newellton	Louisiana	N523BB	Fatal	1	0	0		
2021-01-19	Miami	Florida	N4530T	Minor	0	0	1	The pilot's failure to maintain control of the airplane during takeoff, which resulted in an aerodynamic stall and impact with terrain. Also causal was the flight instructor's inadequate remedial action	Personnel issues – task performance – use of equip/info – aircraft control – pilot, aircraft – aircraft oper/perf/capability – performance/control parameters – directional control – not attained/maintained, personnel issues – action/decision – action – lack of action – Instructor/check pilot
2021-01-19	Leesburg	Virginia	N5880L	None	0	0	0		
2021-01-21	Eagle Point	Oregon	N322TR	Minor	0	0	4		
2021-01-21	Minneola	Florida	N4469Z	Serious	0	1	0		
2021-01-22	Homestead	Florida	N8224H	None	0	0	0	A total loss of power for reasons that could not be determined, as the airplane was not recovered from the water following the ditching	Not determined – not determined – (general) – (general) – unknown/not determined

(continued)

Table 7.1 (Continued)

Event date	City	State	N#	Highest injury level	Fatal injury count	Serious injury count	Minor injury count	Probable cause	Findings
2021-01-22	Mankato	Minnesota	N60BK	None	0	0	0	The pilot's loss of situational awareness while landing, resulting in an impact with power lines	Personnel issues – task performance – workload management – task allocation – pilot, environmental issues – physical environment – object/animal/ substance – wire – awareness of condition
2021-01-22	Queen Creek	Arizona	N90TG	None	0	0	0	The pilot's uncoordinated flight control inputs while entering a hover, which lead to a loss of aircraft control which resulted in the left skid making contact with the ground, causing the helicopter to roll on its side	Personnel issues – task performance – use of equip/ info – aircraft control – pilot, environmental issues – conditions/weather/phenomena - wind – variable wind – effect on equipment, aircraft – aircraft oper/perf/capability – performance/control parameters – prop/rotor parameters – not attained/maintained
2021-01-23	Rockingham	North Carolina	N42714	None	0	0	0	The pilot's improper recovery from a bounced landing, which resulted in a subsequent hard landing and a runway excursion	Personnel issues – task performance – use of equip/ info – aircraft control – pilot, aircraft – aircraft oper/perf/ capability – performance/control parameters – landing flare – incorrect use/operation, personnel issues – action/ decision – info processing/ decision – decision making/ judgment – pilot

2021-01-23	Denton	North Carolina	C-GWAS	None	0	0	0		
2021-01-23	Terrell	North Carolina	N480F	Minor	0	0	1	The pilot's failure to retract the landing gear on the amphibious airplane prior to a water landing, resulting in a nose over and structural damage to the airframe. The mechanic's failure to remove electrical tape from the pitot tube, the pilot's inadequate preflight inspection, and the pilot's failure to visually confirm the landing gear position prior to touchdown were all factors in the accident	Aircraft – aircraft systems – landing gear system – landing gear selector – incorrect use/ operation, personnel issues – task performance – mainte- nance – scheduled/routine maintenance – maintenance personnel, personnel issues – task perfor- mance – inspection – preflight inspection – pilot, personnel issues – task performance – use of equip/info – use of checklist – pilot
2021-01-23	Rio Rancho	New Mexico	N477LB	Fatal	1	0	0		

(continued)

Table 7.1 (Continued)

Event date	City	State	N#	Highest injury level	Fatal injury count	Serious injury count	Minor injury count	Probable cause	Findings
2021-01-24	Boynton Beach	Florida	N266ND	Fatal	1	0	0		
2021-01-24	Anchorage	Alaska	N4265H	None	0	0	0	The instructor pilot's selection of an unsuitable snow-covered runway for landing practice, and the pilot under instruction's failure to maintain directional control during a landing roll, which resulted in a loss of control in unpacked snow and a subsequent nose-over	Environmental issues – physical environment – runway/ land/takeoff/taxi surface – snow/ slush/ice covered surface – contributed to outcome, personnel issues – task performance – use of equip/info – aircraft control – student/instructed pilot, personnel issues – action/ decision – info processing/ decision – decision making/ judgment – instructor/check pilot
2021-01-24	Spotted Bear	Montana	N55BT	None	0	0	0	The pilot's failure to maintain aircraft control during the landing roll on an unimproved snow-covered runway which resulted in a nose-over	Environmental issues – physical environment – runway/ land/takeoff/taxi surface – snow/ slush/ice covered surface – contributed to outcome, personnel issues – task performance – use of equip/info – aircraft control – pilot, personnel issues – action/decision – info processing/decision – decision making/judgment – pilot

Date	City	State	Registration	Severity				Narrative	Findings
2021-01-26	West Palm Beach	Florida	N474AT	None	0	0	0	The examiner's failure to arrest the helicopter descent during a simulated engine failure maneuver and the flight instructor's inadequate remedial action, which resulted in a hard landing	Aircraft – aircraft oper/perf/capability – performance/control parameters – descent rate – not attained/maintained, personnel issues – task performance – use of equip/info – aircraft control – designated examiner, personnel issues – action/decision – action – delayed action – instructor/check pilot
2021-01-26	Salish Sea	Washington	N9114A	Fatal	1	0	0		
2021-01-26	Waller	Texas	N444PZ	Serious	0	1	0		
2021-01-28	Davenport	Iowa	N217US	None	0	0	0	The pilot's failure to maintain directional control of the airplane, and the flight instructor's inadequate oversight during a simulated engine out takeoff that resulted in a runway excursion and impact with a snowbank	Personnel issues – action/decision – action – delayed action – instructor/check pilot, environmental issues – physical environment – object/animal/substance – snow/ice – contributed to outcome, personnel issues – task performance – use of equip/info – aircraft control – pilot, aircraft – aircraft oper/perf/capability – performance/control parameters – directional control – not attained/maintained

(continued)

Table 7.1 (Continued)

Event date	City	State	N#	Highest injury level	Fatal injury count	Serious injury count	Minor injury count	Probable cause	Findings
2021-01-28	Lawrenceville	Georgia	N997MC	None	0	0	0	The pilot's failure to maintain directional control during landing in a gusting crosswind, which resulted in a loss of control and runway excursion	Personnel issues – task performance – use of equip/info – aircraft control – pilot, environmental issues – conditions/weather/phenomena – wind – crosswind – response/compensation, environmental issues – conditions/weather/phenomena – wind – gusts – response/compensation, aircraft – aircraft oper/perf/capability – performance/control parameters – directional control – not attained/maintained
2021-01-29	Ontario	California	N363CM						
2021-01-29	Clare	Michigan	N22101	None	0	0	0	The pilot's failure to maintain directional control during landing that resulted in a runway excursion and impact with terrain	Aircraft – aircraft oper/perf/capability – performance/control parameters – directional control – not attained/maintained, personnel issues – task performance – use of equip/info – aircraft control – student/instructed pilot

Date	Location	State	Registration	Severity				Narrative	Cause factors
2021-01-29	O'Brien	Florida	N9136R	Serious	0	1	1	The flight instructor's improper soft field landing technique while landing on a turf runway, which resulted in a nose over	Aircraft – aircraft oper/perf/capability – performance/control parameters – landing flare – not attained/maintained, personnel issues – action/decision – info processing/decision – decision making/judgment – instructor/check pilot
2021-01-30	Knoxville	Tennessee	N660TN	None	0	0	0	The pilot's failure to maintain directional control of the helicopter, which resulted in rear portion of the right skid contacting the ground and a subsequent dynamic rollover	Personnel issues – task performance – use of equip/info – aircraft control – pilot, aircraft – aircraft oper/perf/capability – performance/control parameters – directional control – not attained/maintained
2021-01-31	Silt	Colorado	N4376Q	Serious	0	3	0		
2021-01-31	Crescent City	California	N291FR	Minor	0	0	2		

- Commercial Aviation
  - o Gimli Glider
  - o Miracle on the Hudson
  - o 737 MAX
  - o De Haviland Comet

My Experience as a Pilot

I had wanted to fly since I was a preteen. I do have a private pilot license, though I am not current. I have had several near misses in my short flying career. The following are some of my flying experiences.

I started working for more than just odd jobs between my sophomore and junior year of high school. This was in the early 1970s. As soon as I had money, I started taking flying lessons. I took a lesson every time I had enough money to pay for the lesson and for car insurance. The plane I was flying was a tail dragging Piper Cub. This was at the Coeur d'Alene, Idaho Airport. However, I had to stop once I started going to the local community college because I had to pay for the tuition and fees.

After I finished my PhD and was hired at the Idaho National Engineering Laboratory (INEL) I stared taking lessons again. This was in Idaho Falls, ID, and the plane I was being trained in was a Cessna 152. It had a very basic instrument suite, with no LORAN-C or Distance measuring equipment. The Skybrary website presents a nice summary of the LORAN-C system:

"The LORAN, or LOng RAnge Navigation, system was developed in the United States during World War II. It was like the British GEE system but used lower frequency radio waves to give it a longer range of as much as 1500 mi. This longer range also resulted in lower accuracies, in the order of tens of miles, but this was deemed acceptable as it was decided that GEE could be used for short range navigation whilst LORAN would be utilized for longer range. LORAN was first used by ships and aircraft in the Atlantic theater but eventually found more extensive use in the Pacific. Acceptance of the longer range, lower accuracy parameters for LORAN demanded a shift to lower frequencies which could reflect off the Ionosphere at night providing 'over the horizon' capability. Over time, advances in technology improved receiver accuracy and reduced unit costs. Corresponding advances in other technologies improved the methodology for synchronizing the signals from the master and secondary sites thus allowing the distances between the beacons of a given chain to be increased. As the accuracy of a hyperbolic system increases with increased baseline distance between the beacons, these changes improved the accuracy of the LORAN system. LORAN used multilateration principles of difference in time of signal arrival from different stations to determine position. In the post-war years, this methodology was combined with technologies that measured the phase shift of those signals which vastly improved the fix accuracy. This improved system was designated Loran-C and the original LORAN system redesignated Loran-A. As the Loran-C system expanded, the original system declined but some chains remained in service until as late as 1980, largely due to a surplus of cheap, discarded Loran-A receivers. The instructor was very good, and I feel I had a great training experience. However, I experienced four near disasters during and just after receiving my private pilot certificate/license. I can freely say that my decisions helped lead to these situations and how I was able to resolve the situation by modifying my actions (Skybrary 2021)."

Radio navigation has been around since right after planes started to fly. Since the 1930s aspects of the Very High Frequency Omni Range (VOR) system has been in available to pilots. Across country there are VOR stations (FAA 2021). Figure 7.1 shows a typical VOR station. VOR is very useful for local navigation within the United States. However, it is not as useful as LORAN-C or GPS.

VOR

Figure 7.1 Very high frequency Omni Range Station. Source: U.S. Department of Transportation (2021).

(Continued)

(Continued)

Idaho Falls is at 4700 ft altitude and, generally, the region is at or above this altitude. For example, the Tetonia/Driggs, Idaho area, at the base of the Teton Mountains is approximately 6000 ft in elevation. The temperatures during any given summer day can vary greatly. It is not uncommon for a day to start out in the 30s or 40s and then in the afternoon to be in the upper 80s or 90s. The Snake River Plain can also be windy. The annual average wind speed in Idaho Falls is approximately 21 mi/h. The average wind speed for the United States is approximately 17 mi/h. The winter weather is better because the daily temperatures can be more consistent. Morning temperatures can be in the 20s or lower and in the afternoon the temperatures can be in the 20s, up to the 40s.

Personal Epiphany

I had a great epiphany during my flight training. It was the first time I used a hood and flew only by instruments. The hood is designed to obscure the student's view out of the front window and force the student to only use the airplane's instruments. Flying with a hood as a student was the best human factors training, I have ever had. I instantly understood the complexity of the task of commercial and military pilots flying in weather. I also realized the very difficult task of landing an aircraft on an aircraft carrier at night or on a foggy or stormy day. I greatly respect modern pilots navigating those conditions. More so, I respect pilots from the past flying aircraft in the 1930s, 1940s, and into the 1950s that had limited instrumentation. During World War II LORAN was implemented and this augmented pilots' ability to navigate. However, navigation by the stars was still used to some extent. Just a note, my flying experience was all in the 1990s, so a while ago.

Cessna 152

The Cessna 152 was manufactured between 1977 and 1985. The engine produced approximately 110 horsepower and the dry weight of the aircraft was approximately 1090 lb, with a gross weight of approximately 1670 lb. Its range was 477 mi and could reach altitudes of 14,700 ft.

Experience 1 – Stuck in an Icy Puddle

The first experience that could have ended in disaster was getting a wheel of the Cessna 152 stuck in an icy puddle. I had flown to Arco, Idaho in the winter. Flying in the cold air of southeast Idaho in winter is rather pleasant. There are not many planes flying, it is dark, so no windshield glare and on many nights no wind. I made this trip as part of the requirement for my long-distance flying time. Arco by air is approximately 65 mi. Arco is very close to the INEL (now INL) and has an elevation of 5325 ft. I don't have record of the temperature that evening. However, it was clear and cold. I made my radio calls and lined up on the runway and landed. That night there was no one at the airport to answer the radio. I tended to land flat, which made my instructor comment about it every time. I landed and taxied to the end of the runway. Then, I made a turn and taxied back to the other far side of the runway. When I made that turn, my port (left) side wheel was just off the tarmac and there was a patch of ice that covered a puddle of water. The wheel broke through the ice and sank down about 8 in. in the icy water. I tried to add power to pull the aircraft out of the puddle, but the aircraft simply wanted to rotate left around the stuck wheel. I sat there and thought what to do next? I decided to leave the aircraft running and then to get out and push on the left strut. This didn't work. I next thought that if I could reach the throttle and push it might come out. That is where I stopped. In my mind I could see the aircraft nose forward into the ground and it set a shiver down my spine. I guess in a way, as an experienced safety guy, I had that bit of clairvoyance that stopped me from doing the stupid thing. At that point I heard on the radio that another plane was going to land. I contacted him and he landed. I shut the aircraft's engine down and we were able to push the plane out of its icy prison. It was happenstance that this person came to my rescue, and I appreciated it greatly. It was a great lesson about ensuring that the aircraft's landing gear stay on the plowed tarmac.

Experience 2 – Density Altitude

The FAA publication FAA-P-8740-2 has thorough discussion of density altitude (FAA 2008).

In this document they discuss that pilots sometimes confuse the term "density altitude" with other definitions of altitude. There are several types of altitude:

- Indicated Altitude is the altitude shown on the altimeter.
- True Altitude is height above mean sea level (MSL).
- Absolute Altitude is height above ground level (AGL).
- Pressure Altitude is the indicated altitude when an altimeter is set to 29.92 in Hg (1013 hectopascal [hPa] in other parts of the world). It is primarily used in aircraft performance calculations and in high-altitude flight.
- Density Altitude is formally defined as "pressure altitude corrected for non-standard temperature variations."

(Continued)

(Continued)

The importance of density altitude is that High Density Altitude = Decreased Performance.

The formal definition of density altitude is certainly correct, but the important thing to understand is that density altitude is an indicator of aircraft performance. The term comes from the fact that the density of the air decreases with altitude. A "high" density altitude means that air density is reduced, which has an adverse impact on aircraft performance. The published performance criteria in the Pilot's Operating Handbook (POH) are generally based on standard atmospheric conditions at sea level (that is, 59 °F or 15 °C and 29.92 in. of mercury). Your aircraft will not perform according to "book numbers" unless the conditions are the same as those used to develop the published performance criteria. For example, if an airport whose elevation is 500 MSL has a reported density altitude of 5000 ft, aircraft operating to and from that airport will perform as if the airport elevation were 5000 ft.

The conditions that impact density altitude are high temperatures, high altitudes, and humidity.

High density altitude corresponds to reduced air density and thus to reduced aircraft performance. There are three important factors that contribute to high density altitude:

1. Altitude. The higher the altitude, the less dense the air. At airports in higher elevations, such as those in the Western United States, high temperatures sometimes have such an effect on density altitude that safe operations are impossible. In such conditions, operations between midmorning and midafternoon can become extremely hazardous. Even at lower elevations, aircraft performance can become marginal and it may be necessary to reduce aircraft gross weight for safe operations.

Density Altitude

2. Temperature. The warmer the air, the less dense it is. When the temperature rises above the standard temperature for a particular place, the density of the air in that location is reduced, and the density altitude increases. Therefore, it is advisable, when performance is in question, to schedule operations during the cool hours of the day (early morning or late afternoon) when forecast temperatures are not expected to rise above normal. Early morning and late evening are sometimes better for both departure and arrival.
3. Humidity. Humidity is not generally considered a major factor in density altitude computations because the effect of humidity is related to engine

power rather than aerodynamic efficiency. At high ambient temperatures, the atmosphere can retain a high water vapor content. For example, at 96 °F, the water vapor content of the air can be eight (8) times as great as it is at 42 °F. High density altitude and high humidity do not always go hand in hand. If high humidity does exist, however, it is wise to add 10% to your computed takeoff distance and anticipate a reduced climb rate.

Check the Charts Carefully

Whether due to high altitude, high temperature, or both, reduced air density (reported in terms of density altitude) adversely affects aerodynamic performance and decreases the engine's horsepower output. Takeoff distance, power available (in normally aspirated engines), and climb rate are all adversely affected. Landing distance is affected as well; although the indicated airspeed (IAS) remains the same, the true airspeed (TAS) increases. From the pilot's point of view, therefore, an increase in density altitude results in the following:

- Increased takeoff distance.
- Reduced rate of climb.
- Increased TAS (but same IAS) on approach and landing.
- Increased landing roll distance.

Because high density altitude has implications for takeoff/climb performance and landing distance, pilots must be sure to determine the reported density altitude and check the appropriate aircraft performance charts carefully during preflight preparation. A pilot's first reference for aircraft performance information should be the operational data section of the aircraft owner's manual or the POH developed by the aircraft manufacturer. In the example given in the previous text, the pilot may be operating from an airport at 500 MSL, but he or she must calculate performance as if the airport were located at 5000 ft. A pilot who is complacent or careless in using the charts may find that density altitude effects create an unexpected – and unwelcome – element of suspense during takeoff and climb or during landing.

So, given all that background from the FAA, I flew from Idaho Falls to the Driggs, Idaho Airport. The airport sits at 6231 ft. On the day I flew there the temperature was approximately 95 °F. I called in to the fixed base operator (FBO) and received no response on the radio, so I lined up and landed. When I taxied to the FBO to use the facilities, there were a group of people who had come in for a fly-in. They were upset because I landed the wrong direction, even though the air was calm and the FBO hadn't responded to my radio call.

(Continued)

(Continued)

So, I used the facilities and prepared to take off. I taxied to the end of the "correct" end of the runway. I did my pre-flight checklist and began to takeoff. Well, the density altitude of that airport at 95 °F was approximately 9500 ft. A Cessna 152 does not perform takeoffs well at that level of density altitude. I did my take off roll and lifted the nose. The plane struggled to get into the air. As I left the end of the runway at probably 30–40 ft elevation, I saw power lines in the distance coming up fast. I gently lifted the nose and hoped (prayed) to miss them. The aircraft continued to climb slowly. High transmission power lines can be from 50 to 150 in height. I cleared them, but just barely. I climbed up to an altitude that I could clear Pine Creek Pass (6780 ft) and headed back to Idaho Falls.

Density altitude is a killer in the west. My mistake was allowing others to influence when I was going to leave the airport. I could have waited a couple hours until it cooled down. The group was so upset, even though they were not actively flying. The mistake made by the FBO was not responding to my radio call in.

Experience 3 – Windshield Glare

This experience is not as involved as the other two but could have ended in disaster just the same. I took off from the Idaho Falls airport on a bright and sunny late fall afternoon. I flew north toward the city of Ashton, Idaho and then started to make my way back. The Idaho Falls airport sits at 4740 ft elevation and has a manned tower. I approached the airport from the north and approximately 25 mi out called in indicating my intention to land. The sun was coming almost directly into my field of vision and the windshield was very scratched. The scratches on the windshield created such a glare I couldn't see anything in front of me. The tower informed me there was another aircraft in the area and to tune to a certain frequency. At this point both the other aircraft and I could not see each other because of the sun conditions and the tower indicated we were close to each other. I made the decision I need to abort. I asked the tower if I was clear to my starboard (Right) and they said yes. I made a right turn and headed out over the desert until the sun set. I then landed with no incident. If I had made the wrong decision I, along the other aircraft, could have crashed into each other.

Experience 4 – High Winds

As part of flight training a student must do a long cross-country flight. I planned mine to fly from Idaho Falls to Twin Falls, Idaho – 167 mi. Twin Falls (Twin) sits at 3743 ft elevation. The next leg was to Logan Utah – 169 mi.

Logan airport sits at 4534 ft. The final leg was back to Idaho Falls – 121 mi. The average wind direction for southern, Idaho and northern, Utah is from the west. On this day, the wind was from the east. I took off and flew to Twin in good time because the wind was at my back. In Twin Falls I didn't need fuel because of the boost from the wind. I took off from Twin and head toward Logan. Here is where the story gets interesting. The wind was so strong I could barely make any progress. I watched the fuel gauge go lower and lower. There is a natural pass into the Cache Valley coming from the west. I watched that pass get closer and closer as the fuel gauge dropped. When I landed I had almost no fuel. The Idaho Falls FBO had called Logan and Twin to see if my plane was down. I filled the tank and took a breath and then headed back to Idaho Falls. My mistake was that I didn't fill up in Twin, even though it appeared I would have enough fuel.

Summary

Inexperienced people make mistakes. What is important is that people learn from their mistakes and or correct actions before a mistake creates a disaster.

7.1 Helicopter Accident

Helicopters are very functional and have a wide range of utility. They are used by militaries for troop transport, weapons platforms, and medical evacuations. In the civilian world they are used for medical evacuations (life flights), construction, inspection of utility lines, logging, firefighting, and site seeing, to name a few. Helicopters have gotten much more reliable over the years. Currently, the fatal accident rate for helicopters is 0.67 per 100,000 air miles (Roskop 2018). In 2016 there were 106 helicopter accidents and 17 of them were fatal. The most common causes of helicopter accidents are:

- Pilot error
- Lack of situational awareness
- Flying into terrain
- Inexperienced pilots
- Overloading
- Faulty or failed parts
- Inadequate maintenance
- Inclement weather
- Air traffic control error

This section will discuss the Liberty Helicopter Crash on March 11, 2018.

7.1.1 Liberty Helicopter Crash March 11, 2018

On March 11, 2018, an aerial photography flight ended in disaster when it lost power and crashed into New York's East River. The five passengers on board all died and the pilot sustained minor injuries. There were numerous contributing factors to this accident, including the failure of the emergency floatation system, the design of the controls, and the FAA regulations for overseeing aerial photography flights. The description in the following text and analyses for this accident primarily focuses on the safety culture of Liberty Helicopters and FlyNyon, human actions, and the contribution of the flight harness and lack of safety management system (SMS). The following accident description was excerpted from the NTSB investigation report (NTSB 2019a).

7.1.1.1 Overview

On March 11, 2018, about 1908 (7:08 p.m.) eastern daylight time, an Airbus Helicopters AS350 B2, N350LH (Figure 7.2), lost engine power during cruise flight, and the pilot performed an autorotative descent and ditching on the East River in New York, New York. The pilot sustained minor injuries, the five passengers drowned, and the helicopter was substantially damaged. The FlyNYON-branded flight was operated by Liberty Helicopters Inc. (Liberty), per a contractual agreement with NYONair; both companies considered the flight to be an aerial photography flight operated under the provisions of Title 14 *Code of Federal Regulations* (*CFR*) Part 91. Visual flight rules (VFR) weather conditions prevailed, and no flight plan was filed for the intended 30-minute local flight, which departed from Helo Kearny Heliport, Kearny, New Jersey, about 1850 (6:50 p.m.).

Figure 7.2 Helicopter involved in the accident. Source: National Transportation Safety Board (2018).

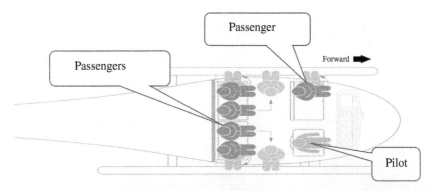

Figure 7.3 Seating configuration. Source: Aircraft Accident Report 2018/Public domain.

Liberty operated the accident flight as a FlyNYON-branded, doors-off helicopter flight that allowed the five passengers (one in the front seat, four in the rear seats) to take photographs of various landmarks while extending their legs outside the helicopter during portions of the flight. Figure 7.3 shows the seating configuration of the six occupants of the helicopter, the five passengers, and the pilot.

For the accident flight (and other FlyNYON flights that Liberty operated), Liberty configured its Airbus AS350 B2 helicopter with the two right and the front left doors removed and the left sliding door locked open. Before departure, each passenger was fitted with a NYONair-provided harness/tether system that NYONair developed with the intent to prevent passengers from falling out of the helicopter. The harness/tether system used on the accident flight consisted of a full-body, workplace fall-protection harness that was secured (with a locking carabiner) to a tether, the other end of which was secured (with another locking carabiner) to an anchor point in the cabin. Each passenger also wore the helicopter's installed, Federal Aviation Administration (FAA)-approved restraints. The pilot (who was seated in the front right seat) wore only an installed, FAA-approved restraint (Figure 7.4). The harness that NYONair provided for each passenger on the accident flight is shown in the black and white Figure 7.5. It is referred to as the "yellow" harness by Liberty and NYONair personnel due to their color and was designed to protect users from workplace fall hazards. The harness included interconnected shoulder straps and leg straps, a chest strap that spanned the two front shoulder straps, and a dorsal D-ring (used on the accident flight for attaching a tether) between the wearer's upper shoulder blades. NYONair attached a pouch containing an emergency cutting tool on either the right or left upper shoulder strap of each harness. Figure 7.5 shows the NYONair system. The tethers for the front passenger's harness is shown in Figure 7.6.

NYONair CX personnel assisted the accident flight passengers with preflight preparations, which included showing the passengers NYONair's 3-minute

Figure 7.4 FAA-approved exemplar four-point restraint with rotary buckle; arrows show direction of release. Source: National Transportation Safety Board (2018).

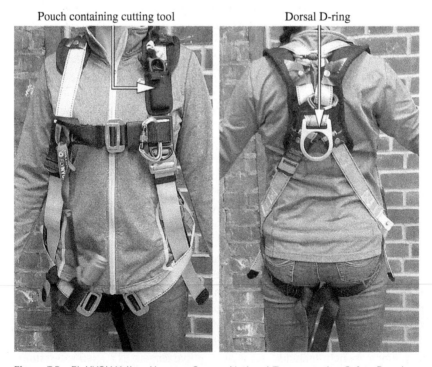

Pouch containing cutting tool   Dorsal D-ring

Figure 7.5 FlyNYON Yellow Harness. Source: National Transportation Safety Board (2018).

Figure 7.6 Front passenger's tether from the accident flight with top locking carabiner inserted through third webbing loop (from floor) with three excess loops hanging as a tail. *Note*: The tether tail's terminating loop is partially obscured in this photograph. Source: National Transportation Safety Board (2018).

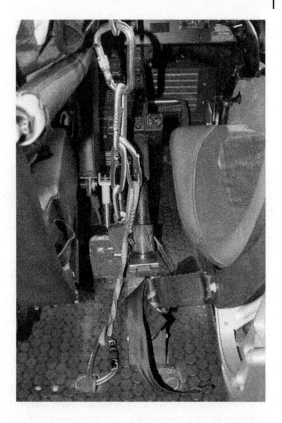

22-second passenger safety video. The video addressed general safety topics (such as harness fitting, use of the headset to hear instructions from the pilot, and securing loose items) and emergency safety procedures (such as harness release, cutting tool use, egress, and location and use of the PFDs and fire extinguisher).

The video narrative for releasing from the harness/tether system in the event of an emergency stated that "the harness can be released by opening the quick release clip in the back of the harness"; the visual that accompanied this narrative showed one person releasing another person's carabiner.

The video also instructed passengers that "a cutter [cutting tool] is also secured to one of your chest straps and will allow you to quickly cut through the harness if you are unable to reach the quick release clip." The visual that accompanied this narrative showed a tether made of a different webbing than the tethers provided to the accident passengers. It showed a person (seated alone in the back of a helicopter) grasping the tether with one hand and applying the cutting tool with the other, successfully cutting through the tether with one pull of the tool (see Figure 7.7).

Figure 7.7 Screen capture from the safety video shown to the passengers before the accident flight showing a person using a cutting tool on a type of tether (NTSB).

After the flight departed, it traveled past various scenic landmarks. Consistent with the standard operating procedures (SOPs) used for FlyNYON flights, the passengers were allowed (when instructed by the pilot) to position themselves to extend their legs outside the helicopter. The two passengers who had been seated in the rear inboard seats removed their installed, FAA-approved restraints and sat on the cabin floor, wearing their harness/tether systems. The passengers seated in the outboard seats were allowed to rotate outboard in their seats. To enable such freedom of movement, the SOPs allowed the passengers to wear their installed, FAA-approved restraint with the lap belt adjusted loosely and the shoulder harness routed under the arm.

A review of radar data and onboard video showed that, when the flight was proceeding northwest over Manhattan toward Central Park at an altitude of 1900 ft MSL, the front passenger, who was facing outboard in his seat with his legs outside the helicopter, leaned back several times to take photographs using a smartphone. The onboard video showed that, each time he leaned back, the tail of the tether attached to the back of his harness hung down loosely near the helicopter's floor-mounted controls (Figure 7.8). At one point, when he pulled himself up to adjust his seating position, the tail of his tether remained taut but appeared to pop upward. Two seconds later, the helicopter's engine sounds decreased, and the helicopter began to descend.

As the pilot performed the emergency procedures to perform an autorotation and address the apparent loss of engine power, he noticed that the fuel shutoff lever (FSOL) was in the shutoff position and that it had been inadvertently moved to that position by the tail of the front passenger's tether, which had become caught on it. The helicopter's FSOL was one of three floor-mounted control levers located

Figure 7.8 Tether loop near floor mounted controls. Source: National Transportation Safety Board (2018).

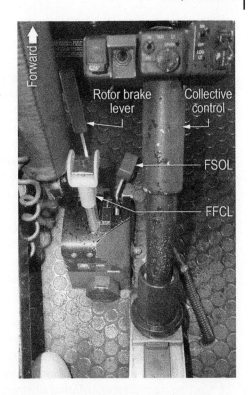

near the base of the pilot's collective control, along with the FFCL and the rotor brake. Each floor-mounted control had a different lever length and head shape; also, the FSOL and rotor brake lever were both red, and the FFCL was yellow (see Figure 7.8). This figure shows the locations of the three levers, noting that the figure is in gray tones.

By design, the FSOL provided a means to close the fuel shutoff valve on the fuel line between the fuel tank and the engine. The rotorcraft flight manual (RFM) for the helicopter specified the closing of the FSOL (by pulling it up and aft) in the event of an emergency landing. The FSOL was equipped with a breakable snap wire routed through a hole in the fuel control plate to keep the FSOL positioned down and forward unless force was applied to the lever to change its position.

Although the pilot pushed the FSOL down to restore fuel flow to the engine and attempted to relight the engine, the helicopter was too low to allow engine power to be restored in time to prevent the emergency landing. The pilot pulled the activation handle to deploy the helicopter's emergency flotation system, and he ditched the helicopter on the East River. However, the helicopter's floats did not fully inflate, and the helicopter rolled right in the water and became fully inverted and submerged about 11 seconds after it touched down.

The pilot was able to release his installed, FAA-approved restraint after he was under water and successfully egress from the helicopter; however, none of the passengers were able to egress, and they all drowned.

The NTSB determines the probable cause of this accident was Liberty Helicopters Inc.'s use of a NYONair-provided passenger harness/tether system, which caught on and activated the floor-mounted engine FSOL and resulted in the in-flight loss of engine power and the subsequent ditching. Contributing to this accident were (i) Liberty's and NYONair's deficient safety management, which did not adequately mitigate foreseeable risks associated with the harness/tether system interfering with the floor-mounted controls and hindering passenger egress; (ii) Liberty allowing NYONair to influence the operational control of Liberty's FlyNYON flights; and (iii) the FAA's inadequate oversight of Title 14 *CFR* Part 91 revenue passenger-carrying operations. Contributing to the severity of the accident were (i) the rapid capsizing of the helicopter due to partial inflation of the emergency flotation system and (ii) Liberty and NYONair's use of the harness/tether system that hindered passenger egress.

The NTSB evaluated the following safety issues:

- Effect of the harness/tether system on the ability of each passenger to rapidly egress from the capsizing helicopter. The investigation found that minimally trained passengers would have great difficulty extricating themselves from the harness/tether system, each of which was equipped with locking carabiners and an ineffective cutting tool, during an emergency requiring a rapid egress.

- Emergency flotation system design, maintenance, and certification issues. The manufacturer of the helicopter's emergency flotation system did not provide information to help operators recognize the presence of unacceptably high pull forces when activating the system; the high pull forces on the accident helicopter's activation system (which resulted from an installation anomaly) contributed to the pilot's mistaken belief that he had taken the necessary action to fully inflate the floats. The FAA's certification review of the emergency flotation system design installed on the accident helicopter did not identify the manufacturer's omission of an activation handle pull-force limitation.

- Ineffective safety management at both Liberty and NYONair. Liberty's managers repeatedly lacked involvement in key decisions related to Liberty-operated FlyNYON flights and allowed NYONair to influence core aspects of the operational control of those flights. Ineffective safety management at both companies allowed foreseeable safety risks to remain unmitigated; these included the potential for passenger interference with the helicopter's floor-mounted controls, partial inflation of the emergency float system, and difficulties passengers would have with the locking carabiners and cutting tools as a means to rapidly release from the harness/tether system.

- Liberty and NYONair's exploitation of the AW/aerial photography exception at 14 CFR 119.1(e) to operate FlyNYON flights under Part 91 with limited FAA oversight. Federal regulations do not define the terms "AW" and "aerial photography" to include only business-like, work-related aerial operations. Both Liberty and NYONair demonstrated deliberate efforts to operate the FlyNYON revenue passenger-carrying flights under Part 91 as aerial photography flights and to avoid any indication that the flights may be commercial air tours, which would be subject to additional FAA requirements and oversight that did not apply to aerial photography flights.

- Lack of policy and guidance for FAA inspectors to perform a comprehensive inspection of Part 91 operations conducted under any of the 14 CFR 119.1(e) exceptions. During the investigation, the FAA determined that the accident flight was a nonstop commercial air tour operated under Part 91 per the 14 CFR 119.1(e)(2) exception. Although an air tour operated under Part 91 is subject to FAA requirements and oversight that exceed what applies to aerial photography flights, the FAA lacks policy and guidance for FAA inspectors to support a comprehensive inspection of Part 91 operations conducted under any of the exceptions in 14 CFR 119.1(e) to ensure that operators are appropriately managing any associated risks.

- Lack of FSOL protection from inadvertent activation. The certification basis for the accident helicopter's FSOL did not require protection from inadvertent activation due to external influences, such as interference from a passenger. However, a design modification that includes protection from external influences could enhance safety.

- Need for guidance and procedures for operators to assess and address passenger intoxication. Although the passenger in the front seat on the accident flight was intoxicated, it was not possible to determine whether alcohol played a role in his inadvertent activation of the FSOL. Despite the existence of an FAA regulation prohibiting the carriage of any passenger who appears to be intoxicated or impaired, neither Liberty nor NYONair had any documented policy or guidance materials, including training, for their employees to identify impaired passengers or for denying boarding of such individuals. While FAA guidance does exist on identifying intoxicated or impaired passengers, operators that conduct revenue passenger-carrying flights under Part 91 or 135 in small aircraft could benefit from guidance specific to their operations, particularly if they have passengers seated in close proximity to the aircraft controls.

- Inadequacy of the review and approval process for supplemental passenger restraint systems (SPRSs) that the FAA implemented after the accident. The FAA's SPRS approval process that it implemented after the accident appears to focus primarily on the SPRS release mechanism without consideration of the expected operational environment or whether the use of an SPRS is warranted.

The NTSB is concerned that, without an assessment of the specific need for and use of an SPRS, the addition of an SPRS may unnecessarily complicate the emergency egress of passengers. Further, without a comprehensive hazard analysis for the use of an SPRS in the operational environment (including aircraft-specific installations), factors that could impede passenger egress, such as the potential for entanglement with headset cords, other equipment, or the SPRS itself; or adversely affect flight safety, such as the potential for the SPRS to interfere with any equipment or controls in a specific aircraft, may be present but go unidentified.

- The tail of the front passenger's tether caught on the FSOL during the flight, which resulted in the inadvertent activation of the FSOL, interruption of fuel flow to the engine, and loss of engine power.
- The pilot autorotated the helicopter successfully and pulled the emergency flotation system activation handle to deploy the floats at an appropriate time; however, the floats inflated partially and asymmetrically.
- Liberty Helicopters Inc.'s and NYONair's decision to use locking carabiners and ineffective cutting tools as the primary means for passengers to rapidly release from the harness/tether system was inappropriate and unsafe.
- The helicopter's landing was survivable; however, the NYONair-provided harness/tether system contributed to the passenger fatalities because it did not allow the passengers to quickly escape from the helicopter.
- The FAA approval process for SPRS that was implemented after the accident is inadequate because it does not provide guidance to inspectors to evaluate any aircraft-specific installations or the potential for entanglement that passengers may encounter during emergency egress.
- Although the crossover hose in the accident helicopter's emergency flotation system design did not perform its intended function to alleviate asymmetric inflation of the floats during a single-reservoir discharge event, buoyancy stability testing showed that even symmetric distribution of the gas from only one reservoir would not enable the helicopter to remain upright in water.

The factors not involved in the accident were:

(1) the pilot's qualifications, which were in accordance with federal regulations and company requirements;
(2) pilot fatigue or medical conditions; and
(3) the airworthiness of the helicopter.

7.1.1.2 Liberty Helicopter's Safety Program

Liberty's CEO said the company's "chain of command," including the COO, DO chief pilot, and director of safety, was responsible for safety. He incorrectly identified Liberty's director of training as the director of safety (a position that was vacant). The COO said the DO and the director of maintenance were responsible for managing safety. The DO said the company did not have a formal safety program or use a flight risk assessment tool, but assessments were made by himself and the chief pilot. He said he and the chief pilot provided safety leadership for the pilots, and the chief pilot said safety was one of his responsibilities.

Regarding the operation of FlyNYON doors-off flights and the use of the NYONair-provided harness/tether system, Liberty's DO said Liberty did not perform a formal risk assessment but instead used an informal process of reviewing the operation through meetings and conversations.

Liberty had a safety officer (a position that was not on the organizational chart) who assumed the safety officer duties (in addition to his line pilot duties) after the previous safety officer left the company about six weeks before the accident. The safety officer said that the position was a holdover from when Liberty participated in the Tour Operators Program of Safety (TOPS). TOPS states their mission and vision are (TOPS 2021):

> Mission
> TOPS works as a collaborative team to develop air tour safety standards and recommended practices that are progressive and best-in-class and exceed regulatory standards.
> Vision
> To be the premier safety standard in the air tour industry.

The COO said Liberty's former director of safety used to provide him with quarterly safety meeting summaries to review. The former director of safety said that, while serving in that role, he had reported to the COO, organized quarterly safety meetings for pilots and heliport personnel, received hazard reports, and updated the company's safety manual. The COO said he expected the company to fill the director of safety position within three months of the former director's departure. When asked why that timeframe was exceeded, he said he was not aware of the extended delay.

Liberty's former safety officer (who had been a line pilot who also served as safety officer from 2014 to January 2018) said he left Liberty because he disagreed with the company's decision to operate FlyNYON flights. He also stated that he thought the authority of Liberty's chief pilot was being undermined by NYONair's CEO.

TOPS, which Liberty left in 2017. Liberty's DO said that, when Liberty was a member of TOPS, Liberty personnel had to attend all the TOPS meetings and undergo audits. Liberty's COO said he decided to cease the company's participation in TOPS as a cost-saving measure until Liberty could develop a new business plan for its downtown sightseeing tours.

Liberty's safety officer said that there was no formal description of his safety officer duties and that he had no training for his role other than the former safety officer having discussed with him the things he should be looking into and what the few responsibilities were as related to the PFDs. He said that, even though the position was informal, he could help the pilots get things addressed by being a second set of eyes on procedures and looking for ways for Liberty to improve their operations. He said Liberty had a hazard reporting system, but it was not being used.

Liberty had a Safety Manual that (according to the manual) served as a "guide for all company personnel in complying with the corporate policy for safety management and mishap prevention." It was not part of Liberty's FAA-approved or -accepted manuals.

Liberty's former director of safety said that he had been responsible for revising the Safety Manual and holding quarterly safety meetings. The manual specified that minutes from the quarterly safety meetings were to be reported to the COO within three days. Liberty's safety officer said he had not yet convened a quarterly safety meeting.

The manual did not describe duties for a "director of safety" position but described a position called the "company safety officer," who reported to the COO and was vested with the authority to prohibit the use of unsafe equipment and suspend unsafe operations. This person was responsible for maintaining the Safety Manual, developing the safety program, evaluating program results, and guiding its implementation through activities that included specifying requirements for internal safety reporting. The specified duties included preparing progress reports of safety activities, safety reports, and safety studies; assisting in providing safety evaluations of new equipment and procedures; and arranging for safety reviews by outside experts, when appropriate.

Per the Safety Manual, the COO was responsible for implementing and establishing support for the safety program and for "integrating safety systems into all areas of operations" through activities that included, in part, the following:

> Delegating specific aviation safety assurance and safety responsibilities, authority, and accountability to the [safety officer] and applicable first-line supervisors.

According to its website, TOPS is an independent, nonprofit organization incorporated in 1996 that promotes collaboration among helicopter air tour operators on the development of voluntary safety standards that exceed regulatory requirements. While a member of TOPS, Liberty was subject to semi-annual audits to verify compliance with TOPS standards. These standards included the requirement to have a safety management program that included top management commitment to safety, a trained safety manager reporting to top management, and procedures (including safety reporting) for identifying hazards. A 2016 TOPS audit of Liberty's safety program determined that it followed these standards. At that time, Liberty had a safety program, a safety manual, a director of safety who convened quarterly safety meetings and reported to top management, and a safety officer who assisted and reported to him.

- Promptly implementing approved recommendations resulting from incident reports, investigations, and other safety improvement projects.

The manual specified that the DO was responsible for the following, in part:

- Supervising, coordinating, and directing flight operations without compromising safety under any circumstance.
- Using measurement and control tools, such as reports and inspections, to enhance safety performance.
- Suspending any operation for safety reasons.

The COO said he did not participate in pilot safety meetings or communicate with the safety officer about operational matters. He said the safety culture at Liberty was such that, if there was any question or concern about the safety of doing something, it was not done.

Liberty's DO said any concerns that any pilots had would be discussed in an intelligent way, and the company would make adjustments, if needed. He said it was a small company, so pilots had the open ability to talk to leadership. The last issue he recalled occurred within the two months before the accident and involved the pilots developing a temperature threshold for cold-weather operations.

The chief pilot said Liberty had a "good safety culture." He said he was a big proponent of safety and believed that the pilots understood that. He said he believed that a good safety culture started at the top and worked its way down and that, for his pilots, he was the face of the company and drove the safety culture at Liberty.

Liberty's safety officer said that, when pilots expressed safety concerns to management, they were addressed. He said that nothing that had been brought to his attention required looking into further or required him to bring it to anyone else's attention, but he would escalate concerns up the chain of command, if needed. He said his biggest safety concerns were that a passenger on a FlyNYON doors-off flight could fall out of the helicopter or that a headset could blow off, break free, and hit the tail rotor.

NTSB determined that before Liberty's development of the cooperative business relationship with NYONair, core aspects of Liberty's safety program, which was not required or approved by the FAA, had begun to deteriorate. For example, Liberty withdrew from participating in TOPS, an air tour industry group, in early 2017. Liberty's director of safety left the company in October 2017, and his position was not backfilled. Liberty's safety officer was new to his position, effective January 2018, after the former safety officer left the company. Liberty's safety officer said his position was informal and that he had no training for his role other than a briefing from the former safety officer. He also reported to Liberty's chief pilot, not the upper management. Although the former director of safety had performed some of the tasks prescribed in the Safety Manual, the manual was not part of Liberty's FAA-approved or -accepted manuals, and there was no indication that the safety officer was tasked to perform the safety actions. Thus, Liberty did not have a robust safety management structure in place before it began integrating the FlyNYON flights into its operation.

As Liberty's air tour business declined due to city restrictions on commercial air tour flights operated out of JRB, the FlyNYON-branded aerial photography flights operated out of 65NJ rapidly increased because they were not subject to the city restrictions, and NYONair was able to generate demand for the flights through social media marketing. As demand for FlyNYON flights surpassed NYONair's ability to operate them using its own fleet, NYONair's CEO approached Liberty's top management with a proposal that Liberty operate some of these flights. The establishment of a contract between Liberty and NYONair for FlyNYON flights provided Liberty with a new revenue stream.

During Liberty's initial operation of FlyNYON flights (late 2017–early 2018), Liberty's chief pilot coordinated closely with NYONair personnel on the execution of these flights, including scheduling helicopters and pilots, developing the SOPs, and handling safety issues. As the volume of Liberty's FlyNYON flight operations increased (and was expected to increase further in the warmer weather months), Liberty's DO, COO, and CEO did not increase any engagement in supervising the day-to-day aspects of Liberty's FlyNYON flight operations. These activities were largely delegated to Liberty's chief pilot, who, along with Liberty's safety officer, worked to address any operational or safety concerns that Liberty's pilots raised.

In contrast, NYONair's CEO and NYONair's director of business operations were closely engaged in the day-to-day operations of FlyNYON flights. A combination of factors, including the NYONair CEO's former employment at Liberty, the close cooperation between personnel from the two companies for Liberty's FlyNYON flights operated out of the NYON Terminal, the "hands-off" approach of Liberty's top management on its day-to-day operations, and the fact that Liberty's DO (who was the father of NYONair's CEO) also served as the DO for East West (NYONair's Part 135 certificate), blurred the roles and responsibilities of personnel from both companies with regard to their ability to influence aspects of Liberty's operation of FlyNYON flights.

The accident pilot said that, during Liberty's initial involvement with FlyNYON flights, he was concerned about whether the cutting tools could cut the tethers. He said the concern was also expressed by pilots "talking amongst themselves" and that the personnel who worked to develop the SOPs were "trying to get a better option." He said he believed an SOP had been developed to address it. Liberty's chief pilot said pilots had asked how tethered passengers could get out if a helicopter goes into the water. He said the answer was that the passengers have knives to cut themselves out, but they should not need them because the helicopter is equipped with floats. He said he had never had a conversation about what to do if the helicopter rolls over because the floats were not working; he said that, in that case, occupants would get the knife and cut themselves out.

Liberty's director of training said that, during a joint training exercise in November 2017, Liberty and NYONair pilots and other personnel had the opportunity to cut the tethers with the cutting tools. He said various participants took between three and ten seconds to cut through the tether, which they held in front of them while the other end was tied to a stationary object. He said the most successful technique involved making an elliptical pattern with the cutting tool and working it back and forth a few times.

Liberty's safety officer said that he felt he could cut through the tether in about five seconds but was concerned about how a typical passenger could cut it, particularly if they were not taking it seriously or had never used the cutting tool before. In a February 8, 2018, e-mail, Liberty's safety officer told Liberty's chief pilot that he was researching new cutting tools and tethers.

The NYONair NYC lead pilot said that she was surprised about the pilots' concerns with the existing cutting tools and tether and that, as soon as the Liberty safety officer brought it to her attention, she added the issue to the agenda for a pilot meeting. She said that, when she tried cutting the tether (using the same type of tether and cutting tool as on the accident flight), she found it difficult, and it took her more than 30 seconds to cut through it.

Pilot meeting minutes dated February 21, 2018, noted that the Liberty safety officer was "researching and procuring" new cutting tools and tethers that "we will be testing shortly." The proposed equipment was discussed again at a pilot meeting on March 7, 2018, that was attended by NYONair's CEO. The minutes from that meeting said, "we are going forward with a bulk inventory purchase which includes the new style tether and [cutting tools] which are much easier to use." NYONair's NYC lead pilot said the reception to the proposal had been enthusiastic and management was accepting of the change. The NYONair chief of staff said that NYONair had decided to purchase the new equipment but had not ordered it before the accident occurred.

A March 7, 2018, pilot meeting resulted in conflict between NYONair's CEO and Liberty's safety officer when he disagreed with a statement by the CEO. Liberty's safety officer said that, after the disagreement, NYONair's CEO included him on a text to Liberty's chief pilot stating that his (the safety officer's) services were no longer required for FlyNYON flights. Liberty's safety officer said that

Liberty operations personnel were informed that he would be flying only tours and charters rather than FlyNYON flights. He said that there was no indication that Liberty's DO either approved of or intervened in this decision and that he supposed it was made solely by NYONair. According to interviews with the Liberty safety officer, NYONair's CEO, and others, the removal was due to a personal conflict, not a safety issue.

Since the departure of Liberty's director of safety in October 2017, Liberty had not had any of its own pilot safety meetings (which the former director of safety had convened quarterly and provided minutes to Liberty's COO). Liberty's safety officer was Liberty's representative for the NYONair-led pilot meetings where logistical, customer service, and safety matters for FlyNYON flights were discussed. These meetings were run by the NYONair NYC lead pilot, who was sometimes assisted by NYONair's CEO, and Liberty's safety officer used them as an opportunity to voice any safety concerns about FlyNYON flight operations from the Liberty pilot group. However, the investigation found that, for a variety of reasons, the meetings were not an effective mechanism for Liberty in managing its pilots' safety concerns.

The NYONair personnel (including the NYC lead pilot) who led and participated in these meetings were not trained in safety management, and NYONair itself did not have a formal safety structure. Although NYONair's CEO said he had appointed the NYONair NYC lead pilot as a safety officer, she did not share that understanding. Additionally, the NYONair NYC lead pilot did not have the authority to enforce safety-related operating policies or to authorize the purchase of safety-related equipment. All such decisions were directly made, or heavily influenced, by NYONair's CEO. However, NYONair's CEO had not established clear lines of accountability through top management or systematic processes for identifying hazards, prioritizing interventions (or mitigations), and mitigating the related risks, which are all key components of a strong safety program.

Liberty pilots were aware of some risks associated with FlyNYON flights, including the potential for entanglement of the harness/tether system with the helicopter's floor-mounted controls; the difficulties passengers would have in accessing the carabiners on their harness/tether system in an emergency (particularly for the FlyNYON Yellow harness); the inadequacy of the passengers' cutting tools to quickly sever their tethers; and the possibility that the emergency flotation system may only partially inflate if the activation handle is not pulled fully aft. Although Liberty pilots took some informal operational measures to address floor-mounted control interference vulnerabilities, and Liberty's training program made pilots aware of the need to pull the float activation handle fully aft, these mitigations were not evaluated at the organizational level, and Liberty did not assess and manage the overall risks in the context of FlyNYON flight operations.

In addition, for FlyNYON flights, due to the previously discussed absence of support from Liberty management (specifically Liberty's DO) in operational and safety-related discussions, Liberty pilots had to advocate for safety improvements themselves.

The culture among Liberty pilots emphasized the responsibility of individual pilots to take the initiative to ensure safe flight operations. Thus, Liberty's safety officer and chief pilot informed NYONair about their concerns about passenger egress, which included their desire for NYONair to provide what they called the blue harnesses for Liberty pilots to use on FlyNYON flights. (Some NYONair pilots were also concerned about passenger egress.)

When Liberty's safety officer and chief pilot first approached NYONair stating that they wanted the blue harnesses for safety reasons (due to better fit on smaller passengers and the presence of a more accessible attachment point), NYONair provided assurances that their concerns would be addressed. However, meeting minutes showed that implementing the request became deprioritized over time, and NYONair management and other employees came to regard the Liberty pilots' concerns about the harness/tether system as superfluous and not urgent.

NYONair's CEO eventually decided that there were no safety issues with the old type of harness and told the pilots to stop asking for the replacement harnesses. Although there is no indication that NYONair performed any risk evaluation to support this decision, there is likewise no indication that Liberty management provided any risk evaluation to counter it. Key decisions related to FlyNYON flight operations were being made by NYONair's CEO, whereas Liberty's top management had little involvement, having delegated these responsibilities to Liberty's chief pilot and Liberty's safety officer. Thus, Liberty's pilots had no one at the top management level to advocate for their concerns. This created impediments to change when the Liberty pilots brought up the need for safety improvements.

Interviews with Liberty and NYONair employees, as well as copies of communications, indicated that the organizational culture at NYONair emphasized goals such as customer satisfaction, expanding the business, and reducing costs. About FlyNYON flights, NYONair's safety goals were focused primarily on ensuring that passengers and equipment did not fall out of the helicopter and did not devote sufficient attention to mitigating other possible risks, like equipment issues that could hinder passenger egress.

When Liberty pilots attempted to take what they considered to be safety action (such as canceling a cold-weather flight and delaying a flight until a passengers could be fitted with a blue harness), NYONair's CEO responded by chastising them for hurting his brand, and he deprioritized their concerns by deciding that they were unrelated to safety. This contributed to a polarization of attitudes and reduced cooperation between the Liberty pilots and NYONair management on safety matters.

As a result, known risks remained unresolved due to blurred lines of authority, lack of safety management expertise, and lack of a formal process at both companies for systematically prioritizing and addressing foreseeable risks. Thus, the NTSB concludes that ineffective safety management at both Liberty and NYONair resulted in a lack of prioritization and mitigation of foreseeable risks.

Both Liberty and NYONair lacked a SMS. According to the FAA's website, "SMS is the formal, top-down business-like approach to managing safety risk, which includes a systemic approach to managing safety, including the necessary organizational structures, accountabilities, policies, and procedures." The goal of SMS is to identify safety hazards, ensure necessary remedial action is implemented to maintain an acceptable level of safety, provide continuous monitoring and regular assessment of the safety level achieved, and continuously improve a company's overall level of safety. When an SMS is implemented, senior management establishes adequate safety resources, develops a safety policy, establishes safety objectives and standards of safety performance, and leads the development of a positive organizational safety culture. Research indicates that this approach to safety leads to improved organizational outcomes (Smith et al. 1978, 5-15; Shannon, Mayr, and Haines 1997, 201-17). Because of the failures of organizational safety management that played a role in this accident, the NTSB believes that adoption of an SMS would enhance safety at both Liberty and NYONair. Therefore, the NTSB recommended that Liberty and NYONair establish an SMS.

In 2015, the FAA required Part 121 air carriers to develop a functioning SMS by March 9, 2018. After determining that inadequate operational safety oversight had been a contributing factor in several accidents involving Part 135 operators, the NTSB recommended in 2016 that the FAA require all Part 135 operators establish an SMS (NTSB 2016). This recommendation is applicable to Part 135 operators such as Liberty and NYONair's East West Helicopters. However, this requirement (if implemented by the FAA) would not apply to commercial air tour operators that operate under Part 91.

Information that the FAA disseminated on the framework of SMS notes that, by design, SMS is scalable to allow the integration of safety management practices into any size operator's unique business model (FAA 2015). The NTSB believes that the safety of commercial air tour operations, regardless of their operating rule, would be enhanced if all air tour operators established an SMS. Therefore, the NTSB recommended that the FAA require all commercial air tour operators, regardless of their operating rule, to implement an SMS.

7.1.1.3 Safety Culture Summary

From the NTSB accident description, a lack of safety culture played a key role in this accident. The facts that Liberty Helicopters pulled out of TOPS and a SMS was not implemented were key factors. Liberty Helicopters director of safety left the company, and the position was not refilled at the time of the accident. In the accident description there was much discussion about who was responsible for safety at the time of the accident. In addition, a risk assessment of the "yellow" harness was not performed. There were two factors associated with the harness that contributed to the accident. One was the fact the tether of the "yellow" harness worn person in the front set seat hooked under the fuel shutoff valve and when the passenger moved forward a harness loop shutoff the fuel to the engine. Second, the harness was not designed to allow easy egress in the case of an accident. The NTSB report stated that the crash was survivable, if the passengers could get out of the harness. A properly designed harness would have mitigated the situation and saved lives. The passenger training on the harness system was also not thorough. The "yellow" harness would be complicated for a novice to get out of. That type of harness is designed to be difficult to get out of because it is designed for fall protection and someone using it for fall protection does not want to slip easily out of it. Having been certified as a Competent Fall Protection person I can attest to the fact that fall protection harnesses are difficult to don and to doff. I can't imagine the terror of the passengers as they tried to get out of the harness. I feel they could not even try to comprehend out to get out of the device in an emergency, let alone find the cutting tool and cut through the harness. I have seen trained, experienced divers panic underwater when their gear didn't work correctly. I had to buddy breath with a friend when his regulator failed. I thought he was going to attack me and take my regulator. So, how would an untrained person without dive gear respond to this sort of underwater situation?

In summary, lack of safety culture on many levels led to this fatal accident. Other helicopter operators should note the failures in this report and improve their operations.

7.2 Commercial Aviation

7.2.1 Successful Landing of Crippled Commercial Airliners

Air Canada Flight 174, nicknamed "Gimli Glider" and US Airways Flight 1549, nicknamed "Miracle on the Hudson" are quite good examples of how catastrophe can be avoided with good training and crew coordination. Each accident had different causes, Flight 174 was initiated by human error and Flight 1549 was initiated by bird strikes. However, each was able to land, though not gracefully, without loss of human life. The following provides in depth examination of these events.

7.2.2 Gimli Glider – Successful Landing of a Crippled Commercial Airliner 1 – July 23, 1983

In the opening chapter reference was made to this accident. It is one of two incidents related to safety culture with a very positive outcome. The following is excerpted from the Final Inquiry Board of Canada report (Government of Canada 1983).

The accident happened on July 23, 1983, when the aircraft was being flown by Captain Pearson in command, assisted by First Officer Quintal. However, to understand the sequence of events that culminated in the emergency landing at Gimli, it is necessary to go back to the previous day when the aircraft was in Edmonton. There were 69 people on board, 61 passengers and 8 crew. The 767-200 aircraft was only four months old, and the crew was not used to a fly-by-wire system. Also, Canada was transitioning from the imperial measurement system to metric (Morely 2018). This accident was made into a movie entitled "Falling from the Sky: Flight 174."

7.2.2.1 Accident Information

During a routine service check, the three fuel quantity indicators, or fuel gauges, situated on an overhead panel between the two pilots, were found to be blank. One gauge is for the center auxiliary tank, one for the left main tank and one for the right main tank. The tanks are in the wings. The center tank has not so far normally been used for flights within Canada. It will, however, necessarily be used when overseas flights begin. These gauges are operated by a digital fuel gauge processor which has two channels. Either one of the channels is normally sufficient to ensure satisfactory operation of the processor to provide fuel indication on the gauges in the cockpit. The processor is located underneath the floor of the aircraft, immediately behind the cockpit. Circuit breakers, one for each channel, are located on a panel on the ceiling of the cockpit, above and slightly to the rear of the pilots' seats.

Mr. Conrad Yaremko, a Certified Aircraft Technician, Category 1 (CAT-1), who had experienced this problem of blank fuel gauges before on the same aircraft, found that he could obtain fuel indication by pulling and deactivating the channel 2 circuit breaker. This rendered channel 2 of the processor inoperative and channel 1 took over. Mr. Yaremko tagged or collared the circuit breaker with a piece of yellow tape which was marked "inoperative." He also stuck another piece of yellow tape above the fuel indicators. This was marked "see logbook." Mr. Yaremko made a note in the journey logbook of the problem with channel 2. Such problems with equipment are commonly referred to as issues. In Edmonton, Mr. Yaremko deferred the issue and entered reference to it in the computerized system which informs online stations of deferred issues. Such an issue is called a deviation because it affects the airworthiness of the aircraft. With this kind of problem an aircraft can only be dispatched after compliance with the conditions of the Minimum Equipment List (MEL). The computerized system referred to is the Deferred File Display (DFD). Mr. Yaremko then dispatched the aircraft after complying with the qualifying conditions of MEL item 28-41-2. Under this item of the MEL, because one of the processor channels was inoperative, the fuel load had to be confirmed and was confirmed using the fuel measuring sticks located under the wings of the aircraft. This procedure is usually called "doing a fuel drip" or "doing a drip check." The MEL is a document developed by Air Canada from the Master Minimum Equipment List (MMEL) issued by the FAA in Washington, USA, and approved by Transport Canada in Ottawa. It lists those circumstances and conditions under which an aircraft may safely be dispatched, even though some of its equipment is inoperative. This is possible because most items of equipment vital to the operation of an aircraft in flight is protected by a system of redundancy. This means that such systems are duplicated so that, if one fails, the other can be used provided that certain stipulated conditions set out in the MEL are met to ensure the safe flight of the aircraft.

The MEL is part of the Boeing 767 Aircraft Operating Manual carried on board the aircraft. It is found in Chapter 1 under the heading "Limitations." Before dispatching the flight, Mr. Yaremko not only made a note in the logbook of the issue with channel 2 of the processor, he also discussed it with Captain John Weir. The latter, as captain of the aircraft, had the final decision as to taking or not taking the aircraft. Captain Weir satisfied himself that it was legal to operate the aircraft under the provisions of the MEL. He did, however, get the impression from his discussion with Mr. Yaremko that the aircraft had arrived in Edmonton from Toronto with the same issue. Mr. Yaremko had, in fact, been referring to a similar problem on a flight from Toronto to Edmonton experienced on July 5, 1983, rather than on July 22. This misunderstanding is important in the light of a conversation that Captain Weir had with Captain Pearson when the aircraft arrived in Montreal on July 23.

Captain Weir and his first officer, Captain Donald Earl Johnson, left Edmonton on the morning of July 23 and flew the aircraft to Montreal via Ottawa. The flight was uneventful. All three fuel gauges operated normally. Upon landing in Montreal, Captain Weir and Captain Johnson met Captain Pearson in the vicinity of the parking lot and discussed briefly the problem relating to the fuel system.

Captain Weir's recollection of the conversation is that they discussed in general terms the fact that there was a problem with the fuel system and that enough fuel should be boarded to go right through to Edmonton. Captain Pearson's recollection of the conversation is that Captain Weir had informed him that the fuel gauges were inoperative, that a fuel drip had to be done to ascertain the amount of fuel on the aircraft and that the aircraft had been operating in that fashion from Toronto to Edmonton and from Edmonton to Ottawa and to Montreal.

Captain Pearson, therefore, received the impression that the issue had been outstanding since the aircraft left Toronto the previous day and that the aircraft had been flown throughout this period with inoperative fuel gauges.

Before the new flight crew arrived on board, Mr. Ouellet entered the cockpit. He was a Certified Avionics Technician, Category 38 (CAT-38), who had been assigned to the aircraft in order to perform the drip check required to satisfy the requirements of MEL item 28-41-2. He noted the entry made in the logbook by Mr. Yaremko. He also noticed the circuit breaker which had been pulled and tagged. He was confused by the entry in the logbook which did not appear to coincide with what he had been taught about the processor in recent training. Because of his confusion, he tried to get to the bottom of the problem by doing what is known as a BITE test on the processor.

BITE stands for Built-in Test Equipment. It refers to the fact that the processor had been designed and built to be able to identify faults within its own system. Before doing the test, he reset the number 2 channel circuit breaker. This caused the fuel gauges in the cockpit to go blank.

Mr. Ouellet was not satisfied with the test and decided that the processor had to be replaced. When he tried to order a new one, he was told that none were available in Montreal, but that one had been ordered to be available that night in Edmonton.

On returning to the flight deck, Mr. Ouellet was distracted by the arrival of the fuel personnel and forgot to pull the number 2 circuit breaker, to deactivate it as Mr. Yaremko had done. Thus, when Captain Pearson arrived on board and saw the blank fuel gauges in the cockpit, this circumstance reinforced his misunderstanding of the conversation with Captain Weir.

When Captain Pearson noticed the collared circuit breaker, he assumed that it had been deactivated and, further, that it was the circuit breaker for the fuel gauges in the cockpit, rather than for the processor which operates them. Both of his assumptions were incorrect. The circuit breaker had not been deactivated, nor was it, strictly speaking, the circuit breaker for the gauges.

However, his assumptions were certainly consistent with his misunderstanding of the conversation with Captain Weir. As a result of the action taken by Mr. Ouellet, the fuel gauges were blank.

When Captain Pearson entered the cockpit, he expected the fuel gauges to be blank. Similarly, the logbook entry made by Mr. Yaremko further confirmed his false assumption about the fuel gauges.

Captain Pearson then consulted the MEL, where he read the provisions of section 28 dealing with the fuel system. Item 28-41-1 clearly indicates that to permit the dispatch of an aircraft, at least two of the three fuel gauges must be working. Item 28-41-2 refers to fuel processor channels. Of the two channels, one is required for dispatch.

Captain Pearson knew that the aircraft was not legal to go with blank fuel gauges. He testified that he had raised the question of legality with one of the attending technicians who assured him that the aircraft was legal to go and that a higher authority, Maintenance Central, now renamed Maintenance Control had authorized the operation of the aircraft in that condition. No such authorization had in fact been given. There is even some question as to whether this conversation took place.

In any event, because of the mistaken assumption already in his mind, Captain Pearson formed the opinion that he could safely take and fly the aircraft, provided the fuel quantity on board the aircraft was confirmed by use of the fuel quantity measuring sticks in the fuel tanks. These are commonly called the drip sticks.

Before dealing with the drip procedure, it should be noted that for some years now all aircraft in Canada have been fueled in liters. That is to say that fuelers deliver fuel in liters and charge for the fuel by the liter. On the other hand, those who calculate the load of the aircraft and those who fly the aircraft do not work in liters, which is a measurement of volume, but rather in a weight measurement.

Prior to the introduction of the Boeing 767 type of aircraft into the Air Canada fleet – 12 had been ordered and 4 delivered at the time in question – weight calculations were made in pounds, an Imperial measurement. When the new aircraft were ordered, a decision was taken, in line with Canadian government policy, to order them with their fuel gauges reading in kilograms, a metric measurement. Similarly, calculations of the take-off weight of the new type of aircraft were to be made in kilograms.

Fuel quantity measuring sticks, or drip sticks, are used to measure the amount of fuel in an Aircraft (Figure 7.9). They are only used in an abnormal situation when the aircraft is being dispatched under the qualifying conditions set out in the section of the MEL relating to fuel. A drip stick is pulled from underneath the aircraft and is so designed that the fuel will not drip out when the stick is pulled. Hence the term drip stick. The stick is also calibrated. Different calibrations are used by different airlines and in Air Canada they vary from one aircraft type to another. Suffice it to say at this point that Air Canada ordered drip sticks for the 767 calibrated in centimeters.

Before leaving Edmonton, Captain Weir and the ground crew complied with MEL item 28-41-2 which at that time stipulated, in part, that one fuel tank quantity processor channel could be inoperative, provided fuel loading was confirmed by use of a fuel measuring stick. In Montreal, before departure, Captain Pearson, First Officer Quintal, and members of the ground crew, all attempted a similar confirmation of the amount of fuel on board the aircraft. The attempt was made even though departure was not permitted by MEL item 28-41-1. This stipulated,

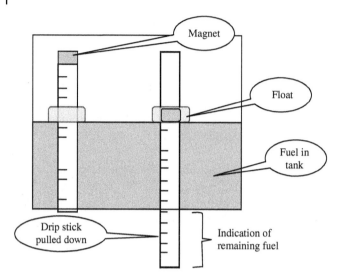

Figure 7.9 Drip stick rendition. Source: Ostrom (2021).

in part, that one left- or right-wing tank fuel gauge might be inoperative, provided fuel quantity in the associated tank was determined by measuring stick. Captain Pearson, who had the final authority on whether to take and fly the aircraft, left Montreal and later left Ottawa with blank fuel gauges, contrary to the provisions of the MEL.

Critical to the determination of the correct fuel quantity by the drip stick method is the conversion from centimeters to liters and from liters to kilograms. The first part is easy because drip tables are provided and kept on board the aircraft. The drip tables, as they existed at the time of the Gimli accident provided a simple means of converting centimeters to liters. On the other hand, converting liters to kilograms involves using a conversion factor. At the time, the conversion factor was called specific gravity. The term, as used to describe the conversion factor of 1.77 lb/l, is a misnomer. The term, however, has been used in the aircraft industry throughout the world for a long time. Unfortunately, the conversion factor, or specific gravity as it was mistakenly called, supplied to those making the calculations in Montreal and Ottawa was 1.77. This is the figure used to convert liters to pounds. The conversion factor to convert liters to kilograms is typically around 0.8.

There are slight variations in any such conversion factor depending on the temperature. There was disagreement among counsel as to whose responsibility it was to make the calculations. Counsel for CALPA submitted that the job had been assigned, according to the provisions of the MEL, to maintenance personnel. Counsel for Air Canada took the position that there was a gap in the system and that the task had not been assigned to anyone. Whoever was supposed to do the calculations, it is clear from the evidence that both in Montreal and Ottawa there was a joint effort on the part of the flight crew and the maintenance personnel to determine the fuel load. On each occasion all concerned used the incorrect conversion factor. As a result, the aircraft left each airport in turn with only half the amount of fuel the flight crew thought they had. Using the conversion factor of 1.77 resulted in a conversion from liters to pounds. To arrive at the correct weight in kilograms, the resulting figure would have had to be divided by 2.2. This was not done. Normally, the aircraft would have been fueled in Montreal and again in Ottawa. However, because of the problem with the gauges, Captain Pearson decided to load enough fuel to go right through to Edmonton with a drip check to be made both in Montreal and in Ottawa. Unfortunately, the same mistake in calculations and conversion was made on both occasions. As a result, the aircraft ran out of fuel in flight.

The first signs of trouble appeared shortly after 8:00 p.m. Central Daylight Time when instruments in the cockpit warned of low fuel pressure in the left fuel pump. The Captain at once decided to divert the flight to Winnipeg, then 120 mi away, and commenced a descent from 41,000 ft. Within seconds, warning lights appeared indicating loss of pressure in the right main fuel tank. Within minutes, the left engine failed, followed by failure of the right engine. The aircraft was then at 35,000 ft, 65 mi from Winnipeg and 45 mi from Gimli. Without power to generate electricity all the electronic gauges in the cockpit became blank, leaving only stand-by instruments, consisting of a magnetic compass, an artificial horizon, an airspeed indicator, and an altimeter. It soon became evident that they would not be able to make it to Winnipeg. Thus, at 8:32 p.m. Central Daylight Time, in consultation with Air Traffic Control, Captain Pearson redirected the aircraft toward Gimli, some 12 mi away on the western shore of Lake Winnipeg. By this time, the flight attendants in the cabin had instructed the passengers in emergency procedures.

Fortunately for all concerned, one of Captain Pearson's skills is gliding. He proved his skill as a glider pilot by using gliding techniques to fly the large aircraft to a safe landing. Without power, the aircraft had no flaps or slats with which to control the rate and speed of descent. There was only one chance at a landing. By the time the aircraft reached the beginning of the runway, it had to be flying low enough and slowly enough to land within the length of the 7200-ft runway.

As they approached Gimli, Captain Pearson and First Officer Quintal discussed the possibility of executing a sideslip to lose height and speed in order to land close to the beginning of the runway. This Captain did on the final approach and touched down within 800 ft of the threshold.

During the descent, First Officer Quintal had tried, without success, manually to lower and lock the nose wheel. As it turned out, his failure to do so helped to slow down the aircraft when it was on the ground, because of the friction caused by contact of the bottom of the nose with the concrete runway. This averted disaster to people at the far end of the runway.

The airfield at Gimli is a disused military base. The far end of the runway from where the aircraft touched down has been adapted for use as a drag racing strip. Just beyond the strip used for racing, drag racing drivers and their families were staying for the weekend in tents and caravans.

Fortunately, the aircraft came to a stop before it reached them. The passengers and crew were safely evacuated. Both fuel tanks were found to be dry.

7.2.2.2 Analysis of the Fuel Problem

The responsibility for the miscalculation of the fuel load on Flight 143 on July 23, 1983, must be borne by both the flight crew and the maintenance personnel involved, particularly those in Montreal.

The issue of corporate responsibility for failing to assign the duty to compute the fuel load in an abnormal fueling situation is a separate issue which is dealt with the following text.

However, no matter whose responsibility it was, both the flight crew and maintenance personnel participated in computing the fuel incorrectly. The evidence indicates that there was a joint effort in Montreal to determine the number of kilograms of fuel on board and the number of kilograms of fuel to be boarded.

There is no evidence to suggest that anyone either shirked or attempted to shirk his responsibility in computing the fuel load on the day in question. Rather, all the individuals involved attempted to assist each other either by providing figures for the calculations, doing the calculations, or checking the calculations.

The discussions about whose responsibility it was, or whose fault it was, arose after the accident with the various individuals and departments involved within Air Canada taking different positions.

This was evident from the testimony of personnel from both Maintenance and Flight Operations.

The calculation of the fuel load by way of the drip procedure is a three-step process:

(1) the drip or measurement of the fuel depth in the tanks in centimeters by use of the measuring or drip sticks;
(2) conversion of the drip stick readings in centimeters to the volume in liters; and
(3) conversion of the volume in liters to the weight in kilograms.

There were three drips done in Montreal. It is important to note that most of the calculations of the fuel load in Montreal were made on the first drip. The first drip was necessary to determine how much fuel was on board the aircraft prior to fueling. This had to be determined in kilograms to compute how much had to be added.

To determine the amount of fuel that had to be put on the aircraft, it was necessary to subtract the amount that was already on board from the total required to fly to Edmonton according to the flight plan. The resulting figure should have been in kilograms and would then have to be converted to liters so that the fueler could be told how many liters to pump.

There is no dispute, and it is quite clear from the evidence, that maintenance personnel, both in Montreal and Ottawa, performed the first step of the drip procedure by pulling the drip sticks and recording the readings in centimeters.

The evidence indicates that the next two steps in the calculation of the fuel by way of this drip procedure, that is, conversion of the readings to liters and conversion of the liters to kilograms, was a joint effort of maintenance personnel and flight crew in Montreal. This was the case, to some extent, in Ottawa as well.

The evidence was at times conflicting and confusing. It does not clearly establish who did what in terms of calculating the fuel load. However, there are common threads in relation to the calculation of the fuel load. One such thread is that each of the principal maintenance personnel in Montreal, Messrs, Ouellet and Bourbeau, and each of the flight crew, Captain Pearson and First Officer Quintal, participated to some degree in calculating the fuel load.

Another common thread indicated by the evidence is that no one involved in making the calculations in Montreal seemed to know how to convert liters to kilograms. Neither the maintenance personnel nor the flight crew had had any training or experience in converting drip stick readings in centimeters to liters and then liters to kilograms.

The following excerpts from the evidence demonstrate this lack of knowledge, training, or experience in converting the drip measurement to liters or the liters to kilograms:

(i) Mr. Rodrigue Bourbeau, a Certified Aircraft Technician, Category l (CAT-I), testified that after Mr. Ouellet gave the drip reading to First Officer Quintal:

> Mr. Quintal says that he never done it before, and we says we haven't done it before either, and we agreed that we are going to do the - our best to convert it, or - so I grabbed the blue book from the aircraft library, there, and I went from the pitch and axis and I end up at 64 and 52.

(ii) First Officer Quintal testified that maintenance personnel did not understand what figures they arrived at when they consulted the drip manual and he pointed out to them that it was liters.

(iii) Mr. Bourbeau testified that he started off himself to make some calculations, but he was not too sure about what he was doing, and he was not going so fast and he therefore gave up on doing the calculations himself.

(iv) First Officer Quintal testified that he believed that by multiplying the number of liters by the specific gravity of 1.77 he would have arrived at kilograms. If First Officer Quintal had been given the correct specific gravity, his calculation for arriving at kilograms may well have been accurate. However, he was given the wrong specific gravity for a conversion directly to kilograms. The correct specific gravity in the metric system for the weight of a liter of fuel is approximately 0.8.

(v) First Officer Quintal testified that he had no training in the use of the drip manual or in fueling procedures.

(vi) Mr. Bourbeau quite frankly pointed out that he had the feeling that if he had carried on with the multiplication he would have "made maybe the same mistake – not dividing by 2.2."

(vii) Mr. Ouellet testified that he started to do some calculations but never finished them because all the figures got "so crowded" that he ran out of paper. He said that he started to "look up" his multiplication and figured that it was not right. He said that he had an idea that the resulting figure would have to be divided by roughly 50% to get kilograms, but he did not have the exact figure. He decided to quit and leave Mr. Bourbeau and First Officer Quintal to do the calculations.

(viii) Captain David Walker, Chief Flying Instructor for the Boeing 767 at Air Canada, testified that no training was provided to pilots with respect to fuel conversion.

(ix) Mr. Roger Morawski, Senior Director of Aircraft Maintenance at Air Canada, with reference to the procedure for converting the drip stick readings to weight, testified:

> I can state it positively that the maintenance people have never performed that function, nor were they trained to perform that function, because always, in Air Canada, ever since I can remember, that function was not assigned to maintenance to perform.

The following excerpts from the evidence indicate that:

(1) the calculation of the fuel load that was on the aircraft prior to fueling done after the first drip;

(2) the calculation as to how much fuel had to be added; and,

(3) the calculation or confirmation of the fuel load after the second drip, were all a joint effort on the part of Captain Pearson, First Officer Quintal, Messrs. Bourbeau and Ouellet:

    (i) Mr. Bourbeau testified that "We agreed that we are going to do the – our best to try to convert it."

(ii) Mr. Bourbeau testified that he himself did some calculations and in fact said that he "had multiplied by 1.77. I don't remember the exact – which figure that I had."

(iii) Mr. Bourbeau testified that First Officer Quintal was doing some calculations.

(iv) Mr. Bourbeau testified, as well, that First Officer Quintal asked the fueler what the specific gravity was and that he was given the wrong conversion factor of 1.

(v) Mr. Bourbeau testified that:

"The first thing that I knew of, they were leaving, the fuel man and Jean – were leaving the cockpit, and they seemed to know how much fuel they were going to board, so I left my calculations, and that's it."

(vi) First Officer Quintal confirmed Mr. Bourbeau's evidence that he, Quintal, made some calculations and requested the specific gravity from the fueler. First Officer Quintal testified: "Oui, j'ai fait la multiplication."

(vii) First Officer Quintal also testified that he offered to help maintenance personnel in their calculations when he saw them having some difficulty with the drip manual when he entered the cockpit.

(viii) Mr. Ouellet indicated that "we got the blue book out" referring to himself and Mr. Bourbeau and went on to testify that "we all worked through it" referring to himself, Mr. Bourbeau and First Officer Quintal.

(ix) Mr. Ouellet also testified that while Mr. Bourbeau and First Officer Quintal were doing some calculations he too started to do some calculations. He did these calculations on his own because he "figured it was a bit crowded to be three guys in the same book, all gathered together on the same page."

(x) Mr. Ouellet went on to describe that he added up the number of liters in the two wing tanks and multiplied by 1.77 on a sheet of paper that he had in his pocket. That piece of paper was subsequently lost. He also pointed out that he did not finish his calculation because all the figures got "so crowded" that he ran out of paper.

(xi) Captain Pearson testified that he participated in the calculation of the fuel load. He testified that "I confirmed the figures from the second drip."

(xii) Captain Pearson testified that he requested the maintenance personnel to show him their figures and that one of the maintenance personnel on board held a sheet of paper between himself and the first officer. Captain Pearson said that he took out his calculator and went over the computations which had been done longhand.

(xiii) Captain Pearson again confirmed that he checked these calculations and in his own words

"I went over these calculations with my Jeppesen computer and found them to be mathematically accurate and I stated that. I said, 'Well, your calculations to me look mathematically correct'."

The evidence referred to earlier indicates that all the individuals in Montreal, both maintenance personnel and flight crew, participated to some degree in determining the fuel load. It also indicates that they really did not know what they were doing in terms of the fuel load calculations and had not been trained for this function. At times, during their testimony, they were uncertain as to what they did. They were in fact uncertain as to what was required to be done.

These excerpts from the evidence also indicate that the wrong specific gravity figure as provided although it may have been what Air Canada requested the fueler to provide.

If the flight crew had consulted their 500 Flight Operations Manual (FOM), they would have found in Chapter 10 at page 20.1 a formula for arriving at kilograms.

Specifically, paragraph 10 of that chapter provides:

"Weight/Volume conversion

$$\text{Pounds loaded} = \text{Imp.gal} \times \text{Spec.wt.}$$
$$= \text{Imp.gal} \times \text{S.G.} \times 10$$
$$= \text{Liters} \times \text{Sp.wt.(1b liter)}$$
$$= \text{Liters} \times \text{S.G.} \times 2.2$$
$$= \text{US gal} \times \text{Sp.wt.(lb.U.S.Gal.)}$$
$$= \text{US gal} \times \text{S.G.} \times 8.33$$

Kilograms loaded = pounds loaded divided by 2.2."

The specific weight, also known as the unit weight, is the weight per unit volume of a material. Jet fuel is approximately 8.35 lb per imperial gallon. Specific gravity (S.G.) of jet fuel is approximately 0.82. A US gallon of jet fuel weighs approximately 6.7 lb/gal. A liter of jet fuel weighs approximately 1.76 lb.

In fairness to the flight crew, however, they would also have found a different specific gravity figure than the one with which they had been provided by the fueler. They would have found in the manual the figure 0.80 for Jet A fuel. It is disconcerting to note that the specific gravity used and accepted by Air Canada in their day-to-day online operations was different than that set out in the manuals.

The crucial miscalculation of the fuel load on Flight 143 came in Montreal, not Ottawa. In Montreal, the fueler was requested by the flight crew to board 50001, one quarter of the required amount. It is interesting to note how this figure was probably arrived at.

In Montreal the drip stick readings after the first drip were 64 and 62 cm. According to the handwritten calculations made by Mr. Ouellet in Exhibit 17, these readings translated to 3924 and 3758 l of fuel in the two tanks, for a total of 7682 l. Multiplying 7682 by 1.77, the figure supplied as the specific gravity one arrives at 13,597. This represented the fuel on board before fueling. It was in pounds, but everyone involved believed it to be kilograms. This figure, if subtracted from the minimum fuel required by the flight plan, 22,300 kg, would produce the figure of 8703. Again it must have been assumed that the latter figure represented the number of kilograms of fuel that had to be added although it was pounds. Since fuel is loaded in liters, the 8703-figure had to be converted to liters. This was probably done by dividing by 1.77. If that division is done, one arrives at the figure of 4916 representing the number of liters to be added. This figure coincides with the evidence of the fueler, Mr. Tony Schmidt, who testified that the 4900 l was rounded to 5000 l at his suggestion.

An illustration of the foregoing calculation is as follows:

STEP I: Computation of the fuel on board:
  Drip stick readings: 62 and 64 cm
  Converted to liters: 3758 and 3924 l
  Total liters on board: 3758 + 3924 = 7682 l.
STEP 2: Conversion of liters on board into kilograms:
  7682 l × 1.77 = 13,597
  Multiplying by 1.77 gave pounds, but everyone involved thought they were kilograms.
STEP 3: Computation of the fuel to be added:
  Minimum fuel required, 22,300 kg – fuel on board, 13,597 assumed to be kilograms = 8703 kg.
STEP 4: Conversion into liters of kilograms to be added:
  8703 kg ÷ 1.77 = 4916 assumed to be liters.

The correct calculation using the minimum required fuel as a base, that is, 22,300 kilograms, as called for by the flight plan, would be as follows:

STEP 1: 3924 + 3758 l from the first drip stick readings of 64 and 62 cm = 7682 l of fuel on board.
STEP 2: 7682 × 1.77 ÷ 2.2 = 6180 kg of fuel on board, prior to fueling.
STEP 3: 22,300 – 6180 = 16,120 kg of fuel to be boarded.
STEP 4: 16,120 ÷ 1.77 × 2.2 = 20,036 l to be boarded.

One can see that if the correct calculation had been made, the amount of fuel that should have been boarded was 20,000 l. This is four times what in fact was added. It is based on the minimum fuel required of 22,300 kg, and not the actual fuel required of 22,600 kg. The latter figure, also set out in the flight plan, includes an estimated 300 kg for taxiing. If the latter figure had been used, then more than 20,000 l should have been boarded.

In any event, the method of calculation set up by Air Canada was a convoluted way of arriving at the number of liters to be boarded and was unduly complicated for a busy flight crew or for equally busy maintenance personnel at departure time.

It must be remembered, however, that it was never intended that this calculation be made because, with inoperative fuel gauges, the aircraft should have been grounded.

As indicated earlier, the crucial miscalculation of the fuel load came in Montreal not Ottawa. The drip procedure in Ottawa was requested as a check. The same error was made in Ottawa, that is, a conversion to pounds that was believed to be kilograms. Since the calculations produced approximately the same figures as in Montreal, allowing for the fuel burned, and since the figures were what the flight crew expected to see, there was no alarm, and the calculations were accepted.

In Ottawa, as in Montreal, there was a joint effort between maintenance personnel and flight crew to determine the fuel load. Maintenance personnel pulled the drip sticks, took the readings and converted them to liters. The figures in liters were then given to Captain Pearson who checked them with his Jeppesen computer after he requested the specific gravity. With respect to what happened in Ottawa concerning the calculation of the fuel load, the following evidence is relevant:

(i) First Officer Quintal, although he did not participate in the calculations in Ottawa, said that Captain Pearson got the number of liters and the specific gravity and multiplied the figures with his Jeppesen computer and then compared them to what maintenance had on a sheet of paper 40.

(ii) Mr. Richard Simpson, a Certified Aircraft Technician, Category 1 (CAT-1), in Ottawa, said that he gave the Captain the total number of liters of fuel that was in both tanks, 11,430. The Captain asked for the conversion factor and received the figure 1.78 from the fueler. He testified that the Captain then took out his circular slide rule and calculated that he had over 20,000 "units of fuel."

(iii) Mr. Simpson did not know how to arrive at kilograms from liters.

(iv) Mr. Robert Eklund, a Certified Aircraft Technician, Category 1 (CAT-1), in Ottawa, testified that he got the drip readings in centimeters and the liter readings for each of the two wing tanks from Mr. Simpson. He said that Mr. Simpson gave the total number of liters to the Captain who took out his Jeppesen computer, asked for the specific gravity of the day, and was given 1.78. The Captain then did some calculations and came up with a figure more than 20,000. The Captain did not say what the figure represented.

Thus, in Ottawa, a similar miscalculation of the fuel load was made. In both Montreal and Ottawa, maintenance personnel and flight crew alike displayed very little knowledge of how to do the calculations. They obviously lacked both training and experience in this respect.

It is instructive to examine what was done in Edmonton before the aircraft left on July 23, 1983 under the command of Captain Weir. Because only one channel of the fuel processor was working properly, and notwithstanding the fact that the actions of Mr. Yaremko had restored fuel indication in the cockpit, the aircraft had to be dispatched under the provisions of MEL item 28-41-2. This is because of the qualifying conditions of the MEL based on the principle of redundancy. Thus, in the case of the fuel processor, an aircraft can be dispatched with only one of its two fuel processor channels working, provided the conditions of MEL item 28-41-2 are complied with.

MEL item 28-41-2 refers to Fuel Tank Quantity Processor Channels. These are the channels of the digital fuel gauge processor. The qualifying conditions relating to that item read:

> (M) One may be inoperative provided fuel loading is confirmed by use of a fuel measuring stick, or by tender uplift, after each refueling and FMC fuel quantity information is available.

These qualifying conditions relating to the fuel processor contemplated that, if one of the two channels were not working, the other would take over and provide normal fuel indications on the gauges in the cockpit.

In compliance with the aforementioned conditions, it was decided to fuel the aircraft and confirm the amount of fuel on board before its departure by doing a drip that is by using the fuel measuring sticks in the wings. Mr. Paul Wood, the fueler, went into the cockpit with Mr. Yaremko to take the readings from the cockpit gauges of the remaining fuel on board in kilograms. He had to do this because, with the fueling panel on the wing open, the gauges at that station were blank. Fueling, therefore, could not be done automatically from the wing position where the hose from the fuel tanker was connected to the nozzle that enables fuel to flow into the wing tanks.

It was not the duty of Mr. Wood to do the conversion from kilograms to liters to know how many liters to board for the flight, but he made the calculations because the captain, first officer and aircraft mechanic were busy.

The fueler had to know the amount of fuel on board the aircraft to calculate how many additional liters to board. He knew how to convert from kilograms to liters.

Asked why he knew how to make such calculations when others seemed to have such difficulty, he said that it was a regular practice of his, so that he would know how many liters he had in his truck and could call dispatch if he was likely to run short. He used a calculator to do the conversions.

When he had completed the fueling, he went back into the aircraft and gave the fuel load to the Captain in liters.

To comply with the provisions of MEL item 28-41-2, a drip was done by Mr. Yaremko assisted by a mechanic, Mr. Robert Serizawa. Mr. Serizawa took down the drip stick readings and gave them to Captain Weir. Captain Weir and his first officer, Captain Johnson, had time to spend about a minute and a half before departure to check the amount of fuel on board, using the measurements from the drip sticks, the drip stick tables and the conversion factor. The drip stick tables are more properly called the fuel measuring stick tables and are commonly referred to as the drip manual or the blue book. Captain Weir knew from previous experience that the correct conversion factor, or specific gravity as it was called at that time, was around 0.8 to convert liters to kilograms.

The fueling referred to earlier was done on the morning of July 23, 1983, at the Edmonton International Airport. It is true that the fuel gauges in the cockpit were working. However, because of the faulty processor, fuel calculations had to be made manually and they were made correctly.

The fueler, Mr. Wood, clearly knew how to make the correct conversion, and so did Captain Weir and his first officer, Captain Johnson.

7.3 Illegal Dispatch Contrary to the MEL: Taking Off With Blank Fuel Gauges

The fact that Flight 143 took off from Montreal on July 23, 1983, with blank fuel gauges was a significant cause of the accident. It was an illegal dispatch contrary to the provisions of the MEL.

As noted in Part II of this report, the MEL is that part of the Operating Manual that governs the legality of the dispatch of an aircraft with inoperative equipment. It details which equipment, under what conditions, may be unserviceable and the aircraft still be fit to be dispatched.

Air Canada's MEL, as well as the MEL for every major airline which operates the 767, has as its base the MMEL prepared by the FAA in the United States. The preamble to the MMEL states the purpose of the MMEL to be:

> ...to provide owners/operators with the authority to operate an aircraft with certain items or components inoperative provided the Administrator finds an acceptable level of safety maintained by appropriate operations limitations, by a transfer of the function to another operating component, or by reference to other instruments or components providing the required information.

Neither Air Canada's MEL nor the MMEL include obviously required items such as wings, rudders, flaps, engines, or landing gear.

Item 28-41-1 of Air Canada's MEL, relating to the Fuel Tank Quantity Indication Systems (Flight Deck) at the time of the accident provided:

"One left- or right-wing tank fuel gauge may be inoperative provided:

(a) Fuel quantity in associated tank is determined by measuring stick, or by tender uplift, after each refueling, and
(b) All pumps for associated tank must be operative, and
(c) FMC fuel quantity information is available 54."

Item 28-41-2 of Air Canada's MEL relating to Fuel Quantity Processor Channels provided:

> One may be inoperative provided fuel loading is confirmed by use of a fuel measuring stick, or by tender uplift, after each refueling and FMC fuel quantity information is available.

It is clear that at least two of the three fuel tank gauges must be working before a passenger aircraft can legally be dispatched. If there is more than one fuel tank gauge inoperative, then the aircraft is required to be grounded.

The dispatch of Flight 143 on July 23, 1983, from Montreal was therefore illegal because none of the fuel tank gauges were working. It was contrary to the MEL. Captain Pearson himself acknowledged, after the fact that he had not been legal to operate. He testified that he addressed the topic of the MEL by saying "to the maintenance fellow": "We are not legal to operate in this configuration," pointing to the indicators, "with all of the fuel quantity indicators unserviceable."

He appears to have consulted the MEL in a very cursory way. He testified that he took out the MEL, read the two items that seemed to apply to the situation and then put the manual back.

Captain Pearson further testified:

> "Q. All right. I take it that it had to be a question because you were of the view that you were not, according to the MEL, legal to operate?
> A. That's right."

Captain Pearson, therefore, acknowledged in his testimony that, on the day in question, he was of the view that he was not legal to operate according to the MEL. Unfortunately, at the time, he came to the opposite conclusion in his own mind.

Questions arise as to why Captain Pearson, a professional pilot of exceptional ability, did not pay more attention to the requirements of the MEL, and why he did not check the legality of the dispatch with his superiors or with someone in Maintenance Central.

The same questions apply equally to the conduct of First Officer Quintal and of Messrs.

Bourbeau and Ouellet. These individuals seem to have paid little attention to the requirements of the MEL and to have done virtually nothing to ensure a legal dispatch.

For the answer to these questions, it is necessary to consider the circumstances under which Captain Pearson made his decision to accept the aircraft with inoperative fuel gauges and also to consider his resulting state of mind.

Objectively there was an illegal dispatch, which Captain Pearson had the authority to avoid.

Mr. Bourbeau also had the authority to avoid the dispatch. He too could have grounded the aircraft.

Captain Pearson, however, had the ultimate authority.

Subjectively, Captain Pearson formed the genuine and honest opinion that he was legal to operate. This opinion stemmed from several factors including, among others, the conversation with Captain Weir in the vicinity of the parking lot in Montreal, the logbook entries made both in Edmonton and Montreal, the belief that there was an overriding authority in Maintenance Central to circumvent the MEL, and, finally, because of his understanding that maintenance clearance for departure had been given.

As for Captain Pearson's state of mind, he gave the following evidence under questioning by Mr. Saul, Counsel for Air Canada:

> "Q. So in this case, on the 23rd of July, Captain, you formed an opinion with respect to the legality of the aircraft?
> A. Yes.
>
> Q. And that is what the MEL requires that you do, you form an opinion?
> A. Yes.
>
> Q. The only problem, as I see it, Captain, is that your opinion was not based on the MEL, but it was based upon a series of assumptions which you had accumulated through the afternoon, do you agree with me?
> A. Yes, that's true.
>
> Q. It was, was it not, a long trail of unfortunate incidences which inevitably led you to this conclusion, that the aircraft was legal?
> A. In hindsight, yes. Looking back.
>
> Q. And that was a conclusion which clearly was wrong, also in hindsight of course?

A. Well, absolutely everything made me believe that it was legal: reading the Jog book, maintenance clearance. Right from the time I arrived at the parking lot till we pushed back, every single thing that happened in relation to this problem only reinforced my belief that it was legal.

Of course, if I didn't believe that I would never have considered operating the aircraft... "

An examination of the various factors which prompted Captain Pearson to leave Montreal with inoperative fuel gauges lead to an understanding of his decision, but it does not, and cannot, justify his decision.

In conclusion, there is no doubt that, from an objective perspective, the dispatch from Montreal without fuel gauges was illegal, that is, contrary to the MEL. On the other hand, there is equally no doubt that Captain Pearson, subjectively, based on the factors and assumptions examined earlier, formed the opinion that it was legal to dispatch the aircraft. It is hard to understand, in principle, how such an experienced or, indeed, any qualified pilot could form such an opinion. It is, based on all the surrounding circumstances outlined in the evidence, to a certain extent understandable. It was not, nor could it be, justifiable in any circumstances.

7.4 Summary of Safety Culture Issues

This incident is, unfortunately, a great example of a positive outcome from a potentially tragic set of human actions. If the flight crew had not had the flying experiences, they had there could have been up to 69 deaths. There were so many mistakes made in this accident. The fact that the aircraft was released to fly, even though the fuel gauges were showing blank, was the major and initial error. That error was committed by maintenance personnel and the flight crew and flight operations. The lack of understanding on how to perform fuel calculations was the secondary error. If the plan had been grounded due to the blank fuel gauges, then that error would not have come into play. The misunderstanding of the MEL procedures was also a contributing error. A lack of thorough training on what constituted the need to ground an aircraft and how to properly calculate fuel using the drip sticks was a major contributor to the accident. Also, the lack of communications between the fuelers, flight crew, maintenance, and flight operations were a significant contributor to the accident. The imperial measurement to metric conversion also played into this accident, but it shouldn't have mattered if the aircraft were to be grounded due to the blank fuel gauges. The aircraft was taken out of service in 2008 and was later scrapped (Morley 2018).

7.5 Miracle on the Hudson River – Successful Landing of a Crippled Commercial Airliner 2, January 15, 2009

This accident, like the Gimli Glider, was also widely publicized and the story was made into a movie, "Sully" that was released in 2016. As compared with the Gimli Glider accident, this accident was not initiated by human error. Instead, it was initiated by a flock of birds. It too had a relatively successful ending. In that, there were no fatalities, although four passengers and one flight attendant sustained serious injuries. The following was excerpted from the NTSB report (NTSB 2016).

7.5.1 Accident Information

On January 15, 2009, about 1527 eastern standard time (EST), US Airways flight 1549, an Airbus Industrie A320-214, N106US, experienced an almost total loss of thrust in both engines after encountering a flock of birds and was subsequently ditched on the Hudson River about 8.5 mi from LaGuardia Airport (LGA), New York City, New York. The flight was en route to Charlotte Douglas International Airport (CLT), Charlotte, North Carolina, and had departed LGA about two minutes before the in-flight event occurred. The 150 passengers, including a lap-held child, and five crewmembers evacuated the airplane via the forward and overwing exits. One flight attendant and four passengers received serious injuries, and the airplane was substantially damaged. The scheduled, domestic passenger flight was operating under the provisions of 14 CFR Part 121 on an instrument flight rules flight plan. Visual meteorological conditions prevailed at the time of the accident.

The accident flight was the last flight of a four-day trip sequence for the flight and cabin crewmembers and the second flight of the day in the accident airplane. The flight crew flew from Pittsburgh International Airport (PIT), Pittsburgh, Pennsylvania, to CLT on a different airplane and then flew the accident airplane from CLT to LGA. The flight crew reported that the flight from CLT to LGA was uneventful.

According to the cockpit voice recorder (CVR) transcript, at 1524:54, the LGA air traffic control tower (ATCT) local controller cleared the flight for takeoff from runway 4. Currently, the first officer was the pilot flying (PF), and the captain was the pilot monitoring (PM). According to the accident flight crew and CVR and flight data recorder (FDR) data, the takeoff and initial portion of the climb were uneventful.

At 1525:45, the LGA ATCT local controller instructed the flight crew to contact the New York Terminal Radar Approach Control (TRACON) LGA departure controller. The captain contacted the departure controller at 1525:51, advising him that the airplane was at 700 ft and climbing to 5000 ft. The controller then instructed the flight to climb to and maintain 15,000 ft, and the captain acknowledged the instruction.

According to the CVR transcript, at 1527:10.4, the captain stated, "birds." One second later, the CVR recorded the sound of thumps and thuds followed by a shuddering sound. According to FDR data, the bird encounter occurred when the airplane was at an altitude of 2818 ft AGL and about 4.5 mi north-northwest of the approach end of runway 22 at LGA. At 1527:13, a sound like a decrease in engine noise or frequency began on the CVR recording. FDR data indicated that, immediately after the bird encounter, both engines' fan and core (N1 and N2, respectively) speeds started to decelerate.

At 1527:14, the first officer stated, "uh oh," followed by the captain stating, "we got one rol- both of 'em rolling back." At 1527:18, the cockpit area microphone (CAM) recorded the beginning of a rumbling sound. At 1527:19, the captain stated, "[engine] ignition, start," and, about two seconds later, "I'm starting the APU [auxiliary power unit]." 5 At 1527:23, the captain took over control of the airplane, stating, "my aircraft."

At 1527:28, the captain instructed the first officer to "get the QRH [quick reference handbook] loss of thrust on both engines." 6 At 1527:33, the captain reported the emergency situation to the LGA departure controller, stating, "mayday mayday mayday ... this is ... Cactus fifteen thirty nine hit birds, we've lost thrust in both engines, we're turning back toward LaGuardia." 7 The LGA departure controller acknowledged the captain's statement and then instructed him to turn left heading 220°.

At 1527:50, the first officer began conducting Part 1 of the QRH ENG DUAL FAILURE checklist (Engine Dual Failure checklist), stating, "if fuel remaining, engine mode selector, ignition," and the captain responded, "ignition." The first officer then stated, "thrust levers confirm idle," and the captain responded, "idle." About four seconds later, the first officer stated, "airspeed optimum relight, three hundred knots. we don't have that," and the captain responded, "we don't."

At 1528:05, the LGA departure controller asked the captain if he wanted to try to land on runway 13 at LGA if it was available, and the captain responded, "we're unable. we may end up in the Hudson [River]." The rumbling sound that the CVR started recording at 1527:18 ended at 1528:08. At 1528:14, the first officer stated, "emergency electrical power ... emergency generator not online." At 1528:19, the captain stated, "it's online." The first officer then stated, "ATC [air traffic control] notify." At 1528:25, the captain stated, "The left one's [engine] coming back up a little bit."

At 1528:31, the LGA departure controller stated that it was going to be "left traffic for runway three one," and the captain responded, "unable." At 1528:36, the traffic collision avoidance system (TCAS) on the airplane transmitted, "traffic traffic." 9 At 1528:46, the controller stated that runway 4 at LGA was available, and the captain responded, "I'm not sure we can make any runway. Uh what's over to our right anything in New Jersey maybe Teterboro?" The controller replied, "ok yeah, off your right side is Teterboro Airport [TEB]." Subsequently, the departure controller asked the captain if he wanted to try going to TEB, and the captain replied, "yes."

At 1528:45, while the captain was communicating with ATC, the first officer stated, "FAC [flight augmentation computer] one off, then on." Fifteen seconds later, the first officer stated, "no relight after thirty seconds, engine master one and two confirm." 11 At 1529:11, the captain announced on the public address (PA) system, "this is the Captain, brace for impact." At 1529:14.9, the CVR recorded the ground proximity warning system (GPWS) warning alert, "one thousand." 12 At 1529:16, the first officer stated, "engine master two, back on," and the captain responded, "back on."

At 1529:21, the CVR recorded the LGA departure controller instructing the captain to turn right 280° and stating that the airplane could land on runway 1 at TEB. At the same time, the CVR recorded the first officer asking the captain, "is that all the power you got? (wanna) number one? Or we got power on number one." In response to the controller, the captain stated, "we can't do it." In response to the first officer, the captain stated, "go ahead, try [relighting] number 1 [engine]." FDR data indicated that engine master switch 1 was moved to the OFF position at 1529:27. The departure controller then asked the captain which runway at TEB he would like, and the captain responded, "we're gonna be in the Hudson."

At 1529:36, the first officer stated, "I put it [the engine master switch] back on," and the captain replied, "ok put it back on … put it back on." At 1529:44, the first officer stated, "no relight," and the captain replied, "ok let's go put the flaps out, put the flaps out." At 1529:53, the LGA departure controller stated that he had lost radar contact with the airplane, but he continued trying to communicate with the captain, stating, "you also got Newark airport off your two o'clock in about seven miles." See Figure 7.10 for the flight track of the airplane.

At 1530:01, the first officer stated, "got flaps out," and, at 1530:03, stated, "two hundred fifty feet in the air." He then stated, "hundred and seventy knots … got no power on either one? Try the other one?" The captain responded, "try the other one." At 1530:16, the first officer stated, "hundred and fifty knots," and, at 1530:17, stated, "got flaps two, you want more?" The captain replied, "no, let's stay at two," and then asked the first officer, "got any ideas?" The first officer responded, "actually not."

At 1530:24, the GPWS issued a "terrain, terrain" warning followed by "pull up," which repeated to the end of the CVR recording. At 1530:38, the first officer then stated, "switch?"

The captain replied, "yes." At 1530:41.1, the GPWS issued a 50-ft warning. The CVR recording ended at 1530:43.7., the captain stated, "we're gonna brace."

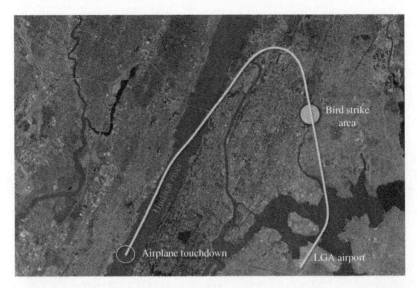

Figure 7.10 Flight track of the airplane. Source: National Transportation Safety Board.

Within seconds after the ditching on the Hudson River, the crewmembers and passengers-initiated evacuation of the airplane. Subsequently, all the occupants were evacuated from the airplane and rescued by area responders. Figure 7.11 shows the airplane occupants on the wings and in the slide/rafts after the evacuation.

Table 7.2 shows the injuries incurred by the occupants of Flight 1549. The airplane was substantially damaged. An examination was conducted on the airplane, and it was found that it had been damaged by Canadian Goose bird strikes. Interesting enough, the Airbus A320-214 now has a new life as an exhibit at the Carolinas Aviation Museum (Ash 2020).

Most of the remainder of the report focuses on the reasons the engines failed from the bird strike. The following discusses the qualifications of the pilots and why they were able to ditch the plane successfully.

7.5.2 Flight Crew and Cabin Crew

The captain, age 57, was hired by Pacific Southwest Airlines on February 25, 1980. Before this, he flew McDonnell Douglas F-4 airplanes for the US Air Force. At the time of the accident, he held a single- and multi-engine airline transport pilot (ATP) certificate, issued August 7, 2002, with type ratings in A320, Boeing 737, McDonnell Douglas DC-9, Learjet, and British Aerospace AVR-146 airplanes. The captain held a first-class FAA airman medical certificate, dated December 1, 2008, with no limitations.

Figure 7.11 A photograph showing the airplane occupants on the wings and in the slide/rafts after the evacuation. Source: National Transportation Safety Board.

Table 7.2 Injuries on Flight 1549.

Injuries	Flight crew	Cabin crew	Passengers	Other	Total
Fatal	0	0	0	0	0
Serious	0	1	4	0	5
Minor	0	0	95	0	95
None	2	2	51	0	55
Total	2	3	150	0	155

Source: NTSB (2010)/Public domain.

According to US Airways records, the captain had accumulated 19,663 total flight hours, including 8930 hours as pilot-in-command, 4765 hours of which were in A320 airplanes. He had flown 155, 83, 39, and 5 hours in the 90, 60, and 30 days, and 24 hours, respectively, before the accident flight. The captain's last A320 line check occurred on December 27, 2007; his last recurrent ground training occurred on February 19, 2008; and his last proficiency check occurred on February 21, 2008. A search of FAA records revealed no accident or incident history, enforcement action, pilot certificate or rating failure, or retest history. A search of the National Driver Register found no record of driver's license suspension or revocation.

The captain stated that he was in excellent health, that he was not taking any prescription medications at the time of the accident, and that he had not taken any medications that might have affected his performance in the 72 hours before the accident. He stated that he drank occasionally but that he had not drunk any alcohol in the week and a half before the accident. The captain reported no major changes to his health, financial situation, or personal life in the last year. A US Airways first officer who had flown with the captain on a six-leg trip sequence in December 2008 described him as exceptionally intelligent, polite, and professional.

7.5.3   The Captain's 72-Hour History

From December 31, 2008, to January 11, 2009, the captain was off duty at his home in the San Francisco, California, area. The captain stated that, when he was off duty, he typically went to sleep about 2300 and woke about 0700 Pacific Standard Time (PST). He stated that he typically needed about eight hours of sleep to feel rested.

On January 12, the captain began a four-day trip sequence with the first officer. He stated that they departed CLT at 1806 and arrived at San Francisco International Airport (SFO), San Francisco, California, at 2119 PST. He stated that he spent the evening at home and that he went to sleep about 2300 PST.

On January 13, the captain awoke about 0700 PST and ate breakfast. He stated that he left his house about 1100 PST and arrived at SFO about 1220 PST. The flight departed SFO about 1315 PST and arrived at PIT about 2100. The captain stated that the total layover time in Pittsburgh was about 10 hours. He added that he did not recall what time he went to bed.

On January 14, the captain awoke about 0510 and ate breakfast. He stated that the quality of his sleep the previous night was good or average and that, although he did not get eight hours of sleep, he was "ok" and felt "normal." The flight crew flew from PIT to LGA and then back to PIT. The captain stated that the total layover time in Pittsburgh was long. He added that he went for a walk around town, ate dinner, answered some e-mails, and went to bed about 2200.

On January 15, the captain awoke about 0640. He stated that the quality of his sleep the previous night was good and that he felt rested. The captain ate breakfast at PIT. The captain's first flight departed PIT at 0856 and arrived at CLT at 1055, at which point the flight crew changed to the accident airplane. The captain stated that he did not get anything to eat in CLT. The flight departed CLT at 1154 and arrived at LGA at 1423. The captain stated that, because they had a quick turnaround at LGA, he purchased a sandwich to eat on the airplane after departure.

7.5.4 The First Officer

The first officer, age 49, was hired by US Airways on April 7, 1986. At the time of the accident, he held a multiengine ATP certificate, issued December 31, 2008, with type ratings in A320, Boeing 737, and Fokker 100 airplanes. The first officer held a first-class FAA airman medical certificate, dated October 7, 2008, with the limitation that he "must wear corrective lenses." The first officer stated during post-accident interviews that he was wearing corrective lenses at the time of the accident.

According to US Airways records, the first officer had accumulated 15,643 total flight hours, including 8977 hours as second-in-command (SIC). The first officer had 37 hours in A320 airplanes, all as SIC. He had flown 124, 55, 37, and 5 hours in the 90, 60, and 30 days, and 24 hours, respectively, before the accident flight. The first officer's last line check on the A320 occurred on January 8, 2009, and his last proficiency check occurred on December 31, 2008. A search of FAA records revealed no accident or incident history, enforcement action, pilot certificate or rating failure, or retest history. A search of the National Driver Register found no record of driver's license suspension or revocation.

The first officer stated that he was in good health, that he was not taking any prescription medications at the time of the accident, and that he had not taken any medications that might have affected his performance in the 72 hours before the accident. He stated that he had not drank alcohol in the last 10 years. The first officer reported no major changes to his health, financial situation, or personal life in the last year. A US Airways check airman who had flown with the first officer for the first officer's operating experience in January 2009 described him as a very good pilot.

7.5.4.1 The First Officer's 72-Hour History

From January 9 through 11, 2009, the first officer was off duty at his home in Madison, Wisconsin. He stated that he typically needed about seven hours of sleep to feel rested.

On January 12, the first officer began the four-day trip sequence with the captain. He stated that, after arriving at SFO from CLT at 2119 PST on January 12, he went to sleep about 2300 PST. The first officer could not recall when he awoke on January 13, but he stated that he felt rested. He stated that the layover in Pittsburgh was less than eight hours and that he did not recall what time he went to bed.

On January 14, the first officer awoke about 0510. He stated that, after flying from PIT to LGA and then back to PIT, the pilots had a long layover in Pittsburgh. He stated that he did not recall when he went to bed. On January 15, the first officer awoke about 0640. He stated that he felt rested and that the quality of his sleep was good. He stated that he did not eat breakfast, which was typical for him.

The first officer stated that, after the flight arrived at CLT, he ate at the airport. He stated that after they arrived at LGA, he got off the airplane and performed a walk around. He stated that they had a quick turn at LGA because the flight had arrived late.

7.5.4.2 The Flight Attendants

Flight attendant A, age 51, was located at the aft-facing, forward jump seat (outboard). She received her initial ground training on June 22, 1982; her initial extended overwater (EOW) 20 training on August 20, 1990; and her last recurrent training on January 31, 2008. Flight attendant B, age 58, was located at the forward-facing, "direct-view" jump seat (aft, center aisle). She received her initial ground training on September 15, 1970; her initial EOW training on September 18, 1989; and her last recurrent training on July 17, 2008. Flight attendant C, age 57, was located at the aft-facing, forward jump seat (inboard). She received her initial ground training on February 27, 1980; her initial EOW training on October 17, 1989; and her last recurrent training on January 31, 2008.

7.5.4.3 Airbus A320-214

Airbus, the manufacturer of the A320 airplane, is headquartered in Toulouse, France. The A320-100/200 series airplanes were type certificated for operation in the United States by the FAA under a bilateral airworthiness agreement between the United States and French governments. The FAA-approved the A320 type certificate (TC) on December 15, 1988. The A320-214 model was approved on December 12, 1996.

The manufacture of the accident airplane was completed by June 15, 1999. The airplane was delivered new to US Airways and was put on its Part 121 operating certificate on August 3, 1999. At the time of the accident, the airplane had accumulated 25,241 total flight hours and 16,299 total cycles. The airplane's last major maintenance inspection was conducted when the airplane had accumulated 24,912 flight hours.

According to the weight and balance manifest provided by US Airways, the airplane departed LGA with a takeoff weight of 151,510 lb, which was below the maximum limitation takeoff weight of 151,600 lb. Assuming a fuel burn of 1500 lb during the climb to 3000 ft and descent at idle thrust, the airplane's weight when it was ditched on the Hudson River was estimated to be about 150,000 lb. The corresponding center of gravity (cg) was 31.1% mean aerodynamic chord (MAC), which was within the takeoff cg limits of between 18.1% and 39.9% MAC.

The airplane was configured with 12 first-class passenger seats; 138 economy-class passenger seats; two cockpit flight crew seats; two cockpit observer seats; and five retractable flight attendant jump seats. Two wall-mounted, aft-facing jump seats were located at the left, forward passenger door (1L); a bulkhead-mounted, forward-facing jump seat was in the aft aisle; and wall-mounted, aft-facing jump seats were located on each side of the aft galley.

The A320 is a "fly-by-wire" airplane, which means that pilot control inputs on the sidesticks are processed by flight control computers that then send electrical signals to the hydraulic actuators that move the pitch and roll flight control surfaces (the ailerons, spoilers, and elevators). The A320 fly-by-wire design incorporates flight envelope protections; the flight computers are designed to prevent exceedance of the safe flight envelope in the pitch-and-roll axes when in "normal law," which is the normal operating mode of the airplane's electronic flight control system. Normal law is one of three sets of control laws (the other two control laws are "alternate law" and "direct law"), which are provided according to the status of the computers, peripherals, and hydraulic generation. The airplane cannot be stalled in normal law. According to the Airbus Flight Crew Training Manual (TM), control law is the "relationship between the PF's input on the sidestick, and the aircraft's response," which determines the handling characteristics of the aircraft.

The A320 flight envelope protections also incorporate a high-AOA protection, which is available from takeoff to landing and is intended, in part, to allow the pilot to pull full aft on the sidestick to achieve and maintain the maximum possible performance while minimizing the risk of stall or loss of control.

Regarding the high-AOA protection, the Airbus Flight Crew Operating Manual (FCOM) stated, in part, that, under normal law, when the AOA becomes greater than a threshold value, the system switches elevator control from flight mode to protection mode, in which the AOA is proportional to sidestick deflection. The AOA will not exceed the allowable maximum even if the pilot gently pulls the sidestick full aft. If the pilot releases the sidestick, the AOA returns to the alpha-protection threshold value and stay there. Under certain conditions, additional features built into the system can attenuate pilot sidestick pitch inputs, preventing the airplane from reaching the maximum AOA.

7.5.4.4 Operational Factors

From April 14 through 16, 2009, the NTSB Operations and Human Performance Group, including members from Airbus, US Airways, the US Airline Pilots Association, and the BEA, conducted flight simulations at the Airbus Training Center, Toulouse, France. The simulations were conducted using an Airbus A320 full-motion, pilot-training simulator and a fixed-base engineering simulator to determine whether the accident airplane could have glided to and landed at LGA or TEB after the bird strike, considering both an immediate return to LGA and a return after a 35-second delay. The simulations were also conducted to evaluate the operational procedures for ditching the airplane within the flightpath and pitch angles assumed during the airplane's ditching certification process.

The simulators were programmed to duplicate as closely as possible the conditions of the accident flight, including winds, temperature, altimeter setting, and weight and balance. The profile flown duplicated as closely as possible the accident profile (the airplane position, thrust setting, altitude at beginning of turns, thrust reduction and cleanup altitudes, speeds, and altitude/speed combination) up to the time of the almost total loss of thrust in both engines. During the simulations, the pilots followed the US Airways Engine Dual Failure checklist after the loss of thrust occurred and relied on their training and experience to complete the test conditions. An observer documented observations and times, and data from the engineering simulator were recorded electronically for later review and analysis.

The pilots were fully briefed on the maneuver before they attempted to perform it in the simulator. The following three flight scenarios were flown: (i) normal landings on runway 4 at LGA, starting from an altitude of 1000 or 1500 ft on approach; (ii) attempted landings at LGA or TEB after the bird strike, starting both from zero groundspeed on takeoff from runway 4 at LGA and from a preprogrammed point shortly before the bird strike and loss of engine thrust; and (iii) ditching on the Hudson River starting from 1500 ft above the river at an airspeed of 200 kts.

During the first flight scenario, all of the pilots were able to achieve a successful landing in both simulators. The flightpath angles at touchdown for these landings ranged from −0.8° to −1.3°. Regarding the second flight scenario, 20 runs were performed in the engineering simulator from a preprogrammed point shortly before the loss of engine thrust in which pilots attempted to return to either runway 13 or 22 at LGA or runway 19 at TEB. Five of the 20 runs were discarded because of poor data or simulator malfunctions. Of the 15 remaining runs, in 6, the pilot attempted to land on runway 22 at LGA; in 7, the pilot attempted to land on runway 13 at LGA; and in 2, the pilot attempted to land on runway 19 at TEB. In eight of the 15 runs (53%), the pilot successfully landed after making an immediate turn to an airport after the loss of engine thrust. Specifically, two of the six runs to land on runway 22 at LGA, five of the seven runs to land on runway 13 at LGA, and one of the two runs to land on runway 19 at TEB immediately after the loss of engine thrust were successful.88 One run was made to return to an airport (runway 13 at LGA) after a 35-second delay, 89 and the landing was not successful.

Regarding the third flight scenario, a total of 14 runs were performed in the engineering simulator in which pilots attempted to touch down on the water within a target flightpath angle of −0.5°, consistent with the structural ditching certification criteria. Two of the 14 runs were discarded because of poor data. Of the remaining 12 runs, 4 were attempted using CONF 2, 4 were attempted using CONF 3, and 4 were attempted using CONF 3/Slats only.

In 11 of the 12 runs, the touchdown flightpath angle ranged between −1.5° and −3.6° (the touchdown flightpath angle achieved on the accident flight was −3.4°).

In 1 of these 12 runs, a −0.2° touchdown flightpath angle was achieved by an Airbus test pilot who used a technique that involved approaching the water at a high speed, leveling the airplane a few feet above the water with the help of the radar altimeter, and then bleeding off airspeed in ground effect until the airplane settled into the water.

7.5.4.5 Flight Crew Training

According to the US Airways Airbus fleet captain, the company has been operating under the Advanced Qualification Program (AQP) since 2002. The AQP is a voluntary program approved and overseen by the FAA that seeks to improve the safety of Part 121 operations through customized training and evaluation. The US Airways AQP included indoctrination training, qualification training (QT), and continuing qualification training (CQT).

Newly hired pilots were required to attend a 9-day indoctrination training course to provide them with an overview of the policies, procedures, and practices at US Airways. After successfully completing the indoctrination training, trainees attended QT, which was a 23-day course including ground school and simulator training. Simulator training occurred in two phases: phase 1 included four days of maneuvers training, during which pilots developed proficiency of core skills and maneuvers and one day of maneuvers validation, and phase 2 included three days of additional simulator training that focused on threat and error management (TEM) and line operations proficiency. After successfully completing QT, trainees completed operating experience in line operations under the supervision of a company check pilot.

US Airways included two PowerPoint presentations on autothrust and AOA protections during ground school. In addition, autothrust, AOA protections, and flight control laws were demonstrated during a simulator session. Information on these topics was also provided to pilots in the US Airways A319/320/321 TM and the A319/320/321 Controls and Indicators Manual.

According to the US Airways AQP manager, the training program was based on a 24-month cycle with a 12-month training evaluation period. Following qualification on an airplane, a crewmember was required to complete annual CQT and quarterly distance-learning modules. The CQT was a three-day course, which included technical ground school; continuing qualification maneuvers observation, consisting of briefings and simulator scenarios; and continuing qualification line-operational evaluation, consisting of simulator sessions like line checks. The CQT was valid for one year and revised annually. The US Airways AQP manager stated that the AQP training program evaluation included a continual review of information collected from integrated data sources and an annual review by an extended review team, which met annually and included members from US Airways and the FAA. During post-accident interviews, the captain stated that company training "absolutely" helped him during the accident event because he was trained on fundamental values to "maintain aircraft control, manage the situation, and land as soon as the situation permits."

According to US Airways training department personnel, bird-strike avoidance training was not included in the ground school curriculum or simulator syllabus. A ground school instructor stated that bird strikes did come up in the lecture environment when pilots asked about "what if" scenarios. He stated that, in these cases, instructors tried to answer the questions to the best of their knowledge.

According to Airbus' vice president of Flight Operations Support and Services, bird-strike hazards were not specifically addressed in its training program but were covered in a flight operations briefing note titled, "Operational Environment: Bird strike Threat Awareness," which was available to all Airbus operators on its website. During training at Airbus on before-takeoff SOPs, exterior airplane lights usage was mentioned as a method to help minimize bird-strike hazards. Although Airbus included engine failure and damage scenarios in its flight-simulator training curriculum, the training did not specifically identify a scenario as being caused by a bird strike.

7.5.4.6 Dual-Engine Failure Training

US Airways QT training included a dual-engine failure in the A320 simulator. According to interviews with US Airways training personnel, instructors ran through the QRH Engine Dual Failure checklist with the crew and provided training on the procedures contained in that checklist before the simulator session.

During the simulator session, the dual-engine failure scenario was initiated at 25,000 ft. The crew was led to attempt to relight the engines by windmilling. The scenario was designed, and the simulator programmed, so that the windmilling relight was not successful, which led the crew to start the APU and attempt an APU-assisted restart of one of the engines. The training scenario was considered completed after the training crew successfully restarted one engine using APU bleed air. During post-accident interviews, a US Airways instructor stated that the scenario was normally completed at an altitude from about 8000–10,000 ft. During the training scenarios, at least one engine is always restarted; therefore, the pilots never reached the point of having to conduct a forced landing or ditching.

The NTSB conducted informal discussions with US operators of A320 airplanes to gather information about flight crew training programs. All the contacted operators indicated that their dual-engine failure training was conducted at high altitudes in accordance with Airbus recommendations and industry practices. The operators revealed that the intent of the training scenarios was to simulate a high-altitude, dual-engine failure scenario and train pilots on the available methods to restart an engine in flight, not to simulate a catastrophic engine failure for which a restart was unlikely.

None of the contacted A320 operators included in their training curricula a dual-engine failure scenario at a low altitude or with limited time available. The A320 operators indicated that the training scenarios generally presented situations for which the course of action and landing location were clear and sufficient time was available to complete any required procedures before landing. The only low-altitude scenarios presented during training were single-engine failures at, or immediately after, takeoff. The A320 operators also indicated that dual-engine failure training was generally only provided during initial, not recurrent, training.

The discussions with A320 operators also indicated that low-altitude, dual-engine failure checklists are not readily available in the industry. One operator stated that it has initiated a review of its procedures because of the US Airways accident to determine if a new checklist needed to be developed. Another operator stated that, although it does not have a low altitude event checklist, in the past, its simulator training had included low-altitude and limited-time dual-engine failures to stimulate pilots' thinking on situational awareness and planning.

7.5.4.7 Ditching Training

US Airways provided ditching training during ground school. The training consisted of a PowerPoint presentation that reviewed the US Airways QRH Ditching checklist, which assumed that at least one engine was running. Ground school also included training on airplane-specific equipment; the use of slides, life vests, and life rafts; and airplane systems related to ditching. According to post-accident interviews and the US Airways Pilot Handbook TM, the function and use of the ditching pushbutton was discussed during ground school, flight simulations, and pre-ground-school scenario-based training in the CQT curriculum.

The US Airways FOM TM included non-airplane-specific guidance on ditching procedures and techniques. In addition, the FOM TM addressed ditching when power was not available and stated the following:

- Power Not Available. If no power is available, a greater than normal approach speed should be used until the flare. This speed margin will allow the glide to be broken early and gradually, decreasing the possibility of stalling high or flying into the water.
- If the wings of the aircraft are level with the surface of the sea rather than the horizon, there is little probability of a wing contacting a swell crest. The actual slope of a swell is very gradual. If forced to land into a swell, touchdown should be made just after the crest. If contact is made on the face of the swell, the aircraft

may be swamped or thrown violently into the air, dropping heavily into the next swell. If control surfaces remain intact, the pilot should attempt to maintain nose up attitude by rapid and positive use of the controls.

Ditching scenarios were not included in either the US Airways or Airbus simulator training curriculum.

The US Airways ditching guidance was similar to military ditching guidance and ditching guidance contained in the FAA Aeronautical Information Manual (AIM), both of which state that, if no power is available, the approach speed used during a ditching should be greater than normal down to the flare to provide the pilot with a speed margin to break the glide earlier and more gradually, thus allowing the pilot time and distance to "feel for the surface." Other literature on this issue also suggested that, when no power is available to the airplane, using flaps may result in the airplane flying at a lower nose attitude and descending more steeply and make it more difficult for the pilot to judge the flare. The NTSB notes that the benefits achieved when using flaps, such as a lower stall speed, should be weighed against the challenges associated with using flaps.

7.5.4.8 CRM and TEM Training

US Airways provided training on crew resource management (CRM) and TEM during basic indoctrination training, CQT, and distance-learning modules. In addition, US Airways integrated CRM and TEM into all aspects of its training, including ground school and flight simulations.

CRM has been defined as "the cognitive, social and personal resource skills that complement technical skills and contribute to safe and efficient task performance" (International Association of Oil and Gas Producers, IOGP 2020).

Anderson (2021) states that "The crash of United Airlines Flight 173 in Portland, Oregon, 1978 highlighted failures in Non-Technical Skills. Analysis of the cockpit voice recordings by the US National Transportation Safety Board (NTSB) found that the plane had run out of fuel while the flight crew were troubleshooting a landing gear malfunction. A contributory factor was the captain's failure to accept input from other flight crew members, as well as the lack of assertiveness from those crew members."

After this crash the NTSB issued a landmark recommendation to require airline crews to have "Cockpit" Resource Management training. The skills mandated by this training were not technical skills, per se, but on what is commonly called, "soft skills." These skills included and still include:

- Situation awareness,
- Decision making,
- Communication,
- Leadership/supervision,

- Teamwork, and
- Awareness of performance shaping factors such as stress and fatigue.

Cockpit Resource Management has evolved into CRM and is used by most international airlines (Anderson 2021).

An American Psychological Association article (APA 2014) presents the argument that CRM has saved thousands of lives. This article cites an article by Helmreich et al. (2002) that every commercial airline flight faces up to four threats. This article states that CRM has helped to manage the threats well and this has significantly reduced major accidents.

The APA article presents two accident scenarios. The first occurred when an Air Florida plane took off from Washington's National Airport (Reagan National Airport) and crashed due to ice in a sensor that caused the speed sensor to read too high. The result was the pilot did not apply enough power. The first officer did not state his concern and the plane crashed killing all but five onboard. This is an excerpt from the CVR:

First Officer: Ah, that's not right.
Captain: Yes, it is, there's 80 [referring to speed].
First Officer: Nah, I don't think it's right. Ah, maybe it is.
Captain: Hundred and twenty.
First Officer: I don't know.

The second example provided concerns the United Airlines DC-10 flight from Denver to Chicago in 1989. The flight crew had been trained in CRM and were able to mitigate the consequences of an event in which the center engine disintegrated and took out the airplane's rudder and aileron control lines. The crew had 34 minutes to figure out a strategy to assess the damage, control the plane, choose a landing site and prepare the passengers for a crash landing. The flight had a three-person flight crew and a fourth pilot, who was a passenger, was also recruited to help. The members of this crew communicated very frequently and all members, senior pilots and junior pilots, expressed their opinions. The airplane landed quite catastrophically. However, 185 of the 296 onboard survived the ordeal.

The article, along with the events of the Flight 1549, demonstrates how well CRM works.

7.5.4.9 FAA Oversight

At the time of the accident, the FAA Certificate Management Office (CMO) for US Airways employed one aircrew program manager (APM) and one assistant APM (who was in training) for the US Airways Airbus fleet. During post-accident interviews, the APM indicated that five aviation safety inspectors assisted him with oversight of US Airways Airbus training and operations. Oversight activity included surveillance and observation of flight crewmembers, instructors, check airmen, and aircrew program designees during training, checking, and line operations. According to interviews, each aviation safety inspector or APM conducted, on average, two or three surveillance activities weekly at US Airways.

The principal operations inspector (POI) for US Airways was stationed at the Coraopolis, Pennsylvania CMO. During interviews, the POI indicated that three assistant POIs were also assigned to the US Airways certificate. The POI stated that he did not normally conduct surveillance activities himself but that he had oversight of the APMs, with whom he interacted daily, and operational programs for each fleet at US Airways. The POI was responsible for approving amendments to training and operational procedures manuals. He stated that he required a review and recommendation by the APMs before he approved any amendments. In addition, any airplane-specific procedures were compared with the manufacturer's recommended procedures, and changes to non-normal procedures were coordinated with the CMO before approval.

7.5.4.10 Summary of Safety Culture Issues

As stated at the start of this accident description, this event was caused by bird strikes and not human error or issues with the safety culture of this crew. The crew used CRM to their advantage and ditched the plane without loss of life. There were five serious injuries, but no one died. In both events, the Gimli Glider and this event, the pilots were able to fly the airplanes successfully and land them without loss of life. The skills of the pilots and the air crew coordination (CRM) saved hundreds of lives.

7.6   737 MAX

Many stories have been written about the issues with the 737 MAX. This case study will focus on the safety culture aspects of the events that led to the two aircraft crashing, with a total of 346 deaths.

7.6.1   Introduction

The safety issues with the 737 MAX are still being addressed as of the writing of this section. They are complicated to say the least and have cost 346 lives. It also greatly tarnished Boeing Airplane Company's (Boeing) reputation. The events leading up to the loss of the 346 lives and the two 737 MAX airplanes demonstrates how an out-of-control organizational culture and poor Federal oversight led to those deaths. There have been news specials on the issues with the 737 MAX, with very in-depth coverage. This section is relatively brief, compared with the magnitude of this complicated case study. The following draws on publicly available information to provide the safety culture aspects of this set of disastrous events.

7.6.2   737 MAX Design and Manufacture

I can't count the number of 737s I have flown on over my over two and half million flight miles. The 737 is the world's best-selling commercial airplane.

According to an article in Aerospace Technology (2021), the 737 MAX aircraft family includes the 737 MAX 7, 737 MAX 8, 737 MAX 9, 737 MAX 10, and 737 MAX 200 variants. The idea behind the new family of 737 aircraft is to deliver improved fuel efficiency for the customers in the single-aisle segment.

Boeing's board of directors approved of the 737 MAX development program in August 2011. The first 737 MAX 7 was launched in May 2013. The first 737 MAX entered service in May 2017, while Boeing rolled out the first 737 MAX 8 in December 2015. The aircraft made its first flight in January 2016 and was first delivered to Malindo Air in May 2017.

The super-efficient design of 737 MAX incorporates new winglet technology, delivering a fuel-burn advantage (Boeing 2021a). The fuselage is strengthened to accommodate new engines and the expanded tail cone and the thickened segment over the flight control surfaces improve steadiness of airflow and avoids the vortex generators on the tail. The modifications further reduce the drag and improve the fuel-efficiency of the airliners. The aircraft delivers an 8% a seat advantage compared with other aircraft in the single-aisle segment (Aerospace Technology 2021).

The flight deck is a glass cockpit with six flat-panel liquid crystal displays (LCDs) (Boeing 2021b).

The 737 MAX is powered by CFM International LEAP-1B engines (GE 2016). The 737 MAX GE engines are designed to reduces CO_2 emissions by an additional ten to 12%.

Boeing has had over 5000 gross orders for 737 MAX since launch (Boeing 2022):

- First delivery of the 737 MAX 8 in May 2017
- First delivery of the 737 MAX 9 in March 2018
- Completed flight testing of the 737 MAX 7 in 4Q19
- 737 MAX began receiving regulatory approval to resume operations and restarted deliveries in 4Q20
- Captured order for 75,737 aircraft from Ryanair in 4Q20
- Secured orders for 100,737 aircraft from Southwest Airlines, 25,737 aircraft from United Airlines, and 23,737 aircraft from Alaska Airlines in 1Q21
- Completed 737 MAX 10 first flight in 2Q21
- Secured orders for 200,737 aircraft for United Airlines and 34,737 aircraft for Southwest Airlines in 2Q21
- There is currently a backlog of 3334 planes

In July 2019, Saudi Arabian airline Flyadeal cancelled its deal with Boeing to buy 30 aircraft because of the fatal crashes in Indonesia and Ethiopia Aerospace Technology (2021).

7.6.3 Accidents

The descriptions of the accidents were excerpted from a report by the NTSB (2019b).

On October 29, 2018, Lion Air flight 610, a Boeing 737 MAX 8, PK-LQP, crashed in the Java Sea shortly after takeoff from Soekarno-Hatta International Airport, Jakarta, Indonesia. The flight crew had communicated with air traffic control and indicated that they were having flight control and altitude issues before the airplane disappeared from radar. The flight was a scheduled domestic flight from Jakarta to Depati Amir Airport, Pangkal Pinang City, and Bangka Belitung Islands Province, Indonesia. All 189 passengers and crew on board died, and the airplane was destroyed. The National Transportation Safety Committee of Indonesia is leading the investigation.

The airplane's digital flight data recorder (DFDR) recorded a difference between the left and right angle of attack (AOA) sensors that was present during the entire accident flight; the left AOA sensor was indicating about 20° higher than the right AOA sensor. During rotation, the left (captain's) stick shaker activated, and DFDR data showed that the left airspeed and altitude values disagreed with, and were lower than, the corresponding values from the right. The first officer asked a controller to confirm the altitude of the airplane and later also asked the speed as shown on the controller radar display. After the flaps were fully retracted, a 10-second automatic aircraft nose-down (AND) stabilizer trim input occurred. After the automatic AND stabilizer trim input, the flight crew used the stabilizer trim switches (located on the outboard side of each control wheel) and applied aircraft nose-up (ANU) electric trim. According to DFDR data, about five seconds after the completion of the pilot trim input, another automatic AND stabilizer trim input occurred. The crew applied ANU electric trim again. DFDR data then showed that the flaps were extended for almost two minutes. However, the flaps were then fully retracted, and automatic AND stabilizer trim inputs occurred more than 20 times over the next six minutes; the crew countered each input during this time using ANU electric trim. The last few automatic AND stabilizer trim inputs were not fully countered by the crew.

During the preceding Lion Air flight on the accident airplane with a different flight crew, the DFDR recorded the same difference between left and right AOA of about 20° that continued until the end of the recording. During rotation, the left control column stick shaker activated and continued for the entire flight, and DFDR data showed that the left airspeed and altitude values disagreed with, and were lower than, the corresponding values from the right. After the flaps were fully retracted, a 10-second automatic AND stabilizer trim input occurred, and the crew countered the input with an ANU electric trim input. After several automatic AND stabilizer trim inputs that were countered by pilot-commanded ANU electric trim inputs, the crew noticed that the airplane was automatically trimming AND. The captain moved the stabilizer trim cutout (STAB TRIM CUTOUT) switches to CUTOUT. He then moved them back to NORMAL, and the problem almost immediately reappeared. He moved the switches back to CUTOUT. He stated that the crew performed three non-normal checklists: Airspeed Unreliable, ALT DIS-AGREE (altitude disagree), and Runaway Stabilizer. The pilots continued the flight using manual trim until the end of the flight. Upon landing, the captain informed an engineer of IAS DISAGREE (indicated airspeed disagree) and ALT DISAGREE alerts, in addition to FEEL DIFF PRESS (feel differential pressure) light problems on the airplane.

On March 10, 2019, Ethiopian Airlines flight 302, a Boeing 737 MAX 8, Ethiopian registration ET-AVJ, crashed near Ejere, Ethiopia, shortly after take-off from Addis Ababa Bole International Airport, Ethiopia. The flight was a scheduled international passenger flight from Addis Ababa to Jomo Kenyatta International Airport, Nairobi, Kenya. All 157 passengers and crew on board died, and the airplane was destroyed. The investigation is being led by the Ethiopia Accident Investigation Bureau.

The airplane's DFDR data indicated that shortly after liftoff, the left (captain's) AOA sensor data increased rapidly to 74.5° and was 59.2° higher than the right AOA sensor; the captain's stick shaker activated. Concurrently, the airspeed and altitude values on the left side disagreed with, and were lower than, the corresponding values on the right side; in addition, DFDR data indicated a Master Caution alert. Like the Lion Air accident flight, a nine-second automatic AND stabilizer trim input occurred after flaps were retracted and while in manual flight (no autopilot). About three seconds after the AND stabilizer motion ended, using the stabilizer trim switches, the captain, who was the PF, partially countered the AND stabilizer input by applying ANU electric trim. About five seconds after the completion of pilot trim input, another automatic AND stabilizer trim input occurred. The captain applied ANU electric trim and fully countered the second automatic AND stabilizer input; however, the airplane was not returned to a fully trimmed condition. CVR data indicated that the flight crew then discussed the STAB TRIM CUTOUT switches, and shortly thereafter DFDR data were consistent with the STAB TRIM CUTOUT switches being moved to CUTOUT.

However, because the airplane remained in a nose-down out-of-trim condition, the crew was required to continue applying nose-up force to the control column to maintain level flight. About 32 seconds before impact, two momentary pilot-commanded electric ANU trim inputs and corresponding stabilizer movement were recorded, consistent with the STAB TRIM CUTOUT switches no longer being in CUTOUT. Five seconds after these short electric trim inputs, another automatic AND stabilizer trim input occurred, and the airplane began pitching nose down.

7.6.4 Design Certification of the 737 MAX 8 and Safety Assessment of the MCAS

The 737 MAX 8 is a derivative of the 737-800 Next Generation (NG) model and is part of the 737 MAX family (737 MAX 7, 8, and 9). The 737 MAX incorporated the CFM LEAP-1B engine, which has a larger fan diameter and redesigned engine nacelle compared with engines installed on the 737 NG family. During the preliminary design stage of the 737 MAX, Boeing testing and analysis revealed that the addition of the LEAP-1B engine and associated nacelle changes produced an ANU pitching moment when the airplane was operating at high AOA and mid Mach numbers. After studying various options for addressing this issue, Boeing implemented aerodynamic changes as well as a stability augmentation function, MCAS, as an extension of the existing speed trim system to improve aircraft handling characteristics and decrease pitch-up tendency at elevated AOA. As the development of the 737 MAX progressed, the MCAS function was expanded to low Mach numbers.

As originally delivered, the MCAS became active during manual flight (autopilot not engaged) when the flaps were fully retracted and the airplane's AOA value (as measured by either AOA sensor) exceeded a threshold based on Mach number. When activated, the MCAS provided automatic trim commands to move the stabilizer AND. Once the AOA fell below the threshold, the MCAS would move the stabilizer ANU to the original position. At any time, the stabilizer inputs could be stopped or reversed by the pilots using their stabilizer trim switches. If the stabilizer trim switches were used by the pilots and the elevated AOA condition persisted, the MCAS would command another stabilizer AND trim input after five seconds.

The FAA's procedures for aircraft type certification require an aircraft manufacturer ("applicant") to demonstrate that its design complies with all applicable FAA regulations and requirements. For transport-category airplanes, as part of this process, applicants must demonstrate through analysis, test, or both that their design meets the applicable requirements under Title 14 *CFR* Part 25. Specifically, 14 *CFR* 25.671 and 25.672 define the requirements for control systems in general and stability augmentation and automatic and power-operated systems, respectively. Title 14 *CFR* 25.1322 addresses flight crew alerting and states, in part, that flight crew alerts must

(1) Provide the flight crew with the information needed to:
    (i) Identify non-normal operation or airplane system conditions, and
    (ii) Determine the appropriate actions, if any.

(2) Be readily and easily detectable and intelligible by the flight crew under all foreseeable operating conditions, including conditions where multiple alerts are provided.

Advisory Circular (AC) 25.1322-1, "Flightcrew Alerting," provides guidance for showing compliance with requirements for the design approval of flight crew alerting functions and indicates that "Appropriate flightcrew corrective actions are normally defined by airplane procedures (for example, in checklists) and are part of a flightcrew training curriculum or considered basic airmanship." Title 14 *CFR* 25.1309 relates to aircraft equipment, systems, and installations, and the primary means of compliance with this section for systems that are critical to safe flight and operations is through safety assessments or through rational analyses; AC 25.1309-1A, "System Design and Analysis," provides guidance for showing compliance with Title 14 *CFR* 25.1309(b), (c), and (d). AC 25.1309-1A explains the FAA's fail-safe design concept, which "considers the effects of failures and combinations of failures in defining a safe design." As part of demonstrating 737 MAX 8 compliance with the requirements in 14 *CFR* 25.1309, Boeing conducted a number of airplane- and system-level safety assessments, consistent with the guidance provided in AC 25.1309-1A.

The NTSB reviewed sections of Boeing's system safety analysis for stabilizer trim control that pertained to MCAS on the 737 MAX. Boeing's analysis included a summary of the functional hazard assessment findings for the 737 MAX stabilizer trim control system. For the normal flight envelope, Boeing identified and classified two hazards associated with "uncommanded MCAS" activation as "major." One of these hazards, applicable to the MCAS function seen in these accidents, included uncommanded MCAS operation to maximum authority. Boeing indicated that, as part of the functional hazard assessment development, pilot assessments of MCAS-related hazards were conducted in an engineering flight simulator, including the uncommanded MCAS operation (stabilizer runaway) to the MCAS maximum authority.

To perform these simulator tests, Boeing induced a stabilizer trim input that would simulate the stabilizer moving at a rate and duration consistent with the MCAS function. Using this method to induce the hazard resulted in the following: motion of the stabilizer trim wheel, increased column forces, and indication that the airplane was moving nose down. Boeing indicated to the NTSB that this evaluation was focused on the pilot response to uncommanded MCAS operation, regardless of underlying cause. Thus, the specific failure modes that could lead to uncommanded MCAS activation (such as an erroneous high AOA input to the MCAS) were not simulated as part of these functional hazard assessment validation tests. As a result, additional flight deck effects (such as IAS DISAGREE and ALT DISAGREE alerts and stick shaker activation) resulting from the same underlying failure (for example, erroneous AOA) were not simulated and were not in the stabilizer trim safety assessment report reviewed by the NTSB. Boeing indicated to the NTSB that, based on FAA guidance, it used assumptions during its safety assessment of MCAS hazards in the engineering flight simulator. Four of these assumptions were the following:

- Uncommanded system inputs are readily recognizable and can be counteracted by overriding the failure by movement of the flight controls "in the normal sense" by the flight crew and do not require specific procedures.
- Action to counter the failure shall not require exceptional piloting skill or strength.
- The pilot will take immediate action to reduce or eliminate increased control forces by re-trimming or changing configuration or flight conditions.
- Trained flight crew memory procedures shall be followed to address and eliminate or mitigate the failure.

Boeing advised that these assumptions are used across all Boeing models when performing functional hazard assessments of flight control systems. These assumptions were consistent with requirements in 14 *CFR* 25.671 and 25.672 and guidance in AC 25-7C, "Flight Test Guide for Certification of Transport Category Airplanes." AC 25-7C stated that short-term forces are the initial stabilized control forces that result from maintaining the intended flightpath after configuration changes and normal transitions from one flight condition to another, "*or from regaining control following a failure. It is assumed that the pilot will take immediate action to reduce or eliminate such forces by re-trimming or changing configuration or flight conditions, and consequently short-term forces are not considered to exist for any significant duration* [emphasis added]." In a 2019 presentation to the NTSB, Boeing indicated that the MCAS hazard classification of "major" for uncommanded MCAS function in the normal flight envelope was based on the following conclusions:

- Unintended stabilizer trim inputs are readily recognized by movement of the stabilizer trim wheel, flightpath change, or increased column forces.
- Aircraft can be returned to steady level flight using available column (elevator) alone or stabilizer trim.
- Continuous unintended nose-down stabilizer trim inputs would be recognized as a stabilizer trim or stabilizer runaway failure and the procedure for stabilizer runaway would be followed.

7.6.5 Assumptions about Pilot Recognition and Response in the Safety Assessment

Functional hazard assessments at the aircraft and systems levels are a critical part of the design certification process because the resulting hazard classifications (severity level) drive the safety requirements for equipment design, flight crew procedures, and training to ensure the hazard effects are sufficiently mitigated. On the basis of Boeing's functional hazard assessment for the MCAS, which assumed timely pilot response to uncommanded MCAS-generated trim input, uncommanded MCAS activation was classified as "major." Boeing was then required to verify that each system that supported MCAS complied with the quantitative and qualitative safety requirements for a "major" hazard, as provided in AC 25.1309-1A, and demonstrate this to the FAA in its aircraft and system safety assessments.

On the Lion Air flight immediately before the accident flight and the Lion Air and Ethiopian Airlines accident flights, the DFDR recorded higher AOA sensor data on the left side than on the right (about 20° higher on the previous Lion Air flight and the Lion Air accident flight and about 59° higher on the Ethiopian Airlines accident flight). As previously stated, the MCAS becomes active when the airplane's AOA exceeds a certain threshold. Thus, these erroneous AOA sensor inputs resulted in the MCAS activating on the accident flights and providing the automatic AND stabilizer trim inputs. The erroneous high AOA sensor input that caused the MCAS activation also caused several other alerts and indications for the flight crews. The stick shaker activated on both accident flights and the previous Lion Air flight. In addition, IAS DISAGREE and ALT DISAGREE alerts occurred on all three flights. Also, the Ethiopian Airlines flight crew received a Master Caution alert. Further, after the flaps were fully retracted, the unintended AND stabilizer inputs required the pilots to apply additional force to the columns to maintain the airplane's climb attitude.

Multiple alerts and indications can increase pilots' workload, and the combination of the alerts and indications did not trigger the accident pilots to immediately perform the runaway stabilizer procedure during the initial automatic AND stabilizer trim input. In all three flights, the pilot responses differed and did not match the assumptions of pilot responses to unintended MCAS operation on which Boeing based its hazard classifications within the safety assessment and that the FAA approved and used to ensure the design safely accommodates failures. Although a number of factors, including system design, training, operation, and the pilots' previous experiences, can affect a human's ability to recognize and take immediate, appropriate corrective actions for failure conditions, industry experts generally recognize that an aircraft system should be designed such that the consequences of any human error are limited. Further, a report on a joint FAA-industry study published in 2002, *Commercial Airplane Certification Process Study: An Evaluation of Selected Aircraft Certification, Operations, and Maintenance Processes*, noted that human performance was still the dominant factor in accidents and highlighted that the industry challenge is to develop airplanes and procedures that are less likely to result in operator error and that are more tolerant of operator errors when they do occur, in particular errors involving incorrect response after a malfunction.

Consistent with this philosophy, the NTSB notes that FAA certification guidance in AC 25.1309-1A that allows manufacturers to assume pilots will respond to failure conditions appropriately is based, in part, upon the applicant showing that the systems, controls, and associated monitoring and warnings are designed to minimize crew errors, which could create additional hazards. While Boeing considered the possibility of uncommanded MCAS operation as part of its functional hazard assessment, it did not evaluate all the potential alerts and indications that could accompany a failure that also resulted in uncommanded MCAS operation. Therefore, neither Boeing's system safety assessment nor its simulator tests evaluated how the combined effect of alerts and indications might impact pilots' recognition of which procedure(s) to prioritize in responding to an unintended MCAS operation caused by an erroneous AOA input. The NTSB is concerned that, if manufacturers assume correct pilot response without comprehensively examining all possible flight deck alerts and indications that may occur for system and component failures that contribute to a given hazard, the hazard classification and resulting system design (including alerts and indications), procedural, and/or training mitigations may not adequately consider and account for the potential for pilots to take actions that are inconsistent with manufacturer assumptions.

Thus, the NTSB concludes that the assumptions that Boeing used in its functional hazard assessment of uncommanded MCAS function for the 737 MAX did not adequately consider and account for the impact that multiple flight deck alerts and indications could have on pilots' responses to the hazard. Therefore, the NTSB recommends that the FAA require that Boeing ensure that system safety assessments for the 737 MAX in which it assumed immediate and appropriate pilot corrective actions in response to uncommanded flight control inputs, from systems such as MCAS, (i) consider the effect of all possible flight deck alerts and indications on pilot recognition and response; and (ii) incorporate design enhancements (including flight deck alerts and indications), pilot procedures, and/or training requirements, where needed, to minimize the potential for and safety impact of pilot actions that are inconsistent with manufacturer assumptions.

Further, because FAA guidance allows such assumptions to be made in transport-category airplane certification analyses without providing applicants with clear direction concerning the consideration of multiple flight deck alerts and indications in evaluating pilot recognition and response, the NTSB is concerned that similar assumptions and procedures for their validation may have also been used in the development of flight control system safety assessments for other airplanes. Therefore, the NTSB recommends that the FAA require that for all other US type-certificated transport-category airplanes, manufacturers (i) ensure that system safety assessments for which they assumed immediate and appropriate pilot corrective actions in response to uncommanded flight control inputs consider the effect of all possible flight deck alerts and indications on pilot recognition and response; and (ii) incorporate design enhancements (including flight deck alerts and indications), pilot procedures, and/or training requirements, where needed, to minimize the potential for and safety impact of pilot actions that are inconsistent with manufacturer assumptions.

Because the FAA routinely harmonizes related standards and guidance with other international regulators who type certificate transport-category airplanes, the NTSB notes that those airplanes may have been designed using similar standards and therefore may also be impacted by this vulnerability. Therefore, the NTSB also recommends that the FAA notify other international regulators that certify transport-category airplane type designs (for example, the European Union Aviation Safety Agency [EASA], Transport Canada, the National Civil Aviation Agency-Brazil, the Civil Aviation Administration of China, and the Russian Federal Air Transport Agency) of Recommendation A-19-11 and encourage them to evaluate its relevance to their processes and address any changes, if applicable.

As early as 2002, the joint FAA-industry study recognized that, while excellent guidance existed for manufacturers on various topics salient to the development of system safety assessments, there were no methods available to evaluate the probability of human error in the operation of a particular system design and that existing qualitative methods for assessing human error were not "very satisfactory." The 2002 study went on to state that the processes used to determine and validate human responses to failure and methods to include human responses in safety assessments needed to be improved. The NTSB notes that a number of human performance research studies have been conducted in the years since the certification guidance contained in AC 25.1309-1A was put in place (in 1988) and this study was conducted and it is likely that more rigorous, validated methodologies exist today to assess error tolerance with regard to pilot recognition and response to failure conditions. The NTSB also believes that the use of validated methods and tools to assess pilot performance in dealing with failure conditions and emergencies would result in more effective requirements for flight deck interface design, pilot procedures, and training strategies. However, we are concerned that such tools and methods are still not commonplace or required as part of the design certification process for functions such as MCAS on newly certified type designs.

Thus, the NTSB concludes that a standardized methodology and/or tools for manufacturers' use in evaluating and validating assumptions about pilot recognition and response to failure condition(s), particularly those conditions that result in multiple flight deck alerts and indications, would help ensure that system designs adequately and consistently minimize the potential for pilot actions that are inconsistent with manufacturer assumptions. Therefore, the

NTSB recommends that the FAA develop robust tools and methods, with the input of industry and human factors experts, for use in validating assumptions about pilot recognition and response to safety-significant failure conditions as part of the design certification process. Further, the NTSB recommends that once the tools and methods have been developed as recommended in Recommendation A-19-13, the FAA revise existing FAA regulations and guidance to incorporate their use and documentation as part of the design certification process, including re-examining the validity of pilot recognition and response assumptions permitted in existing FAA guidance.

A Congressional Committee (Committee) investigated the 737 MAX's design, development, and certification (House Committee, 2020). The major findings of the Committee are listed as follows:

The Committee's preliminary findings identify five central themes that affected the design, development, and certification of the 737 MAX and FAA's oversight of Boeing. Acts, omissions, and errors occurred across multiple stages and areas of the development and certification of the 737 MAX. These themes are present throughout the investigative findings listed next.

(1) Production pressures. There was tremendous financial pressure on Boeing and subsequently the 737 MAX program to compete with Airbus' A320neo aircraft. Among other things, this pressure resulted in extensive efforts to cut costs, maintain the 737 MAX program schedule, and not slow down the 737 MAX production line. The Committee's investigation has identified several instances where the desire to meet these goals and expectations jeopardized the safety of the flying public.

(2) Faulty assumptions. Boeing made fundamentally faulty assumptions about critical technologies on the 737 MAX, most notably with MCAS. Based on incorrect assumptions, Boeing permitted MCAS, software designed to automatically push the plane's nose down in certain conditions, to rely on a single AOA sensor for automatic activation, and assumed pilots, who were unaware of the system's existence in most cases, would be able to mitigate any malfunction. Partly based on those assumptions, Boeing failed to classify MCAS as a safety-critical system, which would have offered greater scrutiny during its certification. The operation of MCAS also violated Boeing's own internal design guidelines established during development.

(3) Culture of concealment. In several critical instances, Boeing withheld crucial information from the FAA, its customers, and 737 MAX pilots. This included hiding the very existence of MCAS from 737 MAX pilots and failing to disclose that the AOA disagree alert was inoperable on the majority of the 737 MAX fleet, despite having been certified as a standard cockpit feature. This alert notified the crew if the aircraft's two AOA sensor readings disagreed, an event that occurs only when one is malfunctioning. Boeing also withheld

knowledge that a pilot would need to diagnose and respond to a "stabilizer runaway" condition caused by an erroneous MCAS activation in seconds or less, or risk catastrophic consequences.

(4) Conflicted representation. The Committee has found that the FAA's current oversight structure with respect to Boeing creates inherent conflicts of interest that have jeopardized the safety of the flying public. The Committee's investigation documented several instances where Boeing authorized representatives (ARs), Boeing employees who are granted special permission to represent the interests of the FAA and to act on the agency's behalf in validating aircraft systems and designs' compliance with FAA requirements, failed to take appropriate actions to represent the interests of the FAA and to protect the flying public.

(5) Boeing's Influence over the FAA's Oversight. Multiple career FAA officials have documented examples to the Committee where FAA management overruled the determination of the FAA's own technical experts at the behest of Boeing. In these cases, FAA technical and safety experts determined that certain Boeing design approaches on its transport category aircraft were potentially unsafe and failed to comply with FAA regulation, only to have FAA management overrule them and side with Boeing instead.

7.7 De Haviland Comet

There have been faulty planes in the past. The best example of a bad commercial airplane is the De Haviland Comet. Sir Geoffrey de Havilland conceived the idea of the DH106 "Comet" in 1943 and design work began in September 1946. The prototype first flew on July 27, 1949 (RAF Museum 2021). Within eighteen months of service two aircraft disappeared. They disappeared within three months of each other.

BOAC Flight 781 was a scheduled British Overseas Airways Corporation (BOAC) passenger flight from Singapore to London. On 10 January 1954, a de Havilland Comet passenger jet operating the flight suffered an explosive decompression at altitude and crashed, killing all 35 people on board. The aircraft, registered G-ALYP, had taken off shortly before from Ciampino Airport in Rome, en route to Heathrow Airport in London, on the final leg of its flight from Singapore. After it exploded, the debris from the explosion fell into the sea near Elba Island, off the Italian coast.

G-ALYP was the third Comet built. The loss of this aircraft marked the second in a series of three fatal accidents involving the Comet in less than twelve months, all caused by structural failures. This airplane crash followed the crash of BOAC Flight 783 near Calcutta, India, in May 1953, and was followed by the loss of South

African Airways Flight 201 in April 1954, which crashed in circumstances similar to BOAC 781 after departing from Ciampino Airport (HMSO 1955). The Secretary of State for Civil Aviation in the United Kingdom ordered a full investigation into the causes of the disappearances. This was carried out by the Royal Aircraft Establishment (RAE) at Farnborough and a court of enquiry was established. A large amount of the wreckage of the Comet G-AYLP was recovered and it was found that the aircraft experienced catastrophic structural damage.

7.8   Summary of Safety Culture Issues

There is not much forgiveness in aircraft crashes. In fact, a six-foot fall is enough to cause death (Matei 2019). The percentage of fatal falls and the associated heights are:

11.7% of fall-related fatalities resulted from falls from heights between 6 and 10 ft
19.7% from falls 11 to 15 ft
17.4% from falls 16 to 20 ft

So, what are the survivability of falls from 100, 1000, 30,000 ft? Not much, if any. It is evident that in all the aviation cases presented, humans played the largest role in the safety of the flight and/or the loss of life. They developed defective designs, made aircraft modifications that decreased (not increased) the possibility of survival, and misinterpreted vital information. Skilled pilots were the saving grace in two of the case studies, but no amount of skill of a pilot can overcome a truly defective aircraft design.

References

Aerospace Technology (2021). Boeing 737 MAX twin-engine airliner. https://www
.aerospace-technology.com/projects/boeing-737-max-aircraft-us/ (accessed
December 2021).
American Psychological Association (2014). Safer air travel through crew resource
management. https://www.apa.org/topics/safety-design/safer-air-travel-crew-
resource-management (accessed November 2021).
Anderson, M. (2021). Non-technical skills ("crew resource management"), human
factors 101. https://humanfactors101.com/topics/non-technical-skills-crm/
(accessed 22 November 2021).
Ash, L. (2020). What happened to the airbus A320 that landed on the hudson?
https://simpleflying.com/miracle-on-the-hudson-aicraft-fate/ (accessed 22
November 2021).

Boeing Airplane Company (Boeing) (2021a). Boeing 373 Max AT winglet. https://www.boeing.com/commercial/737max/737-max-winglets/ (accessed December 2021).

Boeing Airplane Company (Boeing) (2021b). Power on: the new flight-deck displays. https://www.boeing.com/commercial/737max/news/new-flight-deck-displays.page (accessed December 2021).

Boeing (2022). Commercial airplanes fact sheet. https://investors.boeing.com/investors/fact-sheets/default.aspx (accessed April 2022).

Federal Aviation Authority (FAA) (2008). Density altitude, FAA-P-8740-2.

Federal Aviation Administration (FAA) (2015). Safety Management System SMS for 121 Operators. https://www.faa.gov/about/initiatives/sms/specifics_by_aviation_industry_type/121.

Fefderal Aviation Authority (FAA) (2021). Ground-based navigation – very high frequency omni-directional range (VOR). https://www.faa.gov/about/office_org/headquarters_offices/ato/service_units/techops/navservices/gbng/vor (accessed December 2021).

Federal Aviation Regulations (FAR) (2021). 14 CFR 119 – air carriers and commercial operators.

GE (2016). We've got an exclusive look at Boeing's brand new 737 MAX jet. https://www.ge.com/news/reports/weve-got-an-exclusive-look-at-boeing-brand-new-737-max-jet (accessed December 2021).

Government of Canada (1983). Final Report of the Board of Inquiry investigating the circumstances of an accident involving the Air Canada Boeing 767 aircraft C-GAUN that effected an emergency landing at Gimli, Manitoba on the 23rd day of July, 1983.

Helmreich, R.L., Klinect, J.R., Wilhelm, J.A. et al. (2002). *Line Operations Safety Audit (LOSA).* DOC 9803-AN/761. Montreal: International Civil Aviation Organization.

Her Majesty's Stationary Office (HMSO) (1955). Civil Aircraft Accident, Report of the Court of Inquiry into the Accidents of Comet G-ALYP on 10th January 1954 and Comet G-ALYY on 8th April 1954, 1955.

House Committee (2020). Prepared by the Democratic Staff of the House Committee on Transportation and Infrastructure for Chair Peter A. DeFazio, Subcommittee on Aviation Chair Rick Larsen, and Members of the Committee, The Boeing 737 MAX Aircraft: Costs, Consequences, and Lessons from its Design, Development, and Certification. https://www.govinfo.gov/content/pkg/GOVPUB-Y4_T68_2-fb0f3812fefe3515ebcf3f4170fce64b/pdf/GOVPUB-Y4_T68_2-fb0f3812fefe3515ebcf3f4170fce64b.pdf.

International Civil Aviation Organization (ICAO) (2009). Review of the classification and definitions used for civil aviation activities, Montreal.

The International Civil Aviation Organization (ICAO) (2010). Operation of Aircraft, Part 1, Annex 6, Ninth Edition.

International Association of Oil and Gas Producers (2020). Crew Resource Management for Well Operations team. *Report 501*. https://www.iogp.org/bookstore/product/crew-resource-management-for-well-operations-teams/

The International Council of Aircraft Owner and Pilot Associations (IAOPA) (2021). What is general aviation; definition. https://iaopa.aopa.org/about-iaopa#:~:text=General%20aviation%20is%20defined%20by,operations%20for%20remuneration%20or%20hire.%22 (accessed December 2021).

Matei, K. (2019). At what height do falls become deadly?, Safeopedia. https://www.safeopedia.com/at-what-height-do-falls-become-deadly/7/7503 (accessed December 2021).

Morley, D. (2018). Remembering the gimli glider incident on its 35th anniversary, Airline Geeks. https://airlinegeeks.com/2018/07/23/remembering-the-gimli-glider-incident-on-its-35th-anniversary (accessed 23 November 2021).

National Transportation Safety Board (NTSB) (2016). NTSB/AAR-10/03 PB2010-910403, Loss of Thrust in Both Engines After Encountering a Flock of Birds and Subsequent Ditching on the Hudson River US.

National Transportation Safety Board (NTSB) (2019a). Inadvertent Activation of the Fuel Shutoff Lever and Subsequent Ditching Liberty Helicopters Inc., Operating a FlyNYON Doors-Off Flight Airbus Helicopters AS350 B2, N350LH, New York, New York, March 11, 2018, Published December 2019.

NTSB (2019b). Safety Recommendation Report, Accident Numbers DCA19RA017/DCA19RA101. https://www.ntsb.gov/investigations/accidentreports/reports/asr1901.pdf (accessed December 2021).

Roskop, L. (2018). U.S. rotorcraft accident data. http://faahelisafety.org/wp-content/uploads/2018/11/OverviewAccident-Data.pdf (accessed 5 November 2020).

Royal Air Force (RAF) Museum (2021). Comet – the first jet airliner. https://www.rafmuseum.org.uk/research/archive-exhibitions/comet-the-worlds-first-jet-airliner/ (accessed December 2021).

Shannon, H.S., Mayr, J., and Haines, T. (1997). Overview of the relationship between organizational and workplace factors and injury rates. *Safety Science* 26 (3): 201–217. https://doi.org/10.1016/S0925-7535(97)00043-X.

Skybrary (2021). LORAN-C. https://www.skybrary.aero/index.php/LORAN-C (accessed December 2021).

Smith, M.J., Cohen, H.H., Cohen, A., and Cleveland, R.J. (1978). Characteristics of successful safety programs. *Journal of Safety Research* 10 (1): 5–15.

Tour Operators Program of Safety (TOPS) (2021). Safety is TOPS. https://www.topssafety.org/ (accessed November 2021).

8

Nuclear Energy Case Studies

8.0 Introduction

This chapter focuses on cases studies involving radioisotopes and nuclear power. There are four types of case studies presented. These are nuclear power, criticality events, and medical misadministrations. The final case study discusses the aftermath of an abandoned cesium teletherapy treatment machine incident in Brazil.

In Chapter 2 there is an introduction to radiation and radioisotopes. Please refer to that section for an overview.

Please note that this section does not advocate or oppose nuclear power or the use of radioisotopes for industrial and medical purposes. I am just presenting the facts of accidents that have occurred with nuclear materials.

8.1 Nuclear Power

Experimental Breeder Reactor 1 (EBR1) was the first nuclear reactor to generate electricity. It is located on the Idaho National Laboratory (INL) and is now a museum. One of my daughters was a tour guide at the facility for three summers as an intern at the INL. At 1:50 p.m. on December 20, 1951, it became one of the world's first electricity-generating nuclear power plants (NPP)s when it produced sufficient electricity to illuminate four 200-W light bulbs (Michal 2001). Figure 8.1 shows the EBR1 reactor building, and Figure 8.2 shows the interior of the facility. Currently, there are 439 electricity generating NPPs around the world. Figure 8.3 shows the percentage of total electrical generation by country.

NPPs generate electricity the same ways as gas or coal fired plants, the difference is that the heat source to make the steam to turn the turbines comes from the fissioning of fissile materials. There are two basic types of electricity generating reactors in the United States. They are pressure-water reactors and

Impact of Societal Norms on Safety, Health, and the Environment: Case Studies in Society and Safety Culture, First Edition. Lee T. Ostrom.
© 2023 John Wiley & Sons, Inc. Published 2023 by John Wiley & Sons, Inc.

Figure 8.1 Experimental breeder reactor 1 building. Source: Department of Energy.

Figure 8.2 Interior of experimental breeder reactor 1. Source: Department of Energy.

boiling-water reactors. Figure 8.4 is a diagram of a pressure-water reactor (PWR) NPP. Figure 8.5 is a diagram of a boiling-water reactor (BWR).

There has been a general misconception that the cooling towers release something other than steam from an NPP. However, it is truly just steam. Figure 8.6 shows a photo of an NPP and the relationship between the cooling towers and the reactor containment building.

Currently around the world there are 52 reactors under construction (IAEA 2021b). Most of these are PWRs. In my career I have visited 20+ NPPs, mostly

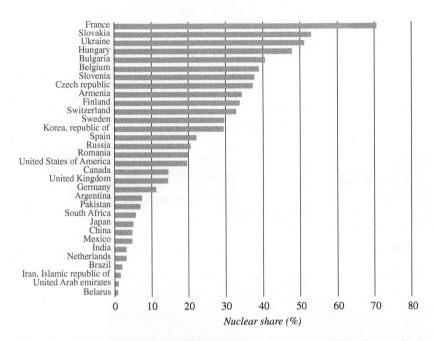

Figure 8.3 Percentage of electricity generated by nuclear power plants. Source: Nuclear Share of Electricity Generation in 2020", International Atomic Energy Agency (IAEA) 2021.

The pressurized-water reactor (PWR)

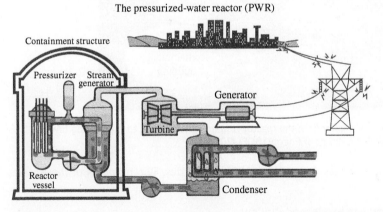

Figure 8.4 Diagram of a pressure-water reactor. Source: NRC (2022)/Public domain.

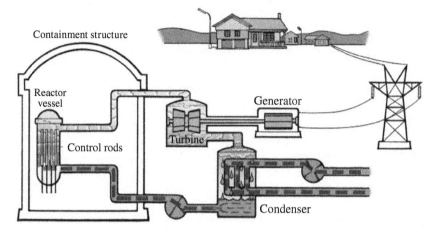

Figure 8.5 Diagram of a boiling-water reactor. Source: NRC (2022)/Public domain.

Figure 8.6 Arrangement of the reactor containment building to the cooling towers. Source: U.S. Department of Energy (DOE) (2022).

in the United States, but a few in other countries. The purposes of these visits were to:

- Investigate an incident or accident.
- Serve on an Nuclear Regulatory Agency (NRC) team examining NPP's Emergency Operating Procedures (EOPs).
- Observing new technologies in the NPP.

The safety culture varies widely from plant to plant and utility to utility. During one visit at an NPP there were guards at every entry point within the plant and

ID checks all the time. However, at one plant I visited I was handed a badge and pointed towards the plant and said to go do my inspection. At that plant I used my cardkey to open a door and in front and below me were the steam lines coming out of containment, going to the turbines. It is a tremendous amount of energy to feel and observe. That feeling lives with me. However, no one came to ask why I opened that door. I was free to wander.

There are many new types of reactors being researched now. These include new sodium cooled reactors and molten salt reactors. Sodium cooled reactors have been tested in the past. In fact, EBR-1 was a sodium/potassium (NaK) cooled reactors. There has been several sodium cooled reactor accidents. Brief overviews of three sodium cooled reactor accidents will be presented. Of course, there are some very famous reactor accidents. These include Chernobyl and Three-Mile Island (TMI). Fukushima-Daichi is the most recent major reactor accident but won't be discussed at this time.

8.1.1 Sodium Cooled Reactors

In Chapter 3 the case study on the T2 accident discussed some of the hazardous properties of sodium (Na). The fact that the metal burns in air and reacts with water to form hydrogen gas (H_2) and sodium hydroxide (NaOH) makes it a relatively hazardous material. Hydrogen gas is, of course, highly flammable and can explode in the presence of oxygen. The reason sodium is a desirable reactor coolant is because, according to Department of Energy, the sodium-cooled fast reactor (SFR) uses liquid metal (sodium) as a coolant instead of water that is typically used in US commercial power plants. This allows for the coolant to operate at higher temperatures and lower pressures than current reactors, thus improving the efficiency and safety of the system (DOE 2021). In a Union of Concerned Scientists (UCS) (2016) article Dr. Lyman of the USC states that "When it comes to safety and security, sodium-cooled fast reactors and molten salt-fueled reactors are significantly worse than conventional light-water reactors." According to the article, SFR such as the Terra Power Natrium reactor system would likely be less "uranium-efficient." They would not reduce the amount of waste that requires long-term isolation in a geologic repository. They also could experience safety problems that are not an issue for light-water reactors. Sodium coolant, for example, can burn when exposed to air or water, and an SFR could experience uncontrollable power increases that result in rapid core melting (UCS 2021).

One thing is for sure; sodium burns in air and reacts with water to create hydrogen. Any coolant leak in a sodium cooled reactor could cause a fire and a possible explosion.

An American Nuclear Society (ANS) Nuclear Newswire article from 2017 (Nuclear Newswire 2017) discusses the US Navy's experiment with sodium cooled reactors. In this article it discusses that the prototype SIR Mark A sodium cooled reactor for the Seawolf SSN-575 (Seawolf) submarine developed sodium leaks in the steam generators and the superheaters. Naval Reactors and Knolls Atomic Power Laboratory tried to fix the leaks. The Seawolf was completed, and the sodium cooled plant and nuclear fuel were installed. The nuclear plant on the Seawolf also developed sodium leaks in the steam generators and superheaters.

Admiral Rickover, the father of the nuclear Navy, was not a fan about the sodium cooled reactors in submarines. He is quoted in the article as having told the Joint Committee on Atomic Energy in 1957: "We went to full power on the Seawolf alongside the dock on August 20 of last year. Shortly thereafter, she developed a small leak. It took us 3 months, working 24 hours a day, to locate and correct the leak. This is one of the serious difficulties in sodium plants." Admiral Rickover was also concerned about radiation exposure to the submarine crew. Another quote in the article from him said: "Sodium becomes 30,000 times as radioactive as water. Furthermore, sodium has a half-life of 14.7 hours, while water has a half-life of about 8 seconds." His other major concern is the hazardous nature of sodium and submarines are surrounded by water.

There has been several sodium cooled reactor accidents. The most notable accidents are:

Santa Susana – Simi Valley California – 1959
Fermi 1 – Near Detroit Michigan – 1966
Monju Reactor – Japan – 1995

8.1.1.1 Santa Susana – 1959

The Santa Susana reactor accident has been written about the most and still has significance today. The Santa Susana Field Laboratory (SSFL) is approximately 30 mi (50 km) from Los Angeles. The owner/stewardship of facility has changed hands over the years. It started out as Rocketdyne, a division of North American Aviation. The Atomics International division of North American Aviation used part of the facility and developed the first commercial NPPs (Places 2022). Liquid metal, for instance sodium, cooled reactor research occurred at the facility for many years. Reactor research was conducted at the facility from 1953 to 1980 (Sapere and Boeing 2005; Department of Energy (DOE), Energy Technology Engineering Center 2022; DOE 2022). Currently, the site is managed by NASA and Boeing.

The following is a synopsis of the accident (DOE 2007). The Sodium Reactor Experiment (SRE), also known as Building 4143, operated from April 1957 to February 1964. It was the highest power research reactor at Santa Susana at 20 MW (or less than 1% of the size of commercial power plants). It was the first facility to supply nuclear powered electricity to a commercial grid and supplied electricity to the city of Moorpark in the late 1950s. Figure 8.7 is an aerial view of the sodium reactor complex.

The SRE accident occurred in July 1959 when there was an accidental partial blockage of sodium coolant in some of the reactor coolant channels. This resulted

Figure 8.7 Aerial view of the sodium reactor complex at Santa Susana. Source: U.S. Department of Energy.

in the partial melting of 13 of the 43 reactor fuel assemblies and the release of some fission products that contaminated the primary reactor cooling system and some of the inside rooms of the facility. All the reactor safety systems functioned properly, and the reactor was safely shut down. The primary pressure vessel, containing the reactor core and sodium coolant, remained intact. Under the oversight of the Atomic Energy Commission (AEC), contamination within the building was cleaned up; the reactor fuel assemblies were then removed, inspected, and stored at the Radioactive Material Disposal Facility. They were later declad in the Hot Lab, and the fuel and cladding were shipped off-site to an AEC approved disposal facility. A second fuel loading was inserted, and operations continued until the reactor was shut down in February 1964 due to termination of the project.

8.1.1.2 Fission Gas Release

A major portion of the radioactivity released from the fuel as a result of the fuel melting was contained in the sodium coolant, but some of the radioactivity was collected in the cover gas in the volume above the sodium coolant inside the reactor vessel. This radioactivity in the cover gas consisted principally of krypton-85, xenon-135, and xenon133 gas and was the same type of radioactivity that collected in smaller quantities during normal operation of the experimental power plant. Other fission products, including iodine-131 and cesium-137 were retained in the primary sodium coolant and were removed during cleanup operations.

During normal routine operations, the cover gas was transferred to large holdup tanks in the SRE facility for the specific purpose of collecting and retaining radioactive gases. After decay, the gas was normally exhausted to the atmosphere through a filtered ventilation system with large quantities of air for dilution of the radioactivity. The releases were always well below those permitted by regulations in existence both then and today. This was done with the approval and oversight of the AEC.

Following the accident, the contaminated reactor cover gas was again transferred to the holding tanks and held long enough for the xenon-135 to decay away (9.1 hour half-life). It was then released to the atmosphere through the stack in a controlled manner over a two month period, in low concentrations that met Federal requirements. This was done with the approval and oversight of the AEC. Based on measurements of the cover gas concentration and volume, approximately 28 curies of krypton-85 (10.7 year half-life) and xenon-133 (5.25 day half-life) was released in this way. This release resulted in a maximum off-site radiation exposure of 0.099 mrem, and an exposure at the location of the nearest resident of 0.018 mrem. These doses are low compared to today's NRC and DOE annual dose limit for unrestricted areas surrounding nuclear facilities (100 mrem/yr) and today's EPA limit for airborne releases (10 mrem/yr). They are also low compared to the average annual background radiation exposure in the United States of 300 mrem/yr. In 1959 the AEC limit for dose to the public in unrestricted areas was 500 mrem/yr.

To summarize, 28 curies of fission gas were released to the environment. This was readily dispersed. No particulate fission products or fuel material (that could potentially "fall-out" onto the ground) were released.

DOE (2022) reports on the cleanup efforts for Building 4143:

> In 1964 the mission of the SRE was completed and the reactor was permanently shut down. Over the next few years, the nuclear fuel and associated reactor equipment were removed and sent to other DOE facilities. Decontamination and Decommissioning of the SRE began in 1974 and continued through 1983. This activity included the removal of the reactor and surrounding soil and concrete, as well as underground structures.
>
> An Environmental Evaluation of the SRE was conducted by DOE in 1983 and found that Building 4143 was acceptably free of contamination and recommended that the facility be released for unrestricted use.
>
> Argonne National Laboratory (ANL) performed an independent verification survey in 1984. The survey found that Building 4143 and its surrounding area was decontaminated to below the limits specified in the draft ANSI Standard N13.12 dated 1982. These levels also met the soil cleanup standards at the time.

There are numerous news articles about the accident that occurred in 1959, countering DOE's claims of site cleanup. An article by Steve Chiotakis (2021) discusses the lingering effects of the contamination caused by the nuclear accident. MSNBC (2021) presents an introduction to a documentary entitled "In the Dark of the Valley" that discusses the aftermath of the accident and the contamination. The California Environmental Protection Agency, Department of Toxic Substances Control (DTSC), Santa Susana Field Laboratory (SSFL) (2022) has a website dedicated to the cleanup of the SSFL. It is located at: https://dtsc.ca.gov/santa_susana_field_lab/, 2022.

8.1.1.3 Fermi 1 – Near Detroit Michigan – 1966

Fermi 1 was a fast breeder reactor power plant cooled by sodium and operated at essentially atmospheric pressure. The reactor plant was designed for a maximum capacity of 430 Megawatt (MW); however, the maximum reactor power with the first core loading (Core A) was 200 MW. The primary system was filled with sodium in December of 1960 and criticality was achieved in August 1963 (NRC 2022b).

The reactor was tested at low power in its first couple years of operation. Power ascension testing above 1 MW commenced in December 1965, immediately following receipt of the high-power operating license. In October 1966, during a power ascension, a zirconium plate at the bottom of the reactor vessel became loose and blocked sodium coolant flow to some fuel subassemblies. Two subassemblies started to melt. Radiation monitors alarmed and the operators manually shut down the reactor. No abnormal releases to the environment occurred. Three years and nine months later, the cause had been determined, cleanup completed, fuel replaced, and Fermi 1 was restarted.

In 1972, the core was approaching the burnup limit. In November 1972, the Power Reactor Development Company made the decision to decommission Fermi 1. The fuel and blanket subassemblies were shipped offsite in 1973. The non-radioactive secondary sodium system was drained and the sodium sent to Fike Chemical Company. The radioactive primary sodium was stored in storage tanks and in 55 gal drums until the sodium was shipped offsite in 1984.

The Monju reactor accident is the most recent, occurring in 1995.

Monju reactor was a Japanese SFR, located near the Tsuruga Nuclear Power Plant, Fukui Prefecture. Construction started in 1986 and the reactor achieved criticality for the first time in April 1994 (STA/NSB 1996). The reactor has been inoperative for most of the time since it was originally built. It was last operated in 2010 and is now closed. Figure 8.8 shows the reactor complex.

On December 8, 1995, Monju, the Japanese prototype fast breeder reactor had an accident in which a thermocouple in a pipe carrying sodium in the secondary plant system had broken, causing approximately 1540 lb (700 kg) of

Figure 8.8 Monju reactor complex. Source: International Atomic Energy Agency (IAEA).

non-radioactive sodium to be spilled. The safety of the reactor, of the plant crew, and of the environment was not jeopardized. However, a fire did occur and there was extensive cleanup of sodium waste. Also, the fire was hot enough to warp the steel structure. A cover up on the nature of the accident also occurred (Pollack 1996; wise-paris.org 2021). One member of the investigating committee even committed suicide.

The plant experienced another accident in August of 2010 when a 3.3-ton "In-Vessel Transfer Machine" fell into the reactor vessel when being removed after a scheduled fuel replacement operation. The Japanese Regulation Authority banned the reactor from restarting in 2013.

8.1.1.4 Safety Culture Summary of Sodium Cooled Reactors
The history of sodium cooled reactors is obviously problematic. Way back in 1957 when Admiral Rickover made his statements to the Joint Committee on Atomic Energy it was clear that sodium leaks will occur. It is also clear that when sodium leaks fires can and do occur. Sodium and water should never mix and, even without water, fires will occur. The sodium from these reactor systems also has long legacies. The sodium is contaminated or radioactive itself and it is a hazardous material. The sodium/potassium (NaK) coolant from EBR-1 was stored at the INL for decades until a safe manner was found to dispose of it. My encounter with it was in the early 1990s when I was with the director of the INEL (INL) at the time doing a safety inspection of a facility. In the facility some engineers were preparing to inject liquid chlorine into the 55-gal drums of NaK in a vessel that

was surrounded with wooden structure. The reaction between sodium and chlorine is extremely exothermic, producing a bright yellow light and a great deal of heat energy (Chemistry 2022). A Ventura County Star (2006) article discussed the sodium burn pit at SSFL that was used to dispose of contaminated equipment. The article interviewed a former employee of SSFL, James Palmer. Mr. Palmer stated that "of the 27 men on Palmer's crew, 22 died of cancers." On some nights Palmer returned home from work and kissed "his [wife] hello, only to burn her lips with the chemicals he had breathed at work."

The industry should really reconsider future sodium cooled reactors.

8.1.2 The Vladimir Lenin Nuclear Power Plant or Chernobyl Nuclear Power Plant (ChNPP) – April 26, 1986

The construction of the ChNPP complex and the town of Pripyat began in 1970. The first reactor was completed in 1977, No. 2 in 1978, No. 3 in 1981, and No. 4 in 1983. The reactors in the complex were RBMK-1000s.

The description of the RBMK reactor design is from IAEA Safety Report Series, No 43 (2005). Accident Analysis for Nuclear Power Plants with Graphite Moderated Boiling Water RBMK Reactors.

RBMKs are boiling water cooled, graphite moderated, channel type reactors (Figure 8.9). The graphite stack together with the fuel and other channels make up the reactor core. The key structural component of the stack is a column made of graphite blocks (parallelepipeds of square cross-section). All graphite blocks have a hole in the center. The central holes of the columns accommodate the tubes of the fuel channels, channels with absorbers placed in them for reactor control and protection, as well as the tubes of other special purpose channels. The thermal contact between the moderating graphite and the channel tubes is

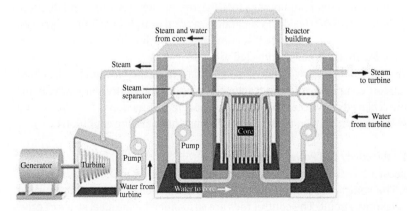

Figure 8.9 RBMK reactor diagram 1. Source: Department of Energy/Public domain.

provided by means of solid contact split rings and sleeves also made of graphite. To improve the thermal contact between the graphite and the tubes, and thus to reduce the graphite temperature, the entire free space of the stack, including the clearances in the block/ring/channel/tube or block/sleeve/channel/tube systems, is filled with a gas mixture consisting mostly of helium. The helium–nitrogen mixture enters the stack from the bottom at a low flow rate and exits from the top through the standpipe of each channel via an individual pipeline. Increases in the moisture and temperature of the mixture provide evidence for detecting coolant leaks from the pressure tubes. The core has top, bottom, and side reflectors. The first two are made up from the same graphite blocks and the latter is formed by the columns.

The core is enclosed in the reactor cavity, formed by the top and bottom plates and a cylindrical barrel hermetically welded to the plates.

The top and bottom plates are pierced by tube lines to house the fuel and other channels. The top plate, mounted on roller supports, rests on the structural components of the reactor building. It takes the weight load from all the reactor channels together with their internal components, as well as part of the weight load from the steam–water pipelines and other service lines (pipes and cables). The bottom plate supports the graphite stack of the core.

Fuel assemblies installed in their channels consist of two subassemblies connected in series. The container type fuel rods are filled with pellets of low enrichment uranium dioxide with the addition of a burnable absorber, erbium. Erbium is a solid element and has an atomic number 68. Fuel claddings are made of zirconium alloy (Zr 1% Neobium, atomic number 41 [Nb]), and channel tubes inside the core are fabricated from another zirconium alloy (Zr 2.5% Nb). Corrosion resistant steel is employed for the inlet and outlet pipelines of the channels outside the core.

The circulation circuit of the reactor is divided into two loops, each including a group of the main circulation pumps (MCPs), the suction, pressure, and distribution group headers (DGHs), drum separators, as well as the downcomers between the drum separators and the MCP suction header. Each of the two circuit loops has half of the fuel channels connected to it.

Figure 8.9 and 8.10 show schematically the layout of the reactor and the circulation circuit components in the reactor building. In the central hall, a refueling machine is placed above the top plate of the reactor. Its function is to unload spent fuel assemblies and to load fresh ones under reactor operating conditions.

8.1.2.1 Reactivity and Power Control

Solid absorber rods are employed to control the reactivity and, thus, the reactor power. The control rods travel on hangers in special channels cooled by water from a separate circuit. The control rods are suspended by the steel strips of drive

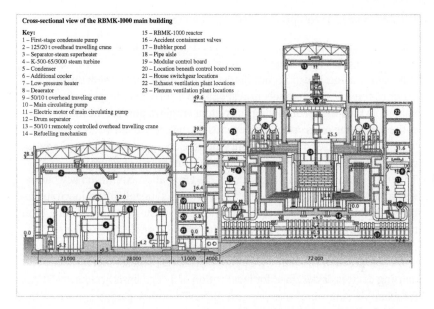

Figure 8.10 RBMK diagram 2. Source: Courtesy of Nuclear Engineering International.

mechanisms mounted on the channel caps. The absorber rods fall into several functional groups.

The related functions at an operating reactor include:

- Monitoring of the neutron power and its rate of increase;
- Automatic maintenance of the specified power in accordance with the signals of the ex-core ionization chambers and in-core sensors;
- Control of the specified radial power distribution;
- Preventive controlled power reduction in response to variations in neutron flux signals;
- Fast controlled power reduction to safe levels;
- Complete shutdown of the reactor by all control rods except the functional group of emergency protection (EP) rods (fast power reduction [FPR] mode);
- Complete shutdown of the reactor by all the rods of the system (emergency protection mode).

All these functions are performed by an integrated monitoring, control, and protection system (CPS) designed in compliance with the requirements specified by the national regulatory authorities. The system has a two-suite arrangement, with either of the two suites being capable of carrying out all of the system's functions. The system was introduced at Kursk-1 in 2002 and was scheduled to be fitted to all RBMK plants at the time of the report.

An RBMK reactor is equipped with the following safety systems:

(a) An emergency core cooling system (ECCS), consisting of two (one fast acting and the other providing long term cooling) subsystems. The fast-acting subsystem, using hydro accumulators, is designed for immediate supply of water to the reactor channels in response to a corresponding emergency signal, while the long-term cooling subsystem employs pumps; the longest delay at the beginning of its operation depends on the time it takes diesel generators to start in response to the emergency protection signal. Both subsystems are capable of feeding water into the circuit at its nominal pressure.

(b) The overpressure protection system of the circulation circuit. Its key components are three groups of main safety valves (MSVs) installed on a loop pipeline that integrates all the ducts collecting steam from the drum separators. From the MSVs, steam flows into a pressure suppression pool (PSP) of the ALS1.

(c) A reactor cavity overpressure protection system (reactor cavity venting system), composed of two parts. One part, consisting of outlet pipelines and condensing devices, is designed for localization of design basis accidents (DBAs), for example rupture of one pressure tube. The other part, equipped with a group of relief devices opening to the atmosphere, is designed to prevent overpressure of the reactor cavity in the case of beyond design basis accidents (BDBAs) associated with multiple pressure tube ruptures (MPTRs).

(d) An accident localization system (ALS) intended for confining accidental coolant releases in leak tight compartments.1 The ALS does not cover the whole circulation circuit. Some pipelines at the top of the circuit, the drum separators, and the steam ducts are located outside the system's hermetic boundary.

(e) An emergency CPS that consists of two independent shutdown systems.

8.1.2.2 Chernobyl Accident

The general population of the Soviet Union was not informed about the Chernobyl accident until several days or weeks after the explosion. I remember watching the news sometime after the accident and the news announcer was showing a satellite photo of the reactor with a glowing red fire in the middle. The CBS news achieve says that the Soviet Union acknowledged the accident on April 28, 1986, because Sweden had detected an atmospheric spike in atmospheric radiation (CBS 2019). My wife was a student in Kiev, Ukraine, at what is now called the Kyiv National University of Technologies and Design at that time. Without knowledge of the reactor accident the students at various schools, including my wife, were instructed to clean all the organic materials from the streets. How contaminated these organic materials were will probably never be known. It is known that the winds took the radioactive cloud northwest, toward Belarus and the Baltic countries, like Sweden. I have seen the reactor building from a distance, but not up close. This accident has and will have a long legacy and might never be cleaned up.

8.1.2.2.1 Accident Sequence

The following accident sequence was developed using the following sources IAEA (1991, 2006a, 2006b), NRC (1987), and DOE (1986). On April 25, 1986, prior to a routine shutdown, the reactor crew at Chernobyl 4 began preparing for an experiment to determine how long turbines would spin and supply power to the main circulating pumps following a loss of main electrical power supply. This test had been carried out at Chernobyl the previous year, but the power from the turbine ran down too rapidly, so new voltage regulator designs were to be tested (IAEA 1991).

Multiple operator actions, including the disabling of automatic shutdown mechanisms, preceded the attempted experiment on the morning of April 26th. By the time that the operator began to shut down the reactor, it was in an extremely unstable condition. The design of the control rods caused a dramatic power surge as they were inserted into the reactor.

The interaction of extremely hot fuel with the cooling water led to fuel disintegration, along with rapid steam production and an increase in reactor pressure. The design characteristics of the reactor were such that substantial damage to even three or four fuel assemblies could, and did, result in the failure of the reactor vessel. Extreme pressure in the reactor vessel caused the 1000-ton cover plate of the reactor to become partially detached. The fuel channels were damaged, and the control rods jammed, which by that time were only halfway down. Intense steam generation then spread throughout the entire core. The steam resulted from water being dumped into the core due to the rupture of the emergency cooling circuit. A steam explosion resulted and released fission products to the atmosphere. A second explosion occurred a few seconds later that threw out fuel fragments and blocks of hot graphite. The cause of the second explosion has been disputed by experts, but it is likely to have been caused by the production of hydrogen from zirconium-steam reactions. Figure 8.11 is a photograph of the damaged reactor and auxiliary buildings.

Two workers died because of these explosions. The graphite (about a quarter of the 1200 tons of it was estimated to have been ejected) and fuel became incandescent and started a number of fires, causing the main release of radioactivity into the environment. A total of about 14 EBq (14×10^{18} Bq) of radioactivity was released, over half of it being from biologically inert noble gases.

About 200–300 tons of water per hour was injected into the intact half of the reactor using the auxiliary feedwater pumps. However, this was stopped after half a day because of the danger of it flowing into and flooding units 1 and 2. From the second to tenth day after the accident, some 5000 tons of boron, dolomite, sand,

Figure 8.11 Damaged Chernobyl reactor. Source: International Atomic Energy Agency (IAEA) (2006b).

clay, and lead were dropped on to the burning core by helicopter in an effort to extinguish the blaze and limit the release of radioactive particles.

It is estimated that all the xenon gas, about half of the iodine and cesium, and at least 5% of the remaining radioactive material in the Chernobyl 4 reactor core (which had 192 tons of fuel) was released in the accident. Most of the released material was deposited close to the reactor complex as dust and debris. Lighter material was carried by wind over the Ukraine, Belarus, and Russia and to some extent over Scandinavia and Europe.

The casualties included firefighters who attended the initial fires on the roof of the turbine building. All these were put out in a few hours, but radiation doses on the first day were estimated to range up to 20,000 millisieverts (mSv), causing 28 deaths, six of which were firemen, by the end of July 1986.

The Soviet Government made the decision to restart the remaining three reactors. To do so the radioactivity at the site would have to be reduced. Approximately 200,000 people ("Liquidators") from all over the Soviet Union were involved in the recovery and clean-up during the years 1986 and 1987. Those individuals received high doses of radiation, averaging around 100 mSv. Approximately 20,000 of the Liquidators received about 250 mSv and a few received 500 mSv. Later, their numbers swelled to over 600,000 but most of these received only relatively low radiation doses.

Causing the main exposure hazard were short-lived iodine-131 and cesium-137 isotopes. Both are fission products dispersed from the reactor core, with half lives of eight days and 30 years, respectively. 1.8 Ebq of I-131 and 0.085 Ebq of Cs-137 were released.) About five million people lived in areas contaminated (above $37\,kBq/m^2$ Cs-137) and about 400,000 lived in more contaminated areas of strict control by authorities (above $555\,kBq/m^2$ Cs-137).

Approximately 45,000 residents were evacuated from within a 10 km radius of the plant, notably from the plant operators' town of Pripyat on May 2nd and 3rd. On 4 May, all those living within a 18-mi radius (30-km), a further 116,000 people from the more contaminated area, were evacuated and later relocated. Approximately 1000 of those evacuated have since returned unofficially to live within the contaminated zone. Most of those evacuated received radiation doses of less than 50 mSv, although a few received 100 mSv or more.

Reliable information about the accident and the resulting contamination was not made available to affected people for about two years following the accident. This led the populace to be distrustful of the Soviet Government and led to much confusion about the potential health effects. In the years following the accident, a further 210,000 people were moved into less contaminated areas, and the initial 30 km radius exclusion zone ($2800\,km^2$) was modified and extended to cover an area of $4300\,km^2$. This resettlement was due to application of a criterion of 350 mSv projected lifetime radiation dose, though in fact radiation in most of the affected area (apart from half a square kilometer) fell rapidly after the accident so that average doses were less than 50% above normal background of 2.5 mSv/yr.

Recent studies have found that the area surrounding the reactors is recovering, though background radiation levels are still above normal background level (IAEA 2006a).

8.1.3 Three Mile Island Accident – March 28, 1979 (NRC 2022a)

The impact of the TMI accident had far reached effects. Not just because of the melting of the core and release of radionuclides but also because of the changes it led to at the NRC and the nuclear industry. It brought about sweeping changes involving emergency response planning, reactor operator training, human factors engineering, radiation protection, and many other areas of NPP operations. It also caused the US Nuclear Regulatory Commission to tighten and heighten its regulatory oversight. Resultant changes in the nuclear power industry and at the NRC had the effect of enhancing safety.

This brief synopsis of the event presents some of those changes.

8.1.3.1 Accident

The accident at the Three Mile Island Unit 2 (TMI-2) NPP near Middletown, Pennsylvania, on March 28, 1979, was the most serious in US commercial NPP

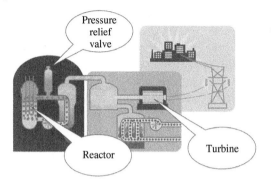

Figure 8.12 Reactor and systems at steady state. Source: NRC (2022)/Public domain.

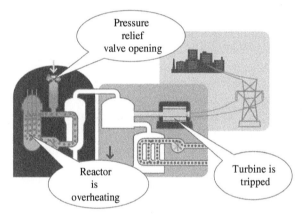

Figure 8.13 Pressure relief valve is open, reactor starting to overheat. Source: NRC (2022)/Public domain.

operating history, even though it led to no deaths or injuries to plant workers or members of the nearby community.

The sequence of certain events, equipment malfunctions, design related problems, and worker errors led to a partial meltdown of the TMI-2 reactor core but only very small off-site releases of radioactivity. Figures 8.12–8.14 are screen shots from a graphic of the accident found at NRC (2022).

8.1.3.2 Summary of Events

The accident began about 4:00 a.m. on March 28, 1979, when the plant experienced a failure in the secondary, non-nuclear section of the plant. The main feedwater pumps stopped running, caused by either a mechanical or electrical failure, which prevented the steam generators from removing heat. First the turbine, then the reactor automatically shut down. Immediately, the pressure in the primary system (the nuclear portion of the plant) began to increase. To prevent that pressure from becoming excessive, the pilot-operated relief valve (PORV) (a valve located at the

Figure 8.14 Fuel is not covered with water and melting. Source: NRC (2022)/Public domain.

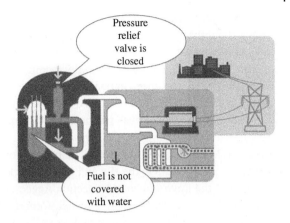

top of the pressurizer) opened. The valve should have closed when the pressure decreased by a certain amount, but it did not. Signals available to the operator failed to show that the valve was still open. As a result, cooling water poured out of the stuck-open valve and caused the core of the reactor to overheat.

As coolant flowed from the core through the pressurizer, the instruments available to reactor operators provided confusing information. There was no instrument that showed the level of coolant in the core. Instead, 'the operators judged the level of water in the core by the level in the pressurizer, and since it was high, they assumed that the core was properly covered with coolant. In addition, there was no clear signal that the pilot-operated relief valve was open. As a result, as alarms rang and warning lights flashed, the operators did not realize that the plant was experiencing a loss-of-coolant (LOCA) accident. At the height of the event most alarm tiles were flashing and gave the operators more confusion as to which alarm to address first. Malone et al. (1980) discuss the human factor engineering aspects of the event. Their team's findings are listed at the end of this synopsis.

The operators took a series of actions that made conditions worse by simply reducing the flow of coolant through the core.

Because adequate cooling was not available, the nuclear fuel overheated to the point at which the zirconium cladding (the long metal tubes which hold the nuclear fuel pellets) ruptured, and the fuel pellets began to melt. It was later found that about one-half of the core melted during the early stages of the accident. Although the TMI-2 plant suffered a severe core meltdown, the most dangerous kind of nuclear power accident, it did not produce the worst-case consequences that reactor experts had long feared. In a worst-case accident, the melting of nuclear fuel would lead to a breach of the walls of the containment building and release massive quantities of radiation to the environment. But this did not occur because of the TMI accident.

The accident caught federal and state authorities off-guard. They were concerned about the small releases of radioactive gases that were measured off-site by the late morning of March 28 and even more concerned about the potential threat that the reactor posed to the surrounding population. They did not know that the core had melted, but they immediately took steps to try to gain control of the reactor and ensure adequate cooling to the core. The NRC's regional office in King of Prussia, Pennsylvania, was notified at 7:45 a.m. on March 28. By 8:00, NRC Headquarters in Washington, D.C. was alerted and the NRC Operations Center in Bethesda, Maryland, was activated. The regional office promptly dispatched the first team of inspectors to the site and other agencies, such as the Department of Energy and the Environmental Protection Agency, also mobilized their response teams. Helicopters hired by TMI's owner, General Public Utilities Nuclear, and the Department of Energy were sampling radioactivity in the atmosphere above the plant by midday. A team from the Brookhaven National Laboratory was also sent to assist .in radiation monitoring. At 9:15 a.m., the White House was notified and at 11:00 a.m., all non-essential personnel were ordered off the plant's premises.

By the evening of March 28, the core appeared to be adequately cooled and the reactor appeared to be stable. New concerns arose by the morning of Friday, March 30. A significant release of radiation from the plant's auxiliary building, performed to relieve pressure on the primary system and avoid curtailing the flow of coolant to the core, caused a great deal of confusion and consternation. In an atmosphere of growing uncertainty about the condition of the plant, the governor of Pennsylvania, Richard L. Thornburgh, consulted with the NRC about evacuating the population near the plant.

Eventually, he and NRC Chairman Joseph Hendrie agreed that it would be prudent for those members of society most vulnerable to radiation to evacuate the area. Thornburgh announced that he was advising pregnant women and pre-school-age children within a 5-mi radius of the plant to leave the area.

Within a short time, the presence of a large hydrogen bubble in the dome of the pressure vessel, the container that holds the reactor core, stirred new worries. The concern was that the hydrogen bubble might burn or even explode and rupture the pressure vessel. In that event, the core would fall into the containment building and perhaps cause a breach of containment. The hydrogen bubble was a source of intense scrutiny and great anxiety, both among government authorities and the population, throughout the day on Saturday, March 31. The crisis ended when experts determined on Sunday, April 1, that the bubble could not burn or explode because of the absence of oxygen in the pressure vessel. Further, by that time, the utility had succeeded in greatly reducing the size of the bubble.

8.1.3.3 Health Effects

Detailed studies of the radiological consequences of the accident have been conducted by the NRC, the Environmental Protection Agency, the Department of Health, Education, and Welfare (now Health and Human Services), the Department of Energy, and the State of Pennsylvania. Several independent studies have also been conducted. Estimates are that the average dose to about 2 million people in the area was only about 1 mrem. To put this into context, exposure from a full set of chest X-rays is about 6 mrem. Compared to the natural radioactive background dose of about 100–125 mrem/yr for the area, the collective dose to the community from the accident was very small. The maximum dose to a person at the site boundary would have been less than 100 mrem.

In the months following the accident, although questions were raised about possible adverse effects from radiation on human, animal, and plant life in the TMI area, none could be directly correlated to the accident. Thousands of environmental samples of air, water, milk, vegetation, soil, and foodstuffs were collected by various groups monitoring the area. Very low levels of radionuclides could be attributed to releases from the accident. However, comprehensive investigations and assessments by several well-respected organizations have concluded that despite serious damage to the reactor, most of the radiation was contained and that the actual release had negligible effects on the physical health of individuals or the environment.

8.1.3.4 Impact of the Accident

The accident was caused by a combination of personnel error, design deficiencies, and component failures. There is no doubt that the accident at Three Mile Island permanently changed both the nuclear industry and the NRC. Public fear and distrust increased, NRC's regulations and oversight became broader and more robust, and management of the plants was scrutinized more carefully. The problems identified from careful analysis of the events during those days have led to permanent and sweeping changes in how NRC regulates its licensees, which, in turn, has reduced the risk to public health and safety.

Here are some of the major changes which have occurred since the accident:

- Upgrading and strengthening of plant design and equipment requirements. This includes fire protection, piping systems, auxiliary feedwater systems, containment building isolation, reliability of individual components (pressure relief valves and electrical circuit breakers), and the ability of plants to shut down automatically;
- Identifying human performance as a critical part of plant safety, revamping operator training and staffing requirements, followed by improved instrumentation and controls for operating the plant, and establishment of fitness-for-duty programs for plant workers to guard against alcohol or drug abuse;

- Improved instruction to avoid the confusing signals that plagued operations during the accident;
- Enhancement of emergency preparedness to include immediate NRC notification requirements for plant events and an NRC operations center which is now staffed 24 hours a day. Drills and response plans are now tested by licensees several times a year, and state and local agencies participate in drills with the Federal Emergency Management Agency and NRC;
- Establishment of a program to integrate NRC observations, findings, and conclusions about licensee performance and management effectiveness into a periodic, public report;
- Regular analysis of plant performance by senior NRC managers who identify those plants needing additional regulatory attention;
- Expansion of NRC's resident inspector program – first authorized in 1977 – whereby at least two inspectors live nearby and work exclusively at each plant in the U.S to provide daily surveillance of licensee adherence to NRC regulations;
- Expansion of performance-oriented as well as safety-oriented inspections, and the use of risk assessment to identify vulnerabilities of any plant to severe accidents;
- Strengthening and reorganization of enforcement as a separate office within the NRC;
- The establishment of the Institute of Nuclear Power Operations (INPO), the industry's own "policing" group, and formation of what is now the Nuclear Energy Institute to provide a unified industry approach to generic nuclear regulatory issues, and interaction with NRC and other government agencies;
- The installing of additional equipment by licensees to mitigate accident conditions, and monitor radiation levels and plant status;
- Employment of major initiatives by licensees in early identification of important safety-related problems, and in collecting and assessing relevant data so lessons of experience can be shared and quickly acted upon;
- Expansion of NRC's international activities to share enhanced knowledge of nuclear safety with other countries in a number of important technical areas.

8.1.3.5 Current Status
Today, the TMI-2 reactor is permanently shut down and defueled, with the reactor coolant system drained, the radioactive water decontaminated and evaporated, radioactive waste shipped off-site to an appropriate disposal site, reactor fuel and core debris shipped off-site to the Department of Energy INL facility, and the remainder of the site being monitored. The owner says it will keep the facility in long-term, monitored storage until the operating license for the TMI-1 plant

Table 8.1 Significant events after the TMI accident.

Date	Event
July 1980	Approximately 43,000 curies of krypton were vented from the reactor building.
July 1980	The first manned entry into the reactor building took place.
Nov. 1980	An Advisory Panel for the Decontamination of TMI-2, composed of citizens, scientists, and State and local officials, held its first meeting in Harrisburg, Pa.
July 1984	The reactor vessel head (top) was removed.
Oct. 1985	Fuel removal began.
July 1986	The off-site shipment of reactor core debris began.
Aug. 1988	GPU submitted a request for a proposal to amend the TMI-2 license to a "possession-only" license and to allow the facility to enter long-term monitoring storage.
Jan. 1990	Fuel removal was completed.
July 1990	GPU submitted its funding plan for placing $229 million in escrow for radiological decommissioning of the plant.
Jan. 1991	The evaporation of accident-generated water began.
April 1991	NRC published a notice of opportunity for a hearing on GPU's request for a license amendment.
Feb. 1992	NRC issued a safety evaluation report and granted the license amendment.
Aug. 1993	The processing of accident generated water was completed involving 2.23 million gallons.
Sept. 1993	NRC issued a possession-only license.
Sept. 1993	The Advisory Panel for Decontamination of TMI-2 held its last meeting.
Dec. 1993	Monitored storage began.

Source: NRC (2022)/Public domain.

expires at which time both plants will be decommissioned. Table 8.1 contains recovery events that occurred.

8.1.3.6 Human Factor Engineering Findings (Malone et al. 1980)

These are the human factor engineering findings from the analysis of the TMI accident.

- Information required by the NPP operators is too often non-existent, poorly located, ambiguous, or difficult to read.

- Annunciators are poorly organized, are not color coded, are often difficult to read and are not arranged in priority order.
- At TMI there are 1900 displays located on the vertical panels. Of these, 503, or 26% cannot be seen by a 5th percentile operator standing at the front panels.
- Labeling of controls and displays is in many cases inadequate or ambiguous, as indicated by the 800 changes made by the operators to the labels provided.
- The Pressurizer and Secondary System sub-panels of Panel 4, a total of 68% of applicable human engineering criteria for labels were not met.

At the time the Metropolitan Edison, the utility that owned TMI, training program was in full compliance with government-imposed standards concerning training. However:

- TMI-2 training was deficient in that it was not directed at the skills and knowledges required of the operators to safely job requirements.
- The essential operator skill is to be able to diagnose what is happening in the plant. The most effective training method of acquiring this skill is simulation. Only 5% of training time is used for simulation training.
- Training in emergency operating procedures (EOPs) was deficient.
- Training at TMI-2 failed to provide for measurement of operator capabilities.
- Training at TMI-2 was deficient in its training of instructors.
- Training at TMI-2 was based on an archaic approach to learning.
- Training at TMI-2 was not closely associated with procedures.
- Training at TMI-2 generally ignored the fact that operators are dealing with a slowly responding system*.
- The training program at TMI-2 did not provide for formal updating and upgrading of methods, materials, and course content.
- Training at TMI- 2 failed to establish in the crew the readiness necessary for effective and efficient performance.

8.1.3.7 Human Engineering and Human Error
- Human engineering planning at TMI-2 was virtually nonexistent
- NRC and the nuclear industry have virtually ignored concerns for human error.
- Where operator – oriented control panel design bases were used (Calvert Cliffs and Oconee) the result was more effective man-machine integration.

8.1.3.8 Procedures
A detailed assessment of EOPs were deficient in content and format.

- There IS little consistency between nomenclature used in procedures and that used on panel components.
- Instructions for control actions seldom provide an indication of the correct (or incorrect) system response.

- Procedures place an excessive burden on operator short-term memory.
- Charts and graphs are not integrated with the text.
- It is not clear which procedures apply to which situations.
- There is no formal method for getting operator inputs into updates of procedures.
- Procedures were grossly deficient in assisting the operators in diagnosing the feedwater system, diagnosing the PORV failure, determining when to override HPI, and determining when to go to natural circulation.

The primary conclusion reached based on this investigation was that the human errors experienced during the TMI incident were not due to operator deficiencies but rather to inadequacies in equipment design, information presentation, emergency procedures, and training.

This general conclusion is supported by several more specific conclusions which are:

- TMI-2 was designed and built without a central concept or philosophy for man-machine integration.
- Lack of a central man-machine concept resulted in lack of definition of the role of operators during emergency situations.
- In the absence of a detailed analysis of information requirements by operator tasks, some critical parameters were not displayed, some were not immediately available to the operator because of location, and the operators were burdened with unnecessary information.
- The control room panel design at TMI-2 violates several human engineering principles resulting in excessive operator motion, workload, error probability, and response time.
- The emergency procedures at TMI-2 were deficient as aids to the operators primarily due to a failure to provide a systematic method of problem diagnosis.
- Operator training failed to provide the operators with the skills necessary to diagnose the incident and take appropriate action.
- Conflicting implications between instrument information, training, and procedures precluded timely diagnosis of and effective response to the incident.

In conclusion, while the major factor that turned this incident into a serious accident was inappropriate operator action, many factors contributed to the action of the operators, such as deficiencies in their training, lack of clarity in their operating procedures, failure of organizations to learn the proper lessons from previous incidents, and deficiencies in the design of the control room.

Therefore, whether operator error explains this particular Case, given all the aforementioned deficiencies, we are convinced that an accident like Three Mile Island was eventually inevitable (Malone et al. 1980).

While at the INL, I was part of NRC teams sent out to evaluate EOPs at numerous NPPs. We also examined the plant's NPP simulators. In general, there was quite a range of the quality and types of EOPs and the quality of the simulators.

Derivan (2014) says that an event at Davis Besse NPP in September 1977 could have turned out the same way because the operators were following their training. Mr. Derivan had an "Ah-ha" moment that turned a potential TMI type event into nothing more than "footnote," as he states.

TMI was a very significant event for nuclear power and human factors engineering.

8.2 Nuclear Criticality

Nuclear criticality accidents are very interesting to me. I have an academic certificate in nuclear criticality safety (NSC). The whole idea of NSC is to prevent the minimum amount of mass of a fissile material in the right geometry to cause a runaway reaction. NSC is achieved by controlling one or more factors of the system within subcritical limits (Tripp 2018).

What is so interesting about criticality events is that it is the geometry of the mass fissile materials that can lead to criticality. The exact definition of a nuclear criticality is:

> Criticality is the state of a nuclear reactor when enough neutrons are created by fission to make up for those lost by leakage or absorption such that the number of neutrons produced in fission remains constant.
>
> (Nuclear Energy Agency 2021).

The case studies presented next have occurred in several different countries, primarily the United States and the Soviet Union. The most recent event that will be presented occurred in Japan. All these events have a human component and are not primarily due to an equipment failure. Chapter 2 contains basic information about radioisotopes and radiation.

The events that will be presented are:

- Mayak Production Association, December 10, 1968, Soviet Union
- National Reactor Testing Station, January 3, 1961
- JCO Fuel Fabrication Plant, September 30, 1999

8.2.1 Mayak Production Association, 10 December 1968 (LANL 2000)

Mayak Production Association is in Ozyorsk, Chelyabinsk Oblast, Russia, formerly the Soviet Union. The accident occurred in a building where various chemical and metallurgical operations with plutonium were performed. Operations were conducted on four, six–hour, shifts per day. The accident occurred on the 19:00 to 01:00 (December 10–11) shift. An unfavorable geometry vessel was being used in an improvised and unapproved operation as a temporary vessel for storing plutonium organic solution. Two independent handling operations with this same vessel and same contents less than one hour apart led to two prompt critical excursions, each one resulting in the severe exposure of a worker. A weak excursion occurred between the two energetic ones, when there were no personnel present.

A small-scale research and development operation had been set up in a basement area to investigate the purification properties of various organic extractants. As originally built, the equipment and piping configuration of this research operation precluded these organics from reaching a set of two 1000 l tanks used for the collection of very lean aqueous solutions (<1 g Pu/l) that were also in the basement. Due to a combination of factors (changes to the piping), organics had inadvertently and unknowingly migrated in significant volume to one of these large aqueous solution tanks in a nearby basement room.

Figures 8.15 and 8.16 show plan and elevation views of these tanks, and the other vessels involved in the accident by their location in the basement room.

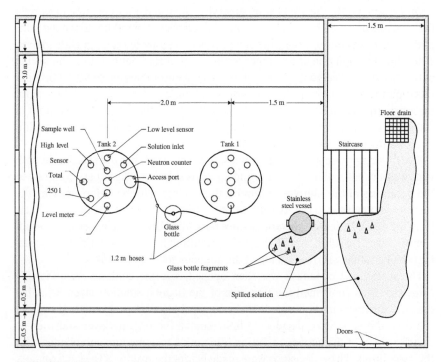

Figure 8.15 Plan view of the tanks involved in the accident.

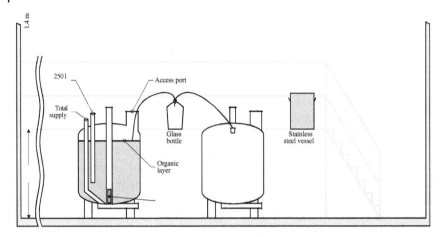

Figure 8.16 Elevation view of the tanks involved in the accident.

Each tank had an operating volume of 800 l. Each was equipped with neutron detectors located near the bottom to monitor the plutonium concentration and to detect any sediment accumulation.

On December 10, the 19:00 shift supervisor instructed an operator to sample the contents of tank 2 before transferring it to recovery operations. Since the sampling device was out of order, the sample was taken by lowering a glass vial on a thin line through one of the upper-level sensor ports (Figure 8.15). The results of the analysis indicated that the plutonium concentration was ~0.6 g/l. Since the total volume of solution in the tank was approximately 800 l, the total plutonium mass was approximately 480 g, which exceeded the criticality safety mass limit of 400 g. According to regulations, the shift supervisor then ordered that two additional confirmatory samples be taken.

When taking these additional samples, it was noticed that the solution in both vials was a combination of organic and aqueous solution. The supervisor ordered the decanting of the samples to remove the organic solution before sending the samples for analysis. In fact, two different organic extractants had been used heavily in the nearby research operations since tank 2 had been last cleaned. As a result, a layer of organic solution with properties resulting from the two different extractants had gradually formed on top of the aqueous solution in tank 2.

The access port of tank 2 was then opened, and the contents were visually inspected. This confirmed the presence of the organic solution layer. Knowing that the downstream equipment was not capable of properly treating organic solution, the supervisor decided to first remove the organic layer and then to transfer a part of the aqueous solution into tank 1 to come into compliance with the mass limits. These decisions were made before the results of the confirmatory sample analyses arrived.

The temporary arrangement of equipment used to remove the organic solution from tank 2 is shown in Figures 8.15 and 8.16. Two reinforced rubber hoses, 1.2 m in length and 13 mm in internal diameter, were fixed into the neck of a 20-l glass bottle (usually used for chemical reagents) with a cloth plug. One hose was connected to the vacuum line on tank 1 and the other was lowered into tank 2 through its access port. If the 20-l bottle became full, it was decided that the contents would be poured into a 60-l vessel usually used for the collection of very low concentration wastes before reprocessing. The bottle and vessel were placed on the platform above tanks 1 and 2. The use of these types of temporary, improvised setups and the use of unfavorable geometry vessels were strictly prohibited by existing regulations.

In the presence of and under instructions from the shift supervisor, two operators used the improvised setup to begin decanting the dark brown (indicative of high plutonium concentration) organic solution. The shift supervisor then left to tend to other duties. After having filled the bottle with approximately 17 l, the operators noted that there was still some amount of organic solution remaining in tank 2. The bottle was then emptied by pouring the contents into the 60-l vessel. During the second filling of the bottle, a mixture of aqueous and globules of organic solution were being drawn into the bottle. As a result, the operators stopped filling the bottle. One operator then went on to other duties while the other went to the shift supervisor for further instructions.

Under instructions from the supervisor, the second operator resumed the decanting of the solution from tank 2 to the bottle. By carefully adjusting the depth of immersion of the hose from the access port the operator was again able to fill the bottle, this time to nearly 20 l. Having disconnected the bottle from the hoses, the operator then poured its contents in the 60-l vessel for a second time. After almost all the solution had been poured out of the bottle, the operator saw a flash of light and felt a pulse of heat. Startled, the operator dropped the bottle, ran down the stairs and from the room. The bottle, which had only a small amount of solution remaining, broke, splashing the remaining contents around the base of the 60-l vessel.

At the instant of the excursion (22:35), the criticality alarm sounded in the room above the tanks. All personnel promptly evacuated to the assigned location (an underground tunnel connecting two adjacent buildings). A similar criticality alarm system in a building approximately 50 m away also sounded almost immediately thereafter, but only for three to five seconds. After the first excursion, the radiation control supervisor on duty informed the plant managers of the accident. He then directed the operator to a decontamination and medical facility, collected the dosimeters (film badges) from all personnel, and strictly warned them not to enter the building where the accident had occurred.

A second excursion was recorded at 23:50, possibly due to cooling of the solution or the release of gas due to a chemical reaction within the solution. This excursion was clearly weaker than the first as it was detected only by thermal neutron detectors within 15 m of the accident. It most likely did not attain prompt criticality or lead to the ejection of solution from the vessel. It occurred when all personnel were at the emergency assembly location.

The shift supervisor insisted that the radiation control supervisor permit him to enter the work area where the accident had occurred. The radiation control supervisor resisted, but finally accompanied the shift supervisor back into the building. As they approached the basement room where the accident had occurred, the γ–radiation (gamma radiation) levels continued to rise. The radiation control supervisor prohibited the shift supervisor from proceeding. Despite the prohibition, the shift supervisor deceived the radiation control supervisor into leaving the area and entered the room where the accident had occurred.

The shift supervisor's subsequent actions were not observed by anyone. However, there was evidence that he attempted to either remove the 60-l vessel from the platform, or to pour its contents down the stairs and into a floor drain that led to a waste receiving tank. (Solution was found on the floor near the drain and around the 60-l vessel at the top of the stairs.) What- ever his actions, they caused a third excursion, larger than the first two, activating the alarm system in both buildings. The shift supervisor, covered in Pu organic solution, immediately exited the room and returned to the underground tunnel. The shift supervisor was then sent to the decontamination and medical facility. At 00:45 the site and building managers arrived. Based on an analysis of available documentation, radiation monitor readings, and interviews with personnel, a recovery plan was developed. By 07:00 the solution in the 60-l vessel had been transferred to several favorable geometry containers. A long handled, large radius of curvature hose and a portable vacuum pump were used to transfer the solution out of the vessel.

Both severely exposed personnel were flown to Moscow for treatment on 11 December. Samples of their blood showed very high ^{24}Na (Sodium 24) activities. Adjusted to the instant of the exposure, they were 5000 decays/min/ml (83 Bq/cm^3) for the operator and 15,800 decays/min/ml (263 Bq/cm^3) for the shift supervisor. The operator received an estimated absorbed dose of about 700 rem and the shift supervisor about 2450 rem. The operator developed acute, severe radiation sickness; both his legs and one hand were amputated. He was still living 31 years after the accident. The shift supervisor died about one month after the accident.

The remaining personnel underwent medical evaluations on the day of the accident. Their dosimeters (film badges) which had just been issued to the personnel on November 20 and November 21, 1968, were used to estimate their doses. The dosimeters indicated that only 6 out of the remaining 27 personnel received doses exceeding 0.1 rem. Their doses were estimated to be 1.64 rem, 0.2 rem, and four with less than 0.15 rem. The operator's dosimeter was overexposed, and the shift supervisor's dosimeter, taken by the radiation control supervisor after the first excursion, indicated a dose 0.44 rem.

Altogether, 19.14 l of solution were recovered from the 60-l vessel. This was a mixture containing 12.83 l of organic solution with a plutonium concentration of 55 g/l and 6.31 l of aqueous solution with a plutonium concentration of 0.5 g/l. Therefore, the plutonium mass that remained in the 60-l vessel after the last excursion was about 709 g. The volume of organic solution and plutonium mass spilled or ejected from the vessel could only be estimated as 16 l and 880 g, respectively.

The number of fissions in the two prompt critical excursions was estimated from (i) the doses received by the operator and shift supervisor and (ii) the measured exposure rate from fission product gamma–rays (1.5 mR/s at 3 m from the vessel, one hour after the last excursion). The number of fissions in the first excursion was estimated at 3×10^{16}, and in the last excursion about 10^{17} fissions. The vessels containing the solution were placed in an isolated room until the radiation levels decayed to an acceptable level, at which time the solution was reprocessed.

The investigation identified several contributing factors to the accident:

- The shift supervisor's decision to take actions that were improvised, unauthorized, and against all regulations, to recover from the plutonium mass limit excess in tank 2.
- Changes to the original piping system that had precluded organic solution transfers to the aqueous solution tanks. As a result of these changes, organic solution could be sent to these tanks in three different ways. These included, (i) the operation of certain valves out of the proper sequence, (ii) through the vacuum and vent lines in the event of stop-valve failures, and (iii) through the purposeful transfer of aqueous solution containing held up organic solution from the extraction facility
- A transfer from tank 1 to tank 2 of about 10 l solution with an unknown plutonium content on 10 December 1968 between 07:00 and 13:00.
- The small-scale organic solution research and development operation was discontinued in this building because of the accident.

The accident caused one fatality, one serious exposure.

8.2.1.1 Safety Culture Issues

The first issue is it was an unapproved operation, with unapproved and unsafe equipment for working with fissile materials. The person or group within the organization that authorized the operation are not identified. It was not clear if the shift supervisor authorized the activity on his own. It is clear he tried to cover up the mistake made, with fatal results. The writeup said that the amount of plutonium in the vessel was 480 g, which exceeded the safe amount of 400 g. Criticality events that occur near workers are quite frequently fatal. These events are unforgiving. Working around any fissile or radioactive material requires strict attention to safety protocols. This event was exactly the opposite of that. Criticality safety was not considered. The worker who lost his legs and one hand and still lived 31 years after the accident was remarkable.

8.2.2 National Reactor Testing Station – January 3, 1961 (LANL 2000)

SL-1 reactor; aluminum-uranium alloy; water moderated; single excursion; three fatalities. Near Idaho Falls, Idaho.

The SL-1 reactor (originally known as the Argonne Low Power reactor) was a direct-cycle, boiling water reactor of 3 MW gross thermal power using enriched uranium fuel plates clad in aluminum, moderated, and cooled by water. Figure 8.17 shows the reactor building. Because the reactor was designed to operate for three years with little attention, the core was loaded with excess ^{235}U (Uranium 235). To counter- balance the excess of ^{235}U, a burnable poison [(Boron 10) (^{10}B)] was added to some core elements as aluminum–^{10}B–nickel alloy. Because the boron plates had a tendency to bow (and, apparently, to corrode, increasing reactivity), some of them were replaced in November 1960 with cadmium strips welded between thin aluminum plates. At that time the shutdown margin was estimated to be 3% (about 4 dollars [$]) compared to the

Figure 8.17 SL-1 reactor building. Source: US Atomic Energy Commission/Wikimedia/ Public domain.

initial value of 3.5–4%. The concept of a using dollar ($) or cents for reactivity is because it is a dimensionless value, specific for reactor conditions at a point in time (Stacey 2001).

The cruciform control rods, which tended to stick, were large cadmium sheets sandwiched between aluminum plates. The nuclear accident was probably independent of the poor condition of the core. After having been in operation for about two years, the SL-1 was shut down December 23, 1960, for routine maintenance; on January 4, 1961, it was again to be brought to power. The three-man crew on duty the night of 3 January was assigned the task of reassembling the control rod drives and preparing the reactor for startup. Apparently, they were engaged in this task when the excursion occurred.

The best available evidence (circumstantial, but convincing) suggests that the central rod was manually pulled out as rapidly as the operator was able to do so. Possibly because it had been stuck and when they pulled hard it became free and came up fast. This rapid increase of reactivity placed the reactor on about a 4-ms period; the power continued to rise until thermal expansion and steam void formation quenched the excursion. The peak power was about 2×10^4 MW, and the total energy release was 133 ± 10 MJ.

The subsequent steam explosion destroyed the reactor and killed two men instantly; the third died two hours later because of a head injury. The reactor building and especially the reactor room were very seriously contaminated by the reactor water, which carried fission products with it. Initial investigations were hindered by the high radiation field (500–1000 R/h) in the reactor room. Despite the large radioactivity release from the core, very little escaped from the building, which was not designed to be airtight.

The SL-1 the reactor core was enclosed, and the water apparently was accelerated upward more or less as a single slug. The energy acquired by the water was sufficient to lift the entire reactor vessel some 9 ft before it fell back to its normal position. If you know where to look, as you drive on Highway 20 near the INL you can see the site where the SL-1 reactor was. Three people died in this accident.

8.2.2.1 Safety Culture Issues

The chain of events leading up to this accident is not clearly understood, other than the fact that most likely a control rood was pulled up too fast and a steam explosion occurred. At that time reactors were still a new technology and many activities like pulling control rods were done manually. The operators might not have been aware that an uncontrolled criticality could occur if a control rod were pulled up too fast. Whether this was a lack of training, a lack of knowledge, or a nefarious act is not known. The reason the steam explosion occurred because a control rod was pulled up too fast is because one of the operators was pinned to the ceiling of the reactor building with a control rod. This book points numerous times that people do not always give new technologies the respect they need before they are thoroughly understood. After this accident control rod changes were done using drive motors and not manually.

The AEC 1962 contains a much more detailed analysis of the accident (AEC 1962).

8.2.3   JCO Fuel Fabrication Plant – September 30, 1999 (LANL 2000)

The accident occurred in the Fuel Conversion Test Building at the JCO company site in Tokaimura, Ibarakin, prefecture, Japan. The building housed equipment to produce either uranium dioxide powder or uranyl nitrate solution from source materials such as uranium hexafluoride or U_3O_8. This building was one of three on site that was licensed to operate with fissile materials. The other two housed large-scale production equipment for the conversion of UF_6–UO_2 for commercial light water reactors and handled only uranium enriched to 5% or less. The Fuel Conversion Test Building was much smaller and was used only infrequently for special projects. It was authorized to handle uranium in enrichments up to 20%. At the time of the accident, Uranium (U) (18.8) fuel processing was underway, with the product intended for the Joyo experimental breeder reactor at the Oarai site of the Japan Nuclear Cycle Development Institute (JNC). The small size ($\sim 300 \times 500$ m) and inner-city location of the JCO Tokai site contributed to a unique aspect of this accident; this was the first process criticality accident in which measurable exposures occurred to off-site personnel (members of the public).

The operation required the preparation of about 16.8 kg of U(18.8) as 370 g/l uranyl nitrate that was to be shipped, as solution, offsite for the subsequent manufacture of reactor fuel. The process was being performed in separate batches to comply with the criticality controls. Procedures specified different uranium mass limits for different enrichment ranges. For the 16–20% range the limit was 2.4 kg uranium. A simplified depiction of the main process equipment and material flow for preparing and packaging the uranyl nitrate, as specified in the license between the JCO Company and the federal government, is shown in Figure 8.18.

Three operators had begun the task on September 29, the day prior to the accident, but were operating according to the procedure indicated in Figure 8.18. There were basically two deviations from the license-authorized procedure that were associated with the actual operations. First, the company procedure that the operators were to have followed, specified that the dissolution step was to be conducted in open, 10-l, stainless steel buckets instead of the dissolution vessel indicated. This change was known to have saved about one hour in dissolution time.

The much more serious procedural departure, however, was the transfer of the nitrate solution into the unfavorable geometry precipitation vessel instead of the

(a) Authorized procedure
Batch

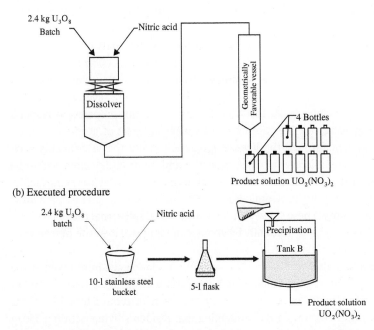

(b) Executed procedure

Figure 8.18 Authorized procedure and the procedure that was performed.

prescribed, favorable geometry columns. This deviation was apparently motivated by the difficulty of filling the product containers from the storage columns. The drain cock below the columns was only about 10 cm above the floor. The precipitation vessel had not only a stirrer to assure a uniform product but greatly facilitated the filling of the product containers.

On September 29 the operators completed the dissolution of four, 2.4 kg batches. The solution was first transferred to a 5-l flask and then hand poured through a funnel into the precipitation vessel. The precipitation vessel was 450 mm diameter by 610 mm high with a capacity of about 100 l.

One operator stood on a ladder to pour the solution. The second operator stood on the floor and held the funnel. Completion of the four batches concluded the three-person team's work for that day.

The next day, September 30, the three operators began dissolving the final three batches that would be required to complete the job. After transferring batches five and six, the pouring of the seventh batch was begun around 10:35. Almost at the end of the pour (183 g of uranium were recovered from the flask) the gamma alarms sounded in this building and in the two nearby commercial fuel buildings.

Workers evacuated from all buildings according to prescribed plans and proceeded to the muster area on site. At this location, gamma ray dose rates far above background were detected and it was suspected that a criticality accident had occurred and was ongoing.

The muster location was then moved to a more remote part of the plant site where dose rates were near to background values. The excursion continued for nearly twenty hours before it was terminated by deliberate actions authorized and directed by government officials. During this time there were several noteworthy aspects of this accident. First, the JCO Company was not prepared to respond to a criticality accident. The gamma alarms were not part of a criticality accident alarm system. In fact, the license agreement stated that a criticality accident was not a credible event. Thus, expertise and neutron detectors had to be brought in from nearby nuclear facilities. Various monitoring devices at the facility as well as the nearby Japan Atomic Energy Research Institute (JAERI), recorded the excursion history. These showed, after a large initial spike, that the power level quasi-stabilized, dropping gradually by about a factor of two over the first approximately 17 hours.

About 4.5 hours after the start of the accident, radiation readings taken at the site boundary nearest to a residential house and a commercial establishment showed combined neutron and gamma ray dose rates of about 5 mSv/hr. At this time the Mayor of Tokaimura recommended that residents living within a 350 m radius of the JCO plant evacuate to more remote locations. After 12 hours, local, Ibaraki-ken, government authorities, recommended that residents within a 10 km radius of the plant remain indoors because of measurable airborne fission product activity.

Shortly after midnight, plans were carried out to attempt to terminate the excursion. It was decided to drain the cooling water from the jacket surrounding the lower half of the precipitation vessel in the recognition that this might remove sufficient reactivity to cause subcriticality. Several teams of three operators each were sent, one at a time, to accomplish this job. The piping that fed the jacket was accessible from immediately outside the building, but it was difficult to disassemble, and the workers were limited to exposures of less than 0.1 Sv each.

When the piping was finally opened at about 17 hours into the accident, not all the water drained from the jacket. This was determined from the various monitoring devices that showed a power drop of about a factor of four and then a leveling off again, indicating that the excursion was not terminated. Complete removal of the water from the jacket was eventually accomplished by forcing argon gas through the piping, again, without entering the building. This led to the shutdown of the reaction at about twenty hours. To assure permanent subcriticality, boric acid was added to the precipitation vessel through a long rubber hose.

A few weeks after the accident, allowing for radiation levels to decay, the solution was sampled from the vessel and analyzed. Based on fission product analysis, it was determined that the total yield of the accident was about 2.5×10^{18} fissions. While there were no radiation detectors that recorded the details of the first few minutes of the excursion history, the operators' exposures and the neutron detector readings at the JAERI-NAKA site provide strong evidence that the reactivity exceeded prompt critical. Experimental results from simulated criticality accidents in solutions would then support a first spike yield of 4 to 8×10^{16} fissions.

The two workers involved in the actual pouring operation were severely overexposed, with estimated doses of 16–20 and 6–10 GyEq, respectively. The third operator was a few meters away at a desk when the accident occurred and received an estimated 1–4.5 GyEq dose. All three operators were placed under special medical care. The operator standing on the floor holding the funnel at the time of the accident died 82 days later. The operator pouring the solution into the funnel died 210 days after the accident. The least exposed operator left the hospital almost three months after the accident.

Factors contributing to the accident, in addition to the stated procedural violations, likely included:

1. A weak understanding by personnel at all levels in JCO of the factors that influence criticality in a general sense, and specifically, a lack of realization that the 45 l of solution, while far sub- critical in the intended storage tanks, could be supercritical in the unfavorable geometry precipitation vessel.
2. Company pressures to operate more efficiently.
3. The mind-set at all levels within JCO and the regulatory authority that a criticality accident was not a credible event; this mind-set resulted in an inadequate review of procedures, plans, equipment layout, human factors, etc. by both the company and the licensing officials.

The Government decided to cancel the license of JCO operations, and the JCO was likely to accept the decision at the time of the printing of this report.

Of the approximately 200 residents who were evacuated from within the 350 m radius, about 90% received doses less than 5 mSv and, of the remaining, none received more than 25 mSv. While there was measurable contamination from airborne fission products on local plant life, maximum readings were less than 0.01 mSv/hr and short-lived.

8.2.3.1 Safety Culture Issues

In many organizations there is still the "Get it Done" mentality. Sometimes it is a sense of loyalty to the company or to a profession. When I worked at the chemical plant, even though safety was the "Number One Priority," operators would say, we are not working for safety today because we have to "get it done." Meaning they had an order to get out. This event is clearly an example of that. However, it cost two people their lives and exposed many others to needless radiation. There is always a chance of a criticality when fissile materials are involved. Solutions are tricky because the exact amount of the fissile material might deviate because the heavier compounds/elements might settle if they are not constantly agitated. In this event, the company lost their license, so many workers were probably displaced, besides those that were killed or injured. Procedures are developed for a reason, and they should be followed.

8.3 Medical Misadministration of Radioisotopes Events

In Chapter 2 there was brief discussion of how radioisotopes are used in medicine. Many times, each year misadministration events occur. Misadministration means giving the radiopharmaceutical to the wrong patient, giving the wrong radiopharmaceutical or wrong activity to the patient, or unjustified examination of pregnant or lactating female patients. Another type of misadministration is to use the wrong route of administration, which includes complete extravascular injections that can result in very high absorbed exposure at the injection site especially if the volume is small, the activity is high, and the radiopharmaceutical has a long retention time. The definition of wrong activity should be made locally. In general, a variation of $\pm 25\%$ from the prescribed activity is regarded as acceptable in diagnostic applications (IAEA 2021a).

IAEA lists what they have determined as the main causes of misadministration of radiopharmaceuticals:

- Communication problems (e.g. insufficient communication, improper labeling of vials and syringes);
- Busy environment, distraction;
- Not knowing local rules;
- Lack of training in emergency situations;
- Responsibilities not clearly defined;
- Inefficient or missing quality assurance (including audits to reveal deficiencies and procedures to deal with emergency situations).

During the early 1990s, I was part of an NRC team that responded to misadministration events around the country. The team then analyzed the events and developed a set of common factors that led the events. The events that were analyzed were (Ostrom et al. 1996, 1994; Ostrom 1993):

Event	Treatment modality and reason for treatment	Type of event	Dose and observed consequences
A	High dose rate brachytherapy for a tumor in the nasal septum	Wrong site	The patient received an unintended dose of 0.76 Gy to the lips. The licensee reported no observed effects.
B	Teletherapy for nonsmall cell lung cancer that had metastasized to the right scapula	Wrong site	The patient received an unintended dose of 6.0 Gy to the lung and spinal cord. The licensee reported no observed effects.
C	Manual brachytherapy for cervical cancer	Wrong site	The patient received an unintended dose of 35.0 Gy to the labial skin. The licensee reported the patient experienced local moist desquamation in this area. The patient also received an unintended dose of 4.5 Gy to the inner aspects of the thighs. There were no observed effects in this area.
D	Diagnostic ^{131}Iodine for the diagnosis of an enlarged thyroid gland	Wrong dose	The patient received a total dose of 45.72 Gy to the thyroid, rather than the intended diagnostic dose! The licensee reported no immediate observed effects.
E	Manual brachytherapy for a tumor obstructing the common bile duct	Wrong site	The patient received an unintended dose of 10.32 Gy to a 1 cm^2 area of the abdominal skin. There were no observed effects. An LPN received an unintended dose of 0.076 Gy to a hand. The LPN experienced no observed effects.
F	Manual brachytherapy for prostate cancer	Wrong dose	The patient received an unintended dose of 56.69 Gy rather than the intended dose of 32.58 Gy. The licensee reported no observed effects.
G	High dose rate brachytherapy for an anal carcinoma	Wrong dose	Probable contributing cause of death

The factors that led to these events were:

- Organizational policy and procedures
- Radiation safety officer oversight
- Training and experience
- Supervision

- Decision errors: errors of judgment
- Decision errors: interpretation errors communications
- Hardware failures

Event G earlier was thoroughly investigated by a large NRC team, including myself. That event occurred in Indiana, Pennsylvania. The following discusses the event.

8.3.1   Loss of Iridium-192 Source at the Indiana Regional Cancer Center (IRCC) – November 1992

8.3.1.1   Introduction

This event involves a medical treatment device that had defects that led to a horrible death of a cancer patient. It occurred almost 30 years ago. However, the fact that a medical device was put into use before it was thoroughly tested is not unique even today. The Edison device that was developed and sold by Theranos is an example of a more recent case (Hartmans, Leskin, and Jackson 2022). Putting aside all the fraud allegations, the fact that the Edison device provided false information is an example of how faulty medical devices can still get into the healthcare system.

8.3.1.2   Event Description

I was part of the NRC team that investigated this event. This investigation was elevated because a death was involved. The following describes the sequence of events and the probably causes of the accident. The following is primarily excerpted from NUREG 1480 (1993).

The event occurred at the Indiana Regional Cancer Center (IRCC) on November 16, 1992, the transportation of the patient from IRCC to Scenery Hill Manor (SHM) nursing home, and the subsequent chronology of events at the SHM nursing home. It also describes the sequence of events involving the loss of the iridium-192 source from IRCC and events associated with the transportation, discovery, and retrieval of the source from the Browning-Ferris Industries (BPI) facility in Carnegie, Pennsylvania (BPI-Carnegie). This section also describes events associated with the second Omnitron International, Inc., (Omnitron) 2000 high dose rate (HDR) afterloader source-wire break at another Oncology Services Corporation (OSC) cancer center in Pittsburgh, Pennsylvania.

8.3.1.3   Patient Treatment Plan

On November 16, 1992, an 82-year-old female patient was treated for anal carcinoma at the IRCC in Indiana, Pennsylvania, using HDR brachytherapy. HDR brachytherapy is a form of internal radiotherapy where an oncologist (Hillman Cancer Center 2022):

- Temporarily implants a catheter, a small plastic tube, or a balloon, in the tumor area.
- Places highly radioactive material inside the body for a sHDRt time and then retracts it using a remote control.
- Removes the catheter after you've completed your entire course of treatment. You may need multiple sessions of HDR brachytherapy.

Figure 8.19 is a photo of the Omnitron device (IAEA 2021b).

The total dose was to be administered in three fractions of 6 Gy (600 rad) each. The HDR treatment plan included the interstitial insertion of the HDR iridium-192 source into five catheters that were strategically implanted in the treatment site. The catheters were implanted on November 13, 1992, and a

Figure 8.19 Omnitron brachytherapy afterloader. Source: United States Nuclear Regulatory Commission (NUREG) (1993).

localization study (simulation) was performed to ascertain the location of the five catheters within the patient.

On November 16, 1992, before the patient was treated, Medical Physicist A generated an Omnitron HDR dosimetry drawing with 40 dosimetric points. The calculated source strength upon initiating the treatment was 1.56 E + 11 Bq (4.219 Ci). The dwell time calculations for the five implanted catheters were generated by computer. On November 16, 1992, at 9:35 a.m., the implanted catheters were connected to the HDR afterloader connecting catheters and patient treatment was initiated. Difficulties were encountered during the source wire insertion into implanted Catheter 5, and Physician A directed termination of treatment without treating the patient through implanted Catheter 5. The patient was disconnected from the HDR afterloader. After the patient was removed from the treatment room, one loose catheter was removed, and the other four catheters were left in the patient for subsequent treatments.

On the basis of the licensee's assumption that the fifth catheter was not treated, Medical Physicist A generated, on November 17, 1992, a second Omnitron dosimetry plan, including different dosimetric drawings and dwell time for each location within each catheter. The fifth catheter was 4.8 Gy (480 rad). The subsequent two fractions were adjusted to 6.6 Gy (660 rad) each to result in a total administered dose of 18 Gy (1800 rad).

This section describes the sequence of events that pertain to the brachytherapy misadministration and accidental loss and transport of an approximate 1.56E + 11 Bq (4.2-Ci) HDR iridium-192 source from the IRCC to BPI, in Warren, Ohio, (BPI-Warren) between November 16 to December 1, 1992. Events preceding the actual loss of the source are also described to explain the incident more fully. Times are approximated.

8.3.1.3.1 Friday, November 13, 1992 At a local hospital, five catheters were surgically inserted adjacent to the patient's rectum to prepare for an HDR treatment. The patient was taken to the IRCC where a simulation study was performed to ascertain the location of the catheters. The patient was subsequently returned to the SHM nursing home. Catheter numbers in this chronology are those taken from the HDR afterloader error log given to NRC by Omnitron.

8.3.1.3.2 Monday, November 16, 1992

7:12 a.m.	The patient departed the nursing home in a local Citizens Ambulance Service ambulance to be transported to the IRCC.
7:30 a.m.	The patient arrived at the IRCC (Figure 2.5).
9:15 a.m.	Medical Physicist A completed the patient treatment plan.

9:20 a.m.	Physician A signed a printout of the patient treatment plan, indicating his approval of the plan.
9:30 a.m.	The patient was moved to the treatment room.
9:35 a.m.	Registered Therapy Technician A (RTT-A), in the presence of Medical Physicist A, connected the patient's five implanted catheters to the HDR afterloader connecting catheters.
9:40 a.m.	RTT-A successfully completed the insertion of the HDR dummy wire into the patient's five implanted catheters.
9:40 a.m.	Medical Physicist A left the IRCC for the day.
9:43 a.m.	RTT-A was present at the HDR afterloader computer console.
	Registered Technologist Radiographer (RTR) was observing the conduct of patient treatment at the HDR computer console. The insertion of the HDR source wire into the patient's catheters was initiated by RTT-A. The computer log indicated the active source wire was successfully inserted into and retracted from the patient's implanted Catheters 1 through 4.

The following error messages were received on the HDR computer monitor for Channel 5.

9:59 a.m.

- Treatment halted due to Error Class 2
- Error Code 1A: Active wire path constriction detected at 82.4 cm;
- Treatment halted due to Error Class 2: Error can be reset by console operator;
- Error Code 55: Console STOP pressed

RTT-A reset HDR computer error message. The following two messages were displayed on the HDR computer monitor:

- Treatment halted due to Error Class 2: Error can be reset by console operator;
- Error Code 6F: Attempt to teat with door open.

RTT-A entered the HDR treatment room and checked the implanted and connecting catheters. RTR entered the HDR treatment room to see to patient's needs. RTI-A and RTR then left the HDR treatment room. RTI-A reset HDR computer error message.

10:01 a.m.	RTI-A attempted to insert the dummy wire into Catheter 5. The following two messages were displayed on the computer monitor: - Treatment halted due to Error Class 2: Error can be reset by console operator; - Error Code 2A: Dummy wire path constriction detected at 90.1 cm; Dummy wire check on Channel 5 failed.

RTT-A reset HDR computer error message.

10:02 a.m.	RTI-A attempted to insert the dummy wire into Catheter 5. RTR was observing the conduct of patient treatment at the HDR computer console. The following messages were displayed on the HDR computer monitor: • Treatment halted due to Error Class 2: Error can be reset by console operator; • Error Code 2A: Dummy wire path constriction detected at 90.0 cm; Dummy wire check on Channel 5 failed.

RTT-A reset HDR computer error message. RTT-A entered the HDR treatment room, disconnected and reconnected Catheter 5 in an attempt to remove constriction. RTR and RTT-B entered the HDR treatment room to assist RTT-A. The three individuals then left the HDR treatment room.

10:07 a.m.	RTT-A attempted to insert the dummy wire into Catheter 5. RTR was observing the conduct of patient treatment at the HDR computer console. The HDR computer monitor indicated the following: • Treatment halted due to Error Class 2: Error can be reset by console operator; • Error Code 2A: Dummy path constriction detected at 90.0 cm; Dummy wire check on Channel 5 failed.

RTT-A reset HDR Computer error message.

10:08 a.m.	RTT-A attempted to insert the dummy wire into Catheter 5. RTR was observing the conduct of patient treatment at HDR computer console. The HDR computer monitor indicated the following: • Treatment halted due to Error Class 2: Error can be reset by console operator; • Error Code 2A: Dummy wire path constriction detected at 90.1 cm; Dummy wire check on Channel 5 failed.

RTR noticed the PrimAlert-10 red light alarm flashing, and RTR informed RTT-A and RTT-B. RTT-A was concerned that the wire was still out of the HDR. RTR was instructed to inform Physician A. RTR informed Physician A, who was in his office in the IRCC, of a problem with inserting the source in Catheter 5. RTT-A observed that the indicator light on the HDR computer monitor was green, indicating that the source was "safe." RTR returned to the HDR treatment room-the door to the room was open-and entered the room but did not walk all the way to the patient. RTR also informed Physician A that the PrimAlert-10 red light alarm was flashing.

Physician A, RTT-A, and RTT-B entered the room while the PrimAlert-10 was flashing. Physician A and RTT-A examined the Catheter 5 connection at the HDR afterloader but observed no source wire. RTT-A disconnected the implanted catheter that he believed to be implanted Catheter 5 from the HDR afterloader connecting Catheter 5, and both Physician A and RTT-A examined the catheter connection at the patient and observed no wire. RTT-B left the HDR treatment room. RTT-A reconnected Catheter 5 to the patient and to the HDR afterloader. Physician A and RTT-A left the HDR treatment room.

Physician A directed RTT-A to again try treating through Catheter 5. RTT-A reset the HDR computer error message. Physician A returned to his office.

10:10 a.m.	RTT-A attempted to insert the dummy wire into Catheter 5. The following messages appeared on the monitor: • Treatment halted due to Error Class 2: Error can be reset by console operator; • Error Code 2A: Dummy wire path constriction detected at 90.1 cm; Dummy wire check on Channel 5 failed.
10:13 a.m.	RTT-A attempted to insert dummy wire into Catheter 5. RTR was observing the conduct of patient treatment at the HDR computer console. The computer monitor indicated the following: • Treatment halted due to Error Class 2: Error can be reset by console operator; • Error Code 2A: Dummy wire path constriction detected at 90.0 cm; Dummy wire check on Channel 5 failed.

RTT-A reset the computer error message.

10:14 a.m.	RTT-A again attempted to insert the dummy wire into Catheter 5 but received the same computer error message previously received on the monitor.
10:16 a.m.	RTT-B informed Physician A of the failure to insert the dummy wire into Catheter 5, and Physician A directed RTT-A to discontinue the treatment.
10:20 a.m.	RTT-A disconnected the patient's implanted catheters from the HDR afterloader. RTT-A removed the patient from the HDR treatment room, and the patient was taken to the stretcher room.

RTT-A unplugged and plugged in again the PrimAlert-10 power supply to reset the alarm sometime during the preceding events.

10:27 a.m.	RTT-A informed Nurse A and Physician A that the stitches were loose on one of the patient's catheters, after which they went to the stretcher room to examine the patient.
10:28 a.m.	The local ambulance was called to transport the patient from the clinic to the nursing home.
10:29 a.m.	Physician A, assisted by Nurse A, removed the loose implanted catheter, which they assumed at that time to be Catheter 5. The local ambulance arrived at the clinic.

10:35 a.m.	Patient A was transferred from the IRCC by ambulance to the SHM Nursing Home. IRCC Nurse A helped two ambulance assistants with placing the patient into the ambulance.
10:48 a.m.	The ambulance arrived at the nursing home and Patient A was transferred to Room 4B.
10:56 a.m.	The ambulance drivers left the nursing home.

8.3.1.3.3 Tuesday, November 17, 1992

3:00 p.m.	Patient A requested that the second radiation therapy treatment scheduled for November 18 be canceled, owing to her inability to tolerate the radiation therapy again. Staff at the SHM nursing home contacted the IRCC and rescheduled the treatment for Monday, November 23, 1992.

8.3.1.3.4 Thursday, November 19, 1992

7:00 p.m.	Certified Nurse Assistant E (CNA-E) was performing perineum care and "removed a piece of gray-black tissue about 1-in. long that was stuck to one of the implants."

8.3.1.3.5 Friday, November 20, 1992

4:30 a.m.	CNA-C noticed that one of the four remaining catheters (later determined to be the one containing the radioactive source) had become dislodged from the patient and protruded approximately 2.54 cm (1 in.) from the body.
6:15 a.m.	During regular patient rounds, Licensed Practical Nurse B (LPN-B) and CNA-C discovered that the catheter that had been protruding earlier on the shift had fallen out of the patient onto the bedding.
	LPN-B picked up the catheter containing the source, placed it in a red bag, which the nursing home uses for medical and biohazardous waste (typically referred to as "red-bag" waste), and transferred this small, red bag to a larger container for medical waste in the soiled utility room where red-bag waste is stored daily. LPN-B was unaware the catheter contained the iridium-192 source.
8:10.a.m.	Registered Nurse-D (RN-D) called the IRCC to determine what method of disposal was needed for the catheter that had fallen out of the patient during the earlier shift. IRCC informed RN-D that what they had done was appropriate and that the catheter could be disposed of in the red-bag waste.
	After discussing the situation with the IRCC, RN-D instructed Maintenance Man A to remove the large red bag of waste from the soiled utility room, which he normally does at 7:30 a.m., and transfer it to the outside waste storage room.
8:30 a.m.	Maintenance Man A took the large red bag of medical waste from inside the soiled utility room to the outside waste storage room, placed it inside a BFI cardboard box and locked the room.

The waste remained in this location for an additional five days, awaiting pickup by BPI from BFI-Carnegie, which usually occurs the last Wednesday of each month.

8.3.1.3.6   Saturday, November 21, 1992

11:10 p.m.	Patient A dies. The remaining three catheters were subsequently removed and disposed of in red-bag waste.

8.3.1.3.7   Wednesday, November 25, 1992

4:30 a.m.	BFI Driver A began picking up medical waste from the first of 22 stops for that day.
9:25 a.m.	BPI Driver A arrived at his 12th stop, SHM Nursing home; picked up three boxes of red-bag waste from this facility; and placed it in a straight truck. Although BPI Driver A had a portable survey meter in the truck, he stated that he did not use it at the SHM nursing home. He continued with his regularly scheduled stops throughout the day, stopping at an additional 10 facilities before returning to BPI-Carnegie.
2:30 p.m.	BPI Driver A arrived at BPI-Carnegie.
3:30 p.m.	BPI Driver A unloaded all the boxes by himself from the straight truck onto BPI Trailer 808, which is a 14.6-m (48-ft) trailer and left for the day. The box containing the source was one of the last ones to be loaded. The box was positioned in the rear left-hand corner of the trailer, approximately 1.8 m (6 ft) off the floor.

8.3.1.3.8   Thursday, November 26, 1992   Because Thursday, November 26, 1992, was Thanksgiving Day, BFI scheduled no transfers to the BPI-Waren Medical Waste Incinerator, and the trailer remained on site until November 27, 1992.

8.3.1.3.9   Friday, November 27, 1992

12:00 Midnight	BPI semitruck Driver B began his routes for the day.
6:15 a.m.	BFI Driver B began a review of paperwork for his second tractor-trailer shipment (808) that day to BPI-Warren. As is customary, Driver B signed and dated all shipping manifests and checked the bottom and top latches at the back of the trailer before placing a padlock on the trailer.
8:30 a.m.	BPI Driver B left BPI Carnegie via 1-79, to I-680N, to I-SOW to Hwy-46N, and on to Hwy-169N. He made no stops between Carnegie and Warren.

8:30 a.m.	Driver B arrived at BPI-Warren, drove over to the unloading area, disconnected Trailer 808, hooked up his cab to an empty trailer, drove to the front office, completed the paperwork for this shipment, and left BPI-Warren. Two fixed radiation monitors inside the facility alarmed, reading above their normal limits of 0.2 µSv (20 µrem) per hour. Employees working that shift began trying to locate the cause of the radiation alarms.
8:50 a.m.	These employees notified both the Plant Manager and the Supervisor on the day shift who also began to attempt to find the source. They surveyed all packages on the conveyor belt and those near the loading dock.
	Because they could not determine the source of the radiation, they began reviewing their shipping records to determine what shipments they had received that morning. They identified two tractor-trailer shipments, 806 and 808, that had arrived that morning at 4 a.m. and 8:30 a.m., respectively, from BPI-Carnegie.
	After identifying the possible cause of the radiation alarms, both the Plant Manager and the Supervisor began to look for the source with their portable survey meters outside of the parked trailers. When they came within approximately 121.9 m (400 ft.) of the trailer, both portable survey meters immediately alarmed and registered at their highest levels over 5 µSv (500 µrem) per hour.
9:45 a.m.	Because the Plant Manager and the Supervisor could not determine which trailer had the radioactive material in it, they decided to drive one trailer at a time behind the main building, using it as a large concrete shield, to see if any of the portable survey meters alarmed as a trailer approached Position B on Figure 8.20.
	Before the tractor-trailer drove behind the main building, the Plant Manager stood at Position C in Figure 8.20 of the facility with his portable survey meter turned on. At this location, there was no indication of radioactivity. As the tractor-trailer came from behind the building, however, the portable survey meter registered its highest level.
10:00 a.m.	As soon as BPI-Warren identified the trailer containing the radioactive material, they parked it as far away on their property as they could from the main building and immediately called BPI-Carnegie to come pick up the trailer as soon as possible. The facility was fenced and secured.
2:30 p.m.	Driver C arrived at BPI-Warren to pick up the tractor-trailer from the Ohio facility, which was then driven to BPI-Carnegie. No radioactive material placards were placed on the tractor-trailer.
4:45 p.m.	BFI Driver C arrived at BPI-Carnegie. Because it was growing late, the driver unhooked and parked the trailer in a back lot where it remained for the rest of the weekend in a fenced, secured area.

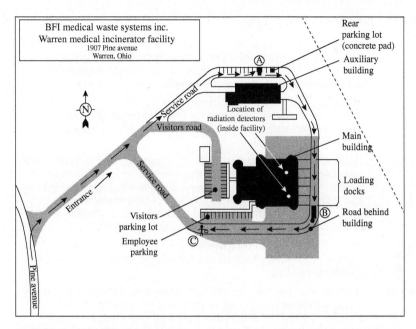

Figure 8.20 Diagram of tractor-trailer route carrying iridium-192 source at BFI-Warren facility. Source: NUREG (1480)/Public domain.

8.3.1.3.10 *Monday, November 30, 1992*

1:00 p.m.	BPI-Carnegie Supervisor A and two safety assistants (Safety Technicians A and B) put gloves on and began to survey boxes from the trailer for radioactivity. When approaching the trailer, each safety technician noticed that the portable survey meter registered its highest level (5 μSv [500 μrem] per hr).
2:30 p.m.	After surveying approximately 40 boxes, they identified the box containing the radioactive material (Figure 2.19). Because the box containing the source had no generator identification labels on the outside, Safety Technicians A and B opened the box and began to go through individual medical red bags looking for information to enable them to identify the originator of the waste. Supervisor A left for an appointment offsite but requested that the safety technicians continue to search for the originator's identification.
3:00 p.m.	One of the two safety technicians found a portion of a prescription in the waste that had an individual's name on it. With this information, they sealed up the box, placed it in a green recycle container, and locked it. They returned to their office with this information and began calling the list of facilities from which waste was picked up on November 25, 1992.
4:00 p.m.	Safety Technician B began calling hospitals and nursing homes to see if any of these facilities could recognize the name they found.
5:00 p.m.	After contacting approximately 15 facilities to identify the location of the individual on the prescription without success, they stopped for the day and went home.

8.3.1.3.11 Tuesday, December 1, 1992

8:30 a.m.	Supervisor A reopened the trailer containing the other medical waste boxes and began to look for the other two containers that had arrived with the box containing the radioactive material.
9:30 a.m.	Supervisor A and Safety Technicians A and B again went through the red bags in these two containers trying to find identifying information. This time, they were successful and found an individual's name associated with the SHM nursing home.
9:45 a.m.	Immediately, Supervisor A called the SHM nursing home to inform them that radioactive material had been discovered in the waste that they had picked up from the home on November 25, 1992. The SHM nursing home staff informed BPI-Carnegie that they did not have any radioactive material at their facility, but they did identify a resident (Patient A) that had recently undergone cancer treatment therapy at IRCC.
10:00 a.m.	The staff at the SHM nursing home immediately called the IRCC and spoke with Nurse A, who contacted Physician A. Physician A suspected a possible source loss.
11:00 a.m.	Nurse A called Medical Physicist A in Johnstown, Pennsylvania, to notify him of the source loss. During the telephone conversation, Medical Physicist A asked RTT-B to evacuate the treatment vault, use remote control to extend the iridium-192 source into a connecting catheter, and observe the PrimAlert-10 radiation monitor to verify the presence of a radiation reading. RTI-B informed Medical Physicist A that the PrimAlert-10 did not detect any radiation levels.
11:40 a.m.	Medical Physicist A arrived at the IRCC and verified the absence of the iridium-192 source by performing an autoradiograph of the source wire, monitoring the PrimAlert-10, and by performing portable survey meter measurements.
11:44 a.m.	Medical Physicist A called BPI-Carnegie to inform them that they would arrive shortly to retrieve the radioactive material.
11:45 a.m.	Medical Physicist A notified OSC's Radiation Safety Officer (RSO) in Harrisburg, Pennsylvania.
11:50 a.m.	Medical Physicist A notified NRC, Region I.
3:15 p.m.	Medical Physicist A and Physician A, arriving in separate vehicles at BPI-Carnegie, were met by Safety Technician A, who had previously supervised and participated in the unloading, identification, and subsequent isolation of the BFI box containing the radioactive material.
	Upon approaching the green recycle container, Safety Technician A unlocked it and stood approximately 2 meters (6.6 feet) away with a portable survey meter turned on. Medical Physicist A also had a portable survey meter and noted that the radiation reading at about 1.5–1.8 meters (5–6 feet) away from the container was above 7.8 mSv (780 mrem) per hour.

Medical Physicist A placed the lead container in which the iridium-192 source was originally shipped on the ground next to the green recycle container. Wearing surgical gloves, Medical Physicist A lifted the box containing the source out of the recycle container, and both Medical Physicist A and Physician A opened the box and began taking plastic bags out. These red plastic bags were removed one at a time and carried toward Medical Physicist A's portable survey meter located approximately 4.6 meters (15 feet) away. The first two bags contained no radioactive material. The third bag contained radioactive material. Using long-handled [about 30 cm (11.8 inches)] forceps, Medical Physicist A opened the red bag and saw several smaller red bags inside. One contained three catheters and one contained a single catheter. The bag with the single catheter was surveyed and indicated the presence of radioactivity. Medical Physicist A quickly walked to the lead container and placed the single catheter containing the source inside.

Medical Physicist A estimated that it had taken approximately 70 seconds from the time they located the red bag containing the source to the time it took to secure the source in its protective shield.

After securing the source, Medical Physicist A placed the container inside his truck and secured its movement by placing rubber-covered sandbags around it. Medical Physicist A surveyed the source container; it read 0.35 mSv (35 mrem) per hour. In addition, he surveyed the cab of his truck and obtained a reading of 6 µSv (0.6 mrem) per hour.

3:45 p.m.	Once the source was secured in place, Medical Physicist A transported it back to the IRCC for storage. No radioactive material placards were placed on the vehicle.
5:10 p.m.	Medical Physicist A arrived at the IRCC and placed the container with the iridium-192 source in the Treatment Room.
5:12 p.m.	Medical Physicist A surveyed the source with his survey instrument and obtained a reading of 0.35 mSv (35 mrem) per hour.

8.3.2   Greater Pittsburgh Cancer Center Incident

On Monday, December 7, 1992, Medical Physicist B, from the Greater Pittsburgh Cancer Center (GPCC), reported that a 1.28 E + 11 Bq (3.45-Ci) iridium-192 sealed source apparently broke off from the end of the source wire while being removed from a patient following a completed endobronchial HDR treatment. The system being used was an Omnitron 2000 HDR afterloader identical to the afterloader involved in the IRCC incident. Further, GPCC was operated by OSC, the same licensee that operated the IRCC at which the November 16, 1992, incident occurred. Please read NUREG 1480 (1993) for the complete writeup on this event.

8.3.3 Omnitron High Dose Rate (HDR) Remote Afterloader System

This section describes the Omnitron International, Inc.'s (Omnitron's) 2000 HDR Remote Afterloader System; the nickel-titanium source wire; the HDR afterloader software; and the manufacturer's quality assurance and quality control (QA/QC) program. In addition, this section discusses the training, operating, and emergency procedures Omnitron gives to its customers.

The Incident Investigation Team (the team) obtained information from Omnitron's brochures and manuals; safety evaluations of the source and device (Registration Certificates); letters provided to the State of Louisiana to support the safety evaluations; and interviews with the persons involved. Additionally, the Food and Drug Administration (FDA) gave the NRC information they obtained from their investigation of the incident at the Indiana Regional Cancer Center (IRCC).

The scope of this investigation included only those mechanical or electrical components that could have contributed to the break in the wire and reflects the team's observations and review of documents and interviews with Omnitron personnel. The afterloader is operated by a computer; however, it is designed so that all low-level safety systems (e.g. hardware, interlocks, watchdog timers) are independent of the computer. The following description summarizes the overall operation of the afterloader, presenting specific details about the areas the team felt could have contributed to the wire break.

8.3.3.1 Description of the Afterloader System
This section discusses the main components of Omnitron's Model 2000 HDR afterloader system, which are the afterloader, the main console, the door status panel, the afterloader system safety features, and the implanted catheters and connecting catheters.

8.3.3.2 High Dose Rate Afterloader
The HDR afterloader (Figures 8.21 and 8.22) contains the mechanical and electrical hardware necessary to execute a treatment. The afterloader contains a microcomputer that communicates with the main console through an RS-422 data link. The unit is approximately 107-cm high, 56-cm wide, and 61-cm deep (42-in. H × 22-in. W × 24-in. D) and weighs approximately 148 kg (325 lb). Casters are mounted on the unit so that it is easy to move. The main components are the friction drive mechanisms, active and inactive wires, optical and mechanical switches, storage, safe treatment channel turret, stepping motors, emergency retract motors, backup power supply, and console interface card.

The turret (Figure 8.23) allows for connection of up to 10 treatment channels. These channels are numbered on the turret head. Each treatment channel can

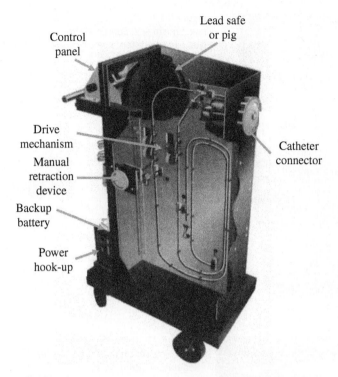

Control panel

Lead safe or pig

Drive mechanism

Manual retraction device

Backup battery

Power hook-up

Catheter connector

Figure 8.21 HDR internal mechanism. Source: United States Nuclear Regulatory Commission (1993).

have up to 20 dwell positions with each position having dwell times of 0.1-second to 3-minute increments. For each channel, the source can be moved in 1.1-cm increments (the minimum) to a maximum treatment distance of 21 cm.

The implanted and connecting catheters are joined to the HDR afterloader by appropriate fittings. The turret rotates to allow the single active wire to extend and enter each implanted catheter for the desired treatment. The computer can determine the location of the turret by use of optical sensors and unique coding on the edge of the turret wheel.

The storage safe provides shielding for the tip of the source wire where the source is located. When the source wire is in the "safe" position (i.e. in the shielded position), the source is located within the center of the safe, and radiation levels at 1 m (39.4 in.) from the surface of the device are below 0.001 mSv/hr (0.1 mR/hr).

Two independent drive systems are used in the afterloader. One drive system drives the source wire used for patient treatment and the other drives the dummy

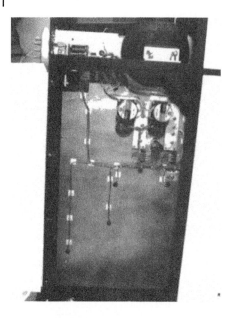

Figure 8.22 HDR internal mechanism – 2. Source: United States Nuclear Regulatory Commission (1993).

Figure 8.23 Catheter connection plate. Source: United States Nuclear Regulatory Commission (1993).

wire used to ensure that the catheter path is not constricted or obstructed. The dummy wire is also used during service and maintenance to ensure that the device is operating and functioning correctly.

Each drive system uses a pinch roller attached to a solenoid. When the solenoid is energized, the pinch roller puts pressure on the wire so that the wire meets the stepping motor's drive roller. The stepping motor moves the wire a predetermined distance (a step) for each signal sent from the computer to the motor. When the drive roller rotates, the wire travels through the guide tubes, owing to the friction between the wire, pinch roller, and drive roller.

An optical encoder mechanism is used for independent wire tracking and is located below the drive mechanism. Pinch rollers that are attached to solenoids are used for each system to keep pressure on the wire so that the wire is in contact with the optical encoder.

During a treatment, the appropriate motors and solenoids for the source-wire path or the dummy wire path are energized causing friction that allows movement of the wire and optical encoder roller.

As a stepping motor drives the source wire forward from the lead safe, a microswitch is tripped (the parked switch, which indicates the safe position, is located at the bottom of the source wire path) that resets a counter to zero (electric pulses are counted). As the wire travels, it reaches a microswitch (home position sensor) just before entering the turret and the distance is tracked by the pulses sent to the computer from the optical encoder.

The source wire travels past the first dwell position and then is pulled back-wards, to remove any slack, to the first dwell position. The wire remains in this position for the prescribed dwell time and then travels to the next dwell position for that prescribed dwell time until all dwell positions and times have been completed. During each movement of the wire the distance is recorded. After all dwell positions and times are completed, the wire is retracted. The number of pulses is counted during the retraction process until the home switch is deactivated. When the wire reaches the safe position, the solenoids are de-energized.

A computer controls the afterloader through an interface board in the computer and a signal multiplexer. Separate circuits on the interface board count the number of pulses in each direction. By using the optical encoders in combination with the stepping motors, the HDR can account for the slippage of the wire. In addition, timers are located on the interface board to time the stepping motors (used for error detection).

An emergency drive system is in the afterloader that is separated from the main drive systems (stepping motors). The emergency drive system consists of a solenoid, pinch roller, and battery-operated dc motor. In the event of an emergency retract, the dc retract motor is energized and power is removed from the solenoids associated with the stepping motors and optical encoders, thus removing any friction forces caused by these systems. The dc motor continues to operate until the end of the source wire (opposite the source-end) contacts the parked switch.

The back of the afterloader contains a status panel, shown in Figure 8.24, which consists of the following:

- The "SAFE," green, light-emitting diode (LED) indicates that the end of the wire opposite the source end has been detected by the sensor (i.e. the parked switch is tripped).
- The "IN PROGRESS," amber, LED indicates that treatment has been initiated.
- The "DUMMY PARKED", 11 green, LED indicates that the end of the wire has been detected by the sensors (i.e. parked switch is tripped).

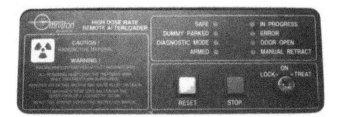

Figure 8.24 Console on the HDR unit. Source: United States Nuclear Regulatory Commission (1993).

- The "ERROR," red, LED indicates that an error has occurred. The error LED is lit whenever an error condition occurs. These error conditions can be caused by malfunctioning equipment, power failures, catheter restrictions, etc., and the cause of the error is displayed on the computer screen.
- The "DIAGNOSTIC MODE," amber, LED indicates that the system is undergoing diagnostics. Only Omnitron personnel are authorized and have access to this mode. Under this mode of operation, direct movement of the source and dummy wires can be controlled through the console.
- The "DOOR OPEN," 11 red, LED indicates that the switch on the treatment room door has been tripped. Either the switch is broken, or the door is open.
- The "ARMED," amber, LED indicates that the system is ready for treatment (i.e. passwords have been entered correctly, keylocks have been set, etc.).
- The "MANUAL RETRACT," red, LED indicates that both the AC stepping motors and the de emergency retract motor failed to retract the wire, and, therefore, a manual retraction is necessary. If this condition occurs, the LED will be lit and an alarm on the console will sound.
- The "RESET," yellow push button has a hinged cover to prevent accidental activation. For resettable errors, pushing the button will erase the error message displayed on the computer console and turn off the "ERROR" LED. For non-resettable errors, a message will appear on the screen that Omnitron needs to be called to fix the error.
- The "STOP," red push button is used to stop treatment. If this button is pushed when the active wire is out, the active wire will automatically retract. Treatment information is saved, and, therefore, treatment can be resumed from the point where it was stopped when this button was pushed.
- A key switch is provided on the console to allow only authorized personnel to control the treatment.

A label containing a radiation symbol and warning appear on the HDR afterloader next to the status panel.

The afterloader is locked, preventing users from accessing the mechanical, electrical, and software maintenance portions of the HDR afterloader system. Entry can only be accessed by the manufacturer.

8.3.3.3 Main Console

The main console consists of a microprocessor, color monitor, printer, disk drive, uninterruptable power supply, and treatment unit interface card (X). The microprocessor is a personal computer (PC) that operates the afterloader and performs first-level safety functions (e.g. dummy wire check, door interlock). The PC controls all wire movement signals and controls. It is used to enter patient data and the patient treatment plan, to initiate treatment, and to perform service diagnostics.

The computer console contains a status panel, shown in Figure 8.25, that consists of the same LEDs, controls, and key switch as those described earlier.

The system printer provides hard copies of treatment printouts, treatment logging and diagnostic reports, and error messages.

Also provided with the control console is an uninterruptable power supply, which allows treatment to continue in the event of a power failure.

8.3.3.4 Door Status Panel

The treatment room door status panel provides additional indication to personnel performing the treatment as to the status of the HDR afterloader. This panel contains the error LEDs as described for the afterloader and control console, an

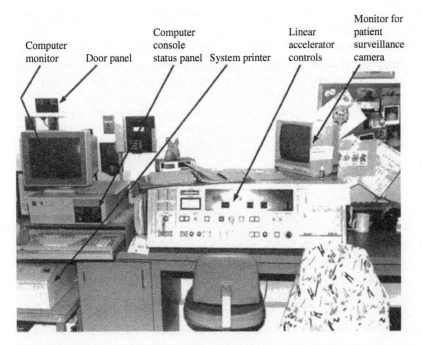

Figure 8.25 HDR operator console. Source: United States Nuclear Regulatory Commission (1993).

emergency stop button to allow the user to halt treatment, and an alarm HDR. The panel is installed near the door of the treatment room.

8.3.3.5 Afterloader System Safety Features

The manufacturer incorporated numerous safety features into the HDR after-loader system. The team observed the performance of those safety features incorporated into the HDR afterloader relevant to the constriction or obstruction of the source wire. Team members requested that Omnitron simulate the constriction condition that occurred at the IRCC with the active source wire. Team members observed that during a constriction in which the dc motor is energized, no audible alarms were activated. However, an error message did appear on the computer console for the entire duration of the wire retraction, and an error message indicating that a constriction occurred was reported to the error log. Other safety features are described in the manufacturer's manuals and brochures.

8.3.3.6 Patient Applicators and Treatment Tubes

Implanted catheters, known as patient applicators or treatment applicators are inserted into the patient during a surgical operation. The active source encased in the cavity of the source wire will reside inside this catheter during treatment time. Typical implant catheters (applicators) include rigid needles, flexineedles, and custom applicators. The flexineedles must be used with appropriate accessories (e.g. needle obturator, treatment tubes).

In the HDR treatment of November 16, 1992, at the IRCC, five 10-cm flexinee-dles were used. Each of the 10-cm flexineedles has a 20-gauge polyester tubing needle shaft, a stainless tip, aluminum coupling, and a nylon suture button. The internal diameter of the 10-cm flexineedle is 0.813 millimeter (mm).

Connecting catheters are known as treatment tubes; these are the tubes between the HDR afterloader and the implanted catheter.

In the HDR treatment of November 16, 1992, at IRCC, five 80-cm connecting catheters were used. Each of the catheters was made of Teflon tubing with aluminum coupling hardware. The nominal coupled length of the inside lumen of the 10-cm flexineedle connected to an 80-cm coupling catheter is 90.34 cm plus or minus 0.2 cm. Figure 3.8 represents a schematic of this arrangement.

8.3.3.7 Description of the Source Wire

The registration certificate for the source wire, Model SL-777 was issued by the Louisiana Radiation Protection Division. The source wire is constructed of Nitinol (nickel-titanium alloy). The source wire was produced in the following way.

Originally the wire comes in a roll from the supplier of the Nitinol wire. Lengths of wire are cut from the roll (cut intentionally long) and sent to another company to produce a cavity in one end of the wire approximately 0.34 mm

(0.014 in.) in diameter and 11-mm (0.43-in.) deep (wall thickness approximately 0.089–0.102 mm [0.0035–0.004 in.]). Previous production required a 13-mm cavity. Instructions are sent with the wire telling the company in which end of the wire to put the cavity.

The wire is sent with a traveler to another company where the cavity is X-rayed. At that time, the X-ray vendor assigns a serial number to the source wire. The wire and videotape of the X-ray are shipped back to Omnitron's Houston, Texas, office.

The wire is inspected at the Houston office for gross defects in material or workmanship and the wire's critical dimensions are checked. The wire is then sent with its traveler to Omnitron's source production facility in Edgerly, Louisiana.

At the Edgerly facility, the critical dimensions are again checked against what is stated on the traveler. If these dimensions are not the same, the discrepancy is either handled over the telephone, the wire was scrapped, or the wire was returned to Houston for rework. Once the personnel are satisfied with the wire, a sample piece of the wire is placed in the welding fixture located in the hot cell and a test weld was performed on a sample piece of the wire.

The weld is inspected for an even-flow, uniform heat distribution (heat ring), and a "shiny" surface. If the weld is satisfactory, the "remaining portion of the wire is cut to length." The wire is placed in the hot cell, and a weld is performed on the end opposite the cavity. All welds are performed, using a tungsten inert gas process. This weld is inspected as mentioned earlier, and if found to be satisfactory, the production continues.

Active iridium wires 5-mm (0.2-in.) long are received from the reactor and are placed in the hot cell. Two active iridium wires are placed in the source wire cavity and a 1-mm plug is placed on top. Previous production required a 3-mm (0.12-in.) plug. This 3-mm (0.12-in.) plug was the size used in the IRCC source wire. This end is welded to form a closure weld. The weld is inspected as previously described, and the outside diameter is measured. If the weld and outside diameter of the wire are acceptable, the wire is leak tested and placed in a transport container. The approximate dimensions of the finished wire are 2201 mm (86.7 in.) in length and 0.58 mm (0.023 in.) in diameter.

The transport container is moved to an HDR afterloader in the laboratory. The wire is loaded in the afterloader and cycled through the connecting catheters (i) to a wipe-test fixture where a wet and dry wipe are taken; (ii) to a critical bend test fixture to ensure the source wire will make the curves without failure; and (iii) to a calibration station where the source is calibrated. The critical bend test consists of two curves, attached together to form a smooth "S" curve. The two radii of the test fixture were chosen to simulate actual patient treatment.

The finished wire is then packaged with the calibration and shipping papers and shipped to the customer. Omnitron staff installed the source at the customer's site. Because of the source's half-life, it is replaced approximately every three months.

8.3.3.8 Prototype Testing Performed on Nickel–Titanium Source Wire

Letters in support of the sealed source certification stated that the manufacturer subjected the source to the tests discussed in this section. Two prototype nickel–titanium wires were tested to the requirements of the International Organization for Standardization (ISO) 2919, "Sealed Radioactive Sources–Classification." This guidance does not provide specific tests for HDR afterloader brachytherapy sources. However, this guidance does provide specific tests, test procedures, and test equipment for other radioactive source usages (e.g. radiography, teletherapy, calibration). The classification of the source defines the testing conditions that the source has met.

Common practice has been to test brachytherapy wire sources to the same specifications as sealed sources used in medical interstitial and intracavitary appliances. The recommended tests for these sources are temperature, pressure, and impact. In addition, common practice has been to perform cycle and tensile tests on source wires where capsules are welded or swaged onto the end of the source as with radiography wires.

In a letter dated January 16, 1992, to the State of Louisiana Radiation Protection Division, Omnitron stated it subjected two prototype sources to the minimum classification recommended for the temperature, pressure, and impact tests. The prototype sources passed the tests in accordance with ISO 2919.

The manufacturer also subjected prototype sources to a bending fatigue test. Catheters connected to an afterloader were curved to represent pathways encountered during treatment. Different curved paths were used during the cycling of the source wire.

8.3.3.9 Description of the Omnitron 2000 Afterloader System Software

Two groups of software programs were available with the Omnitron 2000 afterloader system: (i) the Computerized Treatment Planning System (CTPS) software used to develop a treatment plan (isodose computations) for a patient and (ii) the software required to run afterloader functions.

The Omnitron 2000 HDR afterloader has a dedicated software system used in isodose computations. This software system, called the CTPS, was developed by a medical physics software company independent of Omnitron. Omnitron's medical physics personnel performed specific QA checks of the CTPS software before releasing it to its customers. The CTPS resides on a stand-alone computer hardware system and has a menu-driven interface. Treatment data is transferred to the Omnitron 2000 control console, via a standard 3.5-in. floppy disk drive. The CTPS also automatically corrects for the decay of the source. The CTPS has a full three-dimensional dose optimization feature that plans the HDR brachytherapy doses delivered by the radioactive source. Dose histograms and three-dimensional wire frame displays are produced. Separate programs exist for entering patient data and for entering or editing iridium-192 radioactive source data.

The computer software installed on the Omnitron 2000 control console was developed by Omnitron, with the assistance of a consultant. The Omnitron 2000 computer console at IRCC used Version 3.0 of the software. This software has a menu-driven operator interface and is organized in a modular manner.

The operator must (i) have the Omnitron key in TREAT mode and (ii) enter the Omnitron console program system Access Password before any access to patient or afterloader information is allowed. Some of the software programs in the after-loader console are only accessible to the Omnitron personnel.

The console and afterloader software were designed to monitor all major electrical and mechanical systems, detect the proper operation of these systems, and initiate any afterloader safety functions if necessary. The software is also designed to record and print outpatient treatment information, treatment interruption causes (e.g. "Active wire path constriction detected at XX cm11"), whether the detected condition can be reset by the HDR afterloader operator, and any device check failures detected (e.g. "Dummy wire check on Channel X failed").

The team identified the following features of the Omnitron 2000 hardware and software configuration.

1. The dummy wire executes programmed treatment to test for proper catheter routing, programming, and to ensure that no obstructions or kinks are in the catheters. For patient treatments requiring multiple catheters or applicators (similar to the 5-channel treatment of November 16), the operator can choose between having:
   - dummy wire checks of all programmed channels (i.e. catheters) before source wire treatment of each channel, or
   - dummy wire check of Channel I (one catheter), followed by the source wire entering that channel; dummy wire check of Channel 2, followed by the source wire entering that channel; and so on.

2. If the source wire jams while being extended or while being retracted from the device, one message is generated "Active wire path constriction detected at XX cm." However, the two conditions are acted upon differently by the after-loader. When the source wire jams while retracting and the pull force necessary to retract the wire exceeds that of the stepper motor upper limit, the emergency de retract motor is engaged to retract the source wire. Otherwise, if no wire jam occurs, the stepping motor retracts the wire.

3. When the emergency de motor engages, a message stating that the de motor engaged is displayed momentarily (for approximately 10 seconds) on the console's screen, with no written information being stored in memory or printed out.

4. The software was configured so that when the emergency de retract motor starts retracting the source wire, all devices that monitor and record the wire length information immediately disengage and all wire length information is lost. In addition, activation of the emergency de retract motor is not recorded on the printout.
5. Following an emergency de retract of the source wire, the afterloader operator is allowed to reset this error and restart treatment without taking any corrective actions.

Omnitron International, Inc., located in Houston, Texas, was responsible for the overall QA/QC program. However, although not directly stated, Omnitron's source-wire fabrication facility, located in Edgerly, Louisiana, was responsible for the QA/QC of the source wire once it was received from the Houston office. Therefore, the team separately reviewed the QA/QC program pertaining to (i) the production of the HDR afterloader and the initial fabrication of the nickel-titanium wire and (ii) the assembly of the source wire. Any part of Houston's QA/QC program that pertains to the source wire was considered during the review of the QA/QC program for the Edgerly facility. The NRC does not have specific regulations or guidance that relate to QA/QC programs for vendors of sealed sources and devices. The team used a draft NRC QA/QC guide to review Omnitron's QA/QC program. The team's findings were essentially the same as those reported in the FDA audits as noted in the following text.

The Lafayette, Louisiana, office of the FDA performed an audit of Omnitron's Edgerly, Louisiana, facility during December 1992. Audit findings were presented to Omnitron on December 23, 1992, and documented on FDA Form 483. The inspection report findings included multiple Omnitron deficiencies relating to their (i) quality assurance program, (ii) audits, (iii) device history records, (iv) reworking of components, (v) device master records, (vi) written procedures for in-process and finished device testing, (vii) cleaning and leak testing, (viii) calibration of equipment, (ix) validation of the work process, (x) wire receipt inspection, and (xi) use of white-out on QA documents The Dallas, Texas, office of the FDA performed an audit of Omnitron's Houston, Texas, facility during December 1992 and January 1993. Audit findings were presented to Omnitron on January 11, 1992, and documented on a Form 483. The inspection report findings included multiple deficiencies relating to (i) their QA audits and design changes, (ii) complaint file shortcomings, (iii) medical device reporting, (iv) software (afterloader and console) validation and testing, (v) manufacturing specifications and manufacturing process control, (vi) device history records, and (vii) component receipt inspection. Both FDA audits indicated a lack of procedures, component QA records, and tests. In addition, where procedures, component QA records, and tests existed, numerous examples indicated that standard QA/QC practices were poorly implemented.

Findings documented in the recent FDA audits are similar to the overall team observations. On the basis of FDA audit findings at the Edgerly and Houston offices, the Texas Department of Health, assisted by Houston police, embargoed the Omnitron 2000 HDR afterloader on January 12, 1993. To have the embargo lifted, Omnitron must complete the corrective actions needed to resolve the issues raised during the FDA audits.

Upon request, Omnitron provides a two-day, in-house, or onsite, training course to customers. The course in its current form became available May 1, 1992. In addition, Omnitron provides training at the customer's facility before treatment of the first patient.

Omnitron chose four trainers based on their knowledge and experience. They have not been formally trained or received training to become trainers. Omnitron had (i) no written policies or procedures on what training and experience is required before an individual can become a trainer and (ii) no written procedures specifying who should be trained at the customer's facility. Typically, an individual responsible for training calls or writes the customer's facility and asks the responsible physicist who they feel should be trained.

Typically, the responsible physicist and the person who will most likely perform the treatment are trained. The director of training told the team that Omnitron will periodically call the customers who have not had two people trained to schedule future courses for them.

Omnitron provides the customer with training and reference material. The material includes system specifications, applicator and accessory specifications, treatment planning, HDR software and hardware, an instruction manual, and a one-page emergency procedure. The objectives of the two-day training course are to teach the customer how to perform treatment planning and dose calibration, execute treatment using the HDR afterloader, use emergency procedures in the event of a source retract failure, and discuss uses of accessories and applicators. The emergency procedures, which are provided to the customer, do not address breakage of the wire. Training provided before treatment of the first patient consists of a review of the operation and safety features of the HDR afterloader. Basic radiation safety training is not provided.

According to Omnitron's Sr. Vice President of Research and Development, the safety features and error messages that appear on the console monitor screen are discussed as part of the training; however, these error messages are not addressed in Omnitron's written procedures. Errors that cannot be reset are not discussed in the training, and the customers are told to call Omnitron if an error of this type occurs.

8.3.3.10   Equipment Performance

This section presents (i) the potential analysis of the source-wire breaks, (ii) the immediate cause of the wire break, (iii) the root cause of the wire break, (iv) the failure of the HDR afterloader system to detect the length of the broken wire, and (v) the performance of the PrimAlert-10 area radiation monitor.

The team obtained its knowledge of the source wire and HDR afterloader from Omnitron International, Inc.'s (Omnitron) brochures, visits to the HDR afterloader and source wire facilities, direct observation of the broken wire and afterloader, engineering reference material, material certification, prototype test results, and other information provided by Omnitron. Most of the testing was performed by Southwest Research Institute (SwRI), San Antonio, Texas, under an existing contract with the NRC. The team also reviewed the performance history of the PrimAlert-10 that was installed in the treatment room at the Indiana Regional Cancer Center (IRCC).

8.3.3.11   Failure Analysis Pertaining to the Source Wire

This section describes the failure analysis performed on the broken source wire at IRCC, and the GPCC. It also includes an analysis of other sample wires used for testing purposes and describes (i) possible areas where the source wire failure may have occurred; (ii) observations of equipment used during assembly, testing, and shipping that may have caused wire degradation; (iii) description of the test samples; (iv) test performed of the samples; (v) observations of the test results; (vi) implications of these observations; and (vii) further testing.

8.3.3.12   Possible Failure Areas

On December 4, 1992, the quarantine of IRCC's Omnitron 2000 HDR afterloader was lifted by the NRC, and team members observed the disassembly of the device by Omnitron. The team asked Omnitron personnel specific questions pertaining to the source wire and afterloader.

Upon examination of the inactive portion of the source wire, it appeared that the wire broke just above the bottom of the cavity. The entire source wire length as specified on drawings and the associated traveler was 2201 mm (86.65 in.). The inactive portion of the wire was measured and found to be approximately 2188 mm (86.14 in.). Therefore, the remaining portion of the source wire containing the iridium-192 source was approximately 13 mm (0.51 in.). The cavity produced in the end of the wire was also approximately 13 mm (0.51 in.) in depth.

The team took wipe samples of the active wire and inactive dummy wire and analyzed them on site to determine leakage of radioactivity from the source. The wipe samples were below the regulatory limit of 185 Bq (0.005 μCi) but were sent to Region I for analysis. The wipe samples were divided between Omnitron and NRC, both having representative samples of the dummy and source wires. The results from the gamma spectroscopy test performed in NRC's Region I office indicated a photopeak having an energy corresponding to iridium-192 for both wires. However, the levels obtained were below the regulatory limits of 185 Bq (0.005 µCi).

To examine the HDR afterloader system for conditions that could damage or break the source wire, the afterloader was divided into six areas where the break of the wire could have occurred.

8.3.3.13 Organization of Oncology Services Corporation

The corporate headquarters of OSC is located in State College, Pennsylvania. OSC managed 10 cancer treatment centers that use HDR afterloader systems in five states, including six located in Pennsylvania that are licensed by NRC Region I. Figure 5.1 presents an organization chart for OSC. OSC also managed a number of other cancer centers that do not use HDR afterloaders. All OSC cancer centers used linear accelerators.

OSC corporate headquarters provided support to the cancer centers in clinical and quality assurance areas. The corporate Radiation Safety Officer (RSO) was located at the Harrisburg Cancer Center and was responsible for providing radiological safety oversight to the cancer centers, including radiological safety program development and implementation, radiological safety audits, and radiation safety training.

Physician C was the chief executive officer of the OSC. Physician B was their medical director and primarily dealt with the physicians at the various centers. Physician D was director of all brachytherapy programs. Managers in the corporate office developed policies and expected the various centers to develop the specific procedures needed to follow these policies.

Physician A was the Medical Director of the IRCC and had full authority in all clinical areas and medical care of patients. Physician A was an authorized medical user on OSC's NRC license.

Medical Physicist A was a contract employee who worked two evenings a week at the IRCC to perform treatment planning and review therapy quality control records. Medical Physicist A reported to Physician A and the RSO. During the day he worked at another hospital that had no affiliation with OSC.

Registered Therapy Technician RTI-B was the lead technologist at the IRCC and supervised the other technologists, RTI-A, and the registered technologist radiographer (RTR). As lead technologist, RTT-B assigned and scheduled work for other technologists. Nurse A reported to Physician A.

8.3.3.14 Management Oversight

Indiana Regional Cancer Center, the IRCC staff was headed by Physician A who contracted with OSC. Neither Physician A nor the full-time staff at the IRCC were aware of who the RSO was. Physician A was not certain whether he or the corporation had the responsibility of providing radiation safety training. Physician A stated that he was not aware of any formal radiation safety training that the corporation provided to the technologists and other center staff. During interviews, and after being informed by the team who the RSO was, the staff said that the RSO came to the IRCC about once a year.

The RSO stated that Medical Physicist A was responsible for conducting training at the IRCC. Medical Physicist A said he was not responsible for radiological safety training, and, indeed, the physicist's contract did not explicitly list radiological safety training as one of his duties. However, Medical Physicist A's contract listed technical supervision of technologists as a duty. This responsibility was limited to briefings on individual treatment plans, dosimetry, and/or technical aspects of patient treatment. The medical physicist's contract did not specifically list responsibility for HDR brachytherapy support although Medical Physicist A had verbally agreed with OSC to provide this support. The contract had not been updated to reflect this verbal agreement.

8.3.3.15 Safety Culture
- The team identified the following overall differences in the safety culture of OSC, IRCC, and GPCC:
- OSC staff, including corporate managers, believed that the source wire would not break.
- IRCC staff believed that the source wire would not break. Therefore, the IRCC staff were neither conditioned nor prepared to appropriately respond to a source-wire-break incident. This was demonstrated by IRCC's inadequate response to the November 16, incident.
- Medical Physicist B at the GPCC told the team that after his August 1992 training he believed that a source wire break was a credible, "worst case scenario," accident. He responded appropriately during the source-break incident at the GPCC.
- The IRCC technologists were not familiar with the operation of the portable survey meter.
- Medical Physicist B the GPCC, upon receiving constriction error messages on the computer screen and hearing the audible alarm, entered the treatment room with a portable survey meter. He performed appropriate radiological measurements and assessment and ascertained the location of the source inside the connecting catheter and responded accordingly.
- The IRCC technologists and Physician A either noticed or were informed that the PrimAlert-10 was flashing red. The technologists stated that the PrimAlert-10 had alarmed multiple times without the presence of radiation

in the treatment room, and they, therefore, assumed the PrimAlert-10 was malfunctioning during the November 16, 1992, incident.

- Medical Physicist B at GPCC upon receiving an audible alarm of an error and a visual alarm on the computer screen indicating "Emergency Condition, Manual Retract, Check Source Status," immediately entered the treatment room with a portable survey meter and observed that the PrimAlert-10 was flashing red. Medical Physicist B's assessment of the alarming PrimAlert-10 was that the source was not in its shielded configuration and the physicist confirmed this assessment with portable radiation survey meter measurements and responded accordingly.
- The IRCC technologists had limited knowledge and experience working with radioactive materials. Based on their day-to-day experience working with a linear accelerator, they were conditioned to believe that by turning the linear accelerator off, no radiation would be present. This conditioning and the fact that the HDR afterloader indicated that the source was "safe" caused them to believe that the source was inside the machine and that the PrimAlert-10 alarm was spurious. After completing the dummy wire insertions, Medical Physicist A had left the IRCC and Physician A was not continuously present at the HDR console during patient treatment. IRCC staff had not received a copy of draft procedures: "Oncology Services Corporation, Department of Physics, HDR Treatment Manual."
- Medical Physicist B at GPCC had prior extensive experience working with radioactive materials, including cobalt-60 teletherapy units and an HDR afterloader. Medical Physicist B performed planning and administration of HDR treatments and was continuously present at the HDR console during patient treatment. In addition, Physician E, who is an authorized user, was watching the patient surveillance camera during the treatment. OSC gave draft procedures, "Oncology Services Corporation, Department of Physics, HDR Treatment Manual," to GPCC before the IRCC event.
- Specific aspects of OSC's, IRCC's, and GPCC's safety culture were revealed during the team's personal observations and interviews of personnel.
  Oncology Services Corporation:

- OSC staff believed that the source wire could not break. OSC's Physician D said he assumed the source wire was safe because the Federal Government had licensed it.

Indiana Regional Cancer Center:

- Source Wire. IRCC staff believed the source wire could not break and that if the machine malfunctioned the wire would be stuck outside the machine and would need to be manually retracted. The IRCC technologists knew that if the wire was stuck outside the HDR afterloader that it could be manually retracted by using the wheel on the unit or by just pulling the HDR afterloader away from the patient, effectively pulling the source wire out of the patient.

- Survey Meter. A portable survey meter was positioned close to the HDR afterloader computer console. All the technologists knew that the meter was there and that it detected radiation. None of the technologists routinely used the survey meter. From direct observation of technologists handling the survey meter, the team found that the technologists were not very familiar with its use. Medical Physicist A said he had shown the technologists how to use the survey meter, but this was not documented.

- Alarm Response. RTR noticed the PrimAlert-10 alarm upon the third entry into the treatment room. RTR notified the other two technologists (RTT-A and RTT-B), who assumed the PrimAlert-10 was not working properly rather than that the device was detecting radiation in the room. The PrimAlert-10 had previously alarmed when no radiation was present. Specifically, it would flash for no apparent reason, presumably indicating radiation in the treatment room. On one occasion, over a year before this November 16, 1992, incident, it had flashed all day. On this occasion, RTT-A brought a portable survey meter into the room, as suggested by Medical Physicist A, to ensure that no radiation was in the room. The survey meter detected no radiation, and the technologists concluded that the PrimAlert-10 was not working properly. The IRCC staff stated they believed that the malfunctioning PrimAlert-10 was sent for repair. The PrimAlert-10 vendor stated that the unit currently used at the IRCC had never been returned for service. Neither the IRCC staff nor the team could determine whether the PrimAlert-10 in the IRCC treatment room at the time of this incident was the instrument that caused confusion a year ago or whether it was a different instrument.

- Since the PrimAlert-10 had malfunctioned in the past, the technologists assumed that, if the PrimAlert-10 flashes when the linear accelerator is not turned on or the HDR afterloader is not being used, the monitor was malfunctioning rather than that radiation was present in the treatment room. They, therefore, did not use a portable survey meter upon entering the room after the PrimAlert-10 alarmed on November 16, 1992. RTT-A resolved the problem with the flashing monitor by unplugging the PrimAlert-10 thereby allowing it to reset. RTT-A stated that this had occurred multiple times in the past.

- The RTR had not had this mental conditioning because the RTR had not worked at the IRCC if the other technologists and because the staff at the other facility where the RTR worked believed their radiation monitors were reliable.

- BDR Afterloader Usage. Physician B during an interview discussed the fact that most of the OSC facilities had been linear accelerator facilities until the HDR afterloader systems were brought on site. The IRCC received their HDR after-loader in late 1991 and did not begin using it until February 1992. The IRCC did not have a cobalt-60 teletherapy unit.

- When linear accelerators are turned off no radiation is present, therefore, the IRCC technologists were not used to dealing with a device that contained a radioactive isotope.
- This lack of experience was exacerbated by the fact that the IRCC staff had only performed about 30 treatments on 10 patients since February 1992.
- RTT-A, who treated the patient on November 16, 1992, did not know the activity of the source and did not appear to understand its potential radioactive hazard. RTT-A was unaware of the magnitude of exposure received on entering the treatment room if the HDR afterloader source was unshielded.
- The IRCC staff believed that if the HDR afterloader failed, they would hear an alarm and see an indication on the HDR computer monitor of the failure. Medical Physicist A believed that if the HDR afterloader failed there would be all kinds of "bells and whistles." The HDR afterloader software is designed to produce visible error message indications on the computer console monitor during de emergency retraction of the source wire but is not designed to produce any other record. This visible error message indication should have caused a portion of the upper half of the screen to flash red. However, the afterloader system gives no audible alarm for an emergency retraction. The IRCC technologists did not see the error message indications on the computer console monitor. The IRCC staff revealed that they were not very familiar with the error messages associated with the HDR afterloader because Omnitron had not supplied all the meanings of all the error messages to them.
- IRCC technologists stated that they did not verify the operation of the treatment room door interlock associated with the HDR afterloader before operating the unit. Additionally, IRCC technologists stated they did not use a check source to verify the operability of the PrimAlert-10 area radiation monitor before HDR afterloader operation.
- At the time of the incident at the RTT-A was operating the HDR treatment system from the computer console, and the RTR was also present in the area of the computer console.
- Owing to the placement of the computer console and the patient surveillance camera monitor, it was difficult for one operator to watch both simultaneously. Additionally,
- the PrimAlert-10 was located inside the treatment room and was visible through a window in the treatment room door. Therefore, the operator would have to look away from the console and monitor to observe any alarm on the PrimAlert-10. Physician A was not present at the computer console during the treatment, and Medical Physicist A was off the site.

- The control console panel may be positioned so that it is in as prominent a position as the computer monitor. The panel and the monitor need to be positioned so that when the console panel alarms, the operator would automatically direct attention to the error messages appearing on the computer monitor.
- If the door alarm display panel is used for alarm information, the operator's attention is not automatically directed to error messages appearing on the computer monitor.
- While the emergency de motor is retracting the active source, no alarm is audible, and the error message only stays on the screen untQ. The end of the active wire contacts the park switch. It normally takes approximately 10 seconds for the emergency motor to retract the source. Therefore, the saliency of the alarm indications may be less than optimal.

8.3.3.16 Emergency Operating Procedures

Omnitron gave framed operating procedures to be used in an emergency involving the HDR afterloader to IRCC and GPCC. These procedures were not comprehensive and did not address the incidents that occurred at IRCC or GPCC or suggest an appropriate response to either incident. Each center had posted these procedures by the HDR afterloader console, but they were not used during the incident. The procedures did not provide instructions to workers as to "how to" perform any radiological safety functions, including radiological surveys and precautions.

8.3.3.17 Training

Omnitron Provided Training at the Indiana Regional Cancer Center Omnitron provided the team a letter describing the training they had given the staff at the IRCC from which the following information is summarized:

Medical Physicist A had notified Omnitron that Medical Physicist A, RTT-B, and Physician A would be operating the HDR afterloader. On December 9 and 10, 1991, Trainer A from Omnitron trained Physician A, Medical Physicist A, and RTT-B. The training included:

- instruction in the use of the treatment planning computer to plan patient treatments,
- instruction in operation of the HDR afterloader system,
- a demonstration of the safety features and emergency procedures to be followed,
- a review of the equipment warranty, and support services that Omnitron provides, and
- a demonstration of the use of the various catheters provided with the system.

On February 27, 1992, Trainer A from Omnitron conducted a review of the operation and safety features of the system. The treatment of the first patient was to be conducted on February 28, 1992. From the interviews, the team learned that Physician A, Medical Physicist A, and RTI-B had attended this review and that RTI-A and Nurse A were intermittently present when the review was being conducted.

8.3.3.18   Radiation Safety Training at the Indiana Regional Cancer Center

RTT-A, RTT-B, and RTR stated that during their tenure at IRCC they had not received any formal radiation safety training. All informal training they received was conducted by Medical Physicist A. However, the team could not verify the content of this training. Two technologists stated that the radiation safety training at IRCC was not very good, but the team could not verify whether they had communicated this opinion to their managers. The RSO was under the impression that Medical Physicist A was responsible for training in radiation safety at the IRCC. Medical Physicist A said that the contract with OSC did not explicitly list radiation training or other types of training as one of his duties. The RSO did not consider it appropriate for the RSO to provide radiation safety training to someone with Physician A's qualifications and, further, stated that if Physician A needed this type of training, Physician A should not be approved by the NRC as an authorized user. NRC Region I Inspectors identified similar training weaknesses during a recent inspection at OSC cancer centers in Exton and Mahoning Valley Cancer Centers in Pennsylvania.

OSC's RSO submitted the quality management (QM) program for HDR brachytherapy to the NRC on January 22, 1992. The licensee's QM program was two pages long and addressed the following 10 points:

1. The authorized user was to date and sign each prescription before treatment.
2. The patient's identity was to be verified.
3. The treatment setup was to be verified against the treatment plan before treatment.
4. The staff was required to ask questions concerning the treatment procedure before treatment if they were unsure what to do.
5. Dose calculation checks were required before patient treatment. Two individuals were to verify input data for the HDR afterloader.
6. The physician was required to sign and date a written record after treatment, indicating the dose administered.
7. The staff was allowed to treat patients without checking the dose calculation before treatment if the delay in treatment would jeopardize the patient's health.
8. A "radiologist physicist" was to perform acceptance test on each treatment planning or dose calculating computer program for use with the HDR afterloader system.
9. Periodic reviews were required of the QM program for HDR afterloader treatments.

The team noted that the QM program did not provide, and NRC QM requirements may not have required, guidance or any procedural requirements that could have, upon implementation, prevented the occurrence of this incident. The following problems do not appear to be addressed by the QM program:

- Patient left the treatment room with the source inside. The QM program did not address surveying the patient after treatment. If the program had required a patient survey and the staff had followed the program, the radioactive source left in the patient would have been detected and could have been removed before the patient left the cancer center.
- Identification of the implanted catheters was not accomplished. The QM program did not specifically address the need to identify these catheters and, apparently, the IRCC did not label them. Therefore, the implanted catheters could be incorrectly connected to the HDR afterloader.
- The correct emergency response to most error messages was unclear. The QM program did not specifically address how to respond to error messages from the HDR afterloader system.

8.3.3.19 Summary of Safety Culture Issues

The analysis of this incident was very thorough and the NUREG 1480 document describes in more detail aspects of the event. The investigation team that I was a part of spent a large amount of time on site and at the NRC in Bethesda analyzing the event and writing up the results. As stated at the beginning of this section, the lessons learned from this event are just a pertinent now as they were then. Any new medical device must be analyzed thoroughly, and any device or other nuclear piece of equipment must be designed to be failing safe.

8.4 Goiania, Brazil Teletherapy Machine Incident (IAEA 1988)

About the end of 1985 a private radiotherapy institute, the Instituto Goiano de Radioterapia in Goiania, Brazil, moved to new premises, taking with it a cobalt-60 teletherapy unit and leaving in place a cesium-137 teletherapy unit without notifying the licensing authority as required under the terms of the institute's license. The former premises were subsequently partly demolished. As a result, the cesium-137 teletherapy unit became totally insecure. Two people entered the premises and, not knowing what the unit was but thinking it might have some scrap value, removed the source assembly from the radiation head of the machine. This took the unit home and tried to dismantle.

In the attempt the source capsule was ruptured. The radioactive source was in the form of cesium chloride salt, which is highly soluble and readily dispersible.

Contamination of the environment ensued, with one result being the external irradiation and internal contamination of several persons. Thus began one of the most serious radiological accidents ever to have occurred.

After the source capsule was ruptured, the remnants of the source assembly were sold for scrap to a junkyard owner. The junkyard owner noticed that the source material glowed blue in the dark. Several persons were fascinated by this and over a period of days friends and relatives came and saw the phenomenon. Fragments of the source the size of rice grains were distributed to several families. This proceeded for five days, by which time several people were showing gastrointestinal symptoms arising from their exposure to radiation from the source.

The symptoms were not initially recognized as being due to irradiation. However, one of the persons irradiated connected the illnesses with the source capsule and took the remnants to the public health department in the city. This action began a chain of events which led to the discovery of the accident.

A local physicist was the first to assess, by monitoring, the scale of the accident and took actions on his own initiative to evacuate two areas. At the same time the authorities were informed, upon which the speed and the scale of the response were impressive. Several other sites of significant contamination were quickly identified, and residents evacuated.

Shortly after it had been recognized that a serious radiological accident had occurred, specialists – including physicists and physicians – were dispatched from Rio de Janeiro and Sao Paulo to Goiania. On arrival they found that a stadium had been designated as a temporary holding area where contaminated and/or injured persons could be identified. Medical triage was carried out, from which 20 persons were identified as needing hospital treatment.

Fourteen of these people were subsequently admitted to the Marciho Dias Naval Hospital in Rio de Janeiro. The remaining six patients were cared for in the Goiania General Hospital. Here a whole-body counter was set up to assist in the bioassay program and to monitor the efficacy of the drug Prussian blue, which was given to patients in both hospitals to promote the decorporation of cesium. Cytogenetic analysis was very helpful in distinguishing the severely irradiated persons from those less exposed who did not require intensive medical care.

Decontamination of the patients' skin and dealing with desquamation from radiation injuries and contaminated excreta posed major problems of care. Daily hematological and medical examinations, good nursing care and bioassay of blood cultures contributed to the early detection and therapy of local systemic infections. Four of the casualties died within four weeks of their admission to hospital.

The post-mortem examinations showed hemorrhagic and septic complications associated with the acute radiation syndrome. The best independent estimates of the total body radiation doses of these four people, by cytogenetic analysis, ranged from 4.5 Gy to over 6 Gy. Two patients with similar estimated doses survived. A new hormone-like drug, granulocyte macrophage colony stimulating factor (GMCSF), was used in the treatment of overexposed persons, with questionable results. Within two months all surviving patients in Rio de Janeiro were returned to the Goiania General Hospital, where decorporation of cesium continued until it was safe to discharge them from hospital.

Many individuals incurred external and internal exposure. In total, some 112,000 persons were monitored, of whom 249 were contaminated either internally or externally. Some suffered very high internal and external contamination owing to the way they had handled the cesium chloride powder, such as daubing their skin and eating with contaminated hands, and via contamination of buildings, furnishings, fittings, and utensils.

More than 110 blood samples from persons affected by the accident were analyzed by cytogenetic methods. The frequency of chromosomal aberrations in cultured lymphocytes was determined and the absorbed dose was estimated using in vitro calibration curves. The dose estimates varied from zero up to 7 Gy. Poisson distribution statistical analysis of cells with chromosomal aberrations indicated that some individuals had incurred non-uniform exposures. Highly exposed individuals are still being monitored for lymphocytes carrying cytogenetic aberrations.

Urine samples were collected from all individuals potentially having internal contamination and their analysis was used as a screening method. Urine and fecal samples were collected daily from patients with internal contamination. Intakes and committed doses were estimated with age specific mathematical models. The efficacy of Prussian blue in promoting decorporation of cesium was evaluated by means of the ratio of the amounts of cesium excreted in feces and in urine. A whole-body counter was set up in Goiania, and the effect on the biological half-life of cesium in the organism of the dosage of Prussian blue administered to patients was estimated.

The environment was severely contaminated in the accident. The actions taken to clean up the contamination can be divided into two phases. The first phase corresponds to the urgent actions needed to bring all potential sources of contamination under control, and was in the main completed by 3 October, but elements of this phase persisted until Christmas 1987, when all the mam contamination sites had been dealt with. The second phase, which can be regarded as the remedial phase aiming to restore normal living conditions, lasted until March 1988.

The primary objectives of the urgent response were to prevent high individual radiation doses that might bring about non-stochastic effects; to identify the main sites of contamination; and to establish control over these sites. In the initial response, all actions were aimed at bringing sources of actual exposure under control, and this took three days.

Initial radiation surveys were conducted on foot over the contaminated areas. Seven main foci of contamination were identified, including the junkyards concerned, some of them, with dose levels of up to 2 Sv/h at 3.2 ft (1 m).

An aerial survey by suitably equipped helicopter confirmed that no major areas of contamination had been overlooked. Over a period of two days all of the more than 67 km^2 of urban areas of Goiania were monitored. The extents of the seven known principal foci were confirmed and only one previously unknown, minor area, giving rise to a dose rate of 21 mSv h at 1 m, was discovered.

It was possible for lesser areas of contamination to have been missed, especially in the vicinity of the heavily contaminated areas around the main foci. A complementary system of monitoring covering large areas, although limited to roads, was put into practice. This system used detectors mounted on and in cars, and 80% of the Goiania Road network, over 2090 km, was thus covered. The main foci of contamination were the junkyards and residences where the integrity of the source capsule was breached; these covered an area of about 1 km^2. Action levels in this initial response were set for the control of access (10 μSv/h); for evacuation and prohibited access (2.5 μSv/h and later 10 μSv/h l for houses, and 150 μSv/h for unoccupied areas); and for workers participating in accident management (dose limits and corresponding dose rates per day, week, and month). In total, 85 houses were found to have significant contamination, and 200 individuals were evacuated from 41 of them. After two weeks, 30 houses were free for reoccupation. It should be emphasized that these levels, which correspond roughly to one tenth of the lowest values of the intervention levels recommended by the International Commission on Radiological Protection and the IAEA (non-action levels), were extremely restrictive, owing to political and social pressures.

Subsequently, the dissemination of contamination throughout the area and the hydrographic basin was assessed. A laboratory was set up in Goiania for measuring the cesium content of soils, groundwater, sediment and near water, drinking water, air, and foodstuffs. Countermeasures were only necessary, however, for soil and fruit within a 164 ft (50 m) radius of the main foci.

The subsequent response, consisting mainly of actions undertaken for recovery, faced various difficulties in surveying the urban area and the river basin. These were compounded by the heavy rain that had fallen between 21 and 28 September, which had further dispersed cesium into the environment. Instead of being washed out as expected, radioactive materials were deposited on roofs, and this was the major contributor to dose rates in houses.

Levels of contamination m drinking water were very low. The groundwater was also found to be free of contamination, except for a few wells near the main foci of contamination with concentrations of cesium just above the detection level.

The main countermeasures undertaken during this remedial phase were the decontamination of the main sites of contamination (including areas outside the main foci), of houses, of public places, of vehicles and so on. For decontamination at the main foci, heavy machinery was necessary to remove large amounts of soil and for demolishing houses. Large numbers of various types of receptacles for the waste also had to be constructed. In addition, a temporary waste storage site had to be planned and built. This was done by the middle of November, and decontamination of the main foci and remaining areas was carried out from mid-November up until the end of December 1987.

The investigation levels selected for considering the various actions corresponded to a dose of 5 mSv in the first year and a long-term projected dose of 1 mSv per year in subsequent years. The work included the demolition (and removal) of seven houses and the removal of soil. Areas from which soil was removed were covered with concrete or a soil pad. In less contaminated places, the main source of exposure was contaminated dust deposited on the soil; after removal of the soil layers where necessary, surfaces were covered with clean soil. Of 159 houses monitored, 42 required decontamination. This decontamination was achieved by vacuum cleaning inside and by washing with high pressure water jets outside. Various procedures for chemical decontamination proved to be effective, each adapted to the circumstances, the material concerned and the level of radioactivity.

The action levels for these remedial actions were selected under strong political and public pressures. The levels were set substantially lower than would have resulted from an optimization process. In most cases, they could be regarded as more applicable to normal situations than to an accident recovery phase.

After the Christmas holidays in December 1987 the areas of lower dose rate surrounding the main foci were decontaminated. There was no need for heavy machinery, and optimization procedures were developed and adopted. This stage lasted until March 1988.

From its inception, the response generated large quantities of radioactive wastes. A temporary waste storage site was chosen 12 mi (20 km) from Goiania. Wastes were classified and appropriate types of packaging were used, according to the levels of contamination. The packaging of wastes required 3800 metal drums 55 gal (200 l), 1400 metal boxes (5 tons), 10 shipping containers 1120 ft^3 (32 m^3) and 6 sets of concrete packaging. The temporary storage site was designed for a volume of waste of 140,000–175,000 cubic feet (4000–5000 m^3), encapsulated in about 12,500 drums and 1470 boxes.

The final total volume of waste stored was 122,500 ft^3 (3500 m^3), or more than 275 truckloads. This large volume is directly attributable to the restrictive action levels chosen, both in the emergency period and in the recovery phase. The economic burden of such levels, especially in the latter phase, 1s far from insignificant.

A sampling system was built to monitor the runoff (including rainwater) from the platform on which the waste was placed. The best estimate of the radioactivity accounted for in contamination is around 44 TBq (1200 Ci), compared with the known radioactivity of the cesium chloride source before the accident of 50.9 TBq (1375 C1). No decision has yet been made on the final disposal site for the waste.

8.4.1 Safety Culture Summary

This case study is good example of the problems that can arise when the public is not informed about the hazards associated with an industry, or in this case, a piece of equipment. Providing adequate warnings about the hazards on the piece of equipment or in the facility to clearly state the nature of the hazards could have prevented the deaths and environmental concerns. The organization should never have abandoned the piece of equipment in the first place. They probably did so because they didn't want to pay the disposal cost, but instead it led to deaths, injuries, and an environmental mess.

I have used this example many times. One of our faculty members wanted to buy 100 grams of Uranium and the cost was approximately $1000. The disposal cost for these 100 g was to be about $30,000. I said no, in the loudest voice. Disposing of radioactive materials, no matter how small the activity, is tremendous. Don't buy radioisotopes if you don't have to.

References

Atomic Energy Commission (AEC), (1962), General Electric Company. Idaho Test Station. SL-1 Project; U.S. Atomic Energy Commission. Idaho Operations Office, Division of Technical Information Extension, Government Printing Office.

California Environmental Protection Agency, Department of Toxic Substances Control (DTSC), Santa Susana Field Laboratory (SSFL) (2022). Main page, https://dtsc.ca.gov/santa_susana_field_lab/.

CBS, (2019), Almanac: The Chernobyl nuclear accident, https://www.cbsnews.com/news/almanac-the-chernobyl-nuclear-accident/ (accessed December 2021).

Chemistry (2022). The reaction of sodium with chlorine. https://chem.libretexts.org/Courses/Sacramento_City_College/SCC%3A_CHEM_330_-_Adventures_in_Chemistry_(Alviar-Agnew)/04%3A_Chemical_Bonds/4.03%3A_The_Reaction_of_Sodium_with_Chlorine#:~:text=If%20sodium%20metal%20and%20chlorine,light%20and%20heat%20is%20released (accessed January 2022).

Chiotakis, S. (2021). KCRW, 1959 Santa Susana meltdown still hurts San Fernando Valley community. Why hasn't it been cleaned?. https://www.kcrw.com/news/shows/greater-la/santa-susana.

Department of Energy (DOE) (1986). Report of the US Department of Energy's team analyses of the Chernobyl-4 Atomic Energy Station accident sequence, DOE/NE-0076.

Department of Energy (DOE) (2021). 3 Advanced Reactor Systems to Watch by 2030. https://www.energy.gov/ne/articles/3-advanced-reactor-systems-watch-2030#:~: text=The%20sodium%2Dcooled%20fast%20reactor,and%20safety%20of%20the %20system (accessed January 2022).

Department of Energy (DOE) (2022). Cleanup at the Santa Susana Field Laboratory. https://www.etec.energy.gov/Operations/Major_Operations/SRE.php (accessed January 2022).

Department of Energy (DOE) and Boeing (2007). Sodium reactor experiment (SRE) accident. https://www.etec.energy.gov/Library/Main/Doc._No._28_Short_ Description_of_SRE_Accident_prepared_by_Boeing.pdf (accessed January 2022).

Department of Energy (DOE), Energy Technology Engineering Center (2022). https:// web.archive.org/web/20101118202517/http://www.etec.energy.gov/ (accessed January 2022).

Derivan, M., (2014), TMI operators did what they were trained to do, Nuclear Newswire, https://www.ans.org/news/article-1556/tmi-operators-did-what-they-were-trained-to-do/ (accessed December 2021).

Avery Hartmans, Paige Leskin, and Sarah Jackson (2022). Updated Jan 4, 2022, Business Insider, The rise and fall of Elizabeth Holmes, the Theranos founder who went from being a Silicon Valley star to being found guilty of wire fraud and conspiracy, https://www.businessinsider.com/theranos-founder-ceo-elizabeth-holmes-life-story-bio-2018-4.

Hillman Cancer Center (2022). What is high-dose rate (HDR) brachytherapy? https:// hillman.upmc.com/cancer-care/radiation-oncology/treatment/internal/high-dose-rate-brachytherapy-hdr#:~:text=HDR%20brachytherapy%20is%20a%20form, it%20using%20a%20remote%20control (accessed January 2022).

IAEA (1991). Report INSAG-7 Chernobyl Accident: Updating of INSAG-1 Saftey Series No. 75-INSAG-7.

IAEA (2006a). Chernobyl's Legacy: Health, Environmental and Socio-Economic Impacts and Recommendations to the Governments of Belarus, the Russian Federation and Ukraine, The Chernobyl Forum: 2003–2005.

IAEA (2006b). Environmental consequences of the chernobyl accident and their remediation: twenty years of experience, Report of the Chernobyl Forum Expert Group 'Environment'.

IAEA, (2021a), Misadministrations in diagnostic nuclear medicine, https://www.iaea .org/resources/rpop/healthprofessionals/nuclear-medicine/diagnostic-nuclear-medicine/misadministrations (accessed December 2021).

IAEA (2021b) Nuclear Power Reactors in the World, REFERENCE DATA SERIES No. 2 2021 Edition.

Los Alamos National Laboratory (LANL) (2000). A review of criticality accidents, LA-13638.

Malone, T.B.; Kirkpatrick, M.; Mallory, K.; Eike, D.; Johnson, J.H.; Walker, R.W. (1980). Human Factors Evaluation of Control Room Design and Operator Performance at Three Mile Island - 2 (NUREG/CR-1270).

Michal, R. (2001). Fifty years ago in December: Atomic reactor EBR-I produced first electricity. Nuclear News, American Nuclear Society.

MSNBC (2021). Meet the people exposing the truth about the Santa Susana Field Laboratory. https://www.msnbc.com/msnbc/meet-people-exposing-truth-about-santa-susana-field-laboratory-n1281010 (accessed January 2022).

Nuclear Energy Agency, (2021), Criticality safety, https://www.oecd-nea.org/jcms/c_12776/criticality-safety (accessed December 2021).

Nuclear Newswire (2017). Seawolf tries sodium. https://www.ans.org/news/article-1999/seawolf-tries-sodium/.

Nuclear Regulatory Agency (NRC) (1993). NUREG-1480, Loss of an Iridium-192 Source and Therapy Misadministration at Indiana Regional Cancer Center, Indiana, Pennsylvania, on November 16, 1992. U.S. Nuclear Regulatory Commission, February 1993.

Nuclear Regulatory Agency (NRC) (2022a). Backgrounder on the Three Mile Island accident. https://www.nrc.gov/reading-rm/doc-collections/fact-sheets/3mile-isle.html (accessed January 2022).

Nuclear Regulatory Agency (NRC) (2022b). Fermi – Unit 1. https://www.nrc.gov/info-finder/decommissioning/power-reactor/enrico-fermi-atomic-power-plant-unit-1.html (accessed January 2022).

Nuclear Regulatory Commission (NRC) (1987). NUREG 1250, Report on the accident at the Chernobyl Nuclear Power Station.

Ostrom, L.T. (1993). The human factor in a lethal misadministration. Abstract in the *Proceedings of the 37th Human Factors Society Meeting*, Seattle, Washington, October 1993.

Ostrom, L.T., Leahy, T.J., and Novak, S. (1994). Summary of 1991 and 1992 Misadministration event investigations, EG&G Idaho Formal Report, EG&G-2707, NUREG/CR-6088.

Ostrom, L.T., Rathbun, P., Cumberlin, R. et al. (1996). Lessons learned from investigations of therapy misadministration events. *International Journal of Radiation Oncology, Biology, Physics* 34 (1).

Places, Santa Susana Field Laboratory (2022). https://placeandsee.com/wiki/santa-susana-field-laboratory (accessed January 2022).

Pollack, A. (1996). *Reactor Accident in Japan Imperils Energy Program*. New York Times.

Sapere and Boeing (May 2005). Santa Susana Field Laboratory, Area IV Historical Site Assessment. p. 2. Archived from the original on 12 December 2009. Retrieved 25 January 2010.

STA/NSB (1996). Report on the Investigation into the Sodium Leakage Accident at the Prototype FBR Monju of the PNC, February 9, 1996.

Stacey, W. (2001). *Nuclear Reactor Physics*. Wiley. ISBN: 0-471-39127-1.

Tripp, C. (2018). Nuclear criticality safety standards for nuclear materials outside reactor cores, 83 FR 49956. *Federal Register* 83, 192.

Union of Concerned Scientists (UCS) (2016). Nuclear plant accidents: sodium reactor experiment. htttps://allthingsnuclear.org/dlochbaum/nuclear-plant-accidents-sodium-reactor-experiment/ (accessed January 2022).

Union of Concerned Scientists (UCS) (2021). Report finds that 'advanced' nuclear reactor designs are no better than current reactors—and some are worse. https://ucsusa.org/about/news/report-advanced-nuclear-reactors-no-better-current-fleet (accessed January, 2022).

Ventura County Star (2006). The cancer effect, October, 2006.

Wise-Paris.org, (2021), Two Accidents Have Left Their Traces: Monju and Tokai http://www.wise-paris.org/index.html?/english/ournewsletter/2/page7.html&/english/frame/menu.html&/english/frame/band.html (accessed December 2021).

9

Other Transportation Case Studies

In this chapter there will be several transportation modality case stuadies discussed. These are:

- Large Marine Vessel Accidents
- Duck Boat Accidents
- Recent Railroad Accidents

9.1   Large Marine Vessel Accidents

There are three events described in this section of the chapter. One event involves the collision of an liquefied natural gas (LNG) Carrier Collision with Barge, and two events involve United States Navy Vessels. All three could have had major consequences if conditions were slightly different. However, all three events are significant and involve poor safety culture.

9.1.1   LNG Carrier Collision with Barge

In this incident there were no injuries and no chemical spills or explosions. However, the potential for a major incident was possible. The following accident description is excerpted from the National Transportation Safety Board (NTSB) accident report (2019).

On May 10, 2019, at 1516, the 754-ft-long, 122-ft-wide LNG carrier Genesis River (Figure 9.1) collided with a 297-ft-long tank barge being pushed ahead by the 69-ft-long towing vessel Voyager. On the morning of May 10, 2019, the Genesis River, a 754-ft-long, 122-ft-wide Panama-flagged liquefied gas carrier (see Figure 9.1), was berthed at the Targa Resources Galena Park Marine Terminal, located just east of Houston, Texas, on the upper Houston Ship Channel. Following

Impact of Societal Norms on Safety, Health, and the Environment: Case Studies in Society and Safety Culture, First Edition. Lee T. Ostrom.

Figure 9.1 Genesis River. Source: NTSB Accident Report (2019).

a full onload of liquid propane gas (LPG) cargo, the vessel was scheduled to get under way at noon for an outbound transit of the Houston Ship Channel. With the cargo, the Genesis River had a displacement of 69,249 long tons (70,360 metric tons) and was on an even keel at a draft of 36.8 ft (11.2 m). It has a cargo capacity of 82,400 m^3. The density of **LNG** is 1040 lb/m^3 or 3.94 lb/gal, a little less than one-half the density of water. The energy density of a cubic meter of LNG is 38.140 MJ. A gallon of gasoline has 131.76 MJ. Another conversion that is handy is that a megajoule is approximately equivalent to 278 W. The full cargo of the Genesis River of LNG is approximately 3,142,736 MJ or 870,680.608 W of energy. That is a great amount of energy. By comparison, a 1 kt nuclear weapon develops 4.18×10^9 MJ of energy.

As a result of the collision, two cargo tanks in the barge were breached, spilling petrochemical cargo into the waterway, and a second barge in the Voyager tow capsized.

The Genesis River had been outbound on the Houston Ship Channel when, a few minutes prior to the collision, it met the inbound 740-ft-long, 120-ft-wide liquefied gas carrier BW Oak in the intersection of the Houston Ship Channel and the Bayport Ship Channel, known as the Bayport Flare. After the Genesis River and the BW Oak passed each other port side to port side, the Genesis River approached the southern terminus of the flare and a 16° port turn in the channel. As the Genesis River exited the flare and entered the turn, it crossed over to the opposite side of the Houston Ship Channel and subsequently struck the starboard barge in the Voyager's two-barge tow. The Genesis River's bow penetrated through the barge's double hull and breached its center cargo tanks. The force of the collision capsized the port barge in the tow, and the Voyager heeled considerably before its face wires parted and the vessel righted itself. Over 11,000 barrels of reformate, a gasoline blending stock, spilled into the waterway from the starboard barge's breached cargo tanks.

The Houston Ship Channel was closed to navigation for two days during response operations and did not fully open for navigation until May 15. The total cost of damages to the Genesis River and the barges was estimated at US$ 3.2 million. The cost of reformate containment and cleanup operations totaled US$ 12.3 million. There were no injuries reported.

9.1.1.1 Accident Description

The lower channel transits through Galveston Bay from Morgan's Point to Bolivar Roads near Galveston. This section is comprised of longer, straight segments, with a 530-ft-wide main channel dredged to a project depth of 45 ft. The lower channel also has separate barge lanes located on either side of the main deep-draft vessel channel, each 235 ft wide with a project depth of 12 ft (Figure 9.2). Navigational beacons marking the lower Houston Ship Channel are located to the outside of the barge lanes. As viewed from an outbound vessel, the convention for aid to navigational buoys and beacons, marking the right side of the Houston Ship Channel are green, and those marking the left side are red.

The *Genesis River* had a crew of 28 on board, as well as two pilots required for the transit through the channel. Two pilots were assigned, in accordance with a Houston Pilots policy, to prevent fatigue while handling "wide-bodied vessels" (vessels whose width exceeded 120 ft) for extended periods. During two-pilot jobs, each pilot normally conned the vessel for about half of the transit. Neither pilot nor seniority over the other pilot held authority, and each acted independently while at the conn (unless the conning pilot specifically requested the assistance of the other).

The two pilots boarded the *Genesis River* at the terminal and were escorted to the ship's bridge, arriving at 1148. The pilot who would conn the vessel first (hereafter known as Pilot 1) was given a pilot card – a three-page summary of the ship's particulars, engine speeds, and steering and navigation equipment – at his request. While Pilot 1 reviewed the card, the second pilot (Pilot 2) set up a portable pilot unit (PPU). The master of the vessel arrived about two minutes later and greeted Pilot 1.

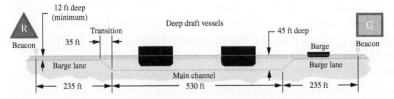

Figure 9.2 Lower Houston Ship Channel profile with navigation beacons as viewed by an outbound vessel. Source: NTSB Accident Report (2019)/National Transportation Safety Board/Public Domain.

In addition to the master, the *Genesis River* bridge team included the officer of the watch (the fourth officer), a helmsman (an able-bodied seaman [AB]), and a cadet. As the bridge team prepared to get under way, Pilot 2 requested that the crew turn off all radar alarms, telling the crew that since the vessel would be passing other vessels at short distances throughout the transit, the alarms indicating closest point of approach would be sounding often and would be a distraction. The safety management system (SMS) for the *Genesis River*'s management company required that both ship's radars be always kept on while in areas of high traffic density and near navigational hazards. However, during a post-accident interview, the master told investigators that the alarms on the vessel's automatic radar plotting aid (ARPA) system that displayed the radar data could not be turned off, so to comply with Pilot 2's request to silence the alarms, he instructed the officer of the watch to put the radars in standby. The SMS also required that one of the vessel's electronic chart display and information systems (ECDISs) be regularly monitored, but the master stated that alarms on these systems likewise could not be silenced, so he told the fourth officer to turn off the online ECDIS as well. The *Genesis River* had two ECDISs on board, which met the *International Convention for the Safety of Life at Sea (SOLAS)* requirements for redundancy. With the ECDISs off, the master instructed the fourth officer to monitor the vessel's position visually by sighting landmarks, navigation buoys, and beacons, and by monitoring the pilot's PPU. Pilot 1 told investigators that he was aware that the ECDIS was off and that the radars were in standby, but he was not concerned because he had the PPU to rely on and had good visibility for seeing navigation aids.

9.1.1.1.1 Transiting the Upper Channel The *Genesis River* took in lines and got under way shortly after noon. Pilot 1 told investigators that within the first few turns in the channel, he determined that the ship had a "small rudder" and that it responded sluggishly to the rudder commands. He stated that he needed to apply 20–30° of rudder to make course changes.

After a short time, Pilot 2 departed the bridge, eventually proceeding to the pilot room (a small lounge and bunkroom located behind the wheelhouse) at 1245. About the same time, the *Genesis River*'s second officer arrived on the bridge to relieve the fourth officer as officer of the watch. When he arrived on the bridge, the second officer noted that the ECDIS was off and that the radars were in standby. The master told him why the equipment was secured and instructed him to continue monitoring the *Genesis River*'s position visually.

At the 6–8 kts ship speed listed in the *Genesis River*'s passage plan; the channel transit was anticipated to take several hours. Sometime after 1300, the master called the chief officer to the bridge to relieve him so that he could eat lunch. The company's SMS normally required the master to be on the bridge while the ship was under pilotage but allowed the chief officer to relieve the master during long navigational transits. Before leaving the bridge, the master briefed the chief officer on the status of the ECDIS and radars and once again instructed the crew to monitor the ship's position visually. The master told the chief officer that he would return to the bridge around 1500.

As the *Genesis River* transited through the upper Houston Ship Channel, Pilot 1 used varying speeds between dead slow ahead and half ahead, combined with large rudder angles, to navigate through the multiple turns in the first half of the voyage. In a straight section of the channel south of the ferry landing at Lynchburg, Texas, the pilot increased speed to full ahead for a brief period "just to see how the ship would respond." The vessel's speed reached 9.6 kts before the pilot slowed the ship in preparation for passing a barge terminal. At 1411, the *Genesis River* met the inbound *Stolt Inspiration*, a 580-ft-long partially laden tanker. After passing the vessel at a speed of 7.1 kts, the *Genesis River* swung to port due to the hydrodynamic forces acting on the vessel in the narrow channel. Pilot 1 told investigators that he had to use hard starboard (35°) rudder to stop the swing. Additionally, he had to use an "engine kick" – a temporary increase in engine revolutions per minute (rpm) – to increase water wash over the rudder to improve its effectiveness.

9.1.1.1.2 *Entering the Lower Channel* Pilot 2 returned to the bridge about 1440 in preparation for taking the conn from Pilot 1 near Morgan's Point. Pilot 2 stated that when he arrived, he listened to the radio communications to get a sense of the vessel traffic. He also observed Pilot 1 as he maneuvered the *Genesis River* past the inbound 600-ft-long tanker *Marvel* at 8.0 kts. Regarding the passing with the *Marvel*, Pilot 1 stated, "Once again, I had to use a kick to get the [*Genesis River*] to stop swinging after I met the ship."

At 1444, Pilot 2 took the conn from Pilot 1. After Pilot 2 issued his first rudder order, he asked Pilot 1, "Y'all over the place?" Pilot 1 responded, "Yup," and added, "She's takin' lotsa wheel…typical Japanese ship, got a little bitty rudder on her." Pilot 1 remained on the bridge for the next 15 minutes, discussing various topics with Pilot 2, including an extended dialogue about the handling characteristics of ships such as the *Genesis River*. The pilots expressed concerns about large ships that were difficult to handle, with Pilot 2 stating, "Yeah, I've sweated a couple times not knowing if they were gonna check-up [stop swinging] after meetin' a wide body there."

At 1446, Pilot 2 ordered the engine to full ahead. A little more than a minute later, as the *Genesis River* was clearing Morgan's Point and steadying on its first long leg of the lower Houston Ship Channel, Pilot 2 asked, "Mate or captain, [do] you have a 10-minute notice we can increase to?" In requesting "10-minute notice," the pilot was asking the crew to increase engine rpm to *sea speed*, which took the engine control system out of maneuvering mode and into navigation full mode.

In navigation full mode, the ability to change speed was limited, and the 10-minute time referred to by the pilot was the amount of prior notice that he would give, under normal circumstances, before requesting another speed change. Pilot 2 told investigators that it was common to increase to sea speed once a vessel entered the lower Houston Ship Channel, and interviews with other pilots confirmed that this was a regular practice among many of them.

The second mate responded yes, and 15 seconds later he asked, "Do you want me to increase now?" Pilot 2 answered, "Yes, that would be great." Following this exchange, an audible tone was captured on the voyage data recorder (VDR) audio, indicating that the *Genesis River*'s engine order telegraph (EOT) lever had been moved, and the telegraph test record registered a change in the EOT to the Nav. Full [navigation full] position. The engine speed, which had been at 60 rpm, then began to slowly increase.

At 1450, the *Genesis River*, transiting at 10 kts, passed the inbound 473-ft-long, 82-ft-wide tanker *Crimson Ray* portside to portside without incident. Nine minutes later, the *Genesis River*, now traveling at 11 kts, passed the inbound 440-ft-long, 73-ft-wide tanker *Nordic Aki*, again without incident.

At 1500, Pilot 1 left the bridge and proceeded to the pilot room. About the same time, an ordinary seaman (OS) requested permission from the second officer to take the helm of the *Genesis River* under the observation of the AB assigned to the helmsman watch. The OS explained to investigators that he was training for promotion to an AB position with the ship's operating company. The second officer gave permission, and the OS took the helm. In a deposition taken in October 2019, the AB stated that he had also requested permission from Pilot 2 to turn over the helm to the OS. However, Pilot 2 told investigators that he was not informed that the OS was at the wheel, and the VDR did not capture audio of the AB or any other crewmember requesting permission from Pilot 2 to change helmsmen. The AB stated that, after turning over the helm, he stood next to the OS and verified that rudder orders were properly executed.

9.1.1.1.3 Meeting with the BW Oak

In the Upper Galveston Bay, the Houston Ship Channel is intersected from the west by the Bayport Ship Channel, which provides access to container, automobile, and petrochemical terminals in Bayport, Texas. Where the two channels intersect, the Bayport channel widens into a funnel shape to allow ships to negotiate the turn from one channel to the other. This area is known as the "Bayport Flare" (Figure 9.3). At the southern terminus of the Bayport Flare, in the vicinity of Five Mile Cut (a shallow channel that extends to the east of the Houston Ship Channel), the Houston Ship Channel makes a 15.7° turn to the east.

At 1505, a pilot on the inbound *BW Oak*, a 740-ft-long, 120-ft-wide liquefied gas carrier, contacted Pilot 2 via very high frequency (VHF) radio channel 13 to

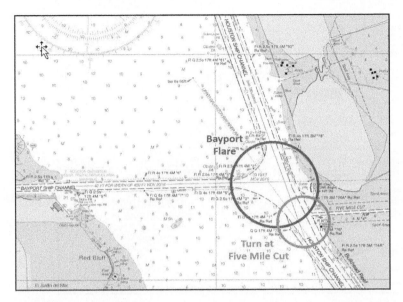

Figure 9.3 Bayport Flare and turn at Five Mile Cut. Source: NTSB Accident Report (2019)/National Transportation Safety Board/Public Domain.

make passing arrangements, and the pilots agreed to a port-to-port passing. At the time, the *Genesis River* was about a mile north of the Bayport Flare. Pilot 2 told investigators that, using the tools on his PPU, he knew that the *Genesis River* would pass the *BW Oak* near the southern part of the Bayport Flare. He stated that the location of the planned passing caused him no concern, as he had met other ships there before. Each of the other Houston Pilots that investigators spoke with stated similarly: they felt comfortable passing in that location or considered it a safe area to pass.

As the *Genesis River* transited south through the channel, its engine speed continued to slowly increase until it reached between 72 and 73 rpm, which crewmembers stated was the programmed rpm setpoint for Nav. Full, between an available range of 60–89 rpm. The vessel's speed over ground was 12 kts.

At 1509, the *Genesis River* entered the Bayport Flare from the north on a course of 161°. Beginning at 1509:22, Pilot 2 set up for the passing with the *BW Oak* by ordering courses to starboard, between 163° and 165°. According to VDR data, the helmsman used up to 25° starboard rudder input and 24° port rudder input to maintain the ordered courses.

As the *BW Oak* entered the turn at Five Mile Cut from the south, the pilot on board the vessel altered course to starboard at 1510:23. Before the turn, the *BW Oak* had been in the center of the channel, and the pilot told investigators that he ordered the turn earlier than he normally would have so that the ship would

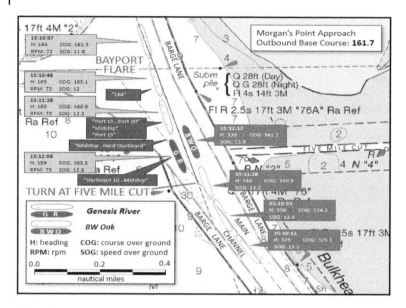

Figure 9.4 Pilot 2 helm orders as Genesis River and BW Oak passed in Bayport Flare. Positions based on automatic identification system (AIS) reporting data from each ship. Source: NTSB Accident Report (2019)/National Transportation Safety Board/Public Domain.

be on the starboard (eastern) side of the channel when it met the *Genesis River* (Figure 9.4).

At 1511:32, when the bows of the *Genesis River* and *BW Oak* were about 0.11 mi apart, Pilot 2 on the *Genesis River* ordered port 15° rudder and then, shortly thereafter, port 20° rudder. Four seconds later, he ordered the rudder to midship. The ship's heading was 164° when he issued the midship order.

At 1511:48, Pilot 2 ordered port 15° rudder, and the rudder moved to the ordered angle as the bow of the *Genesis River* passed the bow of the *BW Oak*. Nine seconds later, the pilot ordered rudder midship, followed almost immediately by hard starboard rudder. The helmsman repeated the command, and the rudder moved to starboard 35° at 1512:07. A second later, with the ship at a speed over ground of 12.6 kts, Pilot 2 issued an order to ease the rudder to starboard 10°, followed two seconds later by an order of rudder midship. The rudder began moving to midship, reaching centerline at 1512:20 just as the stern of the *Genesis River* passed the stern of the *BW Oak*. During this 32-second period, the *Genesis River*'s heading shifted 6° to port, to 158°.

The base course of the channel after the turn at Five Mile Cut (at the southern terminus of the Bayport Flare) was 146°. As the *Genesis River* entered the turn, the vessel was on the western side of the main deep-draft channel. The recorded water

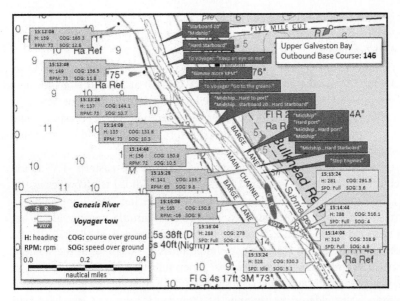

Figure 9.5 Pilot 2 orders and communications before the collision. Source: NTSB Accident Report (2019)/National Transportation Safety Board/Public Domain.

depth under the keel, which had been between 4 and 5 m while the ship transited the Bayport Flare, was now 3 m.

9.1.1.1.4 The Collision At 1512:25, as the ship's heading continued shifting to port and passed 155°, Pilot 2 ordered starboard 20° rudder to stop the swing of the ship. The rudder moved to 20° starboard until the pilot ordered the rudder back to midship at 1512:32 (Figure 9.5). Five seconds later, as the ship's heading passed 151°, Pilot 2 ordered hard starboard rudder. The rudder moved to starboard, reaching 35° at 1512:45, with the ship's heading passing 149°.

After ordering hard starboard rudder, the pilot hailed the towing vessel *Voyager* (Figure 9.5) on VHF radio channel 13. The *Voyager*, with a crew of four, was pushing ahead two tank barges breasted together side by side, with the barge *30015T* to starboard and the barge *MMI3041* to port. Both barges were fully loaded with a cargo of reformate, a gasoline blending stock. With these loads, the barges each had a draft of 10 ft (3.1 m) and a freeboard of 2 ft (0.6 m).

The *Voyager* and its tow were inbound in the barge lane on the eastern side of the Houston Ship Channel, en route from the Buffalo Marine fleeting facility in Texas City, Texas, to the Kirby fleeting facility in Channelview, Texas. Both Voyager's engines were at full throttle, and the tow was making about 5.3 kts speed over ground. The *Voyager*'s relief captain, who had been at the helm of the vessel since taking the watch at noon, answered Pilot 2's radio call. The pilot told him, "[I'm] that ship lookin' at you. Trying to check this thing up. Just keep an eye on me." The relief captain responded, "Roger, Roger." Currently, the *Genesis River* and the tow were 0.7 mi apart.

The *Genesis River*'s heading continued to swing to port. When the heading reached 143° (3° to port of the channel base course) at 1513:07 – a little less than three minutes before the collision – Pilot 2 ordered the mate on watch to "gimme more rpm" and repeated the order a few seconds later. The second mate answered, "Yes, yes, yes." Pilot 2 told investigators that he wanted more engine rpm to improve the rudder's effectiveness.

The *Genesis River*'s bow was now pointed toward the eastern side of the channel, directly at the *Voyager* and its tow. Pilot 2 radioed the *Voyager* again, stating, "She's not checkin' up, *Voyager*." While answering Pilot 2 on the radio, the *Voyager* relief captain moved the towing vessel's throttles to neutral. "What do you need me to do, Captain?" he asked. Pilot 2 responded, at 1513:25, "Go to the greens," meaning the *Voyager* tow should cross the channel to the western side marked by green navigation beacons. When interviewed after the accident, Pilot 2 stated that, because the *Genesis River* was crossing from the western side to the eastern side of the channel, he intended for the two vessels to pass starboard-side to starboard-side once the *Voyager* reached the opposite side of the channel.

The *Voyager*'s relief captain told investigators that when Pilot 2 had first called him, he considered various options in case he needed to maneuver his vessel to avoid the *Genesis River*. He stated that he believed that he could not stop or slow down due to vessel traffic behind him (the towing vessel *Provider*, pushing two barges ahead, was about 0.6 mi astern and transiting at a faster speed) and because he felt that stopping would still leave his vessel in the path of the *Genesis River*. He said he could not turn the tow to starboard because there was a sunken bulkhead just outside the barge lanes (on navigation charts, this bulkhead appears as a dotted line outside the channel on the eastern side, labeled "submerged bulkhead"). He was concerned that hitting the bulkhead with the barges would stop the tow and leave his vessel stranded in the path of the ship bearing down on him. He could not increase speed because the towing vessel had been at full power. With the *Genesis River* pointed directly at him, he felt that his only course of action was to cross the channel to the western side. He stated that the pilot's direction over the radio to do so confirm what he had already determined was the best action, so he immediately increased the *Voyager*'s engine throttles back to full power and put the vessel's rudders over hard to port. Automatic identification system (AIS) data showed that the two vessels were 0.55 mi apart when the head of the tow began turning to port at 1513:35. About the same time, the relief captain sounded the general alarm and radioed the deckhand on watch, instructing him to find the captain to tell him to come to the wheelhouse.

Meanwhile, the *Genesis River*'s engine rpm had remained at the programmed Nav. Full speed of 73 rpm (parametric data from the VDR showed the engine demand signal – the input to the engine control program from the EOT – was also at 73 rpm). Pilot 2 once again asked for more rpm, and the second officer answered, "Yes, sir. Yes, sir. We are going to full." A few seconds later, Pilot 2 told the crew to summon Pilot 1 to the bridge.

At 1513:43, 36 seconds after Pilot 2's initial request for more rpm and 14 seconds after his second request, the *Genesis River* VDR recorded the second officer, speaking in his native language, talking on the ship's phone to the engine control room (ECR). Translated, he said, "Yes, sir, now give us maximum rpm, whatever you can give." After taking the call in the ECR, the first engineer adjusted a fine tuner dial on the side of the ECR EOT lever, changing the Nav. Full speed setting from 73 to 85 rpm. According to the chief engineer, the second officer did not indicate that there was an emergency, so the first engineer and the chief engineer left the ECR to make a round of the engine room, leaving the electrical officer behind in the control room. Parametric data from the VDR showed that the rpm demand signal did not increase after the ECR EOT adjustment. The actual engine speed sporadically registered 74 rpm, but otherwise remained at 73 rpm until just before the collision.

About 1513:46, the *Voyager* captain, who had been exercising in the engine room when the deckhand summoned him, arrived in the wheelhouse. He asked the relief captain what was happening, and the relief captain stated that he responded, "Look!" while pointing out the wheelhouse windows toward the *Genesis River*. After taking a few seconds to survey the situation, the captain asked the relief captain if he wanted the captain to take the conn. The relief captain said that he responded, "No, I got it." The captain remained in the wheelhouse to assist the relief captain.

As the *Voyager* tow turned, its speed over ground slowed, dropping to as low as 3.6 kts. Pilot 2 told investigators that the tow did not turn and cross the channel as quickly as he expected. He again radioed the *Voyager*, stating "You need to go straight to the greens. Take a ninety to the greens, cuz I'm going to go your way again probably." The *Voyager* relief captain responded "Roger, Roger. Straight over."

Meanwhile, as the *Genesis River* crossed the center of the channel toward the eastern side (with the water depth under the keel increasing again to 5 m), the ship's rate of turn to port slowed and then ceased, on a heading of 132°. Then the ship began to swing back to starboard. During this time, Pilot 2 issued a series of rudder orders: first midship, then hard to port, back to midship, then starboard 20°, and finally hard to starboard. The OS at the helm responded when ordered, with the rudder reaching 34° to starboard at 1514:24.

The assigned helmsman on the *Genesis River* (the AB), who was monitoring the OS, recognized that an emergency situation was developing and took back the helm. Seconds later, Pilot 2 ordered the rudder to midship. The rudder returned to centerline at 1514:33. During these rudder movements, Pilot 1, who had returned to the bridge, asked the second mate if both steering pumps were online. The mate responded, "Yeah, already."

Pilot 2 radioed the *Voyager* again, stating, "Go, *Voyager*, go! Go, go, go!" The *Voyager* relief captain responded, "I'm hooked up, hard over, there, brother." At 1514:44, Pilot 2 ordered the rudder hard to port. He told investigators that he knew the ship would swing back across to the western side of the channel, so he ordered the port rudder in an attempt to hold the ship on the eastern side of the channel until it had passed the *Voyager* tow. The helmsman acknowledged the order, and the rudder began swinging to port.

By this time, the *Voyager* tow was crossing the deep draft channel, yet it was still on the eastern side making about 4 kts. The *Genesis River* was still swinging to starboard, and Pilot 2 realized that the port rudder was not going to be effective in holding the ship along the eastern side of the channel. Consequently, he radioed a warning to the *Voyager*. "I'm gonna probably hit ya…sound your general alarm there, *Voy[ager]*…get everybody up." A response from the *Voyager* was not captured on audio recordings.

At 1514:54, the *Genesis River* second officer called the master in his stateroom, telling him to come to the bridge immediately. The master arrived on the bridge about 30 seconds later.

At 1515:00, Pilot 2 ordered the rudder to midship, then immediately ordered the rudder hard to port again. Ten seconds later, Pilot 2 warned the *Voyager* again over the radio, stating, "Wake everybody up on that, uh, *Voyager*." The towing vessel relief captain responded, "We got it, brother. We got 'em. Appreciate it." At 1515:12, Pilot 2 ordered the rudder to midship again, and the helmsman brought the wheel to midship.

The *Genesis River*'s rate of turn back to starboard increased as it approached the bank on the eastern side of the channel. As the ship passed heading 139°, Pilot 2 radioed the *Voyager*, "I'm gonna be swingin' your way real soon. She's comin' your way. You gotta push on it." The *Voyager* relief captain responded, "She's all she's got there, brother; all she's got."

Shortly thereafter, with the *Genesis River*'s rudder at midship, the second mate said, "Go to the port. Go to the port." He told investigators that he was speaking to Pilot 2 at the time, although the pilot did not acknowledge him.

At 1515:29, as the *Genesis River*'s heading passed 143° and back toward the *Voyager*, Pilot 2 ordered the rudder to midship (the rudder was already at midship), then immediately ordered the rudder to hard starboard. The helmsman acknowledged the order, and the rudder began moving to starboard, reaching 35° at 1515:37. About the same time, the second officer said to Pilot 2, "Hard port, sir, hard port." He received no reply. Pilot 2 told investigators that, at that point, he knew that the ship was going to collide with the tow, so he turned to starboard to ensure that the *Genesis River* struck the barges and not the towing vessel. He stated that his principal concern was the people on the *Voyager*. After issuing the starboard rudder order, he radioed the *Voyager*, stating, "You got it hard over there, *Voyager*?…Work with me…We're gonna collide." The *Voyager* relief captain responded, "Roger, Roger. Roger, Roger," while the *Voyager* captain sounded the general alarm.

As Pilot 2 was struggling to control the ship, Pilot 1 ran out to the port bridge wing. He told investigators that from this position, he could see that the *Genesis River* was now in the barge lane on the eastern side of the channel. He said, "As the bow went into the bank on the red [eastern] side, the ship swung into the bank, and the whole ship just rolled up… We touched bottom."

At 1515:43, when the *Genesis River*'s bow was about 600 ft (0.1 mi) from the *Voyager* tow, Pilot 2 ordered "stop engines." The master, who had returned to the bridge, repeated the engine order to the second officer. At 1516:03, seconds before the collision, the master ordered the engine to emergency full astern (also known as "crash astern"). The emergency full astern input to the engine control system overrode the normal propulsion control program and executed an accelerated shift to maximum astern thrust.

At 1516:09, the *Genesis River*'s bow struck barge *30015T* midship on the starboard side (see Figure 9.6), penetrating through the double hull and breaching the no. 2 starboard cargo tank. The gas carrier's bow continued through the barge's hull into the no. 2 port cargo tank. The force of the collision capsized barge *MMI3041*, the tow's port barge, although no tanks were breached. When the *Genesis River* impacted the *30015T*, the port face and long wires securing the

Figure 9.6 Screen capture from wheelhouse video on board the Voyager at the moment that the Genesis River struck barge 30 015T. Source: NTSB Accident Report (2019).

barges to the *Voyager* parted. As the barges were pushed sideways by the *Genesis River*, the *Voyager* pivoted on the starboard face wire until the towing vessel's starboard side contacted the stern of the *30015T*. As the *Voyager* continued to be pulled, it heeled significantly to starboard until the last face and long wires gave way, allowing the relief captain to regain control of the vessel. The loose end of the parted starboard long wire, which had fallen into the water, then fouled the *Voyager*'s starboard propeller, stalling the engine.

Just prior to the collision, the *Voyager* captain had sent the deckhand down to close the engine room door on the starboard-side main deck of the towing vessel. Company policy stated that all main deck doors were to remain closed while the vessel was in operation, but the captain had opened the door to allow air into the engine room while he was exercising. The deckhand reached the door just as the *Genesis River* struck the tow. He was able to close the door, but not before about 200 gal of water entered the engine room.

With the *Genesis River* engine remaining at emergency full astern, the ship began to back away from the barges once the forward motion of the *Genesis River* ceased. Reformate in the breached cargo tanks escaped into the channel from the hole in the *30015T*'s hull. Pilot 2 radioed the Houston Pilots dispatcher, stating, "Bad collision. Shut down the channel." Pilot 1, who had begun communicating with the US Coast Guard Sector Houston–Galveston Vessel Traffic Service (VTS) via his cell phone when the collision was imminent, informed Coast Guard VTS about the accident. The VTS watchstander, who was monitoring vessel traffic in the area and had heard the VHF radio communication between the *Genesis River* and *Voyager* when he received the call from Pilot 1, reported the accident to higher authority and the Coast Guard Command Center, allowing response efforts to commence. VTS watchstanders then advised other vessels in the Houston Ship Channel of the accident and redirected traffic as necessary.

Once the vessels were clear of each other, Pilot 1 radioed the *Voyager* crew to check on their status. The relief captain radioed back, "Everybody's good on the *Voyager*." The *Genesis River* anchored in the channel, awaiting further direction from VTS, while the *Voyager*, reduced in maneuverability due to the loss of the starboard engine, made up to the towing vessel *Provider*. The *Genesis River* was eventually directed to a lay berth in Bayport, and the vessel proceeded to the berth under its own power, mooring at 1736 that evening. The *Voyager* was towed to the Kirby facility at Old River, Texas, arriving at 2120.

9.1.1.1.5 *Result of Collision* Approximately 11,276 barrels (473,600 gal) of reformate spilled into the waterway from the damaged barge *30015T*. When the accident was reported to VTS, the Coast Guard captain of the port closed the Houston Ship Channel to navigation. At 1541, the incident command system was activated, and an incident command post was established at Coast Guard Sector

Houston–Galveston headquarters (the post would later move to a location near Bayport). At the same time, Kirby Inland Marine implemented its spill response plan, hiring various providers to conduct response operations. By 1935, oil spill containment booms had been deployed around barges *30015T* and *MMI3041*, and additional booms were being installed across inlets and other sensitive marine areas around Galveston Bay. Oil skimmers were deployed to recover reformate/water mixture in the vicinity of the accident site, with a total of seven skimmers used during the cleanup.

As cleanup efforts continued, residents in neighborhoods surrounding Galveston Bay reported a petrochemical odor, prompting air quality testing in the areas most affected. Throughout the response, 15,016 air samples measuring for atmospheric flammability and concentrations of benzene and volatile organic compounds were collected. Thirty-nine readings detected benzine or volatile organic compounds at or above 0.5 ppm, but secondary readings for these instances determined that levels were not sustained above actionable levels. Reports indicated that a fish kill impacting between approximately 100 and 1000 fish, shrimp, and crabs occurred on a limited stretch of shoreline, along with other wildlife impacts. Out of 2700 water samples taken between Friday (the accident day) and Sunday, none showed pollution levels requiring action.

At 0400 on May 12, the captain of the port opened the Houston Ship Channel for navigation to outbound traffic only. At 1505 that day, lightering of reformate cargo from barge *30015T* commenced and was completed at 2345 on Tuesday, May 14. In all, 14,000 barrels of pure reformate and 4530 barrels of reformate/water mixture were recovered from the vessel. Once the offload was completed, the barge was towed to a shipyard in Channelview for assessment.

Offloading of cargo from the capsized barge proved more difficult. On Tuesday, May 14, the barge *MMI3041* – still capsized – was towed to a location off the main channel (allowing the channel to reopen for navigation of two-way traffic on May 15) for lightering. Lightering commenced at 1447 and was completed at 1710 the next day. On Sunday, May 26, the *MMI3041* was parbuckled (righted using rotational leverage) and towed to the shipyard in Channelview. All 25,392 barrels of reformate cargo was recovered from the *MMI3041*.

9.1.1.2 Work/Rest of Ships' Crews

Genesis River Pilot 1. Work/rest/sleep records for Pilot 1 were not collected after the accident because he was not at the conn during the collision and therefore did not affect its outcome. Pilot 1 told investigators that he got about eight hours of sleep overnight before the accident voyage, waking at 0930.

Genesis River Pilot 2. According to information submitted to the Coast Guard, Pilot 2 had 23 hours of sleep in the 72 hours before the accident and had slept nine hours the night before the collision. He had worked about 15.5 hours over the same period. He told investigators that he had consumed no alcohol the night before and his sleep had been "really good."

9.1.1.2.1 *Genesis River* Crew

Master. A work/rest log provided by the company showed that the master had 24.5 hours of work and 47.5 hours of rest in the 72 hours prior to the accident. The International Maritime Organization (IMO) standard form used to log work and rest did not specify times of sleep during the rest periods. The master told investigators that the night before the accident, he had gone ashore with the second officer and first-assistant engineer, and, between 2130 and 2230, he consumed three alcoholic drinks (beer). He and other crewmembers returned to the ship about 0400 after being delayed due to the unavailability of taxis, and the master went to sleep immediately after returning. He was awoken between 0930 and 1000 to review cargo documentation. He stated that the quality of his sleep was "good." During the accident voyage, the master went to his cabin after eating lunch. He stated that while in his cabin, he took a nap, sleeping for about 45 minutes and waking when the second officer called him to report the impending accident.

Chief Officer. The work/rest log showed that the chief officer had 27.5 hours of work and 44.5 hours of rest in the 72 hours prior to the accident. He remained aboard the ship the night before the accident and had consumed no alcoholic beverages on May 9 or 10. According to the work/rest log, he had 10 hours of rest between 1430 on the May 9 and 0030 on May 10. He then worked for 1.5 hours, had another rest period for 2.4 hours, then worked from 0430 to 0830. He could not recall the exact number of hours of sleep he had that night but described it as "a good rest." The chief officer stated that he took a 30-minute nap on the morning of the accident, after cargo documentation had been completed and before the vessel got under way, and the work/rest record shows a period of rest between 0830 and 1000. The chief officer worked from 1000 onward through the accident period.

Second Officer. The work/rest log for the second officer showed that he had 28.5 hours of work and 43.5 hours of rest in the 72 hours prior to the accident. The hours of work listed in the form corresponded to a regular watch/duty schedule of 0000–0430 and 1200–1700 daily and included an entry for a 0000–0430 watch on the morning of the accident. However, the second officer told investigators that between about 2200 the night before and 0400 that morning he had been "on shore leave" with the master. He said that he did not consume any alcoholic beverages while ashore. The form notes that the second officer began his day watch on the accident date one hour early, at 1100, prior to getting under way. The second officer stated that, between 0400 and 1100, he had slept. He described the quality of his sleep as "sound."

Able-bodied Seaman. The work/rest log showed that the AB had 24 hours of work and 48 hours of rest in the 72 hours prior to the accident. His hours of work corresponded to a regular watch/duty schedule of 0000–0400 and 1200–1600 daily. The AB stated that he had about six to seven hours of sleep the night before the accident, which he described as "good sleep."

Ordinary Seaman. The work/rest log provided by the company showed that the OS had 27 hours of work and 45 hours of rest in the 72 hours prior to the accident. His hours of work corresponded to a regular watch/duty schedule of 0400–0800 and 1600–2000 daily, as well as additional hours of work beginning at 1200 on the accident date related to deck operations for getting under way and his training watch on the helm. The OS said that he got about six hours of "good sleep" prior to his 0400–0800 cargo watch on the accident date. He also stated that he had napped for about 20–30 minutes after his morning watch, prior to going to his station for getting under way.

9.1.1.2.2 *Voyager Crew* Captain.

A work/rest record provided by the company for the credentialed crewmembers of the *Voyager* was divided by on-watch and off-watch time. The record showed that the captain had 36 hours on watch and 36 hours off watch during the 72 hours prior to the accident, which corresponded to his regular watch schedule of 0500–1200 and 1700–2200 daily. He told investigators that he had slept about 6.5 hours before his morning watch on the accident date, and the quality of his sleep was "good."

Relief Captain. The work/rest record provided by the company showed that the relief captain had 36 hours on watch and 36 hours off watch during the 72 hours prior to the accident, which corresponded to his regular watch schedule of 1200–1700 and 2200–0500 daily. He stated that he had slept about six hours on the morning of the accident before assuming the watch, and the quality of his sleep was also "good."

9.1.1.3 Drug and Alcohol Testing

Pilot 1 and Pilot 2 were administered required postaccident breathalyzer tests for alcohol about 1900 on the accident date after disembarking the *Genesis River*. Results for both pilots were negative. The *Genesis River* pilots also underwent required postaccident urine drug testing, with negative results.

9.1.1.3.1 Genesis River *Crew*

About 1800 on the accident date, after the *Genesis River* had moored in Bayport, all crewmembers on the *Genesis River* were administered postaccident breathalyzer tests for alcohol, and the results were negative. Later that evening, the crew underwent postaccident urine drug testing, and the results were also negative.

9.1.1.3.2 *Voyager Crew*

Between 1800 and 1832 on the accident date, all four members of the *Voyager* crew were administered postaccident breathalyzer tests.

All results were negative. The crew also underwent postaccident urine drug testing, with negative results.

9.1.1.4 Findings

1. Pilot and crew credentialing and experience, use of alcohol or other tested-for drugs, fatigue, and environmental conditions were not factors in the accident.
2. Mechanical and electrical systems on the *Genesis River* and *Voyager* operated as designed, and their functionality was not a factor in the accident.
3. Although the *Genesis River* master's decision to place the vessel's ARPA in standby and turn off the ECDIS deprived the bridge team of critical tools with which to monitor the pilots' actions and ensure that the vessel transited safely, the status of this equipment was not a factor in the accident.
4. The *Genesis River* helmsman properly executed the rudder orders of the pilot and his performance was not a factor in the accident.
5. Although the helmsman in training properly executed the orders of the pilot, placing him at the helm without informing the pilot was contrary to good bridge resource management practice.
6. Maintaining stern trim while under way would have improved the handling characteristics of the *Genesis River*.
7. The combined effect of the speed of the *Genesis River* and the passing of another large vessel in the asymmetrically shaped channel at the southern terminus of the Bayport Flare resulted in an uncontrollable sheer to port by *Genesis River*, initiating a chain of events that led to the collision.
8. The *BW Oak* pilot's maneuvering of his vessel to prepare for the meeting with the *Genesis River* was routine and did not impede the *Genesis River*'s ability to pass.
9. Wide-beam, deep-draft vessels meeting in the Houston Ship Channel in the vicinity of the northern and southern terminuses of the Bayport Flare have a higher risk of loss of control due to complex and varying hydrodynamic forces.
10. Once the *Voyager* and its tow began the turn to port, the collision was unavoidable.
11. An increase in engine rpm to arrest the *Genesis River*'s initial sheer, even if promptly executed after it was ordered by the pilot, would not have prevented the collision.
12. The pilot transiting the wide-beam, deep-draft *Genesis River* at sea speed through the shallow and narrow lower Houston Ship Channel left little margin for error and introduced unnecessary risk.
13. The *Genesis River* pilot's decision not to use emergency full astern or the anchors to avoid the collision was reasonable.

14. The actions of the *Voyager* relief captain to attempt to avoid the collision by crossing the channel were reasonable, given the information available to him at the time he had to make the decision to maneuver.
15. The *Genesis River* pilot's early and frequent communications with the *Voyager* mitigated the impacts of the accident and likely prevented loss of the towing vessel and injuries to its crew.
16. Coast Guard VTS Houston–Galveston's response to the collision was timely and appropriate.
17. The Bayport Flare, as well as other intersections within the Houston–Galveston VTS area, would benefit from regular risk assessments and the consideration of additional vessel routing measures.

9.1.1.4.1   Primary Finding   NTSB determines that the probable cause of the collision between the liquefied gas carrier *Genesis River* and the *Voyager* tow was the *Genesis River* pilot's decision to transit at sea speed, out of maneuvering mode, which increased the hydrodynamic effects of Bayport Flare's channel banks, reduced his ability to maintain control of the vessel after meeting another deep-draft vessel, and resulted in the *Genesis River* sheering across the channel toward the tow.

9.2   Navy Vessel Collisions

There were two very close in time collisions between United States Navy vessels and civilian vessels. There were the first to be presented in the USS FITZGERALD that collided with the Motor Vessel ACX CRYSTAL. The second was the USS JOHN S. MCCAIN and the Motor Vessel ALNIC MC. The descriptions of these accidents are excerpted from the unclassified United States Navy reports on the collisions (US Navy 2017). The photographs are all credited to the US Navy. Note that US Navy ship names are always capitalized. In these reports, the Navy did an excellent job of noting the issues that lead to the collisions, these included the safety culture and human factors issues. Visit Sea Forces (2021) for more information on the Arleigh Burke class Guided Missile Destroyers.

9.2.1   USS FITZGERALD Collided with the Motor Vessel ACX Crystal

USS FITZGERALD collided with Motor Vessel ACX CRYSTAL on June 17, 2017 in the waters of Sagami Wan in vicinity of the approaches to Tokyo Wan.

FITZGERALD is an Arleigh Burke Class Destroyer commissioned in 1995 and homeported in Yokosuka, Japan, as part of the Forward Deployed Naval Forces and Carrier Strike Group FIVE. Approximately 300 Sailors serve aboard

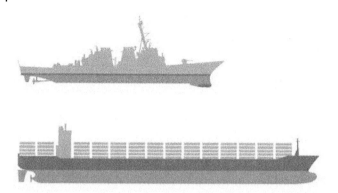

Figure 9.7 Relative size of the USS Fitzgerald.

FITZGERALD. FITZGERALD is 505 ft in length and carries a gross tonnage of approximately 9000 tons.

Figure 9.7 illustrates the relative sizes of the vessels. ACX CRYSTAL (CRYSTAL) is a Philippines flagged container ship built in 2008. CRYSTAL is 728 ft long with a gross tonnage of approximately 29,000 tons.

The collision between FITZGERALD and CRYSTAL resulted in the deaths of seven US Sailors due to impact with FITZGERALD's berthing compartments, located below the waterline of the ship. CRYSTAL suffered no fatalities. US Sailor fatalities were as follows:

GMSN Kyle Rigsby of Palmyra, Virginia, 19 years old.
YN2 Shingo Alexander Douglass, of San Diego, California, 25 years old.
FC1 Carlos Victor Ganzon Sibayan of Chula Vista, California, 23 years old.
PSC Xavier Alec Martin of Halethorpe, Maryland, 24 years old.
STG2 Ngoc Truong Huynh of Oakville, Connecticut, 25 years old.
GM1 Noe Hernandez of Weslaco, Texas, 26 years old.
FCC Gary Rehm, Jr., of Elyria, Ohio, 37 years old.

9.2.1.1 Summary of Findings
The Navy determined that numerous failures occurred on the part of leadership and watchstanders as follows:

Failure to plan for safety.
Failure to adhere to sound navigation practice.
Failure to execute basic watch standing practices.
Failure to properly use available navigation tools.
Failure to respond deliberately and effectively when in extremis (Figure 9.8a,b).

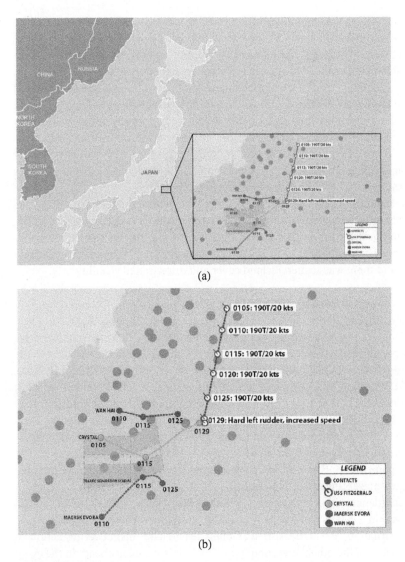

Figure 9.8 (a, b) Illustration map of approximate collision location.

9.2.1.2 Background

On the morning of June 16, 2017, FITZGERALD departed the homeport of Yokosuka, Japan for routine operations. The weather was pleasant with unlimited visibility and calm seas. After a long day of training evolutions and equipment loading operations, FITZGERALD proceeded southwest on a transit to sea from the Sagami Wan operating area at approximately 2300.

FITZGERALD was operating by procedures established for US Navy surface ships when operating at sea before sunrise, including being at "darkened ship." "Darkened ship" means that all exterior lighting was off except for the navigation lights that provide identification to other vessels, and all interior lighting was switched to red instead of white to facilitate crew rest. The ship was in a physical posture known as "Modified ZEBRA," meaning that all doors inside the ship, and all hatches, which are openings located on the floor between decks, at the main deck and below were shut to help secure the boundaries between different areas of the ship in case of flooding or fire. Watertight scuttles on the hatches (smaller circular openings that can be opened or closed independently of the hatch) were left open in order to allow easy transit between spaces.

By 0130 hours on June 17, 2017, the approximate time of the collision, FITZGERALD was approximately 56 nautical miles to the southwest of Yokosuka, Japan, near the Izu Peninsula within sight of land and continuing its transit outbound. The seas were relatively calm at 2–4 ft. The sky was dark, the moon was relatively bright, and there was scattered cloud cover and unrestricted visibility.

9.2.1.3 Events Leading to the Collision

At approximately 2300 local Japan time, both the Commanding Officer and Executive Officer (XO) departed the bridge, the area from which watchstanders drive the ship. As the FITZGERALD proceeded past Oshima Island the shipping traffic increased and remained moderately dense thereafter until the collision. By 0100, FITZGERALD approached three merchant vessels from its starboard, or right side, forward. These vessels were eastbound through the Mikomoto Shima Vessel Traffic Separation Scheme. Traffic separation schemes are established by local authorities in approaches to ports throughout the world to provide ships assistance in separating their movements when transiting to and from ports. The closest point of approach of these vessels and the FITZGERALD was minimal with each presenting a risk of collision.

In accordance with the International Rules of the Nautical Road, the FITZGERALD was in what is known as a crossing situation with each of the vessels. In this situation, FITZGERALD was obligated to take maneuvering action to remain clear of the other three, and if possible, avoid crossing ahead. In the event FITZGERALD did not exercise this obligation, the other vessels were obligated to take early and appropriate action through their own independent maneuvering action. In the 30 minutes leading up to the collision, neither FITZGERALD nor CRYSTAL took such action to reduce the risk of collision until approximately one minute prior to the collision. FITZGERALD maintained a constant course of 190° at 20 kts of speed.

In the several minutes before collision, the Officer of the Deck, the person responsible for safe navigation of the ship, and the Junior Officer of the Deck, an officer placed to assist, discussed the relative positioning of the vessels, including CRYSTAL and whether or not action needed taken to avoid them. Initially, the Officer of the Deck intended to take no action, mistaking CRYSTAL to be another of the two vessels with a greater closest point of approach. Eventually, the Officer of the Deck realized that FITZGERALD was on a collision course with CRYSTAL, but this recognition was too late. CRYSTAL also took no action to avoid the collision until it was too late.

The Officer of the Deck, the person responsible for safe navigation of the ship, exhibited poor seamanship by failing to maneuver as required, failing to sound the danger signal and failing to attempt to contact CRYSTAL on Bridge-to-Bridge radio. In addition, the Officer of the Deck did not call the Commanding Officer as appropriate and prescribed by Navy procedures to allow him to exercise more senior oversight and judgment of the situation.

The remainder of the watch team on the bridge failed to provide situational awareness and input to the Officer of the Deck regarding the situation. Additional teams in the Combat Information Center (CIC), an area on where tactical information is fused to provide maximum situational awareness, also failed to provide the Officer of the Deck input and information (Figure 9.9).

9.2.1.4 Collision

The port (left) side of CRYSTAL's bow, near the top where the anchor hangs, struck FITZGERALD's starboard (right) side above the waterline. CRYSTAL's bulbous bow, under the water, struck FITZGERALD on the starboard side just forward of

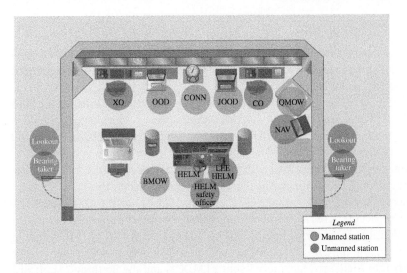

Figure 9.9 Bridge schematic of FITZGERALD.

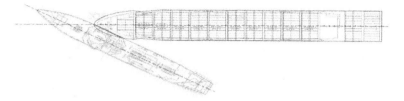

Figure 9.10 Diagram of approximate collision geometry.

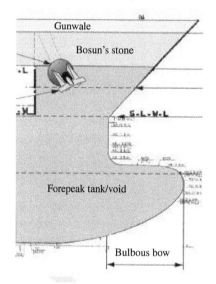

Figure 9.11 Depiction of a bow and bulbous bow.

the middle part of the ship. CRYSTAL's bulbous bow struck the starboard access trunk, an entry space that opens into Berthing 2 through a non-watertight door (Figures 9.10 and 9.11).

The impact of the top of CRYSTAL's bow above the waterline crushed the false bulkhead (a non-structural steel wall that is the shipboard equivalent of a non-load bearing wall in a house, used to divide a space) and non-watertight door within the Commanding Officer's cabin. The Commanding Officer's cabin is composed of two small rooms: an office area with a desk, table, and couch with seating for several people; and his bedroom. The two rooms are separated by this false bulkhead and door. The outer wall was pushed in and torn open by the impact (Figure 9.12).

The impact of CRYSTAL's bulbous bow below the waterline punctured the side of FITZGERALD, creating a hole measuring approximately 13 ft by 17 ft, spanning the second and third decks below the main deck. The hole allowed water to flow directly into Auxiliary Machinery Room 1 (AUX 1) and the Berthing 2 starboard access trunk. The force of impact from CRYSTAL's bulbous bow and resulting

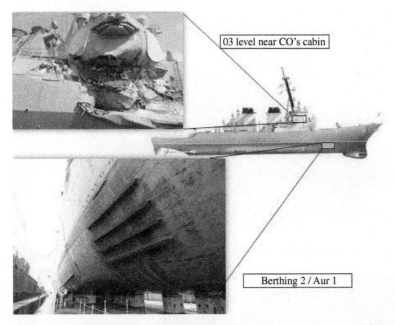

03 level near CO's cabin

Berthing 2 / Aur 1

Figure 9.12 Starboard side of FITZGERALD. Inset Above: Damage to FITZGERALD above the waterline at the following 03 Level near CO's Cabin Inset: FITZGERALD in dry dock with patch over the 13 ft × 17 ft hole at Berthing 2/AUX 1. Source: U.S. Navy, CC by 4.0.

flood of water pushed the non-watertight door between the starboard access trunk and Berthing 2 inward. The wall supporting this door pulled away from the ceiling and bent to a near-90° angle. As a result, nothing separated Berthing 2 from the onrushing sea, allowing a great volume of water to enter Berthing 2 very quickly.

The impact at the moment of collision caused FITZGERALD to list (tilt) a reported 14° to port. FITZGERALD then settled into a 7° starboard list as the sea flooded into Berthing 2 through the starboard access trunk and weighted the ship deeper into the water on the starboard side.

Water poured into the ship from the hole in the hull and flooded into spaces directly connected to areas near the hole or not separated by a watertight barrier, including Berthing 2 and associated spaces; AUX 1 and associated spaces; and other spaces forward of the hole.

Other spaces were flooded due to cross flooding, which is the flooding of spaces that are connected to damaged spaces and that have the ability to be isolated with a watertight barrier, but could not be sealed off in time due to the rapid flooding caused by the large hole in the side of the ship. Additionally, spaces were partially flooded due to a ruptured fire main (large seawater pipes that provide water for

Figure 9.13 Commander's stateroom area-exterior. Source: Bob Collet/Alamy Images.

Figure 9.14 Commander's stateroom-interior.

fighting fires) and ruptured Aqueous Film-Forming Foam (a form of firefighting agent) piping.

Multiple spaces suffered structural damage: Commanding Officer's Cabin (Figures 9.13 and 9.14), Stateroom, and Bathroom; Officer Stateroom; Berthing 2 Starboard Access Trunk; Auxiliary Compartment 1; Repair Locker Number 2 passageway; CIC passageway; multiple Radar and Radar Array rooms; multiple fan rooms; Combat Systems Maintenance Central airlock and ladder-well; Electronic Workshop Number 1; and the Starboard Break.

The collision resulted in a loss of external communications and a loss of power in the forward portion of the ship. Following the collision, FITZGERALD changed the lighting configuration at the mast to one red light over another red light, known as "red over red," the international lighting scheme that indicates a ship that is "not under command." Under International Rules of the Nautical Road, this signifies that, due to an exceptional circumstance such as loss of propulsion or steering, a vessel is unable to maneuver as required.

All US Navy ships are designed to withstand and recover from damage due to fire, flooding, and other damage sustained during combat or other emergencies. Each ship has a Damage Control Assistant, working under the Engineering Officer, to establish and maintain an effective damage control organization. The Damage Control Assistant oversees the prevention and control of damage including control of stability, list, and trim due to flooding (maintaining the proper level of the ship from side to side and front and back), coordinates firefighting efforts, and is also responsible for the operation, care, and maintenance of the ship's repair facilities. The damage control assistant (DCA) ensures the ship's repair party personnel are properly trained in damage control procedures including firefighting, flooding, and emergency repairs. The Damage Control Assistant is assisted by the Damage Control Chief (DCC), a chief petty officer specializing in Damage Control. The officer in charge of damage control efforts, the Damage Control Assistant, called away General Quarters (GQ) to notify the crew to commence damage control efforts.

General Quarters is a process whereby the crew reports to pre-assigned stations and duties in the event of large casualties such as flooding. General Quarters is announced by an alarm that sounds throughout the ship to alert the crew of an emergency or potential combat operations. All crewmembers are trained to report to their General Quarters watch station and to set a higher condition of material readiness against fire, flooding, or other damage. This involves securing additional doors, hatches, scuttles, valves, and equipment to isolate damage and prepare for combat. Navy crews train on Damage Control continuously with drills being run in port and underway frequently to prepare the teams for damage to equipment and spaces. During any emergency condition (fire, flooding, combat operations), the Damage Control Assistant coordinates and supervises all damage control efforts from one of the three Damage Control Repair Lockers.

Damage Control Repair Lockers are specialized spaces stationed throughout the ship filled with repair equipment and manned during emergencies with teams of about 20 personnel trained to respond to casualties. There are three repair lockers on the FITZGERALD: Repair Locker 2, Repair Locker 5, and Repair Locker 3. Repair Locker 2 covers the forward part of the ship, Repair Locker 5 covers the engineering spaces, and Repair Locker 3 covers the aft part of the ship. Each locker is maintained with similar equipment. Personnel assigned to repair lockers are trained and qualified to respond to, and repair damage from, a variety of sources with a specific focus on fire and flooding. Each repair locker can act independently but is also designed to support the others and can take over the responsibilities for any locker if damage prevents that locker's use. The repair lockers are normally unmanned unless the ship sets a condition of higher readiness like General Quarters when they would be manned within minutes.

The Damage Control Assistant and DCC took control of damage control actions immediately after the collision, organizing General Quarters and ship wide efforts.

Reports were received of a 12 ft by 12-ft hole in Auxiliary Compartment 1 and flooding. Reports were also received of a 3 ft by 5-ft hole and flooding in the right side (starboard) passage to Berthings 1 and 2. These two reports were determined to be differing accounts of the same hole caused by the collision that had spread to two spaces. The hole was later determined to be 17 ft by 13 ft.

Based on reports of flooding in Berthing 2 and Auxiliary Compartment 1, supervisors attempted to direct the Repair Locker 2 team, the closest to the report location, in place to combat the flooding. However, due to location of the flooding, the supervisors could not communicate with Repair Locker 2. Efforts were reassigned to Repair Locker 5 as communications were established on internal communications networks still operable, and hand-held radios and sound powered phones, which are phones that do not require electrical power to operate.

Damage control parties used eductors to remove water from flooded spaces. Eductors use a jet of water, typically supplied from the ship's fire-main piping system, to remove water from spaces. FITZGERALD also used three onboard pumps to remove water from the ship. Two of the pumps functioned as designed and a third seized and was inoperable for the duration of the recovery efforts.

Berthing 1 was partially flooded to 5 ft by water entering from the hole created by the impact in Berthing 2 and AUX 1. Damage control parties tried to limit and reduce the flooding with eductors, but with no watertight hatch or door between Berthing 2 and Berthing 1 on the starboard side to act as a barrier against the progressive flooding, water passed freely into Berthing 1 and undermined dewatering efforts. Berthing 1 had water in the space until after the ship returned to Yokosuka and entered dry dock on 11 July Dewatering of Berthing 2 was completed only after the ship entered dry dock. Auxiliary Compartment 1 was dewatered using an additional pump once a patch was installed in Yokosuka.

The Starboard Passageway was partially flooded due to the ruptured fire-main and Aqueous Film Forming Foam (a form of firefighting agent) piping as a direct result of the impact. 1.5 ft of water was reported on deck with 3 ft of firefighting foam. Damage Control Parties used two eductors to effectively counter the flooding. They also used shoring, a material designed to reinforce a structural defect, along the passageway to mitigate the effects of the collision.

Figure 9.15 Non-watertight door frame from Berthing 2 to the ladder going up to Berthing 1 Access Trunk. There is no hatch at the top of this ladder to prevent water from flooding up into Berthing 1.

Radio Central was partially flooded as a result of flooding in space in close proximity. The crew effectively combatted flooding with stuffing compound used in cable way repairs and then dewatered the space using buckets and mops. Main Engine Room 1 had minor flooding due to proximity flooding from Auxiliary Compartment 1's Wastewater Tank and Oily Waste Tank. An eductor was aligned to effectively dewater the space and the water level remained below 3 ft (Figure 9.15).

Sonar Control, Sonar Control Fan Room, Sonar Control Passageway, Sonar Administration Office, and Combat Systems Equipment Room 1 were partially flooded with 5 ft of water on the deck. Subsequent damage control efforts effectively dewatered the space using an eductor and pump.

The Forward 400 Hz, Fan Room, and Power Conversion Room experienced both flooding and white smoke. Progressive flooding through the vents in AUX 1 resulted in 2 ft of water on the deck, which was engaged using eductors, buckets, and swabs. Dewatering efforts continued while water continued to flow in, maintaining 1 ft depth of water or less.

As Sailors reported to their damage control duties and undertook efforts, departments began to account for missing Sailors. Reports were received that three Sailors were trapped in Sonar Control because of the collision. Realizing there was flooding in the spaces above them, the Sailors in Sonar Control radioed for assistance. A team was sent in but initial attempts to reach them were unsuccessful because the passageway was completely obstructed due to damage.

This was also one of the areas that suffered cross-flooding through deck drains that could not be secured before flooding advanced. The team reached the escape hatch above the Sonar Control space, which was topped with a few inches water. They went through the hatch and were able to assist the Sailors trapped inside at approximately 0215.

At 0225, two Sailors were identified as unaccounted for in the Combat Systems Department.

At approximately 0316, four Sailors were reported as unaccounted for. At 0540, a final, accurate all-hands accounting was received in which seven Sailors remained missing.

9.2.1.5 Impact to Berthing 2

Berthing 2 is a crew area containing 42 racks (beds) and spans from one side of the ship across to the other side two decks below the main deck of the ship. It includes its own bathroom, a shower and bathroom space accessed via a non-watertight door. It also includes a lounge filled with sofas, chairs, a table, and a television where the crew in the berthing can relax and recreate. The space is approximately 29 ft long, approximately 40 ft across, and with ceilings approximately 10 ft high. The racks are stacked with a top, middle, and bottom rack, each with a mattress and privacy curtains. Figure 9.16 provides an example of how the racks would have been configured in Berthing 2.

Berthing 2 has three ways out (egress points), of which two are on the port side. The first port side egress is up a ladder through a watertight hatch with a watertight scuttle at the top of the ladder. Figure 9.15 show this egress point on a ship of the same class as FITZGERALD. The second port side egress is an escape scuttle in the

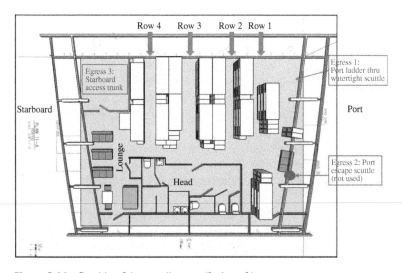

Figure 9.16 Berthing 2 layout diagram (facing aft).

ceiling that leads directly into Berthing 1. Figure 9.16 shows this egress point on a ship of the same class as FITZGERALD. This escape scuttle is usually in the down position, as it would become a trip hazard in Berthing 1 if left in the up position.

The third egress point is on the starboard side, where CRYSTAL's bulbous bow struck FITZGERALD. A non-watertight door leads from Berthing 2 to the starboard access trunk. In that trunk, a ladder leads up one deck into a space just outside of Berthing 1. Figures 9.17–9.19 are examples of the starboard side egress on a ship of the same class as FITZGERALD. There is no hatch separating the starboard access trunk outside Berthing 2 and the space above it. Also in this starboard

Figure 9.17 Sample Berthing 1. Starboard side egress – ladder up. Source: U.S. Navy.

Figure 9.18 Sample Berthing 2. Starboard side egress – ladder up. Source: U.S. Navy.

Figure 9.19 Sample Berthing 3. Starboard side egress – scuttle down to forward IC (no outlet). Source: U.S. Navy.

access trunk is a hatch and watertight scuttle connected to a ladder going down to the Forward Interior Communications (FWD IC) space. FWD IC is not manned underway and has no other exits.

Of the 42 Sailors assigned to Berthing 2, at the time of collision, five were on watch and two were not aboard. Of the 35 remaining Sailors in Berthing 2, 28 escaped the flooding. Seven Sailors perished.

Some of the Sailors who survived the flooding in Berthing 2 described a loud noise at the time of impact. Other Berthing 2 Sailors felt an unusual movement of the ship or were thrown from their racks. Other Berthing 2 Sailors did not realize what had happened and remained in their racks. Some of them remained asleep. Some Sailors reported hearing alarms after the collision, while others remember hearing nothing at all.

Seconds after impact, Sailors in Berthing 2 started yelling, "Water on deck!" and, "Get out!" One Sailor saw another knocked out of his rack by water. Others began waking up shipmates who had slept through the initial impact. At least one Sailor had to be pulled from his rack and into the water before he woke up. Senior Sailors checked for others that might still be in their racks.

The occupants of Berthing 2 described a rapidly flooding space, estimating later that the space was nearly flooded within a span of 30–60 seconds. By the time the third Sailor to leave arrived at the ladder, the water was already waist deep. Debris, including mattresses, furniture, an exercise bicycle, and wall lockers, floated into the aisles between racks in Berthing 2, impeding Sailors' ability to get down from their racks and their ability to exit the space. The ship's 5–7° list to starboard

Figure 9.20 Sample Berthing 2 view from row 3 of racks to the port side (open door on right leading to head). Source: U.S. Navy.

increased the difficulty for Sailors crossing the space from the starboard side to the port side. Many of the Sailors recall that the battle lanterns were illuminated. Battle lanterns turn on when power to an electrical circuit is out or when turned on manually. Battle lanterns are shown hanging from the ceiling in Figure 9.20.

Sailors recall that after the initial shock, occupants lined up in a relatively calm and orderly manner to climb the port side ladder and exit through the port side watertight scuttle. Figure 9.20 provides an example of the route Sailors would have taken from their racks to the port side watertight scuttle on a ship of the same class as FITZGERALD. They moved along designated floor and turned left at the end to access the ladder. Figure 9.20 provides an example and sense of scale. Even though the Sailors were up to their necks in water by that point, they moved forward slowly and assisted each other. One Sailor reported that FC1 Rehm pushed him out from under a falling locker. Two of the Sailors who already escaped from the main part of Berthing 2 stayed at the bottom of the ladder well (see Figure 9.15) in order to help their shipmates out of the berthing area.

The door to the Berthing 2 head (bathrooms and showers) was open and the flooding water dragged at least one person into this area.

Exiting from the head during this flood of water was difficult and required climbing over debris.

As the last group of Sailors to escape through the port side watertight scuttle arrived at the bottom of the ladder, the water was up to their necks. The two Sailors who had been helping people from the bottom of the ladder were eventually forced to climb the ladder as water reached the very top of the Berthing 2 compartment. They continued to assist their shipmates as they climbed, but were eventually forced by the rising water to leave Berthing 2 through the watertight scuttle themselves. Before climbing the ladder, they looked through the water and did not see any other Sailors. Once through the watertight scuttle and completely out of the Berthing 2 space (on the landing outside Berthing 1) they continued to search, reaching into the dark water to try to find anyone they could. From the top of the ladder, these two Sailors were able to pull two other Sailors from the flooded compartment. Both of the rescued Sailors were completely underwater when they were pulled to safety.

The last Sailor to be pulled from Berthing 2 was in the bathroom at the time of the collision and a flood of water knocked him to the deck (floor). Lockers were floating past him and he scrambled across them toward the main berthing area. At one point he was pinned between the lockers and the ceiling of Berthing 2, but was able to reach for a pipe in the ceiling to pull himself free. He made his way to the only light he could see, which was coming from the port side watertight scuttle. He was swimming toward the watertight scuttle when he was pulled from the water, red-faced and with bloodshot eyes. He reported that when taking his final breath before being saved, he was already submerged and breathed in water.

After the last Sailor was pulled from Berthing 2, the two Sailors helping at the top of the port side watertight scuttle noticed water coming into the landing from Berthing 1. They remained in case any other Sailors came to the ladder. Again, one of the Sailors stuck his arms through the watertight scuttle and into the flooded space to try and find any other Sailors, even as the area around him on the landing outside of Berthing 1 flooded. Berthing 1, with no watertight door between it and the landing, began to flood.

Another Sailor returned with a dogging wrench, a tool used to tighten the bolts, on the hatch to stave off flooding from the sides of the hatch. The three Sailors at the top of the ladder yelled into the water-filled space below in an attempt to determine if there was anyone still within Berthing 2. No shadows were seen moving and no response was given.

Water began shooting up and out of the watertight scuttle into the landing. Finding no other Sailors, they tried to close the watertight scuttle to stop the flood of water. The force of the water through the hatch prevented closing the watertight scuttle between Berthing 2 and Berthing 1. The scuttle was left partially open. They then climbed the ladder to the Main deck (one level up from the Berthing 1 landing), and secured the hatch and scuttle between Berthing 1 and the Main deck. In total, 27 Sailors escaped Berthing 2 from the port side ladder.

One Sailor escaped via the starboard side of Berthing 2. After the collision, this Sailor tried to leave his rack, the top rack in the row nearest to the starboard access trunk, but inadvertently kicked someone, so he crawled back into his rack and waited until he thought everyone else would be out of the Berthing 2. When he jumped out of his rack a few seconds later, the water was chest high and rising, reaching near to the top of his bunk.

After leaving his rack, the Sailor struggled to reach the starboard egress point through the lounge area. He moved through the lounge furniture and against the incoming sea. Someone said, "go, go, go, it's blocked," but he was already underwater. He was losing his breath under the water but found a small pocket of air. After a few breaths in the small air pocket, he eventually took one final breath and swam. He lost consciousness at this time and does not remember how he escaped from Berthing 2, but he ultimately emerged from the flooding into Berthing 1, where he could stand to his feet and breathe. He climbed Berthing 1's egress ladder, through Berthing 1's open watertight scuttle and collapsed on the Main Deck. He was the only Sailor to escape through the starboard egress point.

The flooding of Berthing 2 resulted in the deaths of seven FITZGERALD Sailors. The racks of these seven Sailors were located in Rows 3 and 4, the area closest to the starboard access trunk and egress point and directly in the path of the onrushing water, as depicted in Figure 9.21.

After escaping Berthing 2, Sailors went to various locations. Some assembled on the mess decks to treat any injuries and pass out food and water. Others went to their General Quarters stations to assist with damage control efforts. Another Sailor went to the bridge to help with medical assistance. One Sailor later took the helm and stood a 15-hour watch in aft steering after power was lost forward.

9.2.1.6 Findings

Collisions at sea are rare and the relative performance and fault of the vessels involved is an open admiralty law issue. The Navy is not concerned about the mistakes made by CRYSTAL. Instead, the Navy is focused on the performance of its ships and what we could have done differently to avoid these mishaps.

In the Navy, the responsibility of the Commanding Officer for his or her ship is absolute.

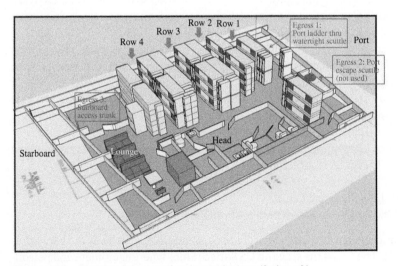

Figure 9.21 Berthing 2 layout of racks and lockers (facing aft).

Many of the decisions made that led to this incident were the result of poor judgment and decision making of the Commanding Officer. That said, no single person bears full responsibility for this incident. The crew was unprepared for the situation in which they found themselves through a lack of preparation, ineffective command and control, and deficiencies in training and preparations for navigation.

9.2.1.7 Training

FITZGERALD officers possessed an unsatisfactory level of knowledge of the International Rules of the Nautical Road.

Watch team members were not familiar with basic radar fundamentals, impeding effective use.

9.2.1.8 Seamanship and Navigation

The Officer of the Deck and bridge team failed to comply with the International Rules of the Nautical Road. Specifically:

FITZGERALD was not operated at a safe speed appropriate to the number of other ships in the immediate vicinity.

FITZGERALD failed to maneuver early as required with risk of collision present.

FITZGERALD failed to notify other ships of danger and to take proper action in extremis.

Watch team members responsible for radar operations failed to properly tune and adjust radars to maintain an accurate picture of other ships in the area.

Watchstanders performing physical look out duties did so only on FITZGERALD's left (port) side, not on the right (starboard) side where the three ships were present with risk of collision.

Key supervisors responsible for maintaining the navigation track and position of other ships: Were unaware of existing traffic separation schemes and the expected flow of traffic.

Did not utilize the Automated Identification System. This system provides real time updates of commercial ship positions through use of the Global Positioning System (GPS).

FITZGERALD's approved navigation track did not account for, nor follow, the Vessel Traffic Separation Schemes in the area.

9.2.1.9 Leadership and Culture

The following are the Navy's findings from the investigation:

- The bridge team and CIC teams did not communicate effectively or share information. The CIC is the space on US Surface Ships where equipment and personnel combine to produce the most accurate picture of the operating environment.

- Supervisors and watch team members on the bridge did not communicate information and concerns to one another as the situation developed.
- The Officer of the Deck, responsible for the safe navigation of the ship, did not call the Commanding Officer on multiple occasions when required by Navy procedures.
- Key supervisors in the CIC failed to comprehend the complexity of the operating environment and the number of commercial vessels in the area.
- In several instances, individual members of the watch teams identified incorrect information or mistakes by others, yet failed to proactively and forcefully take corrective action, or otherwise highlight or communicate their individual concerns.
- Key supervisors and operators accepted difficulties in operating radar equipment due to material faults as routine rather than pursuing solutions to fix them.
- The command leadership did not foster a culture of critical self-assessment. Following a near-collision in mid-May, leadership made no effort to determine the root causes and take corrective actions in order to improve the ship's performance.
- The command leadership was not aware that the ship's daily standards of performance had degraded to an unacceptable level.

9.2.1.10   Fatigue

- The command leadership allowed the schedule of events preceding the collision to fatigue the crew.
- The command leadership failed to assess the risks of fatigue and implement mitigation measures to ensure adequate crew rest.

9.2.1.11   Timeline of Events

June 16, 2017

0001	FITZGERALD is moored at Commander, Fleet Activities Yokosuka (CFAY) Pier 12
0600	Liberty expires for crew
0900	Navigation briefing held to prepare crew for underway and anchoring evolution
1030	Stationed "Sea and Anchor" Detail to get underway
1130	Underway from Pier 12
1210	Anchored in preparation for ammunition on-load
1545	Stationed the Sea and Anchor Detail to get underway from anchor. The Sea and Anchor Detail provides additional personnel experienced in navigating in restricted waters

1624	Underway from anchor
1736	Conducted helicopter deck landing qualifications and aviation certifications
1747	Stationed Modified Navigation Detail due to proximity of shoal water while transition to Sagami Wan. The Modified Navigation detail provides additional personnel when navigating in close proximity to shallow water
1835	The Modified Navigation Detail was secured
1859	Sunset was observed and navigation lights were energized and dimmed to support flight operations
2111	Flight operations were secured
2116	The Modified Navigation Detail was stationed while closing land in preparation for small boat operations to return Afloat Training Group Western Pacific personnel ashore
By 2200	All watchstanders assigned to the 2200–0200 watch time period were on duty
Approximately 2300	Small Boat Operations secured and FITZGERALD proceeded southwest from Sagami Wan to sea
2300	The Commanding Officer left the Bridge
Approximately 2305	The Executive Officer and Navigator left the Bridge
2311–2345	FITZGERALD maintained 16 kts due to high traffic density
2330	Moonrise at 69% illumination
2345	FITZGERALD maneuvered to course 230 and increased speed to 20 kts
2350	FITZGERALD overtook a contact on the left (port) side within three nautical miles and no report was made to the Commanding Officer as required by his Standing Orders procedures. No course and speed determinations were made for this vessel by watchstanders

June 17, 2017

0000	FITZGERALD approached the Vessel Traffic Separation Scheme (VTSS) north of Oshima Island
0000	FITZGERALD was in vicinity of four commercial vessels, two of which were within three nautical miles and no report was made to the Commanding Officer as required by his Standing Orders procedures. No course and speed determinations were made for this vessel by watchstanders
0015	FITZGERALD was passing two commercial vessels, one of which was within three nautical miles and no report was made to the Commanding Officer as required by his Standing Orders procedures. No course and speed determinations were made for this vessel by watchstanders

0022	FITZGERALD altered course to 220 and remained at 20 kts
0033	FITZGERALD altered course to 215
0034	Four vessels passed down the left (port) side with closest point of approach at 1500 yd. The Commanding Officer was informed. No course and speed determinations were made for these vessels. Radar contact on them was not held
0054	FITZGERALD altered course to 190 while remaining at 20 kts
0058	FITZGERALD was in the vicinity of five commercial vessels. Three of these passed on the left (port) side within three nautical miles and no report was made to the Commanding Officer as required by his Standing Orders procedures. No course and speed determinations were made for this vessel by watchstanders
0100	FITZGERALD remained on course 190 at 20 kts
0108	FITZGERALD crossed the bow of a ship at approximately 650 yd, passed a second vessel at two nautical miles, and a third vessel at 2.5 nautical miles. No reports were made to the Commanding Officer as required by his Standing Orders procedures. No course and speed determinations were made for this vessel by watchstanders
0110	FITZGERALD continued on course 190, speed 20 kts. CRYSTAL was ahead on FITZGERALD's starboard side at a distance of 11 nautical miles
0110	Watchstanders unsuccessfully attempted to initiate a radar track on the CRYSTAL
0115	CRYSTAL was closing FITZGERALD's intended track at a high rate of speed.
0117	The FITZGERALD Officer of the Deck plotted a radar track on a vessel thought to be CRYSTAL and calculated that CRYSTAL would pass 1500 yd from FITZGERALD on the right (starboard) side. It is unknown if the officer of the deck (OOD) was tracking the CRYSTAL or another commercial vessel.
0120	The watch stander responsible for immediate support to the Officer of the Deck, the Junior Officer of the Deck, reported sighting CRYSTAL visually and noted that CRYSTAL's course would cross FITZGERALD's track. The Officer of the Deck continued to think that CRYSTAL would pass at 1500 yd from FITZGERALD
0122	The Junior Officer of the Deck sighted CRYSTAL again and made the recommendation to slow. The Officer of the Deck responded that slowing would complicate the contact picture.
0125	CRYSTAL was approaching FITZGERALD from the right (starboard) side at three nautical miles. FITZGERALD watchstanders at this time held two other commercial vessels in addition to CRYSTAL. One was calculated to have closest approach point at 2000 yd and the other was calculated to risk collision. No contact reports were made to the Commanding Officer and no additional course and speed determinations were made on these vessels

0125	The Officer of the Deck noticed CRYSTAL rapidly getting closer and considered a turn to 240T
0127	The Officer of the Deck ordered course to the right to course 240T, but rescinded the order within a minute. Instead, the Officer of the Deck ordered an increase to full speed and a rapid turn to the left (port). These orders were not carried out
0129	The Boatswain's Mate of the Watch, a more senior supervisor on the bridge, took over the helm and executed the orders
As of 0130	Neither FITZGERALD nor CRYSTAL made an attempt to establish radio communications or sound the danger signal
As of 0130	FITZGERALD had not sounded the collision alarm
0130:34	CRYSTAL's bow struck FITZGERALD at approximately frame 160 on the right (starboard) side above the waterline and CRYSTAL's bulbous bow struck at approximately frame 138 below the waterline

9.2.2 Collision of USS JOHN S MCCAIN with Motor Vessel ALNIC MC

9.2.2.1 Introduction

USS JOHN S MCCAIN collided with Motor Vessel ALNIC MC on August 21, 2017 in the Straits of Singapore.

JOHN S MCAIN is a Flight 1 Arleigh Burke Class Destroyer, commissioned in 1994 and homeported in Yokosuka, Japan, as part of the Forward Deployed Naval Forces and Carrier Strike Group FIVE. Approximately 300 sailors serve aboard MCCAIN. MCCAIN is 505 ft in length and carries a gross tonnage of approximately 9000 tons.

ALNIC MC is a Liberia flagged oil and chemical tanker built in 2008. ALNIC MC is approximately 600 ft long and has a gross tonnage of approximately 30,000 tons.

The collision between JOHN S MCCAIN and ALNIC resulted in the deaths of 10 US Sailors due to impact with MCCAIN's berthing compartments, located below the waterline of the ship. ALNIC suffered no fatalities. US Sailor fatalities were the following:

ETC (electronics technician) Charles Nathan Findley of Amazonian, Missouri, 31 years old.
ICC Abraham Lopez of El Paso, Texas, 39 years old.
ET1 Kevin Sayer Bushell of Gaithersburg, Maryland, 26 years old.
ET1 Jacob Daniel Drake of Cable, Ohio, 21 years old.
ITl Timothy Thomas Eckels Jr. of Baltimore, Maryland, 23 years old.
ITl Corey George Ingram of Poughkeepsie, New York, 28 years old.
ET2 Dustin Louis Doyon of Suffield, Connecticut, 26 years old.

ET2 John Henry Hoagland III of Killeen, Texas, 20 years old.
IC2 Logan Stephen Palmer of Harristown, Illinois, 23 years old.
ET2 Kenneth Aaron Smith of Cherry Hill, New Jersey, 22 years old.

9.2.2.2 Summary of Findings
The Navy determined the following causes of the collision:

Loss of situational awareness in response to mistakes in the operation of the JOHN S MCCAIN's steering and propulsion system, while in the presence of a high density of maritime traffic.

Failure to follow the International Nautical Rules of the Road, a system of rules to govern the maneuvering of vessels when risk of collision is present.

Watchstanders operating the JOHN S MCCAIN's steering, and propulsion systems had insufficient proficiency and knowledge of the systems (Figures 9.22 and 9.23).

9.2.2.3 Background
JOHN S MCCAIN departed its homeport of Yokosuka, Japan on May 26, 2017 for a scheduled six-month deployment in the Western Pacific, which at the time of the collision had included operations in the East and South China Seas, and port visits in Vietnam, Australia, Philippines, and Japan. On the morning of August 21, JOHN S MCCAIN was 50 nautical miles east of Singapore, approaching the Singapore Strait and Strait of Malacca, in transit to a scheduled port of call at Changi Naval Base, Singapore. These Straits form a combined ocean passage that is one of the busiest shipping lanes in the world, with more than 200 vessels passing through the straits each day. JOHN S MCCAIN was transiting through the southern end of the Strait. See Figure 2.22. In the predawn hours of August 21, 2017, the

Figure 9.22 Relative size of USS JOHN S MCCAIN.

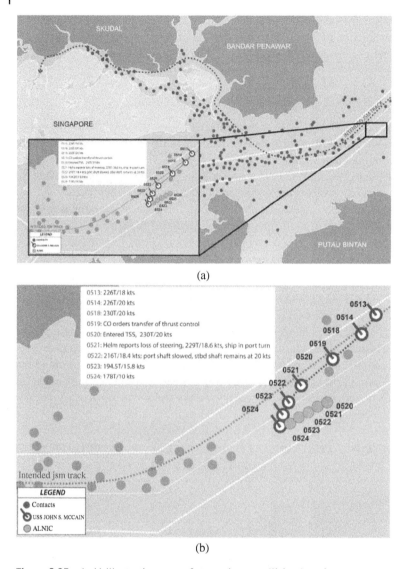

(a)

(b)

Figure 9.23 (a, b) Illustration map of approximate collision location.

moon had set and the skies were overcast. There was no illumination, and the sun would not rise until 0658. Seas were calm, with 1–3 ft swells. All navigation and propulsion equipment was operating properly.

At 0418, JOHN S MCCAIN transitioned to a Modified Navigation Detail due to approaching within 10 nautical miles from shoal water. This detail is used by the

Navy when in proximity of water too shallow to safely navigate as occurs when entering ports. This detail supplemented the on-watch team with a Navigation Evaluator and Shipping Officer, providing additional personnel and resources in the duties of Navigation and management of the ship's relative position to other vessels.

JOHN S MCCAIN was scheduled to enter the Singapore Strait Traffic Separation Scheme less than an hour later. Traffic separation schemes are established by local authorities in approaches to ports throughout the world to provide ships assistance in separating their movements when transiting to and from ports. The Commanding Officer had been physically present on the bridge since 0115, a practice common for operations with higher risk, such as navigating in the presence of busy maritime traffic at night. The XO reported to the bridge at 0430 to provide additional supervision and oversight to enter port.

Although JOHN S MCCAIN entered the Middle Channel of the Singapore Strait (a high traffic density area) at 0520, the Sea and Anchor Detail, a team the Navy uses for transiting narrower channels to enter port, was not scheduled to be stationed until 0600. This Detail provides additional personnel with specialized navigation and ship handling qualifications.

JOHN S MCCAIN was operating by procedures established for US Navy surface ships when operating at sea before sunrise, including being at "darkened ship." "Darkened Ship" means that all exterior lighting was off except for the navigation lights that provide identification to other vessels, and all interior lighting was switched to red instead of white to facilitate crew rest. The ship was in a physical posture known as "Modified ZEBRA," meaning that all doors inside the ship, and all hatches, which are openings located on the floor between decks, at the main deck and below were shut to help secure the boundaries between different areas of the ship in case of flooding or fire. Watertight scuttles on the hatches (smaller circular openings that can be opened or closed independently of the hatch) were left open to allow easy transit between spaces.

9.2.2.4   Events Leading to the Collision

At 0519, the Commanding Officer noticed the Helmsman (the watchstander steering the ship) having difficulty maintaining course while also adjusting the throttles for speed control. In response, he ordered the watch team to divide the duties of steering and throttles, maintaining course control with the Helmsman while shifting speed control to another watchstander known as the Lee Helm station, who sat directly next to the Helmsman at the panel to control these two functions, known as the Ship's Control Console. See Figures 9.24 and 9.25. This unplanned shift caused confusion in the watch team, and inadvertently led to steering control transferring to the Lee Helm Station without the knowledge of the watch team. The CO had only ordered speed control shifted. Because he did not know that

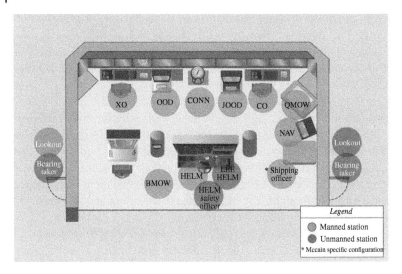

Figure 9.24 Bridge schematic of JOHN S MCCAIN.

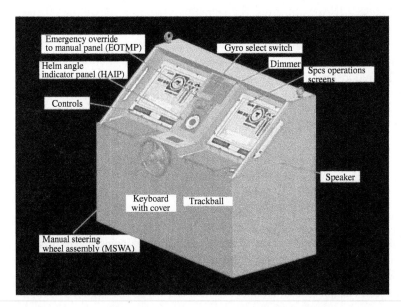

Figure 9.25 Illustration of ship control console on JOHN S MCCAIN.

steering had been transferred to the Lee Helm, the Helmsman perceived a loss of steering.

Steering was never physically lost. Rather, it had been shifted to a different control station and watchstanders failed to recognize this configuration. Complicating this, the steering control transfer to the Lee Helm caused the rudder to go amidships (centerline). Since the Helmsman had been steering 1–4° of right rudder to maintain course before the transfer, the amidships rudder deviated the ship's course to the left.

Additionally, when the Helmsman reported loss of steering, the Commanding Officer slowed the ship to 10 kts and eventually to 5 kts, but the Lee Helmsman reduced only the speed of the port shaft as the throttles were not coupled together (ganged). The starboard shaft continued at 20 kts for another 68 seconds before the Lee Helmsman reduced its speed. The combination of the wrong rudder direction, and the two shafts working opposite to one another in this fashion caused an un-commanded turn to the left (port) into the heavily congested traffic area in close proximity to three ships, including the ALNIC. See Figure 9.23b.

Although JOHN S MCCAIN was now on a course to collide with ALNIC, the Commanding Officer and others on the ship's bridge lost situational awareness. No one on the bridge clearly understood the forces acting on the ship, nor did they understand the ALNIC's course and speed relative to JOHN S MCCAIN during the confusion.

Approximately three minutes after the reported loss of steering, JOHN S MCCAIN regained positive steering control at another control station, known as Aft Steering, and the Lee Helm gained control of both throttles for speed and corrected the mismatch between the port and starboard shafts. These actions were too late, and at approximately 0524 JOHN S MCCAIN crossed in front of ALNIC's bow and collided. See Figure 9.26.

Despite their close proximity, neither JOHN S MCCAIN nor ALNIC sounded the five short blasts of whistle required by the International Rules of the Nautical Road for warning one another of danger, and neither attempted to make contact through Bridge-to-Bridge communications.

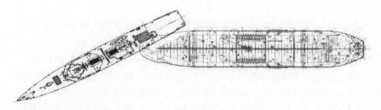

Figure 9.26 Approximate geometry and point of impact between USS JOHN S MCCAIN and ALNIC MC.

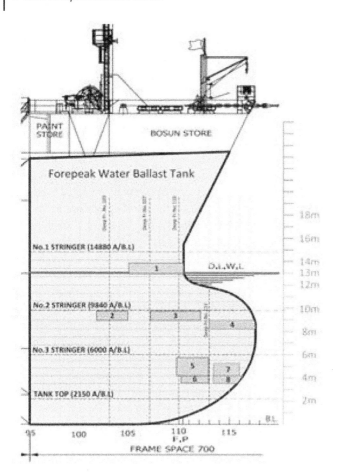

Figure 9.27 Bulbous bow of ALINIC MC and damage to hull from bow to stern.

9.2.2.5 Results of Collision

The bulbous bow of ALNIC MC impacted JOHN S MCCAIN on the port (left) aft side. The impact created a 28-ft diameter hole both below and above the waterline of the JOHN S MCCAIN. See Figures 9.27–9.29.

The point of impact was centered on Berthings 3 and 5 as noted in Figure 9.29. All significant injuries occurred to Sailors that were in Berthing 3 at the time of the impact. All 10 of the fallen Sailors were in Berthing 5 at the time of impact.

ALNIC MC and JOHN S MCCAIN initially remained attached to each other after the collision. Sailors describe this as lasting up to a couple of minutes. The prolonged contact kept the ship from taking a list (tilt to one side) immediately. Sailors on the bridge and on the external deck of the ship immediately after the collision could see ALNIC MC's bow (front of the ship) still lodged into the side of JOHN S MCCAIN. However, within 15 minutes JOHN S MCCAIN had developed a four-degree list to port as the ship flooded.

Figure 9.28 Point of impact on JOHN S MCCAIN from ALINIC MC. Source: Military Collection/Alamy Images.

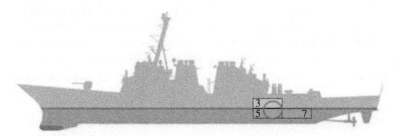

Figure 9.29 Depiction of approximate location of point of impact.

The collision was felt throughout the ship. Watchstanders on the bridge were jolted from their stations momentarily and watchstanders in aft steering were thrown off their feet. Several suffered minor injuries. Some Sailors thought the ship had run aground, while others were concerned that they had been attacked. Sailors in parts of the ship away from the impact point compared it to an earthquake. Those nearest the impact point described it as like an explosion.

As required by Navy procedures, the XO ordered the collision alarm sounded to alert personnel to begin damage control efforts. The Commanding Officer remained on the bridge and the XO departed to the CIC and eventually to Berthing 3 to provide oversight in damage control efforts. The Command Master Chief, the senior assigned enlisted Sailor onboard, went to the area where damage control efforts, known as the Central Control Station, were managed, and then moved about the ship, assisting damage control efforts. After the situation on the bridge stabilized, the Commanding Officer then proceeded to Central Control Station to check on the status of the damage control efforts.

The CO ordered the watch team to announce the collision on the Bridge-to-Bridge radio, which alerted other ships in the area to the collision and the damages. At 0530, JOHN S MCCAIN requested tugboats and pilots from Singapore Harbor to assist in getting the ship to Changi Naval Base.

JOHN S MCCAIN changed its lighting configuration at the mast to one red light over another red light, known as "red over red," the international lighting scheme that indicates a ship that is "not under command." Under the International Rules of the Nautical Road, this warns other ships that, due to an exceptional circumstance, a vessel is unable to maneuver as required.

Most of the electronic systems on the bridge were inoperable until the two ships parted. Main communications systems on the bridge stopped working after the collision and the bridge began using handheld radios to communicate with aft steering. Sound powered phones, which do not require electrical power to transmit communications, and handheld radios were the main means of communication from the bridge. Aft Internal Communications, a space adjacent to Berthing 5 with communications control equipment, quickly flooded and was likely responsible for the loss of bridge communications.

All US Navy ships are designed to withstand and recover from damage due to fire, flooding, and other damage sustained during combat or other emergencies. Each ship has a Damage Control Assistant, working under the Engineering Officer, to establish and maintain an effective damage control organization. The Damage Control Assistant oversees the prevention and control of damage including control of stability, list, and trim due to flooding (maintaining the proper level of the ship from side to side and front and back), coordinates firefighting efforts, and is also responsible for the operation, care, and maintenance of the ship's repair facilities. The Damage Control Assistant ensures the ship's repair party personnel are properly trained in damage control procedures including firefighting, flooding, and emergency repairs. The Damage Control Assistant is assisted by the DCC, a chief petty officer specializing in Damage Control. The officer in charge of damage control efforts, the Damage Control Assistant, called away General Quarters to notify the crew to commence damage control efforts.

General Quarters is a process whereby the crew reports to pre-assigned stations and duties in the event of large casualties such as flooding. General Quarters is announced by an alarm that sounds throughout the ship to alert the crew of an emergency or potential combat operations. All crewmembers are trained to report to their General Quarters watch station and to set a higher condition of material readiness against fire, flooding, or other damage. This involves securing additional doors, hatches, scuttles, valves, and equipment to isolate damage and prepare for combat. Navy crews train on Damage Control continuously, with drills being run in port and underway frequently to prepare the teams for damage to equipment and spaces. During any emergency condition (fire, flooding, combat operations), the Damage Control Assistant coordinates and supervises all damage control efforts from one of the three Damage Control Repair Lockers.

Damage Control Repair Lockers are specialized spaces stationed throughout the ship filled with repair equipment and manned during emergencies with teams of about 20 personnel trained to respond to casualties. There are three repair lockers on the JOHN S. MCCAIN: Repair Locker 2, Repair Locker 5, and Repair Locker 3. Repair Locker 2 covers the forward part of the ship, Repair Locker 5 covers the engineering spaces and Repair Locker 3 covers the aft part of the ship. Each locker is maintained with similar equipment. Personnel assigned to repair lockers are trained and qualified to respond to and repair damage from a variety of sources with a specific focus on fire and flooding. Each repair locker can act independently but is also designed to support the others and can take over the responsibilities for any locker if damage prevents that locker's use. The repair lockers are normally unmanned unless the ship sets a condition of higher readiness like General Quarters when they would be manned within minutes.

Sailors began to locate, report and track flooding, fire, and structural damage to the ship immediately. Significant damage was reported throughout the ship in the moments after the collision, including flooding, internal fuel leakage, loss of ventilation and internal communications, and degradation of many of the ship's other systems.

JOHN S MCCAIN began the process of accounting for all crew members immediately after the collision. This process continued even as the crew made emergency repairs, battled flooding, and helped each other out of damaged spaces. The damage control efforts made confirming the location of personnel difficult. Varying reports of missing Sailors were made in the first minutes after the collision. However, by the submission of the third complete report, there was reasonable confidence that the crew had been accounted for was correct because all of the 10 missing Sailors had been consistently reported missing and all lived in Berthing 5, a space that was inaccessible and flooded.

9.2.2.6 Impact to Berthing 5

Berthing 5 is located aft (near the back of the ship) on the port side. See Figure 9.30. It is approximately 25 ft by 15 ft and has 18 racks, stacked as bunk beds three-high. Each row of racks has a locker for Sailors' belongings. There is a lounge with seats, a small table, and a wall-mounted television. There is a head with one toilet, one shower stall, and one sink.

There are two means to exit Berthing 5: the primary egress (ladderwell) through a hatch with a scuttle (Figure 9.31) and an escape scuttle into Berthing 3 on the deck above (Figure 9.32).

During Modified ZEBRA, the hatch is closed, but the scuttle is open for ease of access. The escape scuttle is normally always closed, as it was at the time of the

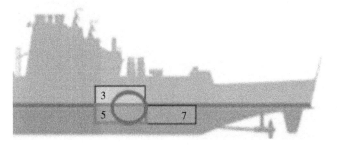

Figure 9.30 Relative positions of Berthings labelled 3, 5, and 7, is point of impact.

Figure 9.31 Primary egress from Berthing 5. Left: from within Berthing 5. Right: Above on the deck outside Berthing 3.

impact. The collision knocked debris in Berthing 3 on top of the escape scuttle connecting Berthing 3 to Berthing 5 below it. This would have made any attempts to open and exit through the escape scuttle very difficult.

Most of the Berthing 5, a space that is normally 15 ft wide, was compressed by the impact to only 5 ft wide. There were 17 Sailors assigned to Berthing 5. At the time of the collision, all were aboard the ship and five were on watch or outside the space. Based on the size of the hole, and the fact that Berthing 5 is below the waterline, the space likely fully flooded in less than a minute after the collision.

Two Sailors who were in Berthing 5 at the time of the collision escaped from the space. The first Sailor was on the second step of the ladder-well leading to the deck above when the collision occurred. The impact of the collision knocked him to the

Figure 9.32 Escape scuttle from Berthing 5. Left: From within Berthing 5. Right: Above in Berthing 3.

ground, leaving his back and legs bruised. Fuel quickly pooled around him, and he scrambled up and back onto the ladder.

The Sailor climbed out of Berthing 5 through the open scuttle, covered in fuel and water from the near instantaneous flooding of the space. He did not see anyone ahead of or behind him as he escaped. He reported seeing two other Sailors in the lounge area, one preparing for watch duties and another standing near his rack. Both Sailors were lost, along with the eight shipmates who were in their racks to rest at the time of the collision.

The second Sailor who escaped from Berthing 5 heard the crashing and pushing of metal before the sound of water rushing in. Within seconds, water was at chest level. The passageway leading to the ladder-well was blocked by debris, wires, and other wreckage hanging from the overhead. From the light of the battle lanterns (the emergency lighting that turns on when there is a loss of normal lights due to power outage) he could see that he would have to climb over the debris to get to the ladder-well.

As he started his climb across the debris to the open scuttle, the water was already within a foot of the overhead, so he took a breath, dove into the water, and swam toward the ladder-well. Underwater, he bumped into debris and had to feel his way along. He was able to stop twice for air as he swam, the water higher each time, and eventually used the pipes to guide him toward the light coming from the scuttle. The Sailor found that the blindfolded egress training, a standard that requires training to prepare Sailors for an emergency and was conducted when he reported to the command, was essential to his ability to escape.

One Sailor was alerted by the first Sailor who escaped Berthing 5 that others were still inside the space, and he went to assist them. When he first reached the closed hatch and open scuttle, the water in Berthing 5 was at the top of the third rung. He tried to enter the space, but was forced back up the ladder by the pressure of the escaping air and rising water, which within seconds had risen to within a foot of the hatch. He saw a Sailor swimming toward the exit and pulled him out of the water through the scuttle between the two decks. This was the second and last Sailor to escape from Berthing 5. His body was scraped, bruised, and covered with chemical burns from being submerged in the mixture of water and fuel.

An additional Sailor who came to assist observed the rescue and, looking down into the berthing, saw "a green swirl of rising seawater and foaming fuel" approaching the top of the scuttle. As the final Sailor to leave Berthing 5 was pulled to safety, the Sailors at the top of the scuttle checked to see if there was anyone behind him. They did not see anybody. By then, so much water was already coming up through the scuttle that it was difficult to close and secure. The fuel mixed in with the water made one of Sailor's hands so slippery that he cut himself while using the wrench designed to secure the scuttle, but the two were able to secure it to stop the rapid flooding of the ship.

9.2.2.7 Impact on Berthing 3

Berthing 3 is immediately above Berthing 5 but spans the width of the ship. There are two points of egress from each side of Berthing 3; on the port side there is a ladder-well leading down into the center of the berthing and an escape scuttle that is in the forward section of the space leading up to the next deck. There were 71 Sailors assigned to Berthing 3.

At 0530, the DCA began receiving reports of a ruptured fire main and water and fuel flooding into Berthing 3. The port side of Berthing 3 suffered substantial damage, including a large hole in the bulkhead. See Figure 9.33. Racks and lockers

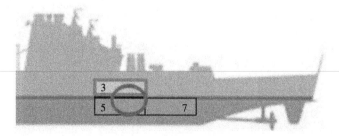

Figure 9.33 Relative positions of Berthings 3, 5, and 7 and point of impact.

detached from the walls and were thrown about, leaving jagged metal throughout the space. Cables and debris hung from the ceiling.

A Sailor from Berthing 3, who was later medically evacuated from the ship, sustained his injuries as the wall next to him blew apart in the collision and threw him to the ground. Water and fuel quickly pooled around him in the short time he was on the ground, and he began crawling over debris to escape. Another Sailor went to him and helped pull him to the lounge area and toward the ladder. On the way, the Sailor who was being assisted fell on the slippery floor and hit his head. Two other Sailors, also injured, helped him reach the flight deck.

Limited lighting guided the remaining Sailors as they left the berthing space. Sailors had to climb over lockers and other debris to escape, using the high vantage point to also minimize the risk of electrocution from traveling through the rising water. Some escaped in only their underwear, and many were bruised and bloodied from injuries sustained in the collision and covered in fuel. At least one Sailor attempted to move the metal rack pinning a trapped shipmate and realized that he could not move it alone. The Sailors who escaped Berthing 3 provided some of the first reports to command climate specialist (CCS) that the space was severely damaged, that it was rapidly taking on water, and that Sailors were trapped inside.

Hearing reports that Sailors were trapped in Berthing 3, the Executive Offer and a group of Sailors, including some who evacuated Berthing 7, went to check on their shipmates. Several Sailors were pinned in their racks as a result of the collision, but, as the two ships pulled apart, the twisted metal shifted and most of the Sailors in Berthing 3 were able to escape as the debris moved. One of these Sailors was pinned in his rack underneath two racks that had collapsed and several lockers that became dislodged during the collision. He was able to escape after ALNIC MC detached. See Figures 9.34 and 9.35.

Figure 9.34 Berthing 3 facing port.

Figure 9.35 Berthing 3 facing port after collision.

However, two Sailors remained pinned in their racks even after the ships separated. Four members of the crew entered Berthing 3 through the jagged metal and rising water to rescue them. The first of these rescuers heard Sailors shouting for help from inside Berthing 3 and tried to enter on the port side; however, the door was blocked by debris, so he ran to the entrance on the starboard (right) side of the berthing.

One of the Sailors trapped in Berthing 3 had been asleep at the time of the collision and was awoken by it. When he opened his eyes, he understood that he was pinned in his rack, with one of his shoulders stuck between his rack and the rack above. He felt both air and water moving around him. He could hear shouting and began shouting himself, which alerted his others that he was trapped. Only his hand and foot were visible by those outside of the rack. The one battle lantern in the area provided the only light for rescuers to find the trapped Sailor. Water was already at knee level when rescuers reached him. The debris was too heavy for the rescuers to move, and a Portable Electric All-Purpose Rescue System, a "jaw of life" cutting device, was required to cut through the metal, separate the panels of the rack, and pull the panels out of the way. After approximately 30 minutes, these efforts allowed the trapped Sailor to pull his arm free. Moments later, the rescuers pulled him from between the racks by his foot. Stretcher bearers came to Berthing 3 and carried this Sailor to the Mess Decks to receive medical treatment.

The second Sailor was in a bottom rack in Berthing 3. His rack was lifted off the floor because of the collision, which likely prevented him from drowning in the rising water, and he was trapped at an angle between racks that had been pressed together. Light was visible through a hole in his rack, and he could hear the water and smell the fuel beginning to fill Berthing 3.

He attempted to push his way out of the rack, but every time he moved the space between the racks grew smaller and he was unable to escape. His foot was outside the rack, and he could feel water. It was hot in the space and difficult to breathe, but he managed to shout for help and banged against the metal rack to get the attention of other Sailors in the berthing space. The Sailors who entered Berthing 3 to rescue others heard this and began assisting him, but he was pinned by more debris than the first Sailor freed.

It took approximately an hour from the time of the collision to free the second Sailor from his rack. Rescuers used an axe to cut through the debris, a crowbar to pull the lockers apart piece by piece, and rigged a pulley to move a heavy locker in order to reach the Sailor. Throughout the long process, his rescuers assured him by touching his foot, which was still visible. Once freed, the Sailor was the last person to escape Berthing 3. Everything aft of his rack was a mass of twisted metal. He had scrapes and bruises all over his body, suffered a broken arm, and had hit his head. He was unsure whether he remained conscious throughout the rescue.

At least one scuttle to Berthing 3 was shut during damage control efforts. The space was electrically isolated and, at 0608, the fire main valves were closed, reducing the amount of flooding. Dewatering efforts began and succeeded in removing the water from Berthing 3 prior to JOHN S MCCAIN's arrival at Changi Naval Base.

9.2.2.8 Impact on Berthings 4, 6, and 7

Berthings 5 and 7 are next to each other on the port side of the ship, mirrored by Berthings 4 and 6, respectively, on the starboard side of the ship. Berthings 4 and 5 are connected across the ship through "cross flooding ducts," designed to distribute water from port to starboard side (or vice versa) to keep the ship level if it takes on water. Berthings 6 and 7 are similarly connected. A 6-ft-long crack in the wall between Berthings 5 and 7, created by the collision, allowed water to move between the spaces.

All Sailors in Berthing 7 were able to evacuate, but water was at approximately knee level as they exited the space. At 0530 there was report of a ruptured pipe in Berthing 7, which added to the flooding caused by the cracked wall separating Berthings 5 and 7. By 0605, Berthing 7 was reported as lost, meaning that it was fully flooded and secured to prevent the flooding from spreading to the rest of the ship.

All Sailors in Berthing 4 were able to evacuate. At 0544, Sailors reported 4 in. of water on the deck in Berthing 4. Sailors in Berthing 4 were thrown about their racks by the force of the collision. By 0627, the berthing space was lost. See Figures 9.36 and 9.37.

All Sailors in Berthing 6 were able to evacuate. At 0546 flooding was reported in Berthing 6, which is across from Berthing 7 on the starboard side of the ship. Despite the crew's dewatering efforts, the space was declared lost at 0627.

Figure 9.36 Scuttle and hatch showing the space completely flooded.

Figure 9.37 Berthing 4 dewatering.

At approximately 0630, because of crew's resiliency and successful damage control and engineering repair efforts, JOHN S MCCAIN was able to proceed under its own power toward Changi Naval Base, Singapore, at an average speed of 3 kts. JOHN S MCCAIN's navigation equipment was degraded because of the collision. While most electronic navigational aids on the bridge were operational, multiple warnings and alerts were illuminated, reducing the navigation team's confidence that the information was reliable. Because of the degraded information, the team relied on "seaman's eye" to stay on track while returning to port. Lack of ventilation across the ship raised concerns based on the amount of fuel that had spilled and the risks posed by rising temperatures inside the ship. The temperatures also drove many Sailors to the flight deck to escape the heat (Figures 9.38 and 9.39).

Figure 9.38 Port side of JOHN S MCCAIN post-collision.

Figure 9.39 Closeup of port side damage. Source: AB Forces News Collection/Alamy Images.

By approximately 1435, JOHN S MCCAIN was moored, and divers were in the water looking for places to enter the hull of the ship. The hole in the port side penetrated not only the hull, but an internal fuel tank as well. The fuel in the water created several hazards to divers and required them to proceed cautiously.

On a second dive at approximately 1500, divers were able to enter the hull of the ship to do initial safety assessments. Many of the conditions they found led to a cautious approach to assure the safety of the divers. The large amount of debris and structural damage required the divers to move slowly about the ship, even cutting holes through racks to access parts of the space. Visibility in Berthing 5 was very poor given the debris and lack of light. The divers had to move about the space almost exclusively by feel. The dive team conducted nearly continuous dive operations over a period of seven days until all ten of the Sailors in Berthing 5 were recovered.

9.2.2.9 Findings

- Collisions at sea are rare and the relative performance and fault of the vessels involved is an open admiralty law issue. The Navy is not concerned about the mistakes made by ALNIC. Instead, the Navy is focused on the performance of its ships and what we could have done differently to avoid these mishaps.
- In the Navy, the responsibility of the Commanding Officer for his or her ship is absolute.
- Many of the decisions made that led to this incident were the result of poor judgment and decision making of the Commanding Officer. That said, no single person bears full responsibility for this incident. The crew was unprepared for the situation in which they found themselves through a lack of preparation, ineffective command and control and deficiencies in training and preparations for navigation.

9.2.2.10 Training

- From the time when the CO ordered the Helm and Lee Helm split, to moments just before the collision, four different Sailors were involved in manipulating the controls at the sea combat commander (SCC).
- Because steering control was in backup manual at the helm station, the offer of control existed at all the other control stations (Lee Helm, Helm forward station, Bridge Command and Control station and Aft Steering Unit). System design is such that any of these stations could have taken control of steering via drop down menu selection and the Lee Helm's acceptance of the request. If this had occurred, steering control would have been transferred.
- When taking control of steering, the Aft Steering Helmsman failed to first verify the rudder position on the After Steering Control Console prior to taking control. This error led to an exacerbated turn to port just prior to the collision, as the indicated rudder position was 33° left, vice amidships. As a result, the rudder had a left 33° order at the console at this time, exacerbating the turn to port.
- Several Sailors on watch during the collision with control over steering were temporarily assigned from USS ANTIETAM (CG 54) with significant differences between the steering control systems of both ships and inadequate training to compensate for these differences.

- Multiple bridge watchstanders lacked a basic level of knowledge on the steering control system, in particular the transfer of steering and thrust control between stations. Contributing, personnel assigned to ensure these watchstanders were trained had an insufficient level of knowledge to effectively maintain appropriate rigor in the qualification program. The senior most officer responsible for these training standards lacked a general understanding of the procedure for transferring steering control between consoles.

9.2.2.11 Seamanship and Navigation

- Much of the track leading up to the Singapore Traffic Separation Scheme was significantly congested and dictated a higher state of readiness. Had this occurred, maximum plant reliability could have been set with a Master Helmsman and a qualified Engineering Lee Helm on watch.
- If the CO had set Sea and Anchor Detail adequately in advance of entering the Singapore Strait Traffic Separation Scheme, then it is unlikely that a collision would have occurred.
- The plan for setting the Sea and Anchor Detail was a failure in risk management, as it required watch turnover of all key watch stations within a significantly congested traffic separation scheme (TSS) and only 30 minutes prior to the Pilot pickup.
- If JOHN S MCCAIN had sounded at five short blasts or made Bridge-to-Bridge VHF hails or notifications in a timely manner, then it is possible that a collision might not have occurred.
- If ALNIC had sounded at least five short blasts or made Bridge-to-Bridge VHF hails or notifications, then it is possible that a collision might not have occurred.

9.2.2.12 Leadership and Culture

- The Commanding Officer decided not to station the Sea and Anchor detail when appropriate, despite recommendations from the Navigator, Operations Officer, and XO.
- Principal watchstanders including the Officer of the Deck, in charge of the safety of the ship, and the Conning Officer on watch at the time of the collision did not attend the Navigation Brief the afternoon prior. This brief is designed to provide maximum awareness of the risks involved in the evolution.
- Leadership failed to provide the appropriate amount of supervision in constructing watch assignments for the evolution by failing to assign sufficient experienced officers to duties.
- The Commanding Officer ordered an unplanned shift of thrust control from the Helm Station to the Lee Helm station, an abnormal operating condition, without clear notification.

- No bridge watchstander in any supervisory position ordered steering control shifted from the Helm to the Lee Helm station as would have been appropriate to accomplish the Commanding Officer's order. As a result, no supervisors were aware that the transfer had occurred.
- Senior officers failed to provide input and back up to the Commanding Officer when he ordered ship control transferred between two different stations in proximity to heavy maritime traffic.
- Senior officers and bridge watchstanders did not question the Helm's report of a loss of steering nor pursue the issue for resolution.

This assessment of USS John S. McCain is not intended to imply that ALNIC mistakes and deficiencies were not also factors in the collision.

9.2.2.13 Timeline of Events

August 20, 2017

1300	Navigation Brief to prepare the crew for the Singapore Strait transit and entering Sembawang, Singapore
Approximately 1326	Rudder swing checks were completed verifying satisfactory operation of the rudder
1730	The Commanding Officer retired to his cabin to rest before reporting to the bridge at 0115 the next morning
1904	JOHN S MCCAIN energized Navigation Lights
2115	Modified Condition Zebra was verified. As explained in the report, this condition maximizes the ability of the ship to gain a watertight status in the event of collision

August 21, 2017

0000	JOHN S MCCAIN is en route to Singapore
0001	Log entries reported that one surface search radar was non-operable
Approximately 0100–0101	Navigation watchstanders began to verify the ship's position at more frequent intervals (15 minutes)
0115	The Commanding Officer arrived on the Bridge
Between approximately 0127 and 0204	Key supervisory watch stations changed personnel
0216	Watchstanders shifted propulsion operations to what is termed split plant, a condition in which different gas turbines drive each of the two shafts separately
0300	Currents were running at a speed of 2.7 kts requiring steering adjustment

0315	Watchstanders report visual sighting of land
0418	Additional watchstanders reported for duties as the Modified Navigation Detail
0426	Navigation watchstanders began determining the ship's position at more frequent intervals (five minutes)
0427	JOHN S MCCAIN turned to avoid surface contacts in the area
0430	The Executive Offer arrived on the bridge
0436	The Commanding Officer ordered steering modes shifted from automatic control to backup manual control
Approximately 0436	Personnel responsible for tracking contacts on radar secured the auto-tracking feature on the SPS-67 radar and began manually tracking surface contacts
Starting at 0437	The bridge ordered various rudder orders to avoid shipping. None of these maneuvers were logged
0444	JOHN S MCCAIN turned to port and steadied on course 227T. On this course, the ship was aligned to enter the westbound Singapore Strait Traffic Separation Scheme
Approximately 0454	Radar contact was gained on the ALNIC nearly ahead of JOHN S MCCAIN on the port side, within eight nautical miles. ALNIC was in the center of a group of three other contacts traveling in the same general direction as JOHN S MCCAIN. Watchstanders did not discuss maneuvering intentions with respect to these contacts
Approximately 0457	JOHN S MCCAIN increased speed to 17 kts
0459	JOHN S MCCAIN reduced speed to 16 kts
0500	Reveille was announced to wake the crew for entering port. The Navigator informed the OOD that previous course changes to the North to avoid surface traffic had put JOHN S MCCAIN behind on its intended track and timeline and recommended an increase in speed to make 18 kts
0500–0524	JOHN S MCCAIN overtook several vessels just north of the eastern entrance to the Singapore Strait Traffic Separation Scheme. The closest point of approach during these passages was as close as 600 yd
0509	JOHN S MCCAIN altered course to 226T
0513	JOHN S MCCAIN increased speed to 18 kts and was steady on course 226T
0514	JOHN S MCCAIN increased speed to 20 kts and was steady on course 226T
0518	JOHN S MCCAIN turned starboard to course 230T, speed 20 kts. The Helmsman was compensating for the effects of currents with between 1° and 4° of right rudder to stay on course 230T

Approximately 0519	The Commanding Officer noticed the Helmsman was struggling to maintain course while simultaneously adjusting throttles. The CO ordered steering
	control separated from propulsion control, with duties divided between the Helm and Lee Helm watch stations. Splitting of the Helm and Lee Helm was not previously discussed at the Navigation Brief or at any time prior to the CO ordering it
Approximately 0520	Supervisory watch stations reported that the Automated Identification System (AIS) representation of contacts was cluttered and "useless." Commercial traffic routinely reports positions via this system, enabling other vessels to use GPS satellite information to accurately determine their positions
05:20:03	JOHN S MCCAIN was overtaking motor vessel GUANG ZHOU WAN. JOHN S MCCAIN was making 18.6 kts over ground. JOHN S MCCAIN closed range from behind ALNIC on ALNIC's starboard side
0520:39	The Lee Helm station took control of steering in computer assisted mode. The shift in steering locations caused the rudder to move amidships
0520:47	Lee Helm took control of the port shaft. Port and starboard shafts were both at 087 RPM/100.1% pitch
Just before 0521	The Helm reported to his immediate supervisor that he had lost steering control. The supervisor informed the Helm to inform the officer in charge of ship safety and navigation, the Officer of the Deck
0521	The Helm reported loss of steering to the Officer of the Deck. The rudders were amidships. JSM was on course 228.7T, engines were all ahead full for 20 kts. JSM was making 18.6 kts over ground and turning to port at 0.26° per second. ALNIC was on course 230T, speed 9.6 kts, and was bearing 164T at a range of approximately 582 yd from JSM
0521	The Conning Officer, the person responsible for issuing steering orders, ordered the Helm to shift steering control to the offline steering units, 1A and 2A.
0521	A loss of steering casualty on the ship's general announcing circuit was announced and After Steering was ordered manned. After Steering is an auxiliary station that has the ability to take control of steering in the event of a problem or casualty to the ship's primary control stations
0521:13	Steering units on the port rudder were shifted as ordered
0521:15	Steering units on the starboard rudder were shifted as ordered
0521:55	The first watchstander reported to After Steering. JOHN S MCCAIN did not have a complete delineated list of personnel to man After Steering in the event of a casualty or problem
0522	JOHN S MCCAIN was on course 216.3T, speed 18.4 kts and was turning to port at a rate of approximately 0.2° per second. Bridge watchstanders followed the Commanding Officer's order to change the lighting configuration to indicate a vessel not under command by the International Rules of the Nautical Road

Approximately 0522:04	The Lee Helm took control of the starboard shaft. The port and starboard shafts remained at a speed of 087 RPM and 100.1% pitch. The Lee Helm did not match the port and starboard throttles that control the speed of the shafts
	JOHN S MCCAIN was on course 216.1 T and turning to port at a rate of approximately 0.25° per second. Rudders were amidships
Approximately 0522:05	The Commanding Officer ordered the ship slowed with a reduction in speed to 10 kts
0522:07	The command to the port shaft lowered speed to 44 RPM and 100.1% pitch. The starboard shaft remained at a speed of 87 RPM and 100.1%. Rudders were amidships. No bridge watchstanders were aware of the mismatch in thrust and the effect on causing the ship's turn to port
0522:40	JOHN S MCCAIN was on course 204.4T, speed 16.6 kts and was turning to the left at a rate of approximately 0.41° per second
0522:45	The Executive Officer noticed the ship was not slowing down as quickly as expected and alerted the Commanding Officer. In response, the Commanding Officer ordered 5 kts. This order was echoed by the Conning Officer. The CO did not announce that he had taken direct control of maneuvering orders as required by Navy procedures
0523:00	The Conning Officer ordered right standard rudder. JOHN S MCCAIN was on course 194.5T at a speed of 15.8 kts. ALNIC was on course 229.8T, 9.6 kts, and was bearing 097T from JOHN S MCCAIN at a range of approximately 368 yd
0523:01	After Steering took control of steering in backup manual mode
0523:06	The port shaft continued to slow. The starboard shaft was ahead at a speed of 87 RPM and 100.1% pitch. The port shaft order at this time was 32 RPM at 81.1% pitch. JSM was on course 192T, speed 15.6 kts and turning to the left at a rate of approximately 0.5° per second
0523:16	The Helm took control of steering at the helm station in Backup Manual mode
0523:24	Throttles were finally matched at the Lee Helm station and both shafts were ahead to reach 5 kts. JOHN S MCCAIN was on course 182.8T, speed 13.8 kts, and turning to port at a rate of approximately 0.54° per second
0523:27	Aft Steering Helmsman took control of steering. This was the fifth transfer of steering and the second time the Aft Steering unit had gained control in the previous two minutes
0523:44	JOHN S MCCAIN was on course 177 T, speed 11.8 kts, and was slowly turning to the left port at a rate of approximately 0.04° per second. The ordered and applied right 15° rudder checked JOHN S MCCAIN's swing to port and the ship was nearly on a steady course
0523:58	ALNIC's bulbous bow struck JSM between frames 308 and 345 and below the waterline

0524:12	After Steering still had control of steering at the ASU in CAM but the rudders moved amidships
0524:24	JSM engines answered "all stop" and the shafts came to idle speed. The ship was on course 138.6T, speed 5.7 kts, and the ship was turning to port at a rate of approximately 1.4° per second
0526	JSM set General Quarters and the Damage Control Assistant assumed responsibility for all DC efforts from CCS

9.2.2.14 Summary of Safety Culture Issues

These three accidents all have similarities. That is that the personnel controlling the ships performed operational maneuvers that were not in accordance with standard practices. A questioning attitude was not evident in any of the three events. All the ships involved were large vessels and could not make course or speed changes quickly. In these circumstances, operational personnel need to be thinking way ahead of where they are in time and place. Personnel who are assisting the personnel driving the ship, so to speak, need to speak up and provide their opinions on what is happening to help avoid and/or mitigate a tragic event. The Genesis River had a Safety Management System (SMS), but as the report discusses, it was not followed to the letter. Implementation of elements of a Crew Resource Management (CRM) type system could help avoid or mitigate accidents like those discussed in this section.

9.3 Stretch Duck 7 July 19, 2018

9.3.1 Introduction

According to Forbes Magazine there were more than 130 duck boats operating in Branson, the Wisconsin Dells, Boston, Seattle, Miami, San Diego, Honolulu, and Washington in 2017. Duck boats are large amphibious vehicles that can seat up to 30 tourists for typical tours that cruise city streets and the surrounding body of water (Goldstein 2018). My family has done a Duck Boat tour in Seattle, and we rather enjoyed it. At the time we took the tour the weather was perfect, without high winds. However, there have been numerous accidents, in which many people perished.

The NTSB report National Transportation Safety Board report (2020), and the Goldstein article discuss accidents that occurred during tours. The most notable were:

- Duc' accident on Lake Hamilton near Hot Springs Arkansas in 1999 that led to 13 people drowning.

- Four people died in 2002 in the Ottawa River when the Lady Duck sank.
- In 2003 a 63-year-old woman fell off a duck boat onto a parking lot and died in Boston.
- An accident in 2010 involved a collision between a stalled duck boat and a tugboat-propelled barge in the Delaware River. Tow Hungarian tourists died in this accident. Another accident near Philadelphia claimed the life of one passenger.

There have also been a series of motor vehicle accidents involving duck boats.

- A 28-year-old woman was run over by a duck boat while riding her scooter, in Boston.
- In 2015 in Seattle, a duck boat collision with a bus, after an axle failed on the duck took the lives of five international college students.

This case study focuses primarily on the Duck accident on Table Rock Lake in 2018.

9.3.2   Accident Description

This case study is excerpted from the NTSB report (2019).

On the evening of July 19, 2018, 17 of the 31 persons aboard the *Stretch Duck 7* died when the amphibious passenger vessel (APV) sank during a high-wind storm that developed rapidly on Table Rock Lake near Branson, Missouri. The *Stretch Duck 7* was built in 1944 as a DUKW landing craft to carry military personnel and cargo during World War II and was then modified for commercial purposes to carry passengers on excursion tours (see Figure 9.40).

DUKW (pronounced "duck") is an acronym that signifies the characteristics of the World War II (WWII) amphibious vessel: $D = 1942$, the year of design; $U =$ utility; $K =$ front-wheel drive; and $W =$ two rear-driving axles.

About 7:08 p.m. central daylight time on July 19, 2018, the 33-ft-long, modified WWII APV *Stretch Duck 7*, part of a fleet of vessels operated by "Ride the Ducks Branson," sank during a storm with heavy winds that developed rapidly on Table Rock Lake near Branson, Missouri. Of the 31 persons aboard, 17 fatalities resulted. Prior to the accident, the National Weather Service (NWS) had issued a severe thunderstorm warning for the area advising of wind gusts of 60 mph. The manager-on-duty advised the captain and driver before departing the shoreside boarding facility to complete the lake portion of the tour before the land tour (which normally occurred first) due to the approaching weather. About five minutes after the vessel entered the water, the leading edge of a storm front, later determined to be a "derecho." A Derecho is a very long lived and damaging thunderstorm. A storm is classified as a derecho if wind damage swath extends

Figure 9.40 Stretch Duck 7 after salvage from Table Rock Lake, July 2018. Source: National Transportation Safety Board.

more than 240 mi and has wind gusts of at least 58 mph or greater along most of the length of the storm's path (National Weather Service 2021). The storm passed through the area generating strong winds and waves reportedly 3- to 5-ft high, with the highest wind gust recorded at 73 mph. The *Stretch Duck 54,* another vessel from the company's fleet that had also been conducting a tour on the lake currently, was able to exit the water after experiencing the severe weather. During its effort to reach land, the *Stretch Duck 7* took on water and sank approximately 250 ft away from the exit ramp. Several first responders, along with the crewmembers and passengers aboard a paddle-wheeler moored nearby, rescued, and triaged 14 passengers, 7 of whom were transported to local hospitals.

The chronology of the events that day were (Professional Mariner 2020):

- 11:20 a.m. (1120) on July 19, 2018: The NWS Storm Prediction Center issues a severe thunderstorm watch for western and central Missouri, including Branson and nearby Table Rock Lake, that was valid until 9 : 00 p.m. (2100).
- 6:00 p.m. (1800): Stretch Duck 7 returns to the company "duck dock" in Branson, about 6 mi from Table Rock Lake, after completing a 90-minute tour on land and water.
- 6:27 p.m. (1827): A manager tells Stretch Duck 7's captain to "go to the water first" for the next tour, but apparently does not explain the reason.
- 6:29 p.m. (1829): Passengers begin boarding the duck boat. There are 29 passengers, a captain and a driver on the vessel.

- 6:32 p.m. (1832): The NWS issues a severe thunderstorm warning for an area that included Table Rock Lake, valid until 1930.
- 6:33 p.m. (1833): Stretch Duck 7 leaves the "duck dock" and begins the 16-minute trip to Table Rock Lake.
- 6:55 p.m. (1855): Stretch Duck 7 enters Table Rock Lake. The captain describes calm winds and a nearly flat surface.
- 6:59 p.m. (1859): The captain alters course to accelerate the vessel's departure from the lake. "Yeah, we're gonna try and beat this weather off the water as fast as we can here," the captain tells the passengers.
- 7:00 p.m. (1900): Strong winds arrive, creating whitecaps on the lake's surface within 15 seconds.
- 7:03 p.m. (1903): Stretch Duck 7's pitching increases in the rough water.
- 7:04 p.m. (1904): Stretch Duck 7's bilge alarm sounds.
- 7:07 p.m. (1907:05): Stretch Duck 54 exits the lake.
- 7:07 p.m. (1907:24): Passengers on Stretch Duck 7 comment about getting wet from incoming water.
- 7:08 p.m. (1908:18): The captain orders passengers to move to the vessel's port side to counteract a starboard list.
- 7:08 (1908:25): The starboard side aft dips below the water's surface and Stretch Duck 7 sinks at the stern seconds later. At about this time, the captain releases the port-side curtain to facilitate passenger escape. The starboard-side curtain is not released. The captain is pushed out of the vessel by water rushing in from the windshield.

Four days after the accident, the *Stretch Duck 7* was salvaged from a depth of 85 ft in the lake. Once the remaining water was pumped out and the vessel was refloated, investigators observed no hull breach. The canopy on the *Stretch Duck 7* was found torn along the center support beam, and the vessel's engine and electronic components were water damaged. Investigators also found that the vessel's side curtain had been released on the port side, but the curtain on the starboard side was closed.

Criminal charges were filed on three employees of the company in July of 2021 (Diaz 2021). The New York Times (NY Times) reported on July 16, 2021, that a local prosecutor charged the boat captain and two other employees over 17 deaths in July 2018 when the Stretch Duck 7 sank. A total of 63 felony charges were filed in Stone County against the Captain Kenneth Scott McKee, of Verona, general manager Curtis Lanham, of Galena, and manager on duty Charles Baltzell, of Kirbyville, the general manager, and the manager on duty the day of the accident for the Ride the Ducks attraction on Table Rock Lake. A federal judge dismissed earlier charges filed by federal prosecutors, concluding they did not have jurisdiction.

Among the dead were nine members of a family from Indianapolis and victims from Missouri, Illinois, and Arkansas. Tia Coleman, a member of that Indianapolis family who lost her husband and three children, said in a statement that her "prayers had been answered" with the charges.

The captain Scott McKee is charged with 17 charges of first-degree involuntary manslaughter and 12 additional charges allege that he endangered child passengers on the boat, five of whom died.

The child-endangerment charges filed over deaths are the most serious, punishable by between 10 and 30 years in prison. The endangerment charges involving children who survived the accident carry a sentence of up to seven years.

An affidavit from a Missouri Highway Patrol sergeant accuses McKee of failing to exercise his duties as a licensed captain by taking his amphibious vehicle onto the lake during a thunderstorm.

The two other employees face 17 charges each of first-degree involuntary manslaughter. They are accused of failing to communicate weather conditions and to cease operations during a severe thunderstorm warning.

The primary cause of this event was trying to operate the vessel during a severe thunderstorm. Whether the captain was made aware of the impending thunderstorm by the managers has not been made clear. However, as shown earlier, the thunderstorm warning was in place by the NWS since 11 a.m. that morning. The contributing causes of the deaths were the canopy and side curtains not breaking away as the vessel sank and that the passengers and not having donned personal floatation devices (PFDs). The NTSB discusses other contributing causes to this and other events as lack of reserve bouncy of the duck boats and the use of the restrictive canopy and side currents. This information is presented next.

9.3.3   1999 Sinking of *Miss Majestic*

In one of the deadliest accidents involving a modified WWII DUKW at the time, 13 passengers lost their lives during an excursion tour aboard the *Miss Majestic* in Lake Hamilton near Hot Springs, Arkansas, shortly before noon on May 1, 1999. About seven minutes after entering the lake, the 31-ft-long vessel (see Figure 9.41) listed to port and rapidly sank by the stern in 60 ft of water. The operator attempted to turn the vessel around, but the water entered the boat rapidly. The operator and most of the 20 passengers were trapped by the vessel's canopy and drawn under water, except one passenger who escaped before it submerged. As the vessel descended to the bottom of the lake, six passengers and the operator were able to escape and, upon reaching the water's surface, were rescued by recreational boaters in the area.

Figure 9.41   Miss Majestic post-salvage, 1999. Source: National Transportation Safety Board.

Vessel maintenance, reserve buoyancy, and survivability – specifically regarding the impediment of the vessel's canopy to the egress of passengers – were among the major safety issues identified by the NTSB's investigation of the accident. Investigators determined that water initially entered the *Miss Maje*stic through the gap between the driveshaft and its housing via an unsecured clamp for the watertight rubber boot. Once the weight of floodwater in the aft portion of the hull reduced the freeboard at the stern to zero, water poured over the immersed transom and into the interior of the *Miss Majestic*, causing it to sink rapidly by the stern. In this event the operator was not at fault. The boat sank due to improper maintenance. Improper maintenance is also a safety culture issue. Throughout this book there are examples of improper maintenance causing or contributing to major accidents.

9.3.4   Types of DUKW Amphibious Vessels

The NTSB has investigated several accidents involving commercial APVs within the United States, including the current investigation of the *Stretch Duck 7*. Whether the incident occurred on land or in the water, in five out of six of these investigations the vessel involved shared similar dimensions and/or characteristics with the World War II-era DUKW amphibious vessel (see Figure 9.42). In total, 37 deaths and 104 injuries resulted from these six DUKW-related accidents.

While there are multiple models of APVs built by various manufacturers, those sharing similar characteristics with the WWII DUKW amphibious vessel have

Figure 9.42 DUKW in military use before conversion to passenger service.

been widely used in tour operations. Further, this subtype exists in several variations that can be distinguished by the following characteristics:

	Original	Stretch	Master jig	Truck
Chassis	DUKW	DUKW	DUKW	M35[a]
Modified	1940s–1990s	Mid-1990s	2003	2005
Exterior	Length: 30′ Beam: 8′	Longer length, similar beam	Raised gunwales,[b] wider beam	Raised gunwales, wider beam
Hull thickness	14 gauge (0.0747″) on side walls	Similar to original design	Increased	Increased
Engine	Gas	Gas	Gas	Diesel

a) M35 military trucks were built in different variations from 1950 to 1999.
b) *Gunwales* are the upper edge of the vessel's sides.

The *Stretch Duck 7*, the *Stretch Duck 54,* and the *Miss Majestic* were all modified WWII DUKWs. However, the *Stretch Duck 7*, as its name intimates, specifically was a "stretch" DUKW, while the *Stretch Duck 54*, which encountered the same severe weather as the *Stretch Duck 7* but was able to exit the water, was a "master jig" DUKW with dimensions similar to a "truck duck." The *Miss Majestic*, a model similar to the *Stretch Duck 7*, was an original DUKW. Based on previous accidents involving modified WWII DUKW vessels, particularly the original and stretch DUKW models, the NTSB focused on these two types of vessels due to their insufficient reserve buoyancy as well as impediments to passenger egress, which compromise not merely the survivability of such vessels but their occupants as well.

NTSB issued 22 safety recommendations related to modified WWII DUKW APVs. At the time of the *Stretch Duck 7* sinking, nine of these safety recommendations had been classified "Closed – Acceptable Action," "Closed – Acceptable Alternate Action," or "Closed – Exceeds Recommended Action," indicating the completion of a response that either complied with, met the objective of, or surpassed what the NTSB recommended, respectively. Four remained pending and were classified "Open – Acceptable Response," indicating a planned action that, when completed, would comply with the safety recommendations. For nine other recommendations, the recipient either disagreed with the recommendation, or otherwise did not plan to satisfy the recommendation, and thus they were classified "Open – Unacceptable Response," "Closed – Unacceptable Action/No Response Received," or "Closed – Unacceptable Action."

Of the total number of safety recommendations related to modified APVs, five were issued nearly two decades ago in response to the sinking of the *Miss Majestic* – the largest number of marine safety recommendations issued because of a single APV accident. The lack of reserve buoyancy on modified DUKW APVs and the dangers of canopies installed on these vessels were identified as important safety issues. Recommendations addressing these issues were directed to operators, manufacturers, and/or those who refurbish DUKW vessels; individual states that had DUKW vessels operating in their jurisdictions; and the Coast Guard. Four out of five were classified "Closed – Unacceptable Action," indicating that the recommendation recipient did not take the recommended action. The one recommendation that was closed acceptably requested the Coast Guard develop and promulgate guidance for all APVs similar to its Navigation and Vessel Inspection Circular (NVIC) 1-01, which was published after the sinking.

The NTSB believed that the failure to implement the safety recommendations related to providing reserve buoyancy for DUKW APVs contributed to the sinking of the *Stretch Duck 7*. Additionally, the failure to implement the recommendation concerning fixed canopies likely increased the number of fatalities that resulted.

Prior to completing its investigation of the *Miss Majestic*, the NTSB assembled key stakeholders in the industry, including the Coast Guard, for a public forum on APV safety in December 1999 and issued an urgent recommendation soon afterward. Having immediate concerns about the risk of flooding and the vulnerability to sinking for all types of APVs, the NTSB issued the following safety recommendation to the operators, manufacturers, and/or businesses that refurbish APVs:

> Without delay, alter your amphibious passenger vessels to provide reserve buoyancy through passive means, such as watertight compartmentalization, built-in flotation, or equivalent measures, so that they will remain afloat and upright in the event of flooding, even when carrying a full complement of passengers and crew.

9.3.5 NTSB Identified Safety Issue No. 1: Providing Reserve Buoyancy

Survivors of the *Miss Majestic* accident confirmed that the vehicle sank by the stern less than a minute after the deck edge was submerged, leaving insufficient opportunity for passengers to escape before the vessel sank. Accordingly, the NTSB issued Safety Recommendation M-02-1 to the Coast Guard, addressing the safety issue of reserve buoyancy to make all APVs more survivable and stable in the event of flooding. However, because the Coast Guard did not concur with this recommendation and did not take any action, Safety Recommendation M-02-1 was classified "Closed – Unacceptable Action" in the NTSB's May 6, 2003, response letter.

DUKW vessels were originally constructed with a low freeboard, an open hull, and no compartmentalization or subdivision, resulting in a design without adequate reserve buoyancy. To reduce the volume of water that could accumulate in these low-freeboard vessels, particularly during beaching and combat operations, the original DUKW design included the installation of a large-capacity bilge pump, referred to as a "Higgins" pump, which was rated at a maximum pumping capacity of about 250 gallons per minute (gpm). Driven by a chain connected to the DUKW's propulsion shaft, the pump would run at a speed proportional to the propeller speed and operate whether the bilges contained water or were dry. To operate the Higgins pump at full capacity, the operator would be required to engage the propeller shaft in the forward direction and operate the engine at full throttle. This action by the operator would be an "active" means of dewatering the vessel, compared to a "passive" safety system, which requires no deliberate action or operation to deploy and generally facilitates fail-safe performance of a vessel.

After the *Miss Majestic* accident, the Coast Guard approved a modification to remove the Higgins pump in DUKW vessels. As an alternative, a sea chest, or watertight containment, was installed around hull penetrations to contain any flooding through them. In addition, these modified vessels were outfitted with high-level alarms and electrically operated bilge pumps that started automatically. On the *Stretch Duck 7*, the Higgins pump had been removed; installed in its place were a sea chest and three 33-gpm bilge pumps. These pumps were found to be in working condition after the accident.

During the waterborne portion of its final voyage, the *Stretch Duck 7* was exposed to high winds and waves estimated at 3–5 ft. The video recorder on board captured bilge alarms sounding four minutes after the vessel encountered severe weather, signaling an ingress of water. About four minutes later, the *Stretch Duck 7* sank. Witness videos showed the vessel pitching in the storm with white-capped waves covering the bow several times. In the forward section of the vessel, the ventilation openings that supplied combustion and cooling air to and from the engine most likely permitted water to enter the engine compartment, from where it would have flowed freely throughout the rest of the hull. The additional water weight would have further lowered the vessel's freeboard and thereby subjected it to more rapid flooding.

Surviving passengers who were interviewed after the accident recalled water quickly rising from under the floorboards of the *Stretch Duck 7*. They described how waves pushed in the starboard-side curtain and water entered from the bottom rail of the curtain where it met the gunwale. Once the water started filling the vessel, it quickly flooded and sank within seconds after covering the passengers' feet. Most passengers recalled the *Stretch Duck 7* had a starboard list in the final moments of the voyage and rapidly sank by the stern. If the *Stretch Duck 7* had been modified to include several subdivided compartments – one approach to a passive safety system – the flooding could have been contained to individual sections of the vessel, thus increasing the vessel's ability to remain afloat. The vessel could have remained afloat and upright indefinitely had it been fitted with built-in flotation or watertight compartmentalization, which can be designed and sized to provide a boat with sufficient reserve buoyancy even when the hull is fully flooded.

9.3.6 Safety Issue No. 2: Removing Canopies and Side Curtains

In the *Miss Majestic* accident report, the NTSB determined that one of the contributing factors to the high number of fatalities was a continuous canopy that entrapped passengers within the sinking vessel. Accordingly, the NTSB issued Safety Recommendation M-02-2 to the Coast Guard, as well as to the states of New York and Wisconsin:

"Until such time that owners provide sufficient reserve buoyancy in their amphibious passenger vehicles so that they will remain upright and afloat in a fully flooded condition (by M-02-1), require the following:

(1) removal of canopies for waterborne operations or installation of a Coast Guard-approved canopy that does not restrict either horizontal or vertical escape by passengers in the event of sinking,

(2) reengineering of each amphibious vehicle to permanently close all unnecessary access plugs and to reduce all necessary through-hull penetrations to the minimum size necessary for operation,

(3) installation of independently powered electric bilge pumps that are capable of dewatering the craft at the volume of the largest remaining penetration to supplement either an operable Higgins pump or a dewatering pump of equivalent or greater capacity,

(4) installation of four independently powered bilge alarms,

(5) inspection of the vehicle in water after each time a through-hull penetration has been removed or uncovered,

(6) verification of a vehicle's watertight condition in the water at the outset of each waterborne departure, and

(7) compliance with all remaining provisions of NVIC 1-01."

The entire passenger and crew space of the *Stretch Duck 7* had been covered by a fixed canopy. Upon recovery of the vessel, the canopy was found torn from front to back (see Figure 9.43). It was peeled back over the starboard side but largely remained intact on the port side. The canopy was constructed of vinyl measuring 0.032 in. thick and pressed into a seam along the horizontal support at the center of the vessel. Underneath the canopy, the *Stretch Duck 7*'s personal flotation devices (PFDs, commonly called lifejackets) were stored above the seating compartment. Of the 56 lifejackets investigators counted post-accident, the majority of them – a total of 41 – were still connected to the vessel's canopy framing by their straps. With the PFDs in their storage locations above the passengers, vertical egress was

Figure 9.43 Torn canopy of the Stretch Duck 7 found during recovery operations. Source: National Transportation Safety Board.

blocked during the sinking, despite the canopy being peeled back over the starboard side.

The canopy framing also created obstructions for clear egress from the vessel. Several surviving passengers recalled hitting various impediments and being pinned against the canopy before they could break through it to escape upward from the submerged vessel. The NTSB believed that some of the fatalities likely resulted from the presence of the canopy and its associated framing.

The NTSB's position on the installation of canopies on modified WWII DUKW APVs has not changed since the *Miss Majestic* sinking. The number of fatalities resulting from the sinking of the *Stretch Duck 7* is further evidence of the continuing, unacceptable risks posed by canopies currently installed on modified WWII DUKW vessels. Given their lack of adequate reserve buoyancy and low freeboard, these vessels are vulnerable to rapid swamping and sinking, leaving passengers and crewmembers little time to evacuate. The NTSB has determined that canopies and their associated supports installed on these vessels impede escape and therefore should be removed before waterborne operations.

The sinking of the *Stretch Duck 7* raised awareness of another impediment to passenger emergency escape. Each side of the *Stretch Duck 7* was outfitted with a clear vinyl side curtain, which was comprised of a continuous sheet of plastic on a reel spanning the entire length of the passenger space. With an electric motor, these two adjustable curtains, designed to be used as protection for passengers during inclement weather, could be lowered (closed) and raised (opened). When lowered, the curtains' bottom rail was held by brackets on the forward and aft sides of the vessel. In an emergency situation, each curtain could be separated from the vessel with manual release levers: the portside curtain could be released from a handle directly above the captain's seat near the top of the portside curtain, and the starboard-side curtain could be released from the corresponding location on the starboard side.

During salvage operations, *Stretch Duck 7*'s portside curtain was found apart from the vessel at the lake's bottom; survivors recalled the captain had manually released it by using a lever above his head just before the vessel sank. However, the starboard-side curtain was found closed; its bottom rail was engaged into the gunwale side brackets, and the lever for releasing the curtain had not been moved into position for that function. Having the side curtain closed created another impediment that prevented emergency escape from the starboard side of the vessel. Although surviving passengers of the *Stretch Duck 7* could not determine whether the curtains, canopy, or other obstructions blocked their escape, the NTSB believes that side curtains employed during waterborne operations further impede egress from the passenger seating area over the gunwale and out the sides of the vessel, especially large curtains that span the length of the vessel.

The NTSB and the Coast Guard have not agreed about many of the NTSB recommendations and the complete report documents these differences.

9.3.7 Findings and Conclusions

The NTSB findings for the Stretch Duck 7 accident are as follows:

1. Having been constructed with a low freeboard and without compartmentalization, or subdivision, the *Stretch Duck 7* lacked adequate reserve buoyancy and therefore quickly sank once water entered the vessel after it encountered severe weather.
2. Both the fixed canopy and a closed side curtain spanning the starboard side of the passenger compartment on the *Stretch Duck 7* impeded passenger escape, which likely resulted in an increased number of fatalities.

9.3.8 Safety Culture Summary Findings

The notification of the thunderstorm came out at 11 a.m. Of course, the strength of storms is not always known until they hit. However, the Stretch Duck 7 event would not have occurred if the captain had not taken the vessel onto the lake with an impending severe thunderstorm, whether it was a derecho or a minor storm. Even a minor thunderstorm produces strong winds, lightening, possibly hail, and heavy rain. The company managers should have curtailed the operation of the Duck Boat tours once the thunderstorm approached. These are both examples of safety culture failures.

The Miss Majestic event could have been prevented if proper maintenance and inspection would have been performed on the vessel.

The increase in reserve buoyancy, removal of the canopies, and side curtains, vessel compartmentalization, and bilge pumps are mitigating factors. They will help to prevent deaths, injuries and damage to the vessels, but they would not have prevented the events.

9.3.9 Other Events

The NTSB report lists in their report other events that have occurred involving APVs.

9.3.9.1 *Minnow,* Milwaukee Harbor, Lake Michigan, September 18, 2000

On September 18, 2000, the *Minnow,* an Alvis Stalwart amphibious vessel, sank in Milwaukee Harbor of Lake Michigan with 19 people on board after experiencing mechanical issues. There were no fatalities. Damage to the vessel was estimated at US$ 170,000. The accident brief was published in the appendix of the report *Sinking of the Amphibious Passenger Vehicle* Miss Majestic, *Lake Hamilton, Near Hot Springs, Arkansas, May 1, 1999,* MAR-02/01. Washington, DC: NTSB.

9.3.9.2 *DUKW No. 1*, Lake Union, Seattle, Washington, December 8, 2001

The *DUKW No. 1*, an original WWII DUKW vessel, sank in Lake Union, Seattle, Washington, on December 8, 2001, when a missing access plug allowed water to flood the hull. No fatalities resulted. Estimated damage was US$ 100,000. The accident brief was published in the appendix of the report *Sinking of the Amphibious Passenger Vehicle* Miss Majestic, *Lake Hamilton, Near Hot Springs, Arkansas, May 1, 1999,* MAR-02/01. Washington, DC: NTSB.

9.3.9.3 *DUKW 34*, Delaware River, Philadelphia, Pennsylvania, July 7, 2010

On July 7, 2010, the tugboat/barge combination *Caribbean Sea/The Resource* collided with the *DUKW 34*, an amphibious vessel modified into a "stretch" DUKW, while carrying 37 persons on board on the Delaware River in Philadelphia, Pennsylvania. Two passengers on board the *DUKW 34* were fatally injured, and several other passengers sustained minor injuries. Damage totaled US$ 130 470. *Collision of TugBoat/Barge* Caribbean Sea/The Resource *with Amphibious Passenger Vehicle* DUKW 34, *Philadelphia, Pennsylvania, July 7, 2010,* MAR-11/02. Washington, DC: NTSB.

9.3.9.4 *DUCK 6*, Seattle, Washington, September 24, 2015

On September 24, 2015, the *DUCK 6*, an amphibious vessel modified into a "stretch" DUKW, crossed the center line into oncoming traffic and struck a motorcoach while traveling on a state bridge in Seattle, Washington. Three other vehicles were damaged. As a result of the crash, five motorcoach passengers died. Seventy-one motorcoach and *DUCK 6* occupants reported injuries. *Amphibious Passenger Vehicle* DUCK 6 *Lane Crossover Collision With Motorcoach on State Route 99, Aurora Bridge, Seattle, Washington, September 24, 2015,* HAR-16/02. Washington, DC: NTSB.

9.4 Recent Railroad Accidents

Railroad accidents happen every day someplace in the world. The railroad accidents from the past were deadly, as shown in the two examples in Chapter 2, and the more recent ones can even be more deadly because of the number of passengers that ride the trains and with the cargo the trains carry. The following two examples illustrate the types of events that happen.

Figure 9.44 Accident scene.

9.4.1 AMTRAK Passenger Train – May 12, 2015

The following accident investigation is excerpted from the NTSB report (National Transportation Safety Board 2015). About 9:21 p.m. eastern daylight time on May 12, 2015, eastbound Amtrak (National Railroad Passenger corporation) passenger train 188 derailed at milepost (MP) 81.62 in Philadelphia, Pennsylvania (Figure 9.44). The train had just entered the Frankford Junction curve – where the speed is restricted to 50 mph – at 106 mph. It was dark and 81 °F with no precipitation; visibility was 10 mi. As the train entered the curve, the locomotive engineer (engineer) applied the emergency brakes. Seconds later, the train, one locomotive and seven passenger cars, derailed. There were 245 passengers, 5 on-duty Amtrak employees, and 3 off-duty Amtrak employees on board. Eight passengers were killed, and 185 others were transported to area hospitals (Table 9.1). Amtrak estimated its damages to be more than US$ 30.84 million; an adjacent Consolidated Rail Corporation (Conrail) track sustained about US$ 330,000 in damages.

9.4.1.1 Accident Scenario

The Amtrak engineer, conductor, and an assistant conductor went on duty at 1:20 p.m. at New York's Pennsylvania Station; they worked on a train that arrived in Washington, DC, at 5:19 p.m. They went off duty for dinner and went back

Table 9.1 Injuries.

Injuries	Number
Fatal	8
Serious	46
Minor	113
None	8
Condition unknown	18
Total	**193**

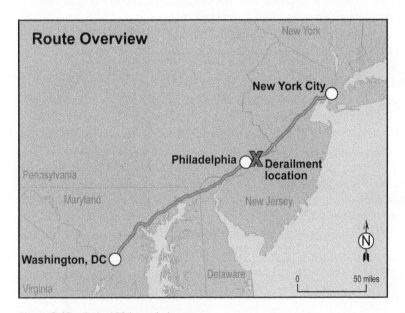

Figure 9.45 Train 188 intended route.

on duty about an hour and 10 minutes later at 6:30 p.m. At that time, another assistant conductor, who had worked on an earlier train from New York, was added to train 188's crew.

Train 188 departed Washington for New York at 7:15 p.m. (Figure 9.45). It made scheduled stops at stations in New Carrollton, Baltimore-Washington International Airport, Baltimore (Pennsylvania Station), and Aberdeen, Maryland, as well as Wilmington, Delaware. Train 188 arrived at Philadelphia's 30th Street Station at 9:06 p.m. where the engineer inspected the locomotive pantograph, as required by Amtrak. The pantograph transmits electrical current from the overhead catenary wire to the train. At 9:10 p.m., the train left the 30th Street Station on time on main track 1.

The engineer, who was alone in the cab, maintained the train speed near the required 30 mph for the next few miles. In postaccident interviews, the engineer said he was monitoring the radio when he heard an eastbound Southeastern Philadelphia Transportation Authority (SEPTA) train engineer and the train dispatcher discussing an incident in an area he was approaching. During the six-minute conversation, the SEPTA engineer said his windshield was shattered near the Diamond Street Bridge – an area where people were known to throw rocks and other objects at passing trains (Figure 9.46). The SEPTA engineer said he had glass in his face and had used the emergency brake to stop his train on main track 1. He requested medical attention.

In this area, four main tracks ran parallel to each other. In the direction train 188 was traveling, the tracks are numbered from right to left, main track 1 through main track 4.

As the Amtrak engineer approached the Diamond Street Bridge, he crossed from main track 1 to main track 2 in preparation for passing the SEPTA train. He accelerated to 67 mph in compliance with the authorized speed and sounded the horn as he approached the SEPTA train, which was stopped to his right at MP 86. The Amtrak engineer broadcasted on the radio that he was about to pass the SEPTA train; he sounded the horn again as he passed in case the SEPTA crewmembers were on the ground inspecting the train for damage as required by operating rules. Despite speculation immediately following the accident that train 188 was also hit by a rock or bullet or other projectile, the engineer did not recall such an incident, nor did FBI testing show any evidence of ballistic material.

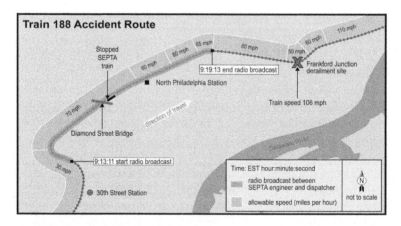

Figure 9.46 Train 188's route through Philadelphia. Source: Amtrak.

The Amtrak engineer continued to comply with authorized speeds, slowing to transit a right-hand curve that had a maximum speed of 65 mph. At MP 83.4, about the time the radio conversation between the SEPTA engineer and the dispatcher ended, the authorized speed increased to 80 mph; the Amtrak engineer moved the throttle to full power, and the train accelerated. The throttle stayed at full power for about 25 seconds, and the train reached about 95 mph. It had reached the 80 mph speed limit near MP 82.3, which was 1.2 mi from the point of derailment. At that point, the Amtrak engineer momentarily reduced the throttle, then returned to full throttle before reducing power about 20 seconds later. The train speed reached 106 mph as it entered the left-hand curve at Frankford Junction. The engineer began emergency braking at 9:20 p.m. Three seconds later, the train had slowed to 102 mph, and the event recorder and forward-facing video stopped recording. The train derailed to the outside of the left-hand curve at MP 81.62.

9.4.1.2 Amtrak

The Rail Passenger Service Act (Public Law 91-518, October 30, 1970), created the National Railroad Passenger Corporation (Amtrak). The act directed that Amtrak develop and operate a modern rail service to meet intercity transportation needs. Amtrak began operations on May 1, 1971. On April 1, 1976, Amtrak acquired its Northeast Corridor property from Conrail. Amtrak's Northeast Corridor is the busiest railroad in North America with about 2200 commuter and freight trains operating on some portion of the Washington-Boston, Massachusetts, route each day (see Figure 9.47). Annually, Amtrak carries more than 24 million passengers and 220 million commuters, either on its trains or on trains running on Amtrak property.

In December 2015, Amtrak completed the implementation of positive train control (PTC) on tracks between New York and Washington, DC, completing installation on most Amtrak property in the Northeast Corridor. PTC has been installed between Boston and New Haven, Connecticut, since 2000. The only exceptions were 7 mi of track located in or adjacent to terminals where trains move slower and automatic train control systems are in service.

Most of the national network of track that Amtrak operates over is owned by other railroads. In fact, 72% of the miles traveled by Amtrak trains is on tracks owned by "host railroads." Host railroads are responsible for PTC installation on their property where passenger trains operate or where poison or toxic-by-inhalation hazardous materials are transported. Amtrak has installed PTC on its locomotives that operate over host railroads.

At the time of the accident, Amtrak trains operating through this territory were authorized by wayside and interlocking signals, as well as cab signals. The signals that authorized the train movements were part of a traffic control system controlled from a dispatching center in Wilmington. All four main tracks were part of

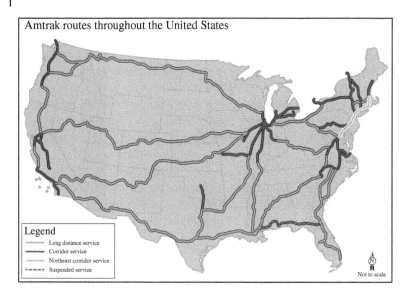

Amtrak routes throughout the United States

Legend
— Long distance service
— Corridor service
— Northeast corridor service
• ■ ■ ■ • Suspended service

Not to scale

Figure 9.47 Amtrak routes throughout the United States. Source: Amtrak.

the traffic control system, and trains were authorized by signals to operate in both directions.

Employees were provided with operating procedures that were part of the Northeast Operating Rules Advisory Committee (NORAC) Operating Rules. Amtrak timetable 5 (effective November 11, 2012) also governed train movements included train speeds at specific locations.

9.4.1.3 Analysis of the Engineer's Actions

The Amtrak engineer told investigators he could not remember what happened immediately preceding the derailment. Specifically, he could not explain why he increased the train's speed to 106 mph as he approached and entered the curve at Frankford Junction where the maximum authorized speed was 50 mph. His last memory until the time of the accident was at the end of the radio conversation (about 9 :19 p.m.) between the disabled SEPTA train engineer and the dispatcher as he negotiated the right-hand curve preceding the derailment. Amtrak records showed the engineer was experienced, certified, and qualified, and he had no previous disciplinary action. He had worked on the Northeast Corridor since 2013 and had traveled through the curve at Frankford Junction hundreds of times.

The engineer maintained a regular work and rest schedule for several days leading up to the accident, and there was no evidence that he suffered from fatigue. There was no evidence of any medical condition that would have impeded his job performance, and postaccident tests showed no evidence that he was impaired during the accident trip by alcohol, other drugs, or any substance. The train 188 conductor and assistant conductors described the engineer in positive terms, including saying that he was a "good engineer" who did "what he was supposed to do."

The engineer's cell phone and records from his cell phone provider showed it was not used during the trip. Specifically, there was no record of any calls, texts, instant messages, or data activity. Amtrak records indicated that the engineer's cell phone did not connect to the train's onboard wireless Internet system on the accident train. Furthermore, an examination of metadata downloaded from the cell phone was consistent with it being powered off during the accident trip.

There were no mechanical or operational issues reported for train 188 on the day of the accident.

NTSB investigators explored the reasons why a qualified, experienced, and apparently alert engineer would accelerate beyond the safe operating speed traveling through the curve at Frankford Junction. Specifically, investigators examined the factors that may have diverted the engineer's attention from train operations before the accident.

Train 188 left Philadelphia 30th Street Station a little after 9 :10 p.m. According to the recorded radio transmissions, the six-minute radio conversation between the SEPTA engineer and the train dispatcher took place between 9 :13 and 9 :19 p.m. The train derailed at 9 :21 p.m., nearly 11 minutes after departing from the 30th Street Station. The Amtrak engineer was very focused on the incident that shattered the windshield of the SEPTA train and sent glass into the face of its engineer as he operated train 188 into the same area. During an interview three days after the accident, the Amtrak engineer accurately recalled the content of the radio transmissions. He said:

> The SEPTA engineer "sounded very upset, and it sounded like the dispatcher was trying to get clear information as to whether or not [the SEPTA engineer] needed medical help". And the SEPTA engineer was not being very clear and so they went back and forth.

During that same interview he also told investigators:

> I was a little bit concerned for my safety. There's been so many times where I've had reports of rocks that I haven't seen anything, that I felt it was unlikely that it would impact me. And I was really concerned for the SEPTA engineer. I had a coworker in Oakland that had glass impact his eye from hitting a tractor-trailer, and I know how terrible that is.

Locomotive engineers are expected to monitor radio transmissions while operating their trains because these communications can be pertinent to the safe operation of their trains. Operationally relevant information might be discussed, including details about issues with the track or signals, deteriorating weather conditions, or issues with trains that might force engineers to make an unplanned stop. In this case, it was important for the Amtrak engineer and other train crews operating in the area to know someone may have thrown an object at a train, and the SEPTA train had made an emergency stop on main track 1. Because the Amtrak engineer was monitoring the radio, he was aware the train crew might be on the tracks inspecting the SEPTA train for damage or setting up protection (such as flags) for it, and he reported this as a concern during his interview on November 10, 2015. This prompted him to sound the train's horn and broadcast a radio advisory that his train was approaching and would be passing the SEPTA train.

As he listened to the six-minute radio conversation between the SEPTA engineer and the dispatcher, the Amtrak engineer continued to operate his train at or below the maximum authorized track speed, and he remained cognizant of the signal indications affecting immediate train operations. The engineer's throttle manipulation that accelerated the train to 106 mph was initiated about 27 seconds after the last radio transmission between the SEPTA engineer and the train dispatcher at 9:19 p.m. Clearly, this action was not appropriate at that time given that the maximum authorized speed was 80 mph and the speed-restricted curve at Frankford Junction was coming up. But if he had traveled another 2 mi, going through the Frankford Junction curve and an adjacent curve at appropriate speeds, he then would have been authorized to operate the train at 110 mph, a speed that he was accustomed to traveling at numerous points along the route.

Investigators used event recorder data to derive the engineer's manipulation of the throttle. The method by which the engineer manipulated the throttle to increase the train's speed to 106 mph was consistent with his description of how he normally made a significant increase in speed when it was appropriate to do so. He told investigators that he typically accelerated with full throttle and then backed off as he approached his target speed. Event recorder data indicated that he did execute this procedure with a slight throttle manipulation when the train reached about 95 mph. Therefore, NTSB concludes that the Amtrak engineer initially accelerated his train to a high rate of speed in a manner consistent with how he habitually manipulated the controls when accelerating to a target speed, suggesting that he was actively operating the train rather than incapacitated moments before the accident.

The NTSB examined the possibility that because of diverting his attention to the extended radio communications between the SEPTA engineer and the dispatcher, the Amtrak engineer may have lost situational awareness. His loss of awareness, combined with the darkness, may have led him either to believe he had already passed the curve at Frankford Junction or to forget about the curve.

9.4.1.4 Loss of Situational Awareness

Situational awareness is defined as "the perception of the elements in the environment with a volume of time and space, comprehension of their meaning, and the projection of their status in the near future (Endsley 1995)." Informally, it is "knowing what's going on."

In this accident, the engineer may have lost awareness of which curve he had traveled through just before he accelerated his train to 106 mph. Immediately following Frankford Junction is a right-hand, 60 mph curve. Two miles before that curve there is a similar right curve near MP 83.5 with a maximum speed of 65 mph. The engineer would have had to operate his train similarly around both right curves. Moreover, each curve is followed by tangent (straight) track that allows the engineer to accelerate to significantly higher speeds. The right curve before the derailment location was followed by tangent track with a maximum operating speed of 80 mph, while the right curve immediately after the derailment location was followed by a 110 mph maximum speed. Given that the two curves were similar, and the engineer's attention was diverted to the radio conversations about the emergency situation with the SEPTA train, he may have confused the right track curve preceding Frankford Junction with the right-hand curve following it. In that case, he would have believed it was appropriate to increase his speed to 110 mph after transiting the first curve.

Furthering the potential for error, the engineer was operating at night when there were fewer visible external cues to help him determine his location. According to the engineer's interview and the forward-facing video, neither the elevated bridge at Frankford Junction (which served as a cue to begin decelerating before the curve) nor the curve itself would have been visible when the engineer began accelerating. If the engineer did not see or focus enough attention on cues indicating that he needed to slow the train, he would be less likely to realize that accelerating the train to 106 mph was an error at that time and location.

The NTSB has investigated other railroad accidents in which the loss of situational awareness resulted when the engineer was engaged in other operational tasks. In its investigation of the 2002 collision of an Amtrak train and a Maryland Area Regional Commuter (MARC) train in Baltimore, the NTSB determined the engineer lost situational awareness because of excess focus on regulating train speed. Because of her excess focus on regulating the train's speed, she failed to see and comply with both the cab and wayside signals indicating she should stop. Additionally, in the 2003 derailment of a Northeast Illinois Regional Commuter Railroad (Metra) train in Chicago, Illinois, the NTSB determined the engineer had lost situational awareness minutes before the derailment because of his preoccupation with paperwork relating to train operations. Because of his preoccupation, he failed to observe and comply with the signal indications.

While diverted attention and the loss of situational awareness can result in errors in identifying one's location, they can also result in errors in executing normal procedures and tasks. Despite the engineer's experience in the area, after operating around the 65 mph curve near MP 83.5, he failed to execute his next significant operating task: decelerating the train as it approached the curve at Frankford Junction, a maneuver he had performed many times. Omitting normal procedural steps is a form of prospective memory error (Dismukes 2006). Prospective memory refers to remembering to perform an intended action at some future time or, more simply, remembering to remember. It typically focuses on when to do something. Failures of prospective memory typically occur when we form an intention to do something later, become engaged with other tasks, and forget the thing we originally intended to do. It is possible, then, that the Amtrak engineer failed to slow his train for the upcoming curve because his attention to the radio communications about the SEPTA train emergency caused him to forget about the impending operation.

The NTSB has investigated accidents where competing information interfered with a crewmember's retention of vital information, which affected the crewmember's future actions. For example, in the 1996 collision between MARC and Amtrak trains in Silver Spring, Maryland, the NTSB determined that the MARC train engineer apparently forgot the most recent signal he had passed and ran his train through the next signal because he was focused on other tasks and information. The engineer was processing competing information that included the mental and physical tasks required to stop the train at an upcoming station; carrying on radio conversations with an engineer on another train; monitoring defect detector broadcasts and disrupted radio broadcasts; and listening to or talking with another crewmember in the cab. Processing these multiple pieces of information interfered with his retention of the signal information.

In addition to this accident, the NTSB has investigated other accidents where experienced crewmembers forgot to complete a normal procedural step, they had successfully performed many times on previous trips. In the 2005 derailment of a Norfolk Southern Railway freight train in Graniteville, South Carolina, the train crew failed to restore a switch to the normal main track position, a task they had routinely performed before the accident. In a number of aviation accidents, experienced crews forgot to perform routine duties such as setting flaps and slats to take-off position; setting hydraulic boost pumps to high position before landing; and arming the spoilers before landing. In many of these accidents, the crewmembers' routine duties were interrupted, and their attention was momentarily diverted. Studies have shown people are vulnerable to forgetting to resume interrupted tasks in a timely manner. When they do resume the interrupted task, they may struggle to mentally reconstruct the point at which they were interrupted, and they are vulnerable to increased errors.

The NTSB concluded that the Amtrak engineer accelerated the train to 106 mph without slowing the train for the curve at Frankford Junction, due to his loss of situational awareness, likely because his attention was diverted to the emergency situation with the SEPTA train.

This accident illustrates that a crewmember's prolonged focus on one area of train operations can take attention away from other critical operations, including those to be performed soon. The NTSB proposed training to help prevent future events. This training included:

- Use locomotive engineer (LE) simulator training to go beyond basic skills and teach strategies for effectively managing multiple concurrent tasks and atypical situations.
- The NTSB also recommended that the Federal Railroad Administration (FRA) develop guidelines for locomotive engineer simulator training programs that go beyond developing basic skills and teach strategies for effectively managing multiple concurrent tasks and atypical situations.
- The FRA contracted for a research study on this topic. Based on the results of that research, a training course was developed by the contractor that teaches strategies to locomotive crews for managing distractions and the importance of sustained attention on the locomotive operating task.

Some of the other strategies relevant to this accident include the following:

- Educating individuals and managers about prospective memory vulnerability and pointing out counter measures individuals can take.
- Minimizing the juggling of multiple tasks concurrently if one of the tasks is vital.
- Pausing to encode an explicit intention to resume an interrupted task after the interruption has ended.
- Analyzing the specific operating environment to identify "hotspots" in which prospective memory and concurrent task demands are high and interruptions are frequent. To the extent possible, redesign procedures and systems to reduce demands, especially when the consequences of memory lapses are serious.
- Designing display and alerting systems for the status of tasks not active where the need for prospective memory is high.

The GPS was in common use in 2015, but even more so today. In this accident, the engineer likely would have benefitted from technology that showed him the location of his train in real time and would have also helped him establish and maintain his situational awareness. The NTSB has advocated the use of memory aids, visual displays, alerting systems, and other strategies and technologies to reduce operator workload and prevent errors. This situational information would assist crews operating in high traffic areas, at night, or in adverse weather conditions. Although there will be less need for such situational information in

PTC-compliant territory, and this technology will be available in some locomotives operating in PTC-compliant territories, there are many areas where PTC will not be implemented. Therefore, the NTSB recommends that the FRA require railroads to install devices and develop procedures that will help crewmembers identify their current location and display their upcoming route in territories where PTC will not be implemented. The system could also alert the engineer of upcoming changes in allowable speed, as do most GPS systems available today on automobiles.

The full NTSB report also discusses signal and train control information.

9.4.1.5 Two-Person Crews

In April 2014, the FRA announced its intention to issue a proposed rule establishing minimum crew size standards for main line freight and passenger rail operations. As stated by the FRA in its 2014 press release, "We believe that safety is enhanced with the use of a multiple person crew—safety dictates that you never allow a single point of failure." On March 15, 2016, the FRA published a notice of proposed rulemaking (NPRM) to establish minimum requirements for the size of train crews depending on the type of operation. The NPRM proposes a minimum requirement of two crewmembers "for all railroad operations, with exceptions for those operations that FRA believes do not pose significant safety risks to railroad employees, the general public, and the environment by using fewer than two-person crews." The proposed rule "would also establish minimum requirements for the roles and responsibilities of the second train crewmember on a moving train, and promote safe and effective teamwork."

There is discussion in the report about whether a two-person crew would reduce the potential for accidents. In so many accidents from the past like The Three-Mile Island Nuclear Power Plant accident and several aviation accidents, multiple person crews can be drawn into a rabbit hole and not find their way out. A two-person engineer crew would also need to use CRM to ensure the crew works effectively together. CRM is discussed in Chapter 7.

9.4.1.6 Factors Not Contributing to This Accident

The locomotive and the passenger cars passed postaccident mechanical inspections. A review of preaccident testing and maintenance records did not reveal any defects. Although it was night, the weather at the time of the accident was clear with good visibility. The engineer did not recall or report the locomotive being struck by an object prior to the derailment, and FBI testing showed no evidence of ballistic material. Postaccident toxicological tests for crewmembers were negative for alcohol and other drugs. Postaccident testing and examination of the engineer did not identify any medical conditions that would have interfered with train operation. There was no evidence of cell phone use by the engineer during the accident trip. Further, the on-duty/off-duty schedule provided adequate time for the employees to obtain rest.

Investigators examined the records for track inspections and maintenance. The undamaged track was examined during the on-scene investigation. The track was inspected and maintained within regulatory standards, and no track anomalies were discovered after the derailment.

The NTSB concludes that none of the following was a factor in this accident: the mechanical condition of the train; a foreign object striking the locomotive; the condition of the track; the weather; medical conditions of the Amtrak engineer; alcohol, other drugs, or any other type of impairment; cell phone use; and fatigue.

NTSB Findings (note that not all of the following findings were presented in this truncated version of the accident report)

1. None of the following was a factor in this accident: the mechanical condition of the train; a foreign object striking the locomotive; the condition of the track; the weather; medical conditions of the Amtrak engineer; alcohol, other drugs, or any other type of impairment; cell phone use; and fatigue.
2. The Amtrak engineer initially accelerated his train to a high rate of speed in a manner consistent with how he habitually manipulated the controls when accelerating to a target speed, suggesting that he was actively operating the train rather than incapacitated moments before the accident.
3. The Amtrak engineer accelerated the train to 106 mph without slowing the train for the curve at Frankford Junction, due to his loss of situational awareness, likely because his attention was diverted to the emergency situation with the SEPTA train.
4. Training focusing on prospective memory strategies for prolonged, atypical situations that could divert crewmember attention may help operating crews become aware of, and take measures to avoid, errors due to memory failure.
5. Cab signal protection to enforce the 50 mph speed restriction in the eastbound direction at Frankford Junction or a fully implemented PTC system would have prevented the accident.
6. The FRA accident database is inadequate for comparing relevant accident rates based on crew size because the information about accident circumstances and number of crewmembers in the controlling cab is insufficient.
7. If the passenger car windows had remained intact and secured in the cars, some passengers would not have been ejected and would likely have survived the accident.
8. Passengers were seriously injured by being thrown from their seats when the passenger cars overturned.
9. Although the passenger equipment safety standards in Title 49 *Code of Federal Regulations* Part 238 provide some level of protection for occupants,

the current requirements are not adequate to ensure that occupants are protected in some types of accidents.

10. Matching patient arrival to hospital capacity in a mass casualty incident is crucial to ensuring optimal care can be provided for all patients.
11. As a result of victims being transported to hospitals without coordination, some hospitals were overutilized while others were significantly underutilized during the response to the derailment.
12. Transportation of injured victims by police or other municipal vehicles early in a mass casualty incident may be a reasonable use of resources.
13. Current Philadelphia Police Department, Philadelphia Fire Department, and Philadelphia Office of Emergency Management policies regarding transport of patients in a mass casualty incident were not, and still are not, integrated.

9.4.1.7 NTSB Probable Cause

The NTSB determines that the probable cause of the accident was the engineer's acceleration to 106 mph as he entered a curve with a 50 mph speed restriction, due to his loss of situational awareness likely because his attention was diverted to an emergency situation with another train. Contributing to the accident was the lack of a PTC system. Contributing to the severity of the injuries were the inadequate requirements for occupant protection in the event of a train overturning.

9.4.1.8 Summary of Safety Culture Issues

The railroad industry, as efficient as it is, does not rely on the most advanced technology to ensure safe passenger travel. Personally, I love train travel. I would like high-speed rail service from West Yellowstone in Montana, through Salt Lake, Las Vegas, to San Diego, with trunk lines to Los Angeles, Phoenix, and Denver. However, the train systems are not up to date and such a system in my lifetime is more than likely not possible. Even with the most advanced technology, operators of complex systems can become overwhelmed and lose situational awareness. However, providing today's state-of-the-art GPS and display technology and staffing with a two-person crew trained in CRM should provide a higher level of safety.

9.4.2 Transportation Safety Board of Canada (2013a)

The description of this accident was excerpted from the Transportation Safety Board of Canada report (2013a). On July 5, 2013, at about 1355, eastward Montreal, Maine & Atlantic Railway (MMA) freight train MMA-002 (the train) departed Farnham (near Brookport, Mile 125.60 of the Sherbrooke Subdivision), Quebec, destined for Nantes (Mile 7.40 of the Sherbrooke Subdivision), Quebec, where it was to be re-crewed and was to continue on to Brownville Junction, Maine.

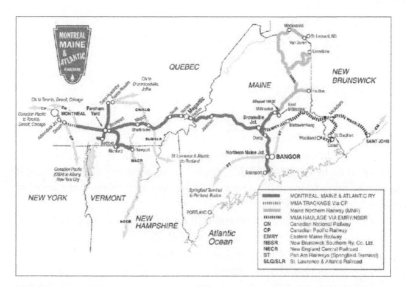

Figure 9.48 Montreal, Maine & Atlantic Railway (MMA) map. Source: MMA, with TSB annotations.

The train's destination was Saint John, New Brunswick (Figure 9.48). The train consisted of 72 tank cars loaded with approximately 7.7 million liters of petroleum crude oil (UN 1267), 1 box car (buffer car), and the locomotive consist (5 head-end locomotives and 1 VB car). The train was controlled by a LE who was operating alone and was positioned in the lead locomotive, MMA 5017. During the trip, the LE reported mechanical difficulties with the lead locomotive, that affected the train's ability to maintain speed. Please note that the complete report contains a great deal of information about the technical aspects of the testing of the brakes and tank cars. Though important to this event, it does not pertain directly to the safety culture associated with this accident.

At around 2250, the train arrived at Nantes, was brought to a stop using the automatic brakes, and was parked for the night on a descending grade on the main track. The LE applied the independent brakes to the locomotive consist. He then began to apply the hand brakes on the locomotive consist and the buffer car (seven cars in total), and shut down the four trailing locomotives. Subsequently, the LE released the automatic brakes and conducted a hand brake effectiveness test without releasing the locomotive independent brakes.

The LE then contacted the rail traffic controller (RTC) responsible for train movements between Farnham and Megantic Station (Megantic), who was located in MMA's yard office in Farnham, to indicate that the train was secured.

The LE then contacted the RTC in Bangor, Maine, who controlled movements of United States crews east of Megantic. During this conversation, the LE indicated that the lead locomotive had continued to experience mechanical difficulties throughout the trip and that excessive black and white smoke was now coming from its smokestack. The LE expected that the condition would settle on its own. It was mutually agreed to leave the train as it was and that performance issues would be dealt with in the morning.

A taxi was called to transport the LE to a local hotel. When the taxi arrived to pick up the LE at about 2330, the taxi driver noted the smoke and mentioned that oil droplets from the locomotive were landing on the taxi's windshield. The driver questioned whether the locomotive should be left in this condition. The LE indicated that he had informed MMA about the locomotive's condition, and it had been agreed upon to leave it that way. The LE was then taken to the hotel in Lac-Mégantic and reported off-duty.

At 2340, a call was made to a 911 operator to report a fire on a train at Nantes. The Nantes Fire Department responded to the call and arrived on site, and the Sûreté du Quebec (SQ) called the Farnham RTC to inform the company of the fire. After MMA unsuccessfully attempted to contact an employee with LE and mechanical experience, an MMA track foreman was sent to meet with the fire department at Nantes. When the track foreman arrived on site, the firefighters indicated that the emergency fuel cut-off switch had been used to shut down the lead locomotive. This shutdown put out the fire by removing the fuel source. Firefighters also moved the electrical breakers inside the locomotive cab to the off position to eliminate a potential ignition source. These actions were in keeping with railway instructions.

Both the firefighters and the track foreman were in discussion with the Farnham RTC to report on the condition of the train. Subsequently, the fire department and the MMA track foreman left the scene.

With no locomotive running, the air in the train's brake system slowly began to be depleted, resulting in a reduction in the retarding force holding the train. At about 0100 (July 06), the train started to roll downhill toward Lac-Mégantic, 7.2 mi away. At about 0115, the train derailed near the center of town, releasing about 6 million liters of petroleum crude oil, which resulted in a large fire and multiple explosions.

The locomotive consist did not derail; rather, it separated from the rest of the train and then further separated into two sections. Data downloaded from the de la Gare Street crossing (located by Megantic Station) showed that the two sections were separated by 104 ft. Both continued traveling eastward onto the Moosehead Subdivision, coming to rest on an ascending grade in the eastern part of town and stopping approximately 475 ft apart.

During the course of this entire sequence, the train passed through 13 level crossings.

Figure 9.49 The Lac-Mégantic derailment site following the accident.

After approximately 1.5 hours, while emergency and evacuation efforts were under way, the leading section of the locomotive consist rolled backwards toward downtown and contacted the trailing section; both sections traveled backwards an additional 106 ft. At approximately 0330, MMA officials secured the locomotive consist of the grade by re-tightening the hand brakes.

As a result of the derailment and the ensuing fires and explosions, 47 people died, and about 2000 people were evacuated. Forty buildings and 53 vehicles were destroyed (Figure 9.49).

The derailed tank cars contained about 6.7 million liters of petroleum crude oil, about 6 million liters of which were released, contaminating approximately 31 ha of land. Crude oil migrated into the town's sanitary and storm sewer systems by way of manholes. An estimated 100,000 l of crude oil ended up in Mégantic Lake and the Chaudière River by way of surface flow, underground infiltration, and sewer systems. About 740,000 l were recovered from the derailed tank cars.

The hydrocarbon recovery and cleanup operation began as soon as the fire was extinguished and the site was stabilized, approximately two days after the derailment. The assessment and remediation of the environment were performed using a combination of monitoring wells and exploratory trenches serviced by vacuum trucks under the guidance of a specialized engineering firm.

Between Nantes and Megantic (Mile 7.40 to the lowest point near Mile 0.00), the average descending grade was 0.94%, and the steepest grade over the length of the train was 1.32% at Mile 1.03 (Figure 9.50). The elevation dropped approximately 360 ft between Nantes and Megantic. For the last 2 mi before the point of

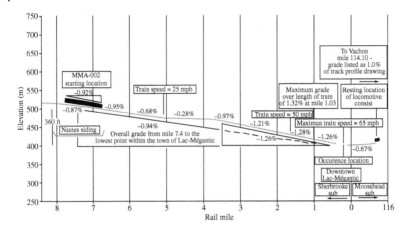

Figure 9.50 Grade and elevation between Nantes and Megantic.

derailment, the track descended at a grade of approximately 1.30%. The maximum horizontal curvature of the track was 4.25°, which was at the derailment location.

Cautionary limits were in effect between Mile 0.82 and Mile 0.00, due to the presence of the yard at Megantic. Movements were to be made in accordance with Canadian rail operating rules (CROR) 94 and 105(c). There was a permanent speed restriction of 10 mph over Frontenac Street (Mile 0.28) until the crossing was fully occupied.

9.4.2.1 Personnel Information

From Farnham to Nantes, MMA-002 was operated by 1 LE positioned in the lead locomotive as per single-person train operations (SPTO) special instructions. The LE was rules-qualified and met fitness and work/rest regulatory requirements. The LE's 2 previous shifts were:

- MMA-002 (eastbound from Farnham to Megantic) on July 2, 2013 from 1230 to 0030, and
- MMA-001 (westbound from Megantic to Farnham) on July 3, 2013 from 0830 to 2030.

Both trips had been performed with a conductor.

On July 5, 2013, the LE awoke at approximately 0530 and reported for duty at 1330 for MMA-002. When the LE was at home in Farnham, he normally slept about eight hours per night. When the LE laid over, he usually slept between five and six hours per night.

The LE was hired by cardiopulmonary resuscitation (CPR) in January 1980, and qualified as an LE in 1986. In September 1996, he transferred to Quebec Southern Railway (QSR) when that company acquired the trackage from CPR. In January

2003, the LE transferred to MMA when QSR was purchased by Rail World, Inc. (RWI), MMA's parent company. During this time, he completed hundreds of trips between Farnham and Lac-Mégantic and was familiar with the territory.

In the 12 months before the accident, the LE completed about 60 eastbound trips on MMA-002. About 20 of these trips were completed as a single-person train operator.

The tank cars originated in New Town, North Dakota, where they were picked up by CPR. At origin, the train consisted of 1 box car (the buffer) and 78 tank cars loaded with petroleum crude oil (UN 1267), a Class 3 flammable liquid. On June 30, 2013, when the train was in Harvey, North Dakota, 1 tank car was removed for a mechanical defect after the train received a safety inspection and a Class I air brake test. This air brake test verifies the integrity and continuity of the brake pipe, as well as the brake rigging, the application, and the release of air brakes on each car.

The petroleum crude oil had been purchased by Irving Oil Commercial G.P. from World Fuel Services, Inc. (WFSI). The shipping documents indicated that the shipper was Western Petroleum Company (a subsidiary of WFSI) and the consignee was Irving Oil Ltd. (Irving).

The cars operated through Minneapolis, Minnesota, Milwaukee, Wisconsin, Chicago, Illinois, and Detroit, Michigan, and arrived in Canada through Windsor, Ontario. The cars traveled to Toronto, Ontario, and underwent a No. 1 air brake test by a certified car inspector on July 4, 2013. The cars departed Toronto as part of a mixed freight train, consisting of 2 locomotives and 120 cars, destined for Montréal. When the train arrived in Montréal, it underwent a routine safety and mechanical inspection in Saint-Luc Yard on July 5, 2013. Mechanical defects were identified on 5 tank cars, which were removed from the train. The remaining tank cars were then interchanged to MMA.

On the morning of July 5, 2013, the cars were taken to Farnham, where they received a brake continuity test and a mechanical inspection by Transport Canada (TC). Minor defects were noted on two cars, and these were corrected. Departing Farnham, the train was approximately 4700 ft long, weighed about 10,290 tons and consisted of the following (Photo 2):

1. lead locomotive MMA 5017, General Electric Company (GE) C30-7;
2. special-purpose caboose (VB car) VB 1;
3. locomotive MMA 5026, GE C30-7;
4. locomotive CITX 3053, General Motors (GM) SD-40;
5. locomotive MMA 5023, GE C30-7;
6. locomotive CEFX 3166, GM SD-40;
7. buffer car CIBX 172032; and
8. 72 tank cars.

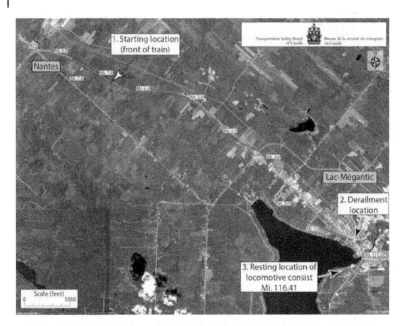

Figure 9.51 The three locations that were the focal points of the investigation: Nantes, downtown Lac-Mégantic, and the location where the locomotives came to a stop (Mile 116.41 of the Moosehead Subdivision). Source: Transportation Safety Board of Canada (2013).

The investigation focused on three locations (Figure 9.51):

- Nantes, where the train was parked;
- downtown Lac-Mégantic, where the train derailed; and
- the ascending grade, east of Megantic, where the locomotive consist came to its final stop (Mile 116.41 of the Moosehead Subdivision).

Railway lines at Nantes are in a rural area where the main track and a siding run parallel and immediately adjacent to public highway 161. The average descending grade on the main track where the train was parked is 0.92%. During site examination, a black oily residue was found on the surrounding vegetation and on the rails where the lead locomotive was parked.

The east siding switch was located at Mile 6.67, and the siding was 7160 ft long. At the time of the accident, several rail cars were being stored there. The siding was equipped with a special derail, located approximately 230 ft west of the switch. A derail is a mechanical safety device that sits on top of the rail and is used to derail runaway equipment. This derail was locked in the derailing position to protect the main track from unintended movements out of the siding.

Switch location for
megantic west turnout

Figure 9.52 Eastward view of the location of the tracks in relation to the first derailed cars: main track (A), yard track 1 (B), yard track 2 (C), yard track 3 (D), and the west and east legs of the wye tracks (E and F). Source: Transportation Safety Board of Canada (2013).

The MMA Megantic Station was located in a commercial district of Lac-Mégantic, where the Sherbrooke and Moosehead Subdivisions met. Frontenac Street, a main thoroughfare, ran through the center of the town. The main track intersected with Frontenac Street just west of the Megantic West turnout and was maintained for a maximum speed of 15 mph. The turnout was located at Mile 0.23, with the switch points facing west.

The derailed equipment covered the main track, three adjacent yard tracks, and the west leg of the wye, which is a triangular arrangement of tracks that can be used for turning rail equipment (Figure 9.52). At the time of the accident, there were box cars parked in yard tracks 1 and 2.

The track and crossing infrastructure was damaged as follows:

- The damage to the main track started approximately 20 ft east of Frontenac Street.
- The main-track turnout, approximately 400 ft of main track, and an additional 2000 ft of yard and wye tracks, including three turnouts, were destroyed.
- Approximately 500 ft from the crossing, the main track was shifted about 4 ft to the north.
- Yard tracks 1 and 2 were demolished from the west-end turnout for about 600 and 500 ft, respectively.
- Rails were curled and twisted, unsettled from tie plates, and moved randomly. Due to the severity of the fire, most track components were badly damaged.

- The Frontenac Street southeast public-crossing cantilever mast and the control box were shattered. Road traffic lights, electrical wires, lighting posts, and other appliances were also damaged.

The derailed equipment at the Lac-Mégantic site consisted of 2 box cars and 63 loaded tank cars.

The derailed equipment came to rest as follows:

- The buffer box car, which had a broken knuckle from a torsional overstress on the leading end, and the first three derailed tank cars were on their sides, jack-knifed, and partially coupled. They came to rest close to each other and came in contact with the seven box cars in yard track 2, derailing 1 of the standing box cars.
- The fourth and fifth derailed tank cars were also on their sides, jackknifed, and resting between yard tracks 2 and 3, about 50 ft north of the main track. They were separated by 125 ft from the preceding cars and had struck a pile of rails stored in the yard.
- The sixth and seventh derailed tank cars, still coupled together, came to rest near yard track 3, about 150 ft north of the main track.
- The eighth derailed tank car was uncoupled and came to rest in a wooded area between yard track 3 and the west leg of the wye.
- All of the remaining derailed tank cars came to rest in a large pileup toward the west leg of the wye, with the last derailed car coming to rest on the Frontenac Street crossing. The ninth and tenth cars stayed coupled and aligned with the roadbed. The next 53 cars came off their trucks, jackknifed, and were severely damaged. The debris from the derailed equipment was confined to the derailment site. Most of the wheel sets and trucks were found on the south side of the pileup, within approximately 400 ft from the Frontenac Street crossing. There were no reports of any pieces of tank cars being projected away from the downtown area.

The last nine tank cars on the train were still coupled to the last derailed car, but did not derail.

Examination of the derailed equipment determined that a hand brake had been applied on the buffer car. No hand brakes were found to have been applied on any of the tank cars.

The locomotive consist came to rest approximately 4400 ft east of the Lac-Mégantic derailment site, at Mile 116.41 of the Moosehead Subdivision (Figure 9.53).

At this location, the track ran parallel to d'Orsennens Street. During site examination, the following was noted:

- There was no damage to the track between the derailment site and the location of the locomotives.

Figure 9.53 Location of the locomotive consist (Mile 116.41 of the Moosehead Subdivision) in relation to the derailment site. The white arrows denote the route of the locomotive consist, which followed the main track. Source: Transportation Safety Board of Canada (2013).

- There was a black oily residue, like the residue observed at Nantes, on the ground adjacent to the lead locomotive, as well as about 600 ft east of where the locomotives came to rest.
- Hand brakes were applied on all five locomotives and the VB car.
- There was severe wear on some of the brake shoes and various degrees of blueing on most of the wheels. Blueing is a blue discoloration of steel surfaces that is indicative of exposure to heat. On railway wheels, tread blueing is caused by the frictional heat generated during a heavy or extended brake application.
- One of the knuckles connecting the second locomotive (MMA 5026) and the third locomotive (CITX 3053) was broken, and a locomotive connector cable had been pinched between the knuckles, indicating that a separation had occurred and had rejoined.
- A broken piece of the knuckle was found under the second locomotive, approximately 15 ft from the coupling. The locomotive knuckle and pin failed in tensile overstress mode, initiating at pre-existing fatigue cracks.

9.4.2.2 Train Brakes

Trains are equipped with two air brake systems: automatic and independent. The automatic brake system applies the brakes to each car and locomotive on the train, and is normally used during train operations to slow and stop the train. Each locomotive is equipped with an independent brake system, which only applies brakes on the locomotives. Independent brakes are not normally used during train operations but are primarily used as a parking brake. Figure 9.54 shows a diagram of the braking system.

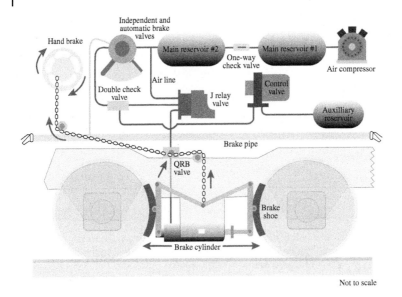

Figure 9.54 Schematic of the locomotive air brake and hand brake. Source: Transportation Safety Board of Canada (2013).

A train's automatic braking system is supplied with air from compressors located on each operating locomotive. The air is stored in the locomotive's main reservoir. This reservoir supplies approximately 90 lb per square inch (psi) of air to a brake pipe that runs along the length of the entire train, connecting to each locomotive and individual car. Air pressure changes within this brake pipe activate the brakes on the entire train.

When an automatic brake application is required, the LE moves the automatic brake handle to the desired position. This action removes air from the brake pipe. As each car's air brake valve senses a sufficient difference in pressure, air flows from a reservoir located on each car into that car's brake cylinder, applying the brake shoes to the wheels.

To release the brakes, the LE moves the automatic brake handle to the release position. This action causes air to flow from the main reservoir on the locomotive into the brake pipe, restoring pressure to 90 psi. Sensing this, each car's brake valve allows air to be released from its brake cylinder, and the shoes are removed from the wheels.

The independent brakes are also supplied with air from the main reservoir. When an independent brake application is required, the LE moves the independent brake handle, which in turn injects up to 75 psi of air pressure directly from the main reservoir into the brake cylinders of the locomotive. This causes the brake shoes to apply to the wheels.

To release the independent brakes, the LE moves the independent brake handle to the release position. This causes air to be released from the locomotive's brake cylinders, and the shoes are removed from the wheels.

A penalty brake application is similar to a full automatic brake application. However, this type of braking further reduces the brake pipe pressure to zero, requiring a moving train to stop and recharge the brake pipe. This type of braking occurs as a result of a "penalty" applied by the system, such as when the reset safety control (RSC) is not reset. This application occurs at a rate that does not deplete all of the air in each car's reservoir.

An emergency brake application is the maximum application of a train's air brakes, during which the brake pipe pressure is rapidly reduced to zero, either from a separation of the brake pipe or operator-initiated action. Following an emergency brake application, a train's entire air system is depleted.

Brake pipe pressure below 40 psi cannot be relied upon to initiate an emergency brake application.

When locomotives are shut down, the air compressors are also shut down and no longer supply air to the train. Given that the system has many connections, which are prone to air leaks, the main reservoir pressure will slowly begin to drop soon afterward.

Because the main reservoir supplies air to the entire system, when its pressure falls to the level of that in the brake pipe, the pressure in both components will thereafter diminish at the same rate. This sequence also occurs when the main reservoir and brake pipe reach the same pressure as that in the brake cylinder, at which point all 3 will lose pressure at the same rate.

As the air in the brake cylinder decreases, the amount of force being applied to the locomotive wheels by the independent brakes is reduced. If the system is not recharged with air, the brakes on the locomotives will eventually become completely ineffective.

In addition to a train's air brake system, all locomotives and rail cars are equipped with at least 1 hand brake, which is a mechanical device that applies brake shoes to the wheels to prevent them from moving or to retard their motion (Figure 9.55). Typically, hand brakes consist of a hand brake assembly, which designates the B-end of each car. When the wheel on the hand brake assembly is tightened, the brakes are applied.

The effectiveness of hand brakes depends on several factors, including hand brake gearing system lubrication and lever adjustment. Also critical is the force exerted by the person applying the hand brake, which can vary widely from one person to another. For example, railway standards are based on an application of 125 lb of force on the outside rim of the hand brake wheel. However, previous Transportation Safety Board (TSB) investigations have noted that, on average, employees apply 80–100 ft-lb of force.

Figure 9.55 Hand brake assembly and wheel at the B-end of a tank car. Source: TSB Canada (2013a).

9.4.2.3 Locomotives

There are no requirements for a locomotive to hold any other equipment when the hand brake is applied. On many locomotives, including the ones in this accident, when the hand brake is applied, only 2 of as many as 12 brake shoes are applied to the locomotive wheels.

For locomotives placed in service after January 4, 2004, the FRA in the United States requires that the hand brake(s) alone be capable of holding a locomotive on a 3% grade. This equates to a net braking ratio of approximately 10%.

Although there were no such requirements prior to 2004, locomotive manufacturers generally designed locomotive hand brakes to meet the 3% holding capacity.

According to Standard S-401 (Brake Design Requirements) of the Association of American Railroads' (AAR) *Manual of Standards and Recommended Practices* (MSRP), the force applied to the wheels by the brake shoes must be equal to about 10% of the car's gross load weight, with 125 lb of force applied to the outside rim of the hand brake wheel.

Unlike hand brakes on many locomotives, hand brakes on cars normally apply all brake shoes (typically 8) to the wheels.

To verify that the hand brakes applied are sufficient to secure the train, crews were required to perform a hand brake effectiveness test, in accordance with CROR 112 (b), to ensure that the equipment will not move. After applying the hand brakes, the test is performed by releasing all of the air brakes and allowing the slack to adjust under gravity, or by attempting to move the equipment slightly with reasonable locomotive force.

If the hand brakes prevent the equipment from moving, then they are determined to be sufficient. If not, additional hand brakes must be applied, and the process repeated until a successful effectiveness test has been completed.

Special instructions of some Canadian railway companies, including MMA, permitted the hand brakes on the locomotive consist to be included in the minimum required number of hand brakes. For example, if a company's special instructions required at least 10 hand brakes to be applied, and the train were operating with 4 locomotives, then only 6 hand brakes were required to be applied on the cars in addition to those on the locomotives.

During an effectiveness test performed with hand brakes applied on the locomotive consist, the LE has to overcome the braking force on the locomotives before moving the rest of the train.

9.4.2.4   Rules and Instructions on Securing Equipment

The CROR are the rules by which Canadian railways under federal jurisdiction operate, which include MMA's Canadian operations. At the time of the accident, CROR 112 stated the following, in part (Canadian Rail Operating Rules 2013b):

(a) When equipment is left at any point a sufficient number of hand brakes must be applied to prevent it from moving. Special instructions will indicate the minimum hand brake requirements for all locations where equipment is left. If equipment is left on a siding, it must be coupled to other equipment if any on such track unless it is necessary to provide separation at a public crossing at grade or elsewhere.

To ensure that there was sufficient retarding force to prevent a train or cars from moving unintentionally, CROR 112 required the effectiveness to be tested when hand brakes were used to secure the equipment. The rule stated:

(b) Before relying on the retarding force of the hand brake(s), whether leaving equipment or riding equipment to rest, the effectiveness of the hand brake(s) must be tested by fully applying the hand brake(s) and moving the cut of cars slightly to ensure sufficient retarding force is present to prevent the equipment from moving [. . .]

In addition to CROR 112, MMA employees were governed by the special instructions in MMA's *General Special Instructions* (GSIs) and *Safety Rules*.

Since MMA operated in former CPR territory, it adopted CPR's General Operating Instructions (GOIs).

The minimum number of handbrakes to be set by MMA's GSIs provided instructions on the minimum number of hand brakes required, and stated in part:

> Crew members are responsible for securing standing equipment with hand brakes to prevent undesired movement. The air brake system must not be depended upon to prevent an undesired movement. [...]

Cars	Handbrakes	Cars	Handbrakes
1–2	1 Hand Brake	50–59	7 Hand Brakes
3–9	2 Hand Brakes	60–69	8 Hand Brakes
10–19	3 Hand Brakes	70–79	9 Hand Brakes
20–29	4 Hand Brakes	80–89	10 Hand Brakes
30–39	5 Hand Brakes	90–99	11 Hand Brakes
40–49	6 Hand Brakes	100–109	12 Hand Brakes

Note: [...] If conditions require, additional hand brakes must be applied to prevent undesirable movement.

The numbers in the table are commonly referred to by MMA employees as the "10% + 2" instruction.

For example, at Sherbrooke, between cautionary limit signs, including the main track and sidings, and at Farnham, the minimum number of hand brakes equated to approximately 10%. For Megantic Yard, the required number was less than 10%.

MMA's Safety Rule 9200 (Sufficient Number – Operating Hand Brakes) stated in part:

> Employees must:
>
> a. Know how to operate the types of hand brakes with which various types of cars are equipped.
>
> [...]
>
> c. Before attempting to operate handbrake, make visual inspection of brake wheel, lever, ratchet and chain.
>
> [...]
>
> f. Be aware of and work within the limits of your physical capabilities and do not use excessive force to accomplish tasks. Past practices that do not conform to the rules are unacceptable.

MMA's Safety Rule 9210 stated in part:

> h. All hand brakes shall be fully applied on all locomotives in the lead consisting of an unattended train.

i. When leaving railway equipment, the minimum number of hand brakes must be applied as indicated in the following chart.19 Additional hand brakes may be required; factors which must be considered are:

Total Number of Cars Empties or Loads Weather Conditions Grade of Track

[...]

k. In reference to the minimum number of hand brakes in the preceding chart, 19 it is acceptable to include the hand brakes applied on locomotives.

[...]

m. There may be situations where all hand brakes should be applied. [...]
o. To ensure an adequate number of hand brakes are applied, release all air brakes and allow or cause the slack to adjust. It must be apparent when slack runs in or out, that the hand brakes are sufficient to prevent that cut of cars from moving. This must be done before uncoupling or before leaving equipment unattended.

Prior to early 2013, CPR's instructions for determining the minimum number of hand brakes were to divide the number of cars to be left unattended by 10, and then add 2. The instructions also included the requirement to secure each locomotive left unattended with its hand brake. When a train was to be left unattended with the locomotive(s) attached, it was acceptable to include the locomotive hand brakes as part of the minimum required number of hand brakes.

Prior to the accident, CPR modified its hand brake instructions, no longer specifying the

minimum number of hand brakes. Crews were responsible for evaluating their train and other operating conditions to determine enough hand brakes and for testing their effectiveness before the equipment were left unattended.

In addition, section 2.0 of CPR's GOIs still stated that on light, heavy, and mountain grades, a specific number of hand brakes (higher than the minimum) was required when a hand brake effectiveness test could not be performed. For example, on grades between 1.0% and 1.29%, hand brakes were required on 25% of the train. Additionally, in some territories, an increased number of hand brakes had to be applied when a movement was stopped on a grade.

At Canadian National (CN), the hand brake instructions in effect at the time of the accident for rail cars left unattended were:

- Divide the number of cars on the train by 10 and add 1 additional hand brake, up to a maximum of 5 hand brakes.
- If the hand brake effectiveness test is not successful, more hand brakes are required to ensure that the movement remains immobilized.

- Certain locations outlined in CN's timetable required double (up to a maximum of 10) the number of hand brakes, depending on the track characteristics.
- Trains with locomotives attached with at least 1 locomotive running can be left on the main track with only 1 locomotive hand brake applied, provided that there is brake continuity throughout the train, the automatic air brakes are fully applied and the independent brakes are applied.

In addition to the aforementioned instructions, CN special instructions for leaving trains or transfers unattended on mountain grade territory were as follows:

- Every effort must be made, including RTC pre-planning, to avoid leaving trains or transfers in steep grades in excess of 0.75%.
- When absolutely necessary, a sufficient number of hand brakes must be applied to prevent any unintended movement caused from possible brake cylinder leak-off.
- The automatic air brakes must not be solely relied upon to secure equipment against undesired movement.
- Stop with the least amount of air brake application possible.
- Leave locomotives attached with brake pipe continuity throughout the train, and do not bleed off cars before applying hand brakes.
- Apply 25% of the train hand brakes on grades between 0.75% and 0.9%, and apply 40% of the train hand brakes on grades up to 1.4%.

Crew members were required to communicate and confirm that they had left the train in accordance with these instructions, and the RTC was to be advised of the number of hand brakes applied.

9.4.2.5 Locomotive Event Recorder

A train's locomotive event recorder (LER) is analogous to a "black box" on an aircraft. The LER monitors and records a number of parameters, including throttle position, time, speed, and distance, as well as pressure within the brake pipe and locomotive brake cylinder.

Changes in the brake pipe pressure cause each car to apply (or release) its air brake. In this accident, because the train was unattended, the LER was instrumental in providing key pieces of data.

Table 9.2 summarizes some important information obtained from the download of the LER on the lead locomotive. Brake pipe pressure is at its maximum at 95 psi (brakes fully released), and locomotive brake cylinder pressure is maximized at 70 psi (full independent brake application). Any drop in brake cylinder pressure indicates a reduction in retarding force.

Table 9.2 Locomotive event recorder information.

Time	mph	Brake pipe pressure (psi)	Locomotive brake cylinder pressure (psi)	Event
July 5, 2013 2249:37	0	82	69	MMA-002 was stopped at Nantes using a 13-psi automatic brake application, and the independent brakes were fully applied
2303:48	0	94	69	The automatic brakes were released. The locomotive independent brakes remained fully applied
2358:42	0	95	69	Lead locomotive MMA 5017 was shut down
July 6, 2013 0005:55	0	94	70	Brake pipe pressure began to decrease, and continued to decrease at an average rate of 1 psi per minute
0013:55	0	79	69	Independent brake cylinder pressure began to decrease at the same rate as the brake pipe pressure
0058:21	1	32	27	MMA-002 began to run away
0115:30	65	16	14	The highest recorded speed of 65 mph was attained
0115:31	65	0	14	Brake pipe pressure dropped to 0 psi as the cars began to derail. The locomotive consist separated into two sections
0117:12	0	0	6	The first section stopped 5016 ft east of the point of derailment, at Mile 116.30 of the Moosehead Subdivision, on a 1% ascending grade
0245:06	1	0	0	The first section of the locomotive consist began to move backwards (west) down the grade toward downtown Lac-Mégantic
0246:23	8	0	0	The first section of the locomotive consist traveled 475 ft west and struck the stationary second section of the consist
0246:42	0	0	0	The two sections rejoined and moved an additional 106 ft west before coming to a final stop

9.4.2.6 Sense and Braking Unit

The sense and braking unit (SBU) is a device placed on the rear of the train and is connected to the train brake pipe. The SBU senses train movement, monitors brake pipe pressure, and sends the information to the locomotive, where it is displayed in the cab. The SBU can also be used to initiate an emergency brake application from the end of the train.

The SBU data from MMA-002 were downloaded. The SBU data and crossing download data were used to corroborate the LER data. An analysis of the SBU data determined that when the SBU first recorded movement (start-to-move) at Nantes, brake pipe pressure at the rear of the train was 29 psi.

Approximately 16 minutes and 40 seconds after the train began to move, the brake pipe pressure at the rear of the train had diminished to 0 psi.

9.4.2.7 Mandatory Off-Duty Times for Operating Employees

The maximum continuous on-duty time for operating employees on a single tour of duty is 12 hours. The *Work/Rest Rules for Railway Operating Employees* specify that operating employees who are off duty after being on duty more than 10 hours at other than the home terminal must have at least 6 continuous hours off duty, with the mandatory off-duty time commencing upon arrival at the accommodations provided by the railway company. In case of an emergency, off-duty employees can be recalled. If a crew's rest is interrupted, the rest time is reset.

The continuous duty time of the LE for MMA-002 was 10 hours and 15 minutes. The LE for MMA-001 was under the same mandatory off-duty time and was lodging at the same accommodations in Lac-Mégantic as the LE for MMA-002.

9.4.2.8 Securement of Trains (MMA-002) at Nantes

With the new train schedule, trains were left at Nantes and at Vachon (the location where the 2 trains could meet, some 10 mi away). By leaving MMA-002 at Nantes, the train could be parked in a location where no crossing would be blocked, where access would be easy for pick-up and drop-off of crews, and where rail access to the siding would be allowed where cars were normally stored. There were no regulations precluding trains, including those carrying dangerous goods (DGs), from being left unattended on a main track. When trains were secured at Nantes, they would be left on the main track with at least 1 locomotive running, the automatic brakes released, the independent brakes applied, and several hand brakes applied.

For two-person crews, train securement was the responsibility of both crew members. Securement consisted of applying several hand brakes and then testing their effectiveness. The conductor would determine the number of hand brakes to be applied and would apply them once the train was brought to a stop. On occasion, LEs would assist in the application of the hand brakes.

With a single-person train operator, the responsibility rested with the LE for both the application and the effectiveness testing of the hand brakes. To ensure that the train would not roll away while the LE was applying the hand brakes, the automatic brakes were applied.

TSB conducted a survey of LEs and conductors to determine train securement practices at Nantes, and it showed that the number of hand brakes applied to trains varied. Two-person crews would consistently apply at least the minimum number of hand brakes specified in MMA's GSIs. Some single-person train operators reported applying less than the minimum number of hand brakes.

To perform a hand brake effectiveness test, some LEs would release the automatic and independent brakes and attempt to move the train, while others would not release the independent brakes and would not attempt to move the train. When a proper hand brake effectiveness test was performed, additional hand brakes would be applied, if required.

For fuel conservation purposes, MMA instructions were to shut down all idling locomotives not equipped with an auto-start. To comply with U.S. regulations (requiring brake testing by qualified employees if a train is off air for more than four hours), the MMA practice was to leave at least 1 locomotive running on U.S.-bound trains. Some crews left all of the locomotives running in all weather.

9.4.2.9 Securement of Trains (MMA-001) at Vachon

Shortly before the accident, MMA-001 was parked in the siding at Vachon by a Brownville Junction single-person train operator who was to be assigned to MMA-002 the following morning. MMA-001, consisting of 5 locomotives and 98 residue tank cars, had been secured with 5 hand brakes, and the independent brakes were applied. The locomotive cab door was not locked, and the train's paperwork along with the reverser were sitting on the locomotive console. The minimum hand brake requirement for a train of this length, as per MMA's instructions, was 11 hand brakes.

9.4.2.10 *Recent Runaway Train History at Montreal, Maine, and Atlantic Railway and Previous* TSB Investigations

According to TSB's Rail Occurrence Database System (RODS), there were five occurrences involving runaway MMA equipment between September 20, 2004 and the date of the accident. All five involved yard-switching movements, one of which involved cars rolling onto the main track.

The TSB has investigated nine runaway train occurrences since 2005; in addition to this accident, five others involved short line railway operations. In all of these occurrences, the investigation into the operations of these railways identified safety deficiencies in training, oversight, and operational practices.

9.4.2.11 *Training and Requalification of Montreal, Maine, and Atlantic Railway Crews in* Farnham

Section 10 of the *Railway Employee Qualification Standards Regulations* (SOR/87-150) states that "a railway company shall, at intervals of not more than three years, have each employee in an occupational category re-examined on the required subjects." CROR General Rule A requires every employee in any service connected with movements to:

> [...]
> (vi) be conversant with and governed by every safety rule and instruction of the company pertaining to their occupation;
> (vii) pass the required examination at prescribed intervals, not to exceed three years, and carry while on duty, a valid certificate of rules qualification;
> [...]

Railways design and administer training and requalification programs according to their needs. The programs usually take place in a classroom setting, where the exam topics are reviewed with an instructor and discussions take place to ensure that the rules are properly understood and applied. Exams vary from knowledge-based to scenario-based, with short-answer questions requiring written responses or with multiple-choice questions.

Knowledge-based exams contain questions that test specific rules or instructions and are typically closed-book. Scenario-based exams require the interpretation and application of CRORs, as well as of special instructions, to frequent scenarios. These exams are usually open-book and promote the development of problem-solving skills while using the company-provided manuals. Instructor feedback is a component of a requalification program. TC has the authority to review companies' training and requalification programs.

MMA delivered training to RTCs, LEs, conductors, and engineering employees. A review of MMA's training and requalification program determined the following:

- MMA's requalification exams tested employees on most CRORs and several MMA special instructions. They were knowledge-based, with short answers and multiple-choice questions.
- Requalification typically consisted of one day to complete the exam and did not always involve classroom training. On many occasions, employees would take the exam home for completion.
- MMA employees did not have the opportunity to review their requalification exam after it was corrected and received no feedback on their mistakes.
- A comparison between multiple requalification exams revealed that, over the years, they had essentially remained the same. However, there was increased use of multiple-choice questions.

- The exams repeated the same question on the minimum number of hand brakes for leaving unattended equipment. They did not have questions on the hand brake effectiveness test, the conditions requiring application of more than the minimum number of hand brakes, nor the stipulation that air brakes cannot be relied upon to prevent an undesired movement.
- Inconsistencies in the correction and grading of exams were noted. On some multiple-choice questions, more than one answer was accepted as correct, and some short-answer questions were answered by writing the applicable CROR number rather than writing the procedure to be followed.
- MMA did not consistently comply with the three-year interval for requalification. For several employees, the deadlines were exceeded by several months, with temporary certificates being issued.

9.4.2.12 Training and Requalification of the Locomotive Engineer

The accident LE had completed a requalification exam in September 2006. The next requalification was completed in December 2009, which was three months beyond the mandatory deadline. The LE received a new certificate of rules qualifications in March 2013, again three months after the expiration of his certificate. In April 2013, the LE completed the requalification exam at home, after having received the new certificate. The LE did not receive feedback on the results of that exam.

The LE's requalification exams in 2006 and 2009 included the same question on the number of hand brakes for a cut of cars left in a siding. In both exams, the LE's answers complied with MMA's hand brake requirements as per the instructions. In 2012, the LE's requalification exam contained two multiple-choice questions on the minimum number of hand brakes required for a cut of cars left unattended. Again, the LE's answers complied with MMA's hand brake instructions.

CROR General Rule A requires every employee in any service connected with movements to:

> (ii) have a copy of this rule book, the general operating instructions, current time table and any supplements, and other documents specified by the company accessible while on duty.

At MMA, the other required documents under CROR General Rule A included its GSIs and *Safety Rules*. At the time of the accident, the LE was not in possession of all of the mandatory documents, including the GSIs and *Safety Rules*.

9.4.2.13 Operational Tests and Inspections at Montreal, Maine, and Atlantic Railway

MMA developed the Operational Tests and Inspections (OTIS) Program for its supervisors to monitor employees' adherence to railway safety rules and instructions. The OTIS program at MMA involved field supervisors observing employees as they performed their work. These observations were to be conducted unannounced. The employee evaluations were based on compliance with GSIs, operating bulletins (OBs), *Safety Rules*, timetables, GOIs, and CROR.

Non-compliance with rules and instructions would be noted, and corrective action could result. Depending on the severity of the infraction, the non-compliance could result in a verbal correction, a letter of reprimand, or a suspension. All observations were entered into the OTIS system with either a "pass" or "failure" evaluation. Employees were notified of the result only if they failed the test.

MMA developed an OTIS manual to aid supervisors in the implementation of the program. The manual outlined the program's objectives, provided guidance on the methods and frequency of test administration, and provided general field instructions on implementing the program. Each supervisor was required to conduct a minimum of 15 OTIS tests per month (that is, 180 per year). Additional guidance provided to the supervisors included:

- instruction in ensuring that observations are conducted at various times and locations so that employees perceive that they may be tested at any time;
- direction in identifying those employees who need remedial rules training or appropriate discipline.
- guidance in periodically advising employees who consistently comply with all operating instructions that they were found to be in compliance with a recent test;
- development of a list of "Core Rules." The 2013 list, on which supervisors were to focus, included CROR 112(a) and (b), and OB 2-133, which covered the application and testing of hand brakes; and
- identification of several rules in which a minimum number of tests per month were to be conducted. For example, CROR 112 was to be tested at least 2 times per month per supervisor.

Supervisors were provided with periodic reports (at least quarterly) on their progress in completing the required number of tests and were reminded of which rules to focus on. Table 9.3 summarizes the number of OTIS observations completed by each supervisor.

For the four-year period from 2009 to 2012, the OTIS results were as follows:

- Of the 3789 tests conducted, 128 of the observations were entered into the system as "Failure."
- Testing on CROR 112 and GSI 112 had been conducted 31 times. There were 2 failures.
- Testing on OB 2-133 had been conducted 35 times. There were five failures.

Table 9.3 Operational Tests and Inspections (OTIS) observations completed per supervisor.

	Supervisor 1	Supervisor 2	Supervisor 3	Supervisor 4	Supervisor 5
2012	197	58	116	89	N/A
2011	208	84	137	216	154
2010	232	181	216	224	260
2009	233	140	199	177	230

Note: Supervisor 5 was no longer employed at MMA after July 2011.

During a test for compliance with hand brakes, supervisors checked the number of hand brakes applied to ensure that the number met the minimum requirement. However, they seldom checked to ensure that a proper hand brake effectiveness test was conducted. To test for a proper hand brake effectiveness test, supervisors had to be at the site, unannounced, when the train arrived. Failing that, supervisors had to review the LER download after the trip. MMA reviewed downloads only after accidents. Since 2009, no employee had been tested on CROR 112(b), which targeted the hand brake effectiveness test. In 2012, US employees had been tested twice on that rule; both tests had resulted in a "Failure."

Since January 1, 2009, the LE had been tested 97 times, and had failed three of these tests. Eight of the tests had been conducted outside of the hours of 0800 to 1800. Of the 97 tests, 70 were conducted within 27 mi of Farnham, and the remaining 27 were conducted in Sherbrooke. Seven of the 97 tests were on CROR 112 or OB 2-133, and the LE had successfully passed. None of the tests targeted the hand brake effectiveness test, and none were performed at Nantes.

9.4.2.14 Implementation of Single-Person Train Operations

At the time of the accident, there were no rules or regulations preventing railways from implementing SPTO. In Canada, there are only two federally regulated railways to have operated using SPTO: MMA, and Quebec North Shore and Labrador Railway (QNS&L).

QNS&L implemented SPTO in 1996, without seeking a Minister's exemption to certain CROR provisions. A collision occurred on the second day of operation. TSB's investigation determined that, without a comprehensive analysis and the implementation of effective compensatory safety measures by the railway, SPTO operation was a contributory factor. As a result, a working group was formed involving TC, QNS&L management and employees, and representatives from industry and labor. Recommencement of SPTO was conditional on arriving at a consensus on minimum operating conditions to ensure a level of safety equivalent to that of two-person operating crews.

In September 1996, rather than requiring railways to obtain exemptions, TC suggested to the Railway Association of Canada (RAC) that rules be developed for SPTO.

In 1997, SPTO was re-implemented at QNS&L with 69 new conditions. Some of the key conditions were to:

- provide LEs and RTCs with 120 to 130 hours of training, including in SPTO emergency procedures, with the training program to be monitored by TC;
- provide increased supervision of LEs; and
- install proximity detection devices (PDDs) on all lead locomotives, track units, and on-track vehicles operating on the main track.

In June 1997, TC acknowledged that the RAC had been developing an SPTO circular for its members, while repeating the expectation that the RAC develop SPTO rules for inclusion in the CROR.

In 1998, the RAC first proposed rule changes to the CROR under Section 20 of the Railway Safety Act (RSA) relating to SPTO. TC rejected these proposed changes, as they did not establish rules for SPTO that would ensure a level of safety equivalent to that of existing crew requirements.

In 2000, the RAC produced SPTO guidelines based on industry review and consultation and made them available to its members. The guidelines were based on the principles of risk assessment, mitigation, and monitoring. They were not approved by TC, nor were they required to be. The guidelines specified the following:

> Railway companies must advise Transport Canada in writing at least 60 days prior to implementing One Person train operations.
>
> [...]
>
> Prior to implementation of One Person operations, the railway company shall identify safety issues and concerns associated with One Person Train Operations, evaluate the risk involved with such an operation, and take appropriate measures to mitigate the risk.
>
> [...]
>
> Each railway company shall develop and institute an appropriate monitoring program for One Person operations that measures the safety performance of the operation.
> This program shall be described to Transport Canada and may be subject to follow-up regulatory review.

9.4.2.15 Canadian Railway Operating Rules (CROR)

TC can order the development of a rule or the amendment of an existing rule. The RAC, in consultation with its member railways, would then draft the rule. Once completed, it must be circulated to employee associations for comment before it is submitted to the Minister for approval. If there are objections to the proposed changes, the RAC can respond to the employee association's objections, and then their comments, along with the RAC response, are provided to the Minister's representatives for consideration. The rules must be approved by the Minister before coming into force. Rules may also be formulated by individual railways on their own, which also requires union consultation and submission to the Minister for approval.

In 2008, a major revision of the CROR by the RAC, approved by TC, introduced General Rule M, which provided in part, "Where only one crew member is employed, operating rules and instructions requiring joint compliance may be carried out by either the locomotive engineer or conductor…" The union consultation period for the rules was 90 days, and a 2-day meeting was held. These rule changes allowed railways to implement SPTO without the need for exemptions from TC to specific CROR rules, such as were required by QNS&L in 1997.

9.4.2.16 Single-Person Train Operations at Montreal, Maine, and Atlantic Railway

In 2003, MMA discussed the implementation of SPTO in Canada with TC. TC advised that MMA should consider QNS&L's SPTO implementation and operation as a Canadian "best practices" model. Between 2004 and 2008, MMA did not pursue SPTO in Canada, as it considered the 69 conditions that had been implemented at QNS&L to be unattainable.

In April 2009, after being informed of MMA's intention to begin SPTO, TC initiated a research project to develop internal guidelines to review SPTO applications. The targeted completion date was October 2009.

In June 2009, MMA submitted its SPTO risk assessment and proposal to TC. MMA advised of its intent to pursue a phased approach to implementing SPTO, using the 23-mi segment between the Maine-Quebec border and Lac-Mégantic as a "test-bed" for further expansion, pending approval by TC. In its risk assessment, MMA stated that a single-person crew member is more attentive when working alone, and cited its previous success on its US network.

In July 2009, TC expressed a number of concerns that centered on deficiencies in MMA operations, including lack of consultation with employees in doing risk assessments, problems managing equipment, problems with remote-control operations, issues with rules compliance, issues with fatigue management, and a lack of investment in infrastructure maintenance. TC reiterated its recommendation that MMA look at the QNS&L consensus-based process as a model in crafting operational conditions.

In May 2010, MMA began its test operation running SPTO. TC was told that MMA's SPTO crews were coming across the border as far as Nantes. However, on a number of occasions, TC became aware that MMA had operated SPTO trains with US crews beyond these limits when there had been weather issues or other operational demands, such as equipment failures. There were no performance indicators identified for tracking, nor was a formal monitoring program established. TC Quebec Region reiterated its concern about MMA's suitability as an SPTO candidate.

In the same month, TC Headquarters and the FRA conducted an informal review of MMA's

US operations, including SPTO. As a result of that review, TC and the FRA identified four areas for subsequent action, including the absence of a written emergency response plan and concerns over employee fatigue, efficacy of company oversight, and rules compliance.

In December 2011, MMA informed TC that, as of January 9, 2012, it was extending SPTO westward to Farnham. TC indicated that it viewed this expansion as a significant change to operations, reiterating that it required a new risk assessment. MMA submitted a revised risk assessment for its SPTO that identified 16 risks. Several mitigation measures were proposed, and where necessary, added to the company's SPTO special instructions, such as informing local authorities, establishing procedures for a single operator when taking control of an unattended train, allowing an SPTO engineer to stop the train for 20-minute naps, and requiring formal communications between the engineer and the RTC at least every 30 minutes.

This risk assessment did not identify or address the specific risks of a single-person train operator performing tasks previously performed by two persons, such as securing a train and leaving it unattended at the end of a shift.

In February 2012, TC met with MMA and the RAC. TC advised MMA that TC did not approve SPTO. MMA only needed to comply with all applicable rules and regulations. TC Quebec Region remained concerned about the safety of SPTO on MMA.

In April 2012, the collective agreement was renegotiated to allow for SPTO. Later in April, TC Quebec Region acknowledged that MMA was going ahead with the expanded use of SPTO to Farnham once the employee consultations were completed and the crews were trained. MMA committed to informing the regulator in advance of the date when SPTO would commence.

The SPTO training plan for LEs (which scheduled training for approximately four hours) was intended to address the new SPTO special instructions.

The actual SPTO training for several LEs, including the accident LE, consisted of a short briefing in a manager's office on the need to report to the RTC every 30 minutes, on the allowance for power naps, and on the need to bring the train to a stop to write clearances. In some cases, training consisted of a briefing within the hour preceding the operator's first SPTO train departure. The training did not cover fatigue management, or a review of tasks normally performed by conductors, such as determining the minimum number of hand brakes. A review of RTC recordings determined that the instruction to communicate to the RTC at least every 30 minutes was not consistently followed.

In July 2012, MMA began operating SPTO between Lac-Mégantic and Farnham. However, no job task analysis with the employees in the territory specific to SPTO was performed, nor were all of the potential hazards associated with those tasks identified. MMA had no plan for further monitoring and evaluating SPTO. MMA did take specific measures, such as:

- extending train radio range to eliminate "dead spots";
- supplying SPTO crews with equipment so that they could operate the locomotive remotely;
- meeting with every community along the track;
- installing mirrors on the left-hand side of its locomotives;
- identifying locations along the track where a helicopter could safely land to evacuate employees; and
- making arrangements with emergency service companies to be on call if an evacuation was needed.

On August 29, 2012, TC became aware that MMA had extended SPTO operations to Farnham. TC did not verify that the mitigation measures identified in MMA's risk assessment were implemented and effective.

In March 2013, TC published an internal guideline to assist in evaluating SPTO applications. The purpose of the guideline was to provide TC regional staff with a guide to review and address SPTO risk assessments provided by railway companies.

9.4.2.17 Review of the Montreal, Maine, and Atlantic Railway Submission and its Relation to the Requirements of Standard CSA Q850

In December 2011, two guidance documents published by TC for filing rule submissions recommended the use of Standard CSA-Q-850-97, *Risk Management Guidelines of Decision-Makers* (October 1997). A comparison was made between MMA's risk assessment on SPTO introduction and the requirements of standard CSA-Q-850-97. There were significant gaps identified in MMA's process. For example, MMA did not quantify safety data to indicate safety trends and to identify some of the possible hazards when major operational changes were being planned.

9.4.2.18 Research into Single-Person Train Operations

MMA's 2009 request prompted TC to renew research into SPTO. TC recognized that it lacked the tools to review an SPTO risk assessment.

TC contracted the National Research Council (NRC) to conduct the research. The report was issued in 2012 and indicated that the safety impact of SPTO is unique to each individual task, and that risk-mitigating countermeasures should be designed accordingly. It stated that "reducing the train crew to one person without appropriate operational changes and technological intervention diminishes safety." The report recommended:

- consolidating human factors knowledge into a best practices resource;
- identifying which technologies are required to fully support SPTO, depending on operational complexity;
- developing an SPTO guide with recommended training and refresher programs for operating personnel;
- developing communication protocols;
- developing a procedures guide to be used to determine if an operation is suitable for SPTO;
- conducting a workshop involving TC, NRC, and railways to review SPTO knowledge and identify one or two specific routes that could be used for a pilot test program; and
- running a pilot test program, complete with detailed monitoring and evaluation, over a two-year period.

In the United States, the FRA conducted a series of cognitive task analyses pertaining to railway operating crews. With respect to the role of the conductor, they found the following:

- Conductors and LEs not only work together to monitor the operating environment outside the locomotive cab, they also work together to plan activities, to solve problems, and to plan and implement risk mitigation strategies.
- Operating in mountain grade territory can significantly alter the complexity of a conductor's duties, introducing additional cognitive demands.
- When the conductor must handle unexpected situations, "these unanticipated situations impose cognitive as well as physical demands on the conductor."
- New technologies, such as PTC, will not account for all of the cognitive support that the conductor provides.

SPTO has been implemented in other parts of the world, including the United States, Europe, Australia, and New Zealand. In many countries, technological advancements were used to mitigate the risks of operating with one less crew member.

9.4.2.19 Safety Culture

All members of an organization, and the decisions made at all levels, have an impact on safety.

TC's *Rail Safety Management Systems Guide: A Guide for Developing, Implementing and Enhancing Railway Safety Management Systems* states:

> An effective safety culture in a railway company can reduce public and employee fatalities and injuries, property damage resulting from railway accidents, and the impact of accidents on the environment.
>
> In simple terms, an organization's safety culture is demonstrated by the way people do their jobs—their decisions, actions and behaviours define the culture of an organization.
>
> The safety culture of an organization is the result of individual and group values, attitudes, perceptions, competencies and patterns of behaviour that determine the commitment to, and the style and proficiency of, an organization's health and safety management system.
>
> Organizations with a positive safety culture are characterized by communications from various stakeholders founded on mutual trust, by shared perceptions of the importance of safety and by confidence in the efficacy of preventive measures.

The Guide also states:

> Experience has shown that a railway company will be markedly more successful in developing a safety culture if employees and their representatives, where applicable, are involved in the development and implementation of the safety management system.

The relationship between safety culture and safety management is reflected in part by the attitudes and behavior of a company's management.

An effective safety culture includes proactive actions to identify and manage operational risk. It is characterized by an informed culture where people understand the hazards and risks involved in their own operation and work continuously to identify and overcome threats to safety. It is a just culture, where the workforce knows and agrees on what is acceptable and unacceptable. It is a reporting culture, where safety concerns are reported and analyzed and where appropriate action is taken. And it is a learning culture, where safety is enhanced from lessons learned.

A company's policies determine how safety objectives will be met by clearly defining responsibilities; by developing processes, structures, and objectives to incorporate safety into all aspects of the operation; and by developing the skills and knowledge of personnel.

Procedures are directives for employees and set management's instructions. Practices are what really happens on the job, which can differ from procedures and, in some cases, increase threats to safety.

The report on the review of the RSA states, "The cornerstone of a truly functioning SMS is an effective safety culture, "and notes that "an effective safety culture is one where past experience is not taken as a guarantee of future success and organizations are designed to be resilient in the face of unplanned events." The RSA review recommended that the TC Rail Safety Directorate and the railway industry "take specific measures to attain an effective safety culture." TC's SMS guide contains a section on achieving an effective safety culture, and TC has published a safety culture checklist for companies to perform a self-assessment on their safety culture.

9.4.2.20 Summary of Safety Culture Issues

The TSB of Canada report lists the major safety culture issues associated with the accident. Whether the safety culture issue is lack of equipment maintenance or operator actions/inactions, the results can be disastrous. In this event, the results were catastrophic, with 47 people dead and millions in property damage. As in the event near Philadelphia, the transport safety organizations debate whether a one-person or two-person crew would have prevented the events. Well-trained two-person crew that are trained in the CRM are much more effective than a one-person crew. The airline industry has found this to be true for years.

References

Diaz, J. (2021). NY times, three men charged in 2018 Missouri duck boat accident. https://www.nytimes.com/2021/07/16/us/missouri-duck-boat-deaths.html (accessed December 2021).

Dismukes, K. (2006, 2006). Concurrent task management and prospective memory: pilot error as a model for the vulnerability of experts. *Proceedings of the Human Factors and Ergonomics Society Annual Meeting* 50 (9): 909–913.

Endsley, M.R. (1995). Toward a theory of situation awareness in dynamic systems. *Human Factors Journal* 37 (1): 32–64.

Goldstein, M. (2018). Forbes, Branson disaster: are duck boat tours unsafe? https://www.forbes.com/sites/michaelgoldstein/2018/07/20/branson-disaster-are-duck-boat-tours-unsafe/?sh=7a8529b4cb39 (accessed December 2021).

National Transportation Safety Board (NTSB) (2015). Derailment of Amtrak passenger train 188 Philadelphia, Pennsylvania May 12, 2015.

National Transportation Safety Board (NTSB) (2019). Collision between liquefied gas carrier genesis river and voyager tow houston ship channel, Upper Galveston Bay, Texas May 10, 2019.

National Transportation Safety Board (NTSB) (2020). Sinking of amphibious passenger vessel stretch duck 7 Table Rock Lake, near Branson, Missouri July 19, 2018.

National Weather Service (2021). Thunderstorm winds and derechos. https://www.weather.gov/safety/wind-thunderstorms-derecho (accessed December 2021).

Professional Mariner (2020). Stretch duck 7: chronology of sinking. https://www.professionalmariner.com/stretch-duck-7-chronology-of-sinking (accessed December 2021).

Sea Forces (2021). Naval information. https://www.seaforces.org/usnships/ddg/Arleigh-Burke-class.htm (accessed December 2021).

Transportation Safety Board of Canada (2013a). Runaway and main-track derailment, montreal, maine & atlantic railway freight train mma-002 mile 0.23, Sherbrooke Subdivision Lac-Mégantic, Quebec 06 July 2013.

Transportation Safety Board of Canada (2013b). Canadian rail operating rules, 112, securing equipment.

Us Navy (2017). Memorandum for distribution: enclosure (1) report on the collision between USS FITZGERALD (DDG 62) and motor vessel ACX CRYSTAL, enclosure (2) report on the collision between USS JOHN S MCCAIN (DDG 56) and motor vessel ALNIC MC 9.

10

Assessing Safety Culture

10.0 Introduction

The eight chapters of case studies and the safety culture implications clearly demonstrate the need to understand the safety culture of organizations. These were just a sampling of the safety culture related accidents that happen each year.

Assessing safety culture is important for several reasons. These include:

1. Of course, to understand the safety culture, norms, and safety climate of the organization.
2. Confirms commitment to the importance of safety for an organization. Conducting a safety culture assessment provides employees and supervisors confirmation that an organization is truly interested in ensuring they make it home safely each day.
3. Educates employees and supervisors on the importance of safety and it is possible to provide additional safety information during the assessment. For instance, if the survey or assessment is properly structured it should not only gather information, but also impart knowledge to the employees and supervisors on safety topics. The assessment should also stimulate employees and supervisors to seek more knowledge on a safety topic.
4. Elicits the underlying foundation of safety within an organization. I have heard this phrase many times in my career; "Today we are working for production, not safety." This indicates that the employees/supervisors in an organization where this is a common phrase are not trying to always be safe and will take shortcuts if a production crunch is happening. Finding this out helps to truly benchmark the level of safety commitment in an organization.
5. Allows employees and supervisors to be heard about their concerns about safety. Employees and supervisors want to be heard about their feelings and experiencing at work. Providing a venue for this allows for better communication. I have always said that employees and supervisors come up with the

Impact of Societal Norms on Safety, Health, and the Environment: Case Studies in Society and Safety Culture, First Edition. Lee T. Ostrom.

best safety and production solutions. Many times, these solutions need to be engineered to be viable, but the idea is sound. Employees need to be listened to.

Chapter 1 of this book discussed safety culture and the interviewees explained their methods of assessing safety culture. This chapter will go more into more depth as to how to assess an organization's safety norms and safety culture informally and formerly. In 1992 we were tasked to assess the safety culture of the Department of Energy's (DOE) Idaho National Engineering Laboratory (INEL) and the affiliated organizations. This was in conjunction with DOE Tiger Team visits to the DOE national laboratories. An initiative involving tiger teams was implemented by the United States DOE under then-Secretary James D. Watkins. From 1989 through 1992 the DOE formed tiger teams to assess 35 DOE facilities for compliance with environment, safety, and health requirements. The concept of Tiger Teams had been created much earlier to examine potential failures in the aerospace industry. Our initial work in safety culture involved developing a safety culture assessment instrument and administering the instrument to several thousand participants. The paper we wrote on the assessment process and the results was published in 1993 in Nuclear Safety (Ostrom, Kaplan, and Wilhelmsen 1993). To date the article has been cited over 350 times.

Since that time numerous organizations and researchers have developed safety culture assessment methods and instruments. These assessment methods and instruments vary widely. The Institute for Nuclear Power Operations (INPO) has developed and performs a very formalized site visit and analysis of a nuclear power plant's (NPPs) safety culture. DuPont De Nemours (DuPont) has developed a validated, standardized survey instrument that DuPont claims can be used to not only assess the safety culture of a facility, but also to compare one facility/organization to another. It is impossible to discuss all these methods and instruments. In the chapter that follows I will provide an overview of a range of safety culture survey instruments organizations and researchers have developed. Though, this is by far not a comprehensive listing.

At the end of the chapter, I will present activities I have been involved with over the years and the instrument we developed in the early 1990s.

10.1 Survey Research Principles

There is a considerable body of research on how to develop survey and how to administer survey instruments. Recent books on the subject include Saris (2014), Fowler (2013), and Esteban-Bravo and Vidal-Sanz (2021).

The following is an overview of survey research. Please consult one of the sources listed for more details on the process.

The topics covered are:

Developing the Survey Instrument
- Developing the questions/statements
- Sampling
- Survey delivery methods
- Reporting results
- Final thoughts

10.1.1 Developing the Survey Instrument

10.1.1.1 Developing the Questions/Statements

There are several questions one needs to ask before beginning the survey development process. These are:

- What exactly do I want to know?
- What do I want to learn from the survey?
- Why do I want to know this information?
- How will I use the information?

There are many types of safety information an organization might wish to be able to understand. Is the focus of the survey to be on the employees' behaviors, on the supervisors' leadership on safety, and/or on the facilities and process hazards? Safety norms are the primary focus of most safety culture surveys. These can be wide ranging. For instance:

- Is personal protective equipment (PPE) available and is it used appropriately?
- Do employees follow procedures, or do they take shortcuts?
- Do supervisors direct employees to perform tasks in an unsafe manner?
- Are tools used in an appropriate way?

The purpose of the survey should be clear and not burdensome to the participants. The survey should be to the point and focused on the information the organization truly wants to collect. I have taken some surveys that drive me crazy. The worst are hotel surveys that indicate they are brief and wind up being 50–100 questions. Many times, I start these and then abandon them. A survey should be able to be performed in a reasonable amount of time. What is reasonable? Well, that depends on the organization. However, the length of an average safety meeting would be appropriate. Say, 30 minutes.

The survey should not be designed to collect superfluous data. Keep to the purpose of the survey and not deviate. For instance, if the organization wishes to know about the employees' safety behaviors, it should not shift into questions about the lunchroom. However, the survey could ask questions about management support for safety.

A survey can be composed of questions or statements. A survey question on safety could be something like:

1. Are safety glasses available in production area A? Yes No I do not know
2. Do you always wear the proper safety equipment for entering the radiation protection zone? Yes No
3. Does your supervisor ensure you always have the proper tools for the task? Yes No
4. Do you know who to ask for safety related questions? Yes No

Statements in surveys are worded so that a range of responses are possible. Usually, a Likert Scale is associated with the statement or a sliding scale on which a participant can provide their level of agreement or disagreement with the statement (Jamieson 2004). Examples of these types of statements are shown in the following text:

1. Management cares about my safety.

Strongly disagree	Disagree	Neither agree or disagree (neutral)	Agree	Strongly agree
1	2	3	4	5

2. My supervisor ensures that I have the tools I need to perform the work I am assigned.

Strongly disagree	Disagree	Neither agree or disagree (neutral)	Agree	Strongly agree
1	2	3	4	5

Indicate on the following scale your adherence to the following statements.

1. I follow the lock and tag procedure.

2. I follow the hot work procedures.

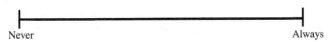

Open-ended questions can be used also, but they require much more work to analyze. The results of these questions can range widely. An example of an open-ended question is:

1. Describe the worst safety hazard in the plant.

The type of analysis needs to be considered prior to the survey development. Surveys using a Likert scale can be quantitatively analyzed. Surveys using Yes/No answers can only provide counts of answers. Open-ended questions are analyzed using qualitative methods.

10.1.1.2 Question/Statement Development

Questions and statements can be developed in many ways. One is of course, questions the organization wishes to know. Like the use of PPE. However, to really get at the heart of the safety culture the questions/statements should be developed with an open mind at the start of the process. Multiple methods can be used to derive questions/statements.

Preliminary surveys can also be sent out to a small number of employees/supervisors/safety professionals/managers with the goal of obtaining some initial questions. The responses from these surveys can then be fed into focus groups or Delphi sessions.

Focus groups made of small groups of employees can be a great way to start the question/statement development process. These groups should be led by a skilled facilitator. We have used this method and Galloway (2010) recommends that these groups be limited to about 10 employees. Focus groups of supervisors and managers can also be used for this purpose. It is wise not to have a combined group of employees and supervisors. In such a group some employees might not provide honest information.

A Delphi session can be used to develop questions/statements as well (Hall et al. 2009).

The Delphi method is a process based on the results of multiple rounds, in this case, queries sent to a panel of experts. In a Delphi session a skilled facilitator presents maybe an issue the organization is having, like minor accidents in a production area. The experts for the purposes of question/statement development are employees, supervisors, safety professionals, or managers. The experts consider the information and provide input to the facilitator. After each round, the experts are presented with an aggregated summary of the last round, allowing each expert to adjust their answers according to the group response. This process combines the benefits of expert analysis with elements of the wisdom of crowds.

- The Delphi method is a process used to arrive at a group opinion or decision by surveying a panel of experts.

- Experts respond to several rounds of questions, and the responses are aggregated and shared with the group after each round.
- The experts can adjust their answers each round, based on how they interpret the "group response" provided to them.
- The ultimate result is meant to be a true consensus of what the group thinks.
- The questions/statements are developed from the results.

Reviews of other safety surveys can be used as a starting point of questions/statements for a new survey. These questions/statements should really be modified to fit the context of the organization.

At the end of the process the draft survey should be tested on a small group of employees to see if it elicits the information desired.

Many organizations, like universities and research labs like National Laboratories have Institutional Review Boards (IRBs) that ensure that any human participants are protected from unethical research practices (HHS 2022). Please communicate with your IRB before any human, or for that matter animal, research. Even the interviews in Chapter 1 were cleared through University of Idaho's IRB.

10.1.1.3 Sampling

The next item to consider is the sample population. Is it a sample or is it the population? A sample is a subset of a population. A population is a complete demographic. The entire population of Sweden is a population. The entire athletic department at a university is population. A statistical sampling method can be used to determine the sample size to ensure that the results have some level of confidence in the data. For instance, what should the sample size be to ensure a 95% confidence level in the data (Lohr 2019). Using a sample, instead of the entire population will reduce the total number of participants required to take the survey to ensure quality data.

However, in some cases sampling the entire population is beneficial. For instance, if the accident rate in a certain department is much higher than in other departments.

10.1.1.4 Demographics

It is important to only collect the demographics needed to provide the least amount of information about who is taking the survey, without the ability to identify a certain individual. That is, if you want the survey to identify Department A personnel, then do not ask items like, years of service, gender, education level, or age. A participant will look at the list or demographic questions and know they could identify him or her. Only ask which department a participant works in. This way the survey is anonymous. Asking too many demographic questions could introduce bias or people will refuse to take the survey. I have not done some surveys because I knew the demographics could identify me, not that I am paranoid or anything (☺).

10.1.1.5 Survey Delivery

Surveys can be delivered in person, by mail, or via the Internet. I have used all three. There are several items to consider when delivering the survey in person. These are:

- Use a location that is quiet and away from the work area.
- The survey should be conducted during normal work hours.
- There should be adequate time to take the survey. Ensure it is not a timed event.
- Ensure the directions are understandable and possibly have someone read the instructions to the group.
- Allow adequate personal space for the participants to take the survey.
- It is optional to provide water and snacks, but this can make survey taking more palatable.

There are many Internet, web-based survey tools available. These tools are relatively easy to use and are reasonably priced. Once again, the directions as to how to take the survey should be understandable and there should be a phone number or email link for the participant to ask questions. An email or flyer should be delivered to the potential participants prior to the survey being available that explains the purpose and goals of the survey. The directions should also explain that the survey is anonymous and how the results will be used. These web-based survey tools usually also provide statistical analyses of the data. This is beneficial because you do not need to run the data through statistics software. However, a statistician might need to be consulted to help interpret the data.

Surveys mailed to the home are discouraged for safety culture surveys.

Consent forms should be provided for all participants where appropriate. These are usually required by IRBs. The completed forms must be maintained. Consult your IRB for proper content for consent forms.

10.1.1.6 Analyzing the Results and Reports

As stated earlier, web-based survey tools usually also provide statistical analyses of the data. In person surveys will require the data to be analyzed using a statistical software package. A statistician might need to be consulted to help analyze and interpret the data. Consult one of the texts cited in this section on how to analyze the data.

A report should be written on what the survey found, and the results communicated at some level to the employees, supervisors and manager. If the results are not communicated to the participants, they can feel the organization is hiding something.

10.1.1.7 Final Thoughts on Developing and Delivering Surveys

- Simplest way to encourage participation in a survey is to make it look simple!
- Survey length is important, do not ask too much.
- Ensure the survey instructions are understandable by your participant population.
- Run the survey through the IRB where appropriate.
- Do not collect too much demographic data. The participants will think Big Brother is watching.
- Use the appropriate statistical analytical methods.

10.1.2 Safety Culture Assessment Methods

10.1.2.1 DuPont (DuPont) De Nemours Sustainable Solutions (DSS)

My first introduction to DuPont's safety products was their Stop® program while I was working for PPG in La Porte, TX. At the time the Stop program was one of a kind. It is one of the precursors to many of the Human Performance Improvement (HPI) programs in use today. The Stop program is still being marketed, with both positive (Noria 2022) and negative (Rebbitt 2015; Monforton and Martinez 2016) reviews.

DSS has developed and markets a safety perception survey. They market that their perception survey can offer an organization the following benefits (DSS 2022):

- Measure and monitor safety culture leading indicators
- Data and analytics help identify potential risk areas
- Understand the safety culture of your organization
- Easy access to results through an online portal
- Reassess and benchmark performance every 12 months
- Results are plotted for strategic safety culture perspective
- Available in 40+ languages

DSS market brochure states that their perception survey was designed to capture perceptions and attitudes toward safety held by a wide cross-section of hourly workers, professionals, supervisors, and managers within an organization. The survey:

- Allows you to uncover employee perception gaps across job roles, locations, and divisions.
- Helps you better understand the strength of your organizational culture across teams and demographics.
- Benchmarks your results against companies in your industry to track and improve your safety performance.

The survey consists of 24 questions and measures and organization's safety culture in three areas of safety management. These are leadership, process and actions, and structure. The DSS survey has been administered in 45 countries and has over two million responses.

A paper by Mike Hewitt who is the Vice President, Global Workplace Safety Practice, DuPont Safety Resources, discusses the development and validation of the survey instrument (2011). The purpose of the survey is to determine whether safety is a core value of the management and employees of an organization. The survey is administered to all levels of an organization, and it seeks to determine whether employees and managers are involved in the safety functions of the organization. The survey also seeks to determine what the attitudes are for off the job safety as well.

DuPont developed a metric called the Relative Culture Strength (RCS). This metric is calculated using an organization's OSHA Total Recordable Injury Rate and results from the survey. DuPont has collected quite a large database of survey respondents and uses these data to do the comparisons between organizations. Organizations have also used the survey to compare their progress over a period of time.

10.1.2.2 Department of Energy Assessment of Safety Culture Sustainment Processes

In a June 2020 report the DOE discusses the assessment of eight of the maturity of the safety culture assessment process (DOE 2020). This report identified the following key areas.

10.1.2.2.1 Cultural Awareness Emphasizing the concept of safety culture in the DOE complex has served to heighten awareness of safety, promote ongoing conversations about culture as a set of organizational competencies that influence long term success in mission accomplishment, and stimulate renewed attention to positive working relationships as the foundation for the safe performance of work. Management initiates communications so that attention to organizational factors that promote safe work performance is a vital, visible topic of discussion throughout the organization. Multiple employee-led committees and teams champion active employee identification of concerns, improvements, and effective practices. A variety of communication channels and media help shape and sustain mutually respectful relationships and collaborative engagement. The use of formal, joint management and employee groups for analyzing and resolving safety issues promotes shared decision-making and responsibility for safe mission accomplishment.

10.1.2.2.2 Relative Maturity of Safety Culture DOE's ability to use culture as a management concept for continuous improvement of safe work performance is in the early stages of maturity. Two factors illustrate the bases for this conclusion:

- All assessed organizations used surveys as the primary method for quantifying safety culture improvements. In three cases, the organizations could demonstrate that their surveys had proven validity and reliability, but the remaining organizations could not. Thus, at those organizations, the validity of measurements can be uncertain and the credibility of the data not understood, reducing the reliability of decisions based on the surveys.
- When asked the question, "What does safety culture mean to you?" the great majority of responses related to worker safety. It is not evident that the term safety culture is understood to apply to the principles of nuclear safety, radiological safety, industrial safety, and environmental safety. This difference in focus limits the factors that are monitored and attended to as aids for decision making.

10.1.2.2.3 Federal Oversight Safety culture is a topic of ongoing discussion between the DOE Federal field offices and contractor management. Despite familiarity with the concepts of safety culture, Federal oversight that focuses specifically on contractor safety culture with defined processes, performance objectives, and criteria, is not evident. The team found through interviews that there is a widely shared perception among DOE and contractor officials that safety culture is not an overarching contractual requirement, and that formal oversight is limited to contract requirements. Over the course of this assessment, there has been some progress toward incorporating safety culture into contract requirements.

DOE identified 10 best practices:

The report summarizes 10 best practices in safety culture sustainment. These best practices are:

- A "shared governance" model to encourage strong relationships between leadership and the workforce.
- A documented description of a reliable and repeatable safety culture monitoring, analysis, and continuous improvement.
- A formal change management model that uses evidence-based decisions to shape the direction of change and social-science based models to design the processes for sustained change over.
- An initiative that combines the principles of Technical Conscience with the HPI tools described in DOE Handbook 1028 to support integrity in developing and maintaining engineering.

- Weekly Safety/Security Shares published on an easily accessed web page, targeted toward specific hazards, with a referenced link to the applicable Mission Success Model principle, that are timely, thought-provoking, and designed to encourage dialogue among.
- A dedicated organizational improvement department to support employee development, performance improvement, and performance.
- A knowledge preservation management program that captures key knowledge from retiring personnel through.
- Adoption of elements of the 2017 EFCOG *Guide to Monitoring and Improving Safety Culture.*
- Use of Integrated Safety Management Surveillance Team results to improve management's awareness of field status, potential safety issues, and worker attitudes toward the work.
- Inclusion of safety culture expectations into recent requests for proposals, contract clauses, and contractor performance evaluation.

The report listed five recommendations:

- Use a standard framework for culture monitoring and reporting to promote consistency, comparability, and credibility.
- Designate safety culture oversight monitors to provide an ongoing picture of cultural factors that warrant management attention and develop culture competencies among Federal staff.
- Develop safety culture competencies throughout the organizations.
- Enhance culture assessment and learning with peer reviews.
- Implement survey techniques that will produce credible safety culture assessment results.

10.1.2.3 Institute for Nuclear Power Operations Safety Culture Assessment

INPO identified eight principles of a strong nuclear safety culture (INPO 2004). These are listed as follows:

1. Everyone is personally responsible for nuclear safety.
   Responsibility and authority for nuclear safety are well defined and clearly understood. Reporting relationships, positional authority, staffing, and financial resources support nuclear safety responsibilities. Corporate policies emphasize the overriding importance of nuclear safety.
2. Leaders demonstrate commitment to safety.
   Executive and senior managers are the leading advocates of nuclear safety and demonstrate their commitment both in word and action. The nuclear safety message is communicated frequently and consistently, occasionally as a stand-alone theme. Leaders throughout the nuclear organization set an example for safety.

3. Trust permeates the organization.

   A high level of trust is established in the organization, fostered, in part, through timely and accurate communication. There is a free flow of information in which issues are raised and addressed. Employees are informed of steps taken in response to their concerns.

4. Decision-making reflects safety first.

   Personnel are systematic and rigorous in making decisions that support safe, reliable plant operation. Operators are vested with the authority and understand the expectation, when faced with unexpected or uncertain conditions, to place the plant in a safe condition. Senior leaders support and reinforce conservative decisions.

5. Nuclear technology is recognized as special and unique.

   The special characteristics of nuclear technology are taken into account in all decisions and actions. Reactivity control, continuity of core cooling, and integrity of fission product barriers are valued as essential, distinguishing attributes of the nuclear station work environment.

6. A questioning attitude is cultivated.

   Individuals demonstrate a questioning attitude by challenging assumptions, investigating anomalies, and considering potential adverse consequences of planned actions. This attitude is shaped by an understanding that accidents often result from a series of decisions and actions that reflect flaws in the shared assumptions, values, and beliefs of the organization. All employees are watchful for conditions or activities that can have an undesirable effect on plant safety.

7. Organizational learning is embraced.

   Operating experience is highly valued, and the capacity to learn from experience is well developed. Training, self-assessments, corrective actions, and benchmarking are used to stimulate learning and improve performance.

8. Nuclear safety undergoes constant examination.

   Oversight is used to strengthen safety and improve performance. Nuclear safety is kept under constant scrutiny through a variety of monitoring techniques, some of which provide an independent "fresh look."

INPO has developed a formalized method of assessing these principles at nuclear facilities. This methodology is contained in a document entitled "Industry Joint Initiative Nuclear Safety Culture Assessment" (NRC 2022). The assessment team includes:

- Team Leader
- Team Executive

- Host Peer
- Team Members
- Administrative Assistants
- Behavioral Scientist (when appropriate)
- NSCA Process Manager

In this methodology the assessment team reviews records and interviews management and employees. The team assesses the strength of the safety culture by comparing the data the team collects against the eight principles earlier.

10.1.2.4 Developing Team Findings

The team meets on a daily basis and discusses what they have learned. By the middle of the week the team is forming consensus on what have observed and learned about the plant. By the end of the week the team has developed preliminary findings and they meet with plant personnel during an exit meeting. The final report is prepared for the plant after approximately four weeks from the end of the site visit.

It is clear that this a very formalized process and provides an in-depth analysis of a nuclear plant's safety culture.

10.1.3   United States Air Force Assessment Tool

The Human Factors Division supports the Department of the Air Force safety mission to safeguard Airmen/Guardians, protect resources, and preserve combat capability by addressing the number one cause of Air and Space Forces mishaps: human error. The division includes experts from medicine, physiology, psychology, aircraft operations, and aircraft maintenance. These experts help investigators examine how human behavior contributes to mishap risks and causes. They further analyze mishap data and provide policy recommendations to commanders. The division administers the Air Force Combined Mishap Reduction System (AFCMRS) survey and can provide commanders an On-site Safety Assessment (OSA). The division trains DAF safety professionals in human factors principles and in the DOD Human Factors Analysis and Classification System. The AFCMRS is a safety culture survey tool that allows commanders to assess the safety of their units by anonymously surveying member's attitudes and perceptions. AFCMRS is a web-based survey that can be conducted at the squadron, group, wing, NAF, and MAJCOM levels. Survey items focus on organizational processes, organizational climate, resources, and supervision. Respondents provide Likert scores or direct comments to survey questions. Results of the surveys allow commanders to compare their unit to other anonymous units and analyze their own data (Air Force Safety Center 2022).

10.2 Assessing Health Care Safety Culture

The Healthcare Foundation is United Kingdom (UK) based organization. They performed a research study to determine which available safety culture survey instruments were applicable for use in the UK healthcare system. They assembled a listing of safety culture assessment tools, a description, and how they should be used (2011). The following presents a synopsis of their work. The organization

Table 10.1 Healthcare safety culture survey tools.

Tool and developer	Usage examples	Psychometric properties	Key strengths	Key weaknesses
Hospital Survey on Patient Safety Culture (AHRQ)	Hospitals in the United States, United Kingdom, Belgium, China, the Netherlands, Turkey, Saudi Arabia, Spain, Lebanon, and others	Psychometric properties have been tested. Issues with staffing scale identified	Can compare with other countries and industries	Focuses only on hospitals Has some validity issues
Manchester Patient Safety Culture Assessment Framework (NPSA)	Hospitals in the United Kingdom, pharmacy in the United Kingdom, hospitals in Canada	No psychometric properties reported in empirical literature	Focuses on the broader notion of safety culture	Little has been published about usage
Safety Attitudes questionnaire (developed from aviation tool)	Hospital, ICU, pharmacy, primary care, long-term care in many countries	Psychometric properties extensively tested and well validated	• Well validated and established • Can compare with other countries and industries	• Not used much in UK • Some think it takes time to complete
Safety Climate Survey (University of Texas and US IHI)	Hospitals in North America	Some validation undertaken but no detailed studies of psychometric properties	• Short and easy to complete • Has been compared with other surveys	• Tested mainly in North America • Developed some time ago
Patient Safety Climate in Healthcare Organizations (Stanford, funded by AHRQ)	Hospitals in the United States	Validation of psychometric properties undertaken	• Studies with large sample sizes have validated the tool	• Has been used mainly by one group of researchers • Tested almost exclusively in US hospitals

scanned numerous research data bases and selected safety culture assessment tools that fit into these criteria:

- be primary research or reviews,
- be readily available online, in print or from relevant (healthcare) organizations,
- be available in abstract, journal article, or full report form,
- address one or more of the core questions listed, and
- be available in English or readily available for translation.

In addition to these criteria, the Healthcare Foundation screened the instruments using several caveats. These caveats included:

First, they acknowledge the search was not exhaustive.

Second, they standardized the language, so that terms like safety climate were all changed to safety culture. They also acknowledged that treatment types vary by country. For instance, primary care in the United Kingdom might differ from primary care in the United States. In addition, the results organizations published might lack specificity.

Third, the results presented are usually limited to one facility and from this it is difficult to determine whether results can be compared from one facility to another. There is little published evidence about how well lesser known tools or unnamed tools may have worked in specific local contexts.

The organization developed a table of five safety culture safety climate survey instruments they felt were the most appropriate for use in the UK healthcare industry. These are found in Table 10.1.

10.3 Seven Steps to Assess Safety Culture

Shawn Galloway (2010) has published an article called "Seven Steps to Assessing Safety Culture." In the article he points out that every organization has a safety culture, but it might not be the culture they would desire. He also points out how complex an organization's can be. He says the influences on a safety culture include location in a certain city or state, leadership, supervisory styles, peer pressure, workplace conditions, and logistics. The goal of his process is to improve the safety culture. His steps are:

1. Review Documentation, Programs, and Policies
   Review all pertinent safety documentation for an organization; also review the organizations processes for communication, safety committees, incident investigations, incentives, and recognition programs. Also, examine the incident reports, the safety organization, and the safety expectations for supervisors

and managers. Doing so will provide an understanding of the current safety cultural foundation, as he puts it.

2. Communicate Prior to Employee Interaction

   Mr. Galloway states that it is critical to communicate the intentions of the safety culture assessment prior to interacting with the employees. If this is not done the employees might not understand why the person doing the assessment is asking questions. He states that it is essential to communicate to employees that their responses will be anonymous, and that the assessment is not a fault finding exercise.

3. Conduct a Location Walk

   Early in the process Mr. Galloway suggests doing a walkaround or tour of the facility to observe the tasks being performed and to get a grasp of the potential safety issues.

4. Leadership Discussion

   This step involves meeting with management to brief them on the safety culture assessment process and to build awareness of the process that will be performed. As with the employees, ensure managers that this is not a faulty finding exercise. Mr. Galloway also emphasizes that the collective bargaining unit should be involved in these early discussions as well.

5. Utilize a Customized Safety Perception Survey

   Mr. Galloway stresses that a customized safety perception survey be used for the assessment, rather than an off-the-shelf perception survey. The organization should develop their own survey.

6. Conduct Group and Individual Interviews

   Mr. Galloway writes that group and individual interviews provide insight into a safety culture. Individual interviews can provide deeper insight into a group's dynamics and can help to explain responses on a safety culture survey. He writes that one-on-one interviews with a complete staff of an organization should be performed for smaller organizations. Group interviews should be conducted with no more than 10 individuals and should be led by a trained facilitator. In all cases the results of these interviews should be anonymous.

7. Provide a Report Focusing on Internally Actionable Items

   This step involves two items. The first is an exit meeting with key management and collective bargaining unit leaders to discuss the findings from the safety culture assessment. The second is to write a report that focuses on the findings that can be acted upon. The report should also suggest a short-term action plan to help keep momentum going after the assessment.

   Mr. Galloway finishes the article with some suggestions about what to do next. His most important suggestion is to include the workers in the next phase and to seek to develop a culture through a "journey," rather than through an event.

10.3.1   A Framework for Assessing Safety Culture

Johan Götvall developed a framework for assessing safety culture as part of his master's thesis in Chemical Engineering from Chalmers University of Technology (Götvall 2014). Besides the safety the framework he developed; the thesis is a good resource for locating additional safety culture assessment instruments. He also did a considerable amount of research on how researchers defined safety culture.

10.3.2   Agency for Healthcare Research and Quality

The Agency for Healthcare Research and Quality's (AHRQ) website contains information on their Surveys of Patient Safety Culture™ (SOPS) (AHRQ 2022). The website also contains links to their database of responses from 9000 providers. AHRQ's definition of Patient Safety Culture is:

> Patient safety culture is the extent to which an organization's culture supports and promotes patient safety. Patient safety culture refers to the beliefs, values, and norms that are shared by healthcare practitioners and other staff throughout the organization that influence their actions and behaviors. Patient safety culture can be measured by determining what is rewarded, supported, expected, and accepted in an organization as it relates to patient safety.

The website says that the SOPS instrument can provide information on the following types of organizations:

- Hospital Survey on Patient Safety Culture
- Medical Office Survey on Patient Safety Culture
- Nursing Home Survey on Patient Safety Culture
- Community Pharmacy Survey on Patient Safety Culture
- Ambulatory Surgery Center Survey on Patient Safety Culture

AHRQ says that the results of the SOPS can:

- Raise staff awareness about patient safety.
- Assess the current status of patient safety culture.
- Identify strengths and areas for patient safety culture improvement.
- Examine trends in patient safety culture change over time.
- Evaluate the cultural impact of patient safety initiatives and interventions.

10.3.3   Graduate Student Safety Culture Survey

The following is a safety culture survey I developed and conducted at the University of Idaho, Idaho Falls Center, in the spring of 2021. The survey was to determine the level of safety knowledge the graduate students who work in laboratories have and the hazards they have the most knowledge about.

The University of Idaho, Idaho Falls Center (UIIF) is co-located with Idaho State University. The center primarily supports the educational needs of the Idaho National Laboratory (INL). The faculty at UIIF conduct research in the areas of nuclear engineering, bioenergy, cybersecurity, material science, artificial intelligence, robotics, and human factors/risk assessment. There are well over 40 graduate students at the center. About half of the students are traditional graduate students. The remainder is part time graduate students, who work fulltime.

The survey was designed to have a limited number of questions to help ensure it would be completed by as many of the current graduate students as possible. The survey was approved by UI's IRB and was distributed electronically to all the graduate students. Sixteen students completed the survey. The following is the survey that was developed and administered:

1. Are you:
   a. A new UIIF graduate student. – first semester
   b. A continuing graduate student.
   c. A graduate student that will complete your degree this semester.
2. Will or have you been conducting laboratory research?
   a. If yes (check all that apply):
      i. Will it be using chemicals?
      ii. Will it involve pressurized equipment?
      iii. Will it involve equipment above room temperatures?
         1. 70–200° F
         2. Above 200° F
      iv. Will there be rotating or vibrating equipment?
3. Did you do undergraduate laboratory research?
4. Did you receive safety training as an undergraduate student?
5. Did you receive training on PPE as an undergraduate?
6. What level of knowledge do you have of the following potential hazards?
   - EARLY-Some
   - DEVELOPING-Elementary
   - PROFICIENT-Intermediate
   - ADVANCED-Professionally Trained
     o High Voltage
     o Toxic Metals
     o Flammable/Combustible Solids
     o Flammable/Combustible Liquids
     o Flammable/Combustible Gases
     o Chemical Waste
     o Pyrophoric Metals

- o Asphyxiation Hazards
- o Confined Spaces
- o Ionizing Radiation
- o Non-ionizing Radiation
- o Compressed Gas cylinders
- o Magnetic Fields
- o Laser
- o High intensity visual light
- o Infrared light
- o Ultraviolet light
- o Pressure vessel
- o Noise/sound, below 16 000 Hz
- o Noise/sound, above 16 000 Hz and into ultrasonic
- o Overhead work
- o Elevated work
- o Cryogenic
- o Pressure vessel
- o Mechanical (moving parts)
- o Ergonomic related
  - Lifting
  - Desk workstation
  - Awkward postures
  - Repetitive motions

7. What learner style best describes you?
   a. Visual (You prefer using pictures, images, and spatial understanding.)
   b. Aural (You prefer using sound and music.)
   c. Verbal (You prefer using words, in both speech and writing.)
   d. Physical (You prefer using your body, hands, and sense of touch.)
   e. Logical (You prefer using logic, reasoning, and systems.)

8. In regard to Question 9, which of these media do you feel would provide you with the best safety information:
   a. Short videos showing safety situations that a person can access anytime
   b. Long videos covering several safety topics
   c. PowerPoint interactive safety presentations
   d. PowerPoint page turner safety presentations
   e. Booklets
   f. Workbooks
   g. Posters
   h. Lectures
   i. One-on-one mentoring
   j. Hand on demonstrations

9. Do you prefer to learn?
   a. In a social situation with other social learners
   b. In a solitary situation
10. Did your major professor provide safety training concerning the laboratory you will or have been working? Yes No
11. What safety information would you have liked to have been provided when you started working in a laboratory? Short answer
12. Do you know how to actuate a fire alarm?
13. Do you know how to get out of the building in the event of a fire alarm or other emergency?
14. Do you know where the closet fire extinguisher is?
15. Have you had fire extinguisher training?
16. Do you know where the AED (automated external defibrillator) in the building is?
17. Are you or do you know who is trained to use the AED?
18. Have you been trained in first aid?

The results of the survey provided very valuable information. The following are the insights the data provided:

1. Most of the students had received safety training as undergraduate students and this training included the use of PPE.
2. The students who will be working in labs understand the type of work they will be performing. However, the level of knowledge of the hazard types varies greatly.
3. Two-thirds of the major professors do not provide specific safety training.
4. All the student responders understand how to exit the buildings in the event of a fire alarm and 90% know how to operate the fire alarm. Less than half of the student responders are familiar with where the AED is in the buildings. However, they are familiar with where the fire extinguisher is located.
5. Almost two-thirds of the students have had some first aid training.
6. Most of the student responders are visual learners and like learning in a group setting.

The results of a safety survey can help focus safety efforts and information can be directed on the potential problem areas. Figure 10.1 is an example of the results from the survey.

10.3.4 Idaho National Engineering Laboratory Survey

The introduction of this chapter discussed the safety culture survey we developed and administered in the early 1990s and that was cited numerous times (Ostrom, Kaplan, and Wilhelmsen 1993). The INL went through several name

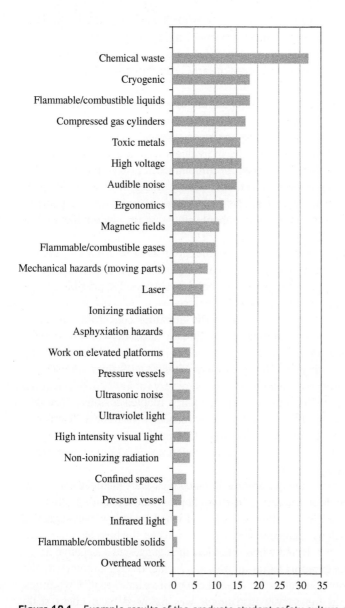

Figure 10.1 Example results of the graduate student safety culture survey.

changes during the last 30 years. In the early 1990s it was called the INEL. Then in the late 1990s it became the Idaho National Engineering and Environmental Laboratory (INEEL). The following discussion of the survey development is a synopsis from our 1993 article. Please see the article for the complete development and analysis procedures.

The development process of the safety culture assessment instrument included three techniques. The first technique involved interviewing 86 EG&G Idaho employees from the INEL, including managers, professionals, office workers, and laborers from various facilities. The individuals were asked three interview questions addressing safety and procedure compliance at EG&G Idaho.

The second technique used to generate survey items involved holding an all-managers meeting in which managers were asked to write down a personal safety credo: what they say they believe about safety that they would like each of their employees to understand.

The third technique involved querying other sources of information, such as previous interview data concerning a recent organizational climate survey, a literature review, and previous personnel opinion surveys, for possible norms. Possible safety norms suggested by these sources were selected for inclusion in the new survey instrument. Review of the literature concerning organizational climate, organizational norms, safety climate, and safety norms provides a conceptual framework into which items might be organized. Of particular importance in this sorting was the research of Litwin and Stringer (1968). The categories of safety norms ultimately selected were very similar to their categories of social norms except that ours were particularly adapted to safety. The data gathered were sorted into the following categories: Safety Awareness, Teamwork, Pride and Commitment, Excellence, Honesty, Communications, Leadership and Supervision, Innovation, Training, Customer Relations, Procedure Compliance, Safety Effectiveness, and Facilities.

A total of 84 statements, divided among the categories, were included in the original survey. Statements on the survey instrument presented had both positive and negative wording. Positive wording was selected when interview data suggested a positive norm, such as "people work safely, even when the boss isn't looking." Negative wording was selected when interview data suggested a negative norm, such as "We hesitate to report minor injuries and incidents." An attempt was also made to have a reasonable balance between both positive and negative wordings.

Each statement in the survey instrument was followed by a Likert scale. The five-point Likert scale allows respondents to indicate the extent to which they agree or disagree with each statement. An example of a scale is shown in Table 10.2.

A seven-point Likert scale be used also, but a five-point provides enough fidelity (Jamieson 2004).

Table 10.2 Example of Likert scale.

Strongly disagree	Disagree	Neither agree or disagree (neutral)	Agree	Strongly agree
1	2	3	4	5

Responses 1, 2, 4, and 5 in Table 10.2 are self-explanatory; however, the third, neither disagree nor agree response, is not as obvious. If an employee responds with a 3, they are saying they are neutral in their response to the statement. This does not mean the item does not pertain to them; they are saying they do not have an opinion either positive or negative concerning an item. This is a legitimate response for an employee to have. The instructions should say that if a statement does not pertain to you then do not answer it. The data generated from individuals not responding to statements are also of significant value. The percent nonrespondents for a statement can give an indication of the employees' assessment of those questions which pertain to them. Sutton (1989) says that nonparticipant data are important because they can give an indication that individuals:

1. Have never been asked to participate in the process being investigated.
2. Cannot or are not willing to participate in the survey process. Therefore, the reasons why individuals did not respond to statements should be in results and to provide input to those trying to draw inferences across many groups and organizations.

The survey was administered during the month of January 1991 to about 4000 employees of DOE-ID and its eight contractors (EG&G Idaho; Rockwell; MSE, Inc.; Chem-Nuclear Geotech; West Valley-Nuclear; Winco; PTI; and MK-Ferguson). Care was taken to administer the survey in a manner to reduce bias and to help ensure honest information from the participants.

A statistical sampling method was used that specified the number of employees needed to be surveyed to have a 95% level of confidence in the data. The results from the survey pointed out both the strengths and weaknesses in the safety cultures of the organizations.

The data collected were tested for reliability using the Cronbach's Alpha (Nunnally 1978). Other statistical tests were conducted on the data to determine the level of correlation between statement results.

Many types of statistical analysis were performed on the data, and a sampling of the results is contained in the original article.

All available forms of data should be collected and analyzed before making judgments about the safety culture of an organization. In addition to the questionnaire itself, data gathered should include accident statistics, safety performance data, records of employee and management concerns, and other measures of product quality and organizational performance.

Other important sources of input to the analysis process are the explanations and interpretations given by those surveyed. Ideally, each group surveyed should be given an opportunity to review and interpret their own.

The following are the original statements from the survey and the categories the statements were associated with:

Safety Awareness

1. In our company, the employees are aware of their part in safety.
2. In our company, people think safety concerns do not relate to office workers.
3. People are aware of the safety hazards in their area and are careful to minimize and avoid them.
4. Around here, people do not think much about safety.

Teamwork

5. Safety professionals in this company tend to be bright and capable people.
6. In this company, people ask for help with safety when they need it.
7. Around here, you'll be better off if you hide your problems and avoid your supervisor.
8. People do go out of their way to help each other work safely.
9. Safety personnel are unavailable when we need help.
10. Around here, employees who have to follow safety and health procedures are seldom asked for input when the procedures are developed or changed.

Pride and Commitment

11. Around here, people take pride in how safely we operate.
12. In this company, people stand up for the safety of their operations when others criticize it unfairly.
13. Around here, people look at the company safety record as their own safety record and take pride in it.
14. In this company, I cannot significantly impact the company's safety record.
15. In this company, people think safety is not their concern – it's all up to their manager and others.
16. Around here, people see safety as the responsibility of each individual.
17. This company cares about the safety of its employees.

Excellence

18. In this company, we have the highest standards for safety performance.
19. Around here, people are always trying to improve on safety performance, even when they are doing well.

20. People are often satisfied with routine and mediocre consideration for safety.
21. Around here, the way we work now is safe enough.
22. In this company, there is no point in trying harder to be safe; no one else is.

Honesty

23. In this company, people work safely, even when the boss is not looking.
24. Around here, people wear safety equipment even when they know they aren't being watched.
25. Around here, people are willing to comply with safety measures and regulations.
26. In this company, people try to get around safety requirements whenever they get a chance.

Communications

27. In this company, we hesitate to report minor injuries and incidents.
28. We do not get adequate information about what is going on with safety in the company.
29. Around here, there's lots of confusion about who to contact for safety concerns.
30. Around here, safety statistics are seldom studied and discussed.
31. In our company, safety hazards are seldom discussed openly.
32. Timely feedback is seldom provided when a safety hazard is reported.
33. In this company, you cannot raise a safety concern without fear of retribution.
34. In this company, we have very few safety signs or posters.
35. Around here, employee ideas and opinions on safety are solicited and used.
36. People who raise safety concerns are seen as troublemakers

Leadership and Supervision

37. It's a tradition; safety matters are given a low priority in meetings.
38. In our company, managers do not show much concern for safety until there is an accident.
39. In this company, the people who make safety decisions do not know what is going on at the workers' level.
40. Around here, work is organized so that you can do the job safely.
41. Around here, managers seldom work with their groups to identify and correct safety concerns or problems.
42. In our company, employees who will implement plans are seldom involved in reviewing their safety implications.
43. Managers/supervisors are often not available to answer health and safety questions.

44. My manager/supervisor discussed safety and health issues in my last employee evaluation.
45. Supervisors are receptive to learning about safety concerns.
46. In this company, people who work safely get no real rewards.
47. Little special recognition is given to safe employees.

Innovation

48. Around here, people are constantly on the lookout for ways of doing things more safely.
49. People tend to hang on to the old ways of doing things without regard to their safety implications.
50. In this company, people are encouraged to express new safety ideas and suggestions.
51. Around here, you get little recognition for new safety ideas.
52. It's a tradition; you do not raise safety ideas that your boss does not have first.

Training

53. People mostly give lip service to safety training; they do little to actively support it.
54. In this company, safety training is compromised in favor of more pressing demands.
55. Around here, managers are not very well trained to identify and address safety concerns.
56. In this company, safety training does not address subjects of real concern.
57. It's a tradition; safety training is done on a regular basis.
58. People in this company are well prepared for emergencies, and everyone knows just how to respond.
59. I know who to talk to when I see a hazard or have health and safety concerns.

Customer Relations

60. Employees here are always looking for ways to satisfy the customers' needs and requirements.
61. Customers here count on our company to do its work safely.

Procedure Compliance

62. In this company, we have a long way to go in improving our compliance.
63. In this company, people are often uncertain about what the safety procedures are for the work they do.
64. In general, people are well acquainted with the safety procedures for their job.

65. In this company, the safety procedures are relevant to employees' particular circumstances.
66. Around here, there are lots of safety procedures that do not really apply to the particular areas or circumstances in which they are supposed to be used.
67. There are so many procedures they interfere with doing a job safely.
68. In this company, area requirements for protective clothing and equipment may not reflect the actual hazards.
69. In this company, employees use their heads and raise lots of questions about why things are being done the way they are.
70. In this company, procedures are too detailed, making compliance a mindless activity.
71. It's a tradition; people carefully follow the written procedures.
72. In this company, people can be confident they are safe when they are following the rules.
73. Around here, you cannot expect praise and recognition for complying with procedures.
74. In this company, following safety procedures is consistently expected.
75. Safety procedures tend to be too vague and general to apply in specific situations.

Safety Effectiveness

76. When it comes down to it, people in this company would rather take a chance with safety than miss a schedule or budget commitment.
77. In this company, people are willing to expend a great deal of effort to get a job done safely.
78. In this company, work is not done that jeopardizes other workers or the public.
79. Employees rarely take the initiative to get safety problems taken care of.
80. Around here, people can report a safety problem several times, yet the problems may remain and not get corrected.
81. Our daily routines do not show that safety is an important value.

Facilities

82. In this company, the physical conditions of work locations inhibit safe work.
83. In this company, facilities are designed with safety in mind.
84. Concern and attention is being given to maintaining good safety conditions in our facilities.
85. People tend to keep their facility neat and orderly.
86. Around here, good housekeeping is not just the janitor's job – people clean up their own areas.

87. In this company, fire and electrical hazards are accepted in some of our facilities.
88. Around here, we really keep on top of the snow and ice problems and prevent them from getting out of hand.

10.4 Chapter Summary

This chapter provided a brief guide on how to develop a safety culture survey and presented several safety culture survey instruments. Each organization is different, just as each person is different. Customizing a survey is the best bet for ensuring the results are tailored to a particular organization. However, a standardized survey can be used to benchmark an organization or department to another organization or department. It is not difficult to develop and administer a survey but following some basic steps can help ensure the best quality results.

References

Agency for Healthcare Research and Quality (AHRQ) (2022). https://www.ahrq.gov (accessed February 2022).

Air Force Safety Center (2022). Human factors. www.safety.af.mil/Divisions/Human-Factors-Division (accessed January 2022).

Department of Energy (DOE) (2020). Assessment of safety culture sustainment processes at U.S. Department of Energy Sites, 2020.

Dupont Sustainable Solutions, (2022) The new DSS Online Safety Perception Surveys™, https://sps.consultdss.com/ (accessed April 2022).

Esteban-Bravo, M. and Vidal-Sanz, J. (2021). *Marketing Research Methods: Quantitative and Qualitative Approaches*. Cambridge University Press.

Fowler, F. (2013). *Survey Research Methods (Applied Social Research Methods)*, Fifthe. SAGE Publications, Inc.

Galloway, S. (2010). Assessing your safety culture in seven simple steps, EHS today. https://www.ehstoday.com/safety/article/21914669/assessing-your-safety-culture-in-seven-simple-steps (accessed February 2022).

Götvall, J. (2014). A framework for assessing safety culture. Master's Thesis in Chemical Engineering. Chalmers University of Technology.

Hall, E., Lentz, C.A., Sandwisch, J.L. et al. (2009). *The Delphi Primer: Doing Real-World or Academic Research Using a Mixed-Method Approach (The Refractive Thinker)*, 1ste. The Lentz Leadership Institute.

Heath and Human Services (HHS) (2022). What are IRBs?. https://www.hhs.gov/ohrp/education-and-outreach/online-education/human-research-protection-training/lesson-3-what-are-irbs/index.html#:~:text=Institutional%20Review %20Boards%2C%20or%20IRBs%2C%20review%20research%20studies%20to %20ensure,and%20adequately%20protect%20research%20participants (accessed February 2022).

Hewitt, M. (2011). *Relative Culture Strength A Key to Sustainable World-Class Safety Performance*. DuPont DeNours.

Institute for Nuclear Power Operations (2004). Principles for a strong nuclear safety culture. https://www.nrc.gov/docs/ML0534/ML053410342.pdf (accessed December 2021).

Jamieson, S. (2004). Likert scales: how to (ab) use them. *Medical Education* 38 (12): 1217–1218.

Litwin, G. and Stringer, R. (1968). *Motivation and Organizational Climate*. Boston: Harvard University Press.

Lohr, S. (2019). *Sampling: Design and Analysis*, 2nde. Chapman and Hall/CRC.

Monforton, C. and Martinez, J. (2016). DuPont must end its safety charade, Austin American-Statesman. https://www.statesman.com/story/news/2016/09/23/ dupont-must-end-its-safety-charade/10093078007/ (accessed February 2022).

Noria (2022). DuPont enhances STOP for each other safety program. Reliable plant. https://www.reliableplant.com/Read/8703/dupont-enhances-stop-for-each-or-safety-program (accessed February 2022).

Nuclear Regulatory Commission (NRC) (2022). Industry joint initiative nuclear safety culture assessment. https://www.nrc.gov/docs/ML0918/ML091810805.pdf (accessed January 2022).

Nunnally, J. (1978). *Psychometric Theory*, Seconde. New York: McGraw Hill.

Ostrom, L.T., Wilhelmsen, C.A., and Kaplan, K. (1993). Assessing safety culture. *Journal of Nuclear Safety* 34 (2).

Rebbitt, D. (2015). The mighty have fallen, the Safety Mag. https://www.thesafetymag .com/ca/news/opinion/the-mighty-have-fallen/187071 (accessed February 2022).

Saris, D. (2014). *Evaluation, and Analysis of Questionnaires for Survey Research, 2nd Edition (Wiley Series in Survey Methodology)*, 2nde. Wiley.

Sutton, R. (1989). Reactions of nonparticipants as additional rather than missing data: opportunities for organizational research. *Hum Relations* 42 (5): 423–439.

The Health Foundation (2011). Measuring safety culture. https://www.health.org.uk/ publications/measuring-safety-culture (accessed December 2021).

Index

Note: Page numbers in *italics* refer to figures; Page numbers in **bold** refer to tables. US spelling is used in this index.

Impact of Societal Norms on Safety, Health, and the Environment: Case Studies in Society and Safety Culture, First Edition. Lee T. Ostrom.
© 2023 John Wiley & Sons, Inc. Published 2023 by John Wiley & Sons, Inc.